VOLUME 1

2000 B.C. to A.D. 699

Science
and
Its
Times

Understanding the

Social Significance of

Scientific Discovery

VOLUME 1
2000 B.C. to A.D. 699

Science
and
Its
Times

Understanding the
Social Significance of
Scientific Discovery

Neil Schlager, Editor

Josh Lauer, Associate Editor

Produced by Schlager Information Group

GALE GROUP

Detroit
New York
San Francisco
London
Boston
Woodbridge, CT

Science and Its Times

VOLUME 1

2000 B.C.
TO A.D. 699

NEIL SCHLAGER, *Editor*
JOSH LAUER, *Associate Editor*

GALE GROUP STAFF

Amy Loerch Strumolo, *Project Coordinator*
Christine B. Jeryan, *Contributing Editor*

Mark Springer, *Editorial Technical Specialist*

Maria Franklin, *Permissions Manager*
Margaret A. Chamberlain, *Permissions Specialist*
Deb Freitas, *Permissions Associate*

Mary Beth Trimper, *Production Director*
Evi Seoud, *Assistant Production Manager*
Stacy L. Melson, *Buyer*

Cynthia D. Baldwin, *Product Design Manager*
Tracey Rowens, *Senior Art Director*
Barbara Yarrow, *Graphic Services Manager*
Randy Bassett, *Image Database Supervisor*
Mike Logusz, *Imaging Specialist*
Pamela A. Reed, *Photography Coordinator*
Leitha Etheridge-Sims, *Image Cataloger*

ISBN: 0-7876-3933-8

Printed in the United States of America
10 9 8 7 6 5 4 3 2 1

Library of Congress Cataloging-in-Publication Data

Science and its times : understanding the social significance of scientific discovery / Neil Schlager, editor.
 p.cm.
Includes bibliographical references and index.
 ISBN 0-7876-3933-8 (vol. 1 : alk. paper) — ISBN 0-7876-3934-6 (vol. 2 : alk. paper) —
ISBN 0-7876-3935-4 (vol. 3 : alk. paper) — ISBN 0-7876-3936-2 (vol. 4 : alk. paper) —
ISBN 0-7876-3937-0 (vol. 5 : alk. paper) — ISBN 0-7876-3938-9 (vol. 6 : alk. paper) —
ISBN 0-7876-3939-7 (vol. 7 : alk. paper) — ISBN 0-7876-3932-X (set : hardcover)
 1. Science—Social aspects—History. I. Schlager, Neil, 1966-
Q175.46 .S35 2001
509—dc21
                   00-037542

Contents

Contents
2000 B.C.
to A.D. 699

Physical Sciences

Technology and Invention

Preface

The interaction of science and society is increasingly a focal point of high school studies, and with good reason: by exploring the achievements of science within their historical context, students can better understand a given event, era, or culture. This cross-disciplinary approach to science is at the heart of *Science and Its Times.*

Readers of *Science and Its Times* will find a comprehensive treatment of the history of science, including specific events, issues, and trends through history as well as the scientists who set in motion—or who were influenced by—those events. From the ancient world's invention of the plowshare and development of seafaring vessels; to the Renaissance-era conflict between the Catholic Church and scientists advocating a sun-centered solar system; to the development of modern surgery in the nineteenth century; and to the mass migration of European scientists to the United States as a result of Adolf Hitler's Nazi regime in Germany during the 1930s and 1940s, science's involvement in human progress—and sometimes brutality—is indisputable.

While science has had an enormous impact on society, that impact has often worked in the opposite direction, with social norms greatly influencing the course of scientific achievement through the ages. In the same way, just as history can not be viewed as an unbroken line of ever-expanding progress, neither can science be seen as a string of ever-more amazing triumphs. *Science and Its Times* aims to present the history of science within its historical context—a context marked not only by genius and stunning invention but also by war, disease, bigotry, and persecution.

Format of the Series

Science and Its Times is divided into seven volumes, each covering a distinct time period:

Volume 1: 2000 B.C. to A.D. 699

Volume 2: 700-1449

Volume 3: 1450-1699

Volume 4: 1700-1799

Volume 5: 1800-1899

Volume 6: 1900-1949

Volume 7: 1950-present

Dividing the history of science according to such strict chronological subsets has its own drawbacks. Many scientific events—and scientists themselves—overlap two different time periods. Also, throughout history it has been common for the impact of a certain scientific advancement to fall much later than the advancement itself. Readers looking for information about a topic should begin their search by checking the index at the back of each volume. Readers perusing more than one volume may find the same scientist featured in two different volumes.

Readers should also be aware that many scientists worked in more than one discipline during their lives. In such cases, scientists may be featured in two different chapters in the same volume. To facilitate searches for a specific person or subject, main entries on a given person or subject are indicated by bold-faced page numbers in the index.

Within each volume, material is divided into chapters according to subject area. For volumes 5, 6, and 7, these areas are: Exploration and Discovery, Life Sciences, Mathematics, Medicine, Physical Sciences, and Technology and Invention. For volumes 1, 2, 3, and 4, readers will find that the Life Sciences and Medicine chapters have been combined into a single section, reflecting the historical union of these disciplines before 1800.

Arrangement of Volume 1: 2000 B.C. to A.D. 699

Volume 1 begins with two notable sections in the frontmatter: a general introduction to science and society during the period, and a general chronology that presents key scientific events during the period alongside key world historical events.

The volume is then organized into five chapters, corresponding to the five subject areas listed above in "Format of the Series." Within each chapter, readers will find the following entry types:

Chronology of Key Events: Notable events in the subject area during the period are featured in this section.

Overview: This essay provides an overview of important trends, issues, and scientists in the subject area during the period.

Topical Essays: Ranging between 1,500 and 2,000 words, these essays discuss notable events, issues, and trends in a given subject area. Each essay includes a Further Reading section that points users to additional sources of information on the topic, including books, articles, and web sites.

Biographical Sketches: Key scientists during the era are featured in entries ranging between 500 and 1,000 words in length.

Biographical Mentions: Additional brief biographical entries on notable scientists during the era.

Bibliography of Primary Source Documents: These annotated bibliographic listings feature key books and articles pertaining to the subject area.

Following the final chapter are two additional sections: a general bibliography of sources related to the history of science, and a general subject index. Readers are urged to make heavy use of the index, because many scientists and topics are discussed in several different entries.

A note should be made about the arrangement of individual entries within each chapter: while the long and short biographical sketches are arranged alphabetically according to the scientist's surname, the topical essays lend themselves to no such easy arrangement. Again, readers looking for a specific topic should consult the index. Readers wanting to browse the list of essays in a given subject area can refer to the table of contents in the book's frontmatter. Finally, readers of Volume 1 should be aware that the volume includes a handful of events—for instance, the building of the pyramids in Ancient Egypt—that occurred before 2000 B.C.

Additional Features

Throughout each volume readers will find sidebars whose purpose is to feature interesting events or issues that otherwise might be overlooked. These sidebars add an engaging element to the more straightforward presentation of science and its times in the rest of the entries. In addition, each volume contains photographs, illustrations, and maps scattered throughout the chapters.

Comments and Suggestions

Your comments on this series and suggestions for future editions are welcome. Please write: The Editor, *Science and Its Times*, Gale Group, 27500 Drake Road, Farmington Hills, MI 48331.

Advisory Board

Amir Alexander
Research Fellow
Center for 17th and 18th Century Studies
UCLA

Amy Sue Bix
Associate Professor of History
Iowa State University

Elizabeth Fee
Chief, History of Medicine Division
National Library of Medicine

Lois N. Magner
Professor Emerita
Purdue University

Henry Petroski
A.S. Vesic Professor of Civil Engineering and
    Professor of History
Duke University

F. Jamil Ragep
Associate Professor of the History of Science
University of Oklahoma

David L. Roberts
Post-Doctoral Fellow, National Academy of
    Education

Morton L. Schagrin
Emeritus Professor of Philosophy and History of
    Science
SUNY College at Fredonia

Hilda K. Weisburg
Library Media Specialist
Morristown High School, Morristown, NJ

Contributors

Amy Ackerberg-Hastings
Independent Scholar

Mark H. Allenbaugh
Lecturer
George Washington University

James A. Altena
The University of Chicago

Peter J. Andrews
Freelance Writer

Kenneth E. Barber
Professor of Biology
Western Oklahoma State College

Bob Batchelor
Writer
Arter & Hadden LLP

Katherine Batchelor
Independent Scholar and Writer

Sherri Chasin Calvo
Freelance Writer

H. J. Eisenman
Professor of History
University of Missouri–Rolla

Ellen Elghobashi
Freelance Writer

Lindsay Evans
Freelance Writer

Loren Butler Feffer
Independent Scholar

Randolph Fillmore
Freelance Science Writer

Richard Fitzgerald
Freelance Writer

Maura C. Flannery
Professor of Biology
St. John's University, New York

Donald R. Franceschetti
Distinguished Service Professor of Physics and
*   Chemistry*
The University of Memphis

Diane K. Hawkins
Head, Reference Services—Health Sciences Library
SUNY Upstate Medical University

Robert Hendrick
Professor of History
St. John's University, New York

James J. Hoffmann
Diablo Valley College

Leslie Hutchinson
Freelance Writer

Joseph P. Hyder
Science Correspondent
History of Mathematics and Science

P. Andrew Karam
Environmental Medicine Department
University of Rochester

Evelyn B. Kelly
Professor of Education
Saint Leo University, Florida

Judson Knight
Freelance Writer

Lyndall Landauer
Professor of History
Lake Tahoe Community College

Josh Lauer
Editor and Writer
President, Lauer InfoText Inc.

Adrienne Wilmoth Lerner
Department of History
Vanderbilt University

Brenda Wilmoth Lerner
Science Correspondent

K. Lee Lerner
Prof. Fellow (r), Science Research & Policy Institute
Advanced Physics, Chemistry and Mathematics,
Shaw School

Eric v. d. Luft
Curator of Historical Collections
SUNY Upstate Medical University

Lois N. Magner
Professor Emerita
Purdue University

Ann T. Marsden
Writer

Kyla Maslaniec
Freelance Writer

William McPeak
Independent Scholar
Institute for Historical Study (San Francisco)

Duncan J. Melville
Associate Professor of Mathematics
St. Lawrence University

Sarah C. Melville
Visiting Assistant Professor of History
Clarkson University

Edith Prentice Mendez
Assistant Professor of Mathematics
Sonoma State University

Leslie Mertz
Biologist and Freelance Science Writer

J. William Moncrief
Professor of Chemistry
Lyon College

Stacey R. Murray
Freelance Writer

Lisa Nocks
Historian of Technology and Culture

Stephen D. Norton
Committee on the History & Philosophy of Science
University of Maryland, College Park

Neil Schlager
Editor and Writer
President, Schlager Information Group

Gary S. Stoudt
Professor of Mathematics
Indiana University of Pennsylvania

Dean Swinford
Ph.D. Candidate
University of Florida

Lana Thompson
Freelance Writer

Todd Timmons
Mathematics Department
Westark College

Philippa Tucker
Post-graduate Student
Victoria University of Wellington, New Zealand

David Tulloch
Graduate Student
Victoria University of Wellington, New Zealand

Stephanie Watson
Freelance Writer

Giselle Weiss
Freelance Writer

Michael T. Yancey
Freelance Writer

Introduction: 2000 B.C. to A.D. 699

Overview

Throughout the course of human history, science and society have advanced in a dynamic and mutual embrace. Regardless of scholarly contentions regarding an exact definition of science, the history of science in the ancient world is a record of the first tentative steps toward a systematic knowledge of the natural world. During the period 2000 B.C. to 699 A.D., as society became increasingly centered around stable agricultural communities and cites of trade, the development of science nurtured necessary practical technological innovations and at the same time spurred the first rational explanations of the vastness and complexity of the cosmos.

The archaeological record provides abundant evidence that our most ancient ancestors' struggle for daily survival drove an instinctive need to fashion tools from which they could gain physical advantage beyond the strength of the relatively frail human body. Along with an innate curiosity about the workings and meanings of the celestial panorama that painted the night skies, this visceral quest for survival made more valuable the skills of systematic observation, technological innovation, and a practical understanding of their surroundings. From these fundamental skills evolved the necessary intellectual tools to do scientific inquiry.

Although the wandering cultures that predated the earliest settlements were certainly not scientifically or mathematically sophisticated by contemporary standards, their efforts ultimately produced a substantial base of knowledge that was fashioned into the science and philosophy practiced in ancient Babylonia, Egypt, China, and India.

While much of the detail regarding ancient life remains enigmatic, the pattern of human history reveals a reoccurring principle: ideas evolve from earlier ideas. In the ancient world, the culmination of early man's intellectual advances ultimately coalesced in the glorious civilizations of classical Greece and Rome, where the paths of development for science and society were clearly fused. Socrates' observation that "The unexamined life is not worth living," expresses an early scientific philosophy that calls thinking people to examine, scrutinize, test, and make inquires of the world. This quest for knowledge and rational thought gave a practical base to the development of modern science and society.

The Formulation of Science

In ancient societies, the natural world was largely explained by the whims of gods or the dreams of man. Against this backdrop, the earliest scientists and philosophers struggled to fashion explanations of the natural world based on observation and reasoning. From a fundamental practice of counting, for example, ultimately evolved Pythagorean arguments about the nature of numbers. From attempts to explain the essential, basic constituents of the material world came Leucippus (c. 440 B.C.), Democritus (c. 420 B.C.), and Epicurus (342-270 B.C.), who argued that matter was composed of extremely small particles called atoms.

The advancement of science was consistently spurred by an increasing need to measure and manipulate the world. It is evident from early mathematical problems embodied in both the Moscow and Rhind papyri that practical mathematics and geometrical reasoning in were well developed in ancient Egypt, especially they as related to the science of building and construction. From these practical roots grew the flower of formal mathematical theory in ancient Greece.

Unfortunately, many of the once-cherished arguments of ancient science ultimately proved erroneous. Despite their flaws, these philosophical statements of logic and mathematics were stepping stones to modern scientific understanding. For example, until swept aside in the Copernican Revolution of the 1500s, errant models of the universe made by Ptolemy (127-145) dominated the Western intellectual tradition for more than a millennium. Although Aristotle's (384-322 B.C.) physics asserted that a moving body of any mass had to be in contact with a "mover," and for all things there had to be a "prime mover," this flawed but testable hypothesis did not yield until brought under the empirical and mathematical scrutiny of Italian astronomer and physicist Galileo Galilei (1564-1642) and English physicist and mathematician Sir Isaac Newton (1642-1727).

Amidst misguided concepts were often found examples of solid scholarship and brilliant insights into natural phenomena. Euclid's *Elements*, a synthesis of proofs, was the seminal mathematical text of the period. Aristarchus of Samos (310-230 B.C.) proposed that Earth rotated around the Sun more then 1,700 years before Polish astronomer Nicolaus Copernicus (1473-1543) defied church doctrine to reassert the heliocentric view. Another example of the depth of intellectual progress in the ancient world can be found in the work of Eratosthenes of Cyrene (276-194 B.C.), who, while working at the great library in Alexandria, Egypt, used elegant deduction and clever empiricism to deduce a reasonable estimate of the circumference of Earth at a time when only the most primitive of maps could be constructed.

Ancient Mesopotamian and Egyptian Science and Mathematics

Reconstructed from the scattered and fragmented remains of paintings and pots, the record of human civilization begins with the early settlements founded along the banks of the Tigris and Euphrates Rivers in about 3500 B.C. Although scholars don't believe that this early civilization invented writing, they did keep records, used a calendar based on the phases of the Moon, and made the first technological use of metals. The Mesopotamian culture that followed used cuneiform writing to detail the ebb and flow of early history, from the Sumerian King Gilgamesh through the collapse of Sumer and the rise of Babylon.

The advancements of science in Mesopotamia are concentrated in two divergent periods, the earlier Babylonian period (1800-1500 B.C.) and the later Seleucid period (400-100 B.C.). It's clear that many of the mathematical techniques and skills used in these societies predate either of these periods. The earliest papyrus and cuneiform writings known show a wide practical application of mathematics, especially as related to building and construction. In an effort to fashion more accurate calendars, particular attention was paid to the seasonal movements of the stars. The Babylonian development of a sexagesimal (base-60) numerical system allowed accurate calculation of the movements of the celestial sphere needed for the advancement of astronomy and the practice of astrology. By the sixth century B.C., Egyptian priests used crude instruments to measure the transits of stars across the night sky, and observations of the Sun allowed for accurate predictions regarding annual Nile flooding.

Writing in the ancient world let people codify and calculate many things. Alongside the laws of Lipit-Ishtar and the Amorite king Hammurabi (the first codes of law in world history) are remnants of ancient religious beliefs and primitive medical practices. Mummies, medicines, and ointments provide first-hand testimony of primitive medical practices in ancient Egypt. In China, the development of acupuncture marked a systematized and well documented integration of anatomy and physiology that persists today. Codified Hebrew dietary laws still reflect early religious practice and practical health concerns.

Mesopotamian mathematicians were able to construct base-60 systems, rudimentary uses of π, quadratic equations, and techniques that foreshadowed the Pythagorean theorem to influence the mathematics of Greece, Rome, Egypt, and China. Advancements in mathematics provided tangible progress. The counting board and abacus became important everyday tools to aid the development of trade. Priestly concern for the development of an algorithmic calendar needed for religious practice also allowed the development of mathematical methods for the accurate apportionment of foodstuffs. The incorporation of the Indian concept of zero provided a much-needed boost for theoretical and practical development in mathematics. Of utilitarian value, these workable mathematical systems utilizing the null or zero concept were nearly duplicated in ancient Chinese and Mayan cultures.

The Science of Greece and Rome

In ancient Greece, the cradle of classical civilization, human understanding of the physical universe and the mathematical laws that governed its behavior reached intellectual heights that would not be scaled again until late in the Renaissance.

Modern atomic theory and the logical divisions of matter trace back to Democritus and the pre-Socratic philosophers. The assertion that matter had an indivisible foundation made the universe finite and knowable within developing systems of logic by Zeno and other Greek philosophers. Early theories of the nature of matter became the subject of intellectual and societal discourse. Ideas of atomism and the nature of the elements were developed and argued in Plato's *Timaeus,* Aristotle's writings, and in the assertions of Epicurean and Stoic philosophers.

Trade contacts and the march of Alexander the Great's armies helped advance knowledge in ancient Greece by bringing scientific knowledge from early Egypt, Babylon, India, and China. In addition, the ancient world had a confluence of intellectual needs that did not require physical contact. The need to develop accurate calendars in China, for example, stimulated the development and use of many of the same astronomical and astrological techniques in Mediterranean cultures. Regardless of the culture, within these societies independent observations of the celestial sphere slowly yielded a firm foundation for the advancement of astronomy.

The assimilation of science and culture also provided a powerful drive in the evolution of cosmological and theological systems that associated the wanderings of the planets with the whims of gods and goddesses. Although the interpretation of celestial events as signs from the supernatural persisted well into eighteenth-century Europe, early myths and legends are replete with references to the prediction and observation of both solar and lunar eclipses. Beyond their importance in local religious festivals, interpretations of the heavens became, if not actual, at least legendary explanations for the birth of kings and the fall of dynasties. The prediction of a 585 B.C. solar eclipse by Thales, for example, is held to have led to the cessation of war between the Medes and the Lydians.

The Foundations of Modern Science

Aristotle's theories regarding chemistry and the four elements (e.g., earth, air, fire, and water) fostered an elusive and futile search for a fifth element (the ether) that would vex scientists until the assertion of relativity theory by German-American physicist Albert Einstein (1879-1955) in the twentieth century.

Until the collapse of the Western Roman civilization, there were constant refinements to physical concepts of matter and form. Yet, for all its glory and technological achievements, the science of ancient Greece remained essentially nothing more than a branch of philosophy. Science would wait almost another 2,000 years for experimental methodology to inject its vigor.

In some very important ways Roman civilization returned science to its Egyptian and Mesopotamian roots. The Romans, like those earlier civilizations, subordinated science to the advancement of architecture and engineering. Accordingly, Roman achievements were tangible: aqueducts, bridges, roads, and public buildings that were the finest and most durable to be built until late in the Renaissance.

Neither were the ethics of science much advanced in the Roman Empire. The very structure of Roman society retarded the growth of science because of continued reliance upon slave labor, a resource that provided little incentive to develop labor-saving technologies. The value of scientific thinking is also put in perspective when considering that although the nature of matter was called into question, the ancient social institutions of slavery remained largely unchallenged.

If science was little more than a handmaiden to the Roman arts of military tactics and weaponry, it was swept from the philosophical stage during the decay and fall of the Roman Empire, the beginning of the Dark Ages, and the rise of Christianity. Objective evidence regarding the universe became increasingly sifted through theological filters that demanded evaluation of observation and fact in theological terms. As new theologies ascended over old, societies that had relied upon ancient unifying political and social structures became fragmented and intellectually isolated. These divisions not only hampered further advancements in science, they led to a loss of much of the intellectual wealth of the classical age. Although Arab scientists managed to preserve a portion of the scientific knowledge and reasoning of the ancient and classical world, the fall of the Roman Empire plunged Western civilization into the Dark Ages and medieval era in which science was to fitfully slumber for seven centuries.

K. LEE LERNER

Chronology: 2000 B.C. TO A.D. 699

c. 3500 B.C. Beginnings of Sumerian civilization and pictographic writing that will evolve into cuneiform; some 400 years later, hieroglyphics make their first appearance in Egypt.

c. 3500 B.C. The wheel is invented in Sumer.

c. 2650 B.C. Half a millennium after the Pharaoh Menes united Lower and Upper Egypt, the Old Kingdom begins with the Third Dynasty, builders of the pyramids.

1200s B.C. Moses leads the nation of Israel out of Egypt, and writes down the Ten Commandments and other laws, laying the foundations of the Judeo-Christian tradition.

612-538 B.C. A series of empires rise and fall in the Near East, as Assyria is replaced by Babylon, which in turn is replaced by Persia; this is also the period of Israel's Babylonian captivity, during which key Judeo-Christian concepts such as the Messiah and Satan are forged.

Late 500s-early 400s B.C. Western philosophy is establishment by Thales of Miletus and others who follow; Eastern philosophies and religions are founded by the Buddha, Confucius, and Lao-tzu;, the beginnings of the Roman Republic (507 B.C.) and of democracy in Greece (502 B.C.).

c. 500 B.C. Greek philosopher and mathematician Pythagoras derives his famous theorem, studies the relationship between musical pitch and the length of the strings on an instrument, and establishes the high-

ly influential idea that the universe can be fully explained through mathematics.

479 B.C. Battle of Mycale, last in a series of engagements between Greeks and Persians, leads to Greek victory. It's followed by 75-year Athenian Golden Age; during this time, Socrates, Pericles, Sophocles, Herodotus, Hippocrates, and many others flourish.

c. 400 B.C. Greek physician Hippocrates and his disciples establish a medical code of ethics, attribute disease to natural causes, and use diet and medication to restore the body.

c. 350 B.C. Aristotle establishes the disciplines of biology and comparative anatomy, and makes the first serious attempt to classify animals.

334-323 B.C. Alexander the Great subdues more territory in less time than any conqueror before or since; as a result of his conquests, Hellenistic civilization spreads throughout much of the known world.

c. 310 B.C. Greek explorer Pytheas sets off on a voyage that takes him to Britain and Scandinavia.

c. 300 B.C. Euclid writes a geometry textbook entitled the *Elements* that codifies all the known mathematical work to its time; it's destined to remain the authority on mathematics for some 2,200 years.

c. 260 B.C. Aristarchus, a Greek astronomer, states that the Sun and not Earth is the center of the universe, and that the planets revolve around it; unfortunately, Ptolemy will later reject this heliocentric view in favor of a geocentric universe, a notion only refuted by Copernicus in the 1500s.

221 B.C. Ch'in Shih-huang-ti unites China, establishes the Ch'in Dynasty with himself as the first Chinese emperor, and orders the building of the Great Wall.

c. 220 B.C. Archimedes discovers the principle of buoyancy, noting that when an object is placed in water, it loses exactly as much weight as the weight of the water it has displaced.

c. 120 B.C. Chang Chi'en, a diplomat in the service of Chinese emperor Han Wu-ti, makes contact with Greek-influenced areas in western Asia; this is the first link between the Far East and the West, and it leads to the opening of the Silk Road.

31 B.C. Thirteen years after the assassination of Julius Caesar, his nephew Octavian defeats his last enemies, Antony and Cleopatra, at the Battle of Actium; this marks the effective beginning of the Roman Empire, and of the two-century-long *Pax Romana.*

c. A.D. 30 Jesus of Nazareth dies on the cross, and six years later, the Pharisee Saul has a vision on the road to Damascus; this leads him to embrace Christianity and help establish a formal religion based on the teachings of Jesus.

A.D. 105 Chinese inventor Tsai Lun perfects a method for making paper from tree bark, rags, and hemp.

A.D. 180 The death of Marcus Aurelius, last of four highly capable Roman emperors who ruled since A.D. 96, signals the end of the *Pax Romana,* and the beginning of the Roman Empire's decline.

A.D. 313 Roman emperor Constantine ends persecution of Christians, leading to adoption of Christianity as official religion of Roman Empire; later, he divides the empire into eastern and western halves, governed at Constantinople and Rome respectively.

A.D. 410 Half a century after the Huns entered Europe, beginning the destruction of the Western Roman Empire, the Visigoths under Alaric sack Rome; 60 years later, the Western Empire officially ends.

c. A.D. 450 Stirrups, brought westward by invading nomadic tribes, make their first appearance in Europe; in the view of many historians, this is one of the most important inventions in history since it makes warfare on horseback effective, and thus opens the way for knights and feudalism.

A.D. 622 The prophet Muhammad and his followers escape from Mecca, marking the beginning of the Muslim calendar—and of a series of Muslim conquests which by 750 will spread Islamic rule from Morocco to Afghanistan, and from Spain to India.

Exploration and Discovery

Chronology

c. 1472 B.C. Egypt's Queen Hatshepsut sends an expedition south to the land of Punt, in the region of modern-day Somalia.

c. 600 B.C. Carthaginian sailors circumnavigate the African continent; around this time, Carthaginians also establish European colonies such as Marseilles and Barcelona, and later a group led by Hanno founds colonies in West Africa.

c. 452-c. 424 B.C. Herodotus journeys throughout much of the known world, collecting material—including geographical information—for the *History*.

325-324 B.C. At the end of Alexander's conquests, which greatly increase Greek knowledge of the world, a general named Nearchus commands a fleet that explores the sea route from the mouth of the Indus River to the head of the Persian Gulf.

c. 310 B.C. Greek explorer Pytheas sets off on a voyage that takes him to Britain and Scandinavia.

c. 120 B.C. Chang Chi'en, a diplomat in the service of Chinese emperor Han Wu-ti, makes contact with Greek-influenced areas in western Asia; this is the first link between the Far East and the West, and it leads to the opening of the Silk Road.

55-54 B.C. Julius Caesar leads the first two Roman expeditions to Britain.

c. A.D. 100 Kanishka, greatest ruler of the Kushan empire centered in Peshawar, India (now Pakistan), establishes a vital link between East and West, helping introduce Hellenistic culture to India, and Buddhism to China.

A.D. 300s-600s A wave of westward migrations, beginning with the Huns in the 300s, forever alters the character of Europe and its awareness of the outside world.

A.D. 399 Fa-hsien, a Chinese Buddhist, travels to India and Ceylon, the first significant contact between China and the Indian subcontinent.

A.D. 563 Saint Columba establishes the first notable monastery in the British Isles, at Iona off the coast of Scotland; this and other sites in Scotland and Ireland will help preserve Western civilization during the dark centuries from 500 to 800.

A.D. 629-664 Chinese Buddhist pilgrim Hsüan-tsang (Xuan Zang) makes two journeys throughout India, collecting extensive geographic and cultural information.

Overview:
Exploration and Discovery 2000 B.C. to A.D. 699

Background

Lost far back in time are the names of the earliest explorers who roamed in search of game and edible plants across land they had not yet learned to cultivate. Many generations later, man drifted across great landmasses, populating Europe, Asia, Africa, and the Americas—some settling in agreeable climates, others becoming nomads forever on the move. No written records remain of these earliest journeys. Man made limited exploration of his world—sometimes from curiosity, but mainly from a constant need to search for food. The earliest explorers in the modern sense of the word, those who left us a written record of their travels, were limited to the peoples and lands of the small "world" around the Mediterranean. These exploits, often fueled by a civilization's desire for military conquest, are the earliest examples of true exploration.

The world's earliest recorded civilizations were those of Egypt and Sumer, followed by Babylon, Assyria, the Minoans on the island of Crete, and the Greeks. The earliest recorded examples of exploration were those by Egyptians who had led expeditions up the Nile River, and the Assyrians who explored the Tigris and the Euphrates Rivers. Around 1492 B.C., Egyptian Queen Hatshepsut sent a number of ships on a trading expedition to the land of Punt. The journey involved crossing 150 miles (241 km) of desert from the Nile to the shores of the Red Sea, then rowing and sailing some 1,500 miles (2,413 km) towards the Arabian Sea. Illustrations of the expedition were sculpted on the walls of the mortuary temple of Hatshepsut in an unparalleled record of such an undertaking.

Early Sea Expeditions

Although the Egyptians left some of the earliest records of their journeys, the first real explorers were the Phoenicians—renowned for their pursuit of trade and colonization in the Mediterranean region and for their crafts such as Tyrian cloth and glassblowing—who ventured from the coast in search of new routes for trade and expansion of their culture. The Phoenicians had a significant effect on human culture, encouraging trade between groups, thus exposing various civilizations and cultures in the Mediterranean

basin to each other and spreading science, philosophy, and other ideas through the ancient world. Many historians even believe that Pharaoh Necho II's Phoenician fleet may have circumnavigated the continent of Africa, thanks to a story about a large three-year Phoenician expedition around 600 B.C. recounted by Greek historian Herodotus in his work *History*.

The cross-cultural trade and exploration of the Phoenicians is best evidenced by the voyage over a century later by the mariner Hanno of Carthage. With the purpose of reinforcing Phoenician colonies and founding new ones, Hanno's expedition sailed through the Straits of Gibraltar, known then as the Pillars of Hercules, and along the north and west coasts of Africa to establish settlements to guard new and expanding trade routes. Hanno's detailed diary of the voyage, an inscribed stele known as the Periplus, is said to be the longest known text by a Phoenician writer. The success of his journey would not be repeated until the golden age of Portuguese exploration some 2,000 years later.

With the exception of Hanno, very few of the adventures of Phoenician explorers were recorded. It would not be until around 330 or 325 B.C. that another adventurer would leave behind an account of his explorations—this time to the north. Voyaging to the north in search of new lands beyond what the Greeks called the "Habitable World" was Pytheas of Massalia, a Greek adventurer, astronomer, and scholar. He followed the Atlantic coast around Spain and on to Brittany, setting course for Britain in search of tin and other items of interest to the Greeks, who were great trade rivals of the Phoenicians. Pytheas was the first to probe the cold arctic regions—possibly as far north as Norway or Iceland—and the first to bring back an account of the frozen sea.

Land Expeditions

Following the example set by sea explorers, important land exploration in antiquity was mostly conducted in a quest for military superiority and control, by soldiers leading their armies in wars waged in the Mediterranean arena, especially by the Greeks and the Persians. Military men such as Athenian officer Xenophon used their knowledge of the geography of the region to guide huge armies into battle before returning to their native lands. The best early example of military exploration is that of Alexander the Great, whose exploration in expanding an empire was so vast that it would be over a thousand years before another civilization, the Vikings, would even come close to its scale of conquest and discovery.

Beginning in 334 B.C., Alexander's Greek forces crossed into Asia Minor, defeated the Persians, then marched into Syria, Phoenicia, and Egypt before leaving the Mediterranean for the heart of the Persian Empire. They seized Babylon, then continued marching northeast to the shores of the Caspian Sea, through Afghanistan, over the Khyber Pass, and across the Indus River into India by 326 B.C. In India, Alexander set an example for mass exploration rarely equaled in history by splitting his returning expedition, sending his best ships under the command of the Greek admiral Nearchus to learn more about the "nature of the sea" by returning home via the Persian Gulf and ordering a separate party to travel overland through southern Persia. Alexander himself led a land party along the Makran coast, where the desert took its toll on his men, before returning triumphant to Babylon in 323 B.C. Over ten years, Alexander's armies had traveled over 20,000 miles (32,180 km), a feat not equaled in antiquity.

A few centuries after Alexander's Greek forces swept across Asia Minor to India, the Roman Empire reached its peak—shortly after Julius Caesar's conquest of Gaul in the Gallic War (58-50 B.C.). Following the Punic Wars (264-201 B.C.), the Roman Empire had rapidly expanded its borders, eventually amassing untold riches and dominating lands to the north as far as Britain (Albion) and to the south as far as the Atlas Mountains in northern Africa—but with a greater interest in colonization, not exploration. The Romans did, however, make possible trade, communication, and travel as never before experienced by the peoples of the known world. It was also the Roman era that spawned the greatest geographer of antiquity, Strabo of Asia Minor, whose monumental work *Geographica* would not be surpassed as a guide to the Western world until late in the Middle Ages.

Asian Exploration

In addition to the conquest and colonization of the empires of antiquity, Asian cultures quested for new opportunities for religious education and conversion—evidenced by the adventures of Chinese monks who journeyed long distances to the West to visit the birthplace of Buddha and to study Buddhist scriptures. Others, such as Fa-Hsien in the early fifth century and Hsüan-tsang in the seventh century, journeyed for many years throughout China and India and influenced widespread acceptance of the Buddhist faith in their homeland of China—and for new routes to commerce, especially for the luxury commodity of silk. The Chinese began venturing westward with this delicate resource—much desired by the wealthy Roman Empire—along the Silk Road, a set of overland routes connecting China to Antioch, Damascus, and other cities of the eastern Mediterranean. The Silk Road venture was initially organized by Han Dynasty emperor Wu-ti, who sent imperial bodyguard Chang Ch'ien and 100 men as emissaries to the West. Ch'ien and his party spent a decade as prisoners of the Hsiung-Nu, better known in the West as the Huns, but eventually escaped to discover Persia, Arabia, and even Rome, gaining a wealth of political, diplomatic, and economic knowledge for the Chinese, who, in turn, eventually established the Silk Road, linking their culture to the West.

Looking Ahead

In the Middle Ages, as the civilizations of the world developed and expanded, man's curiosity about his world developed into a desire to explore and conquer new lands and peoples. Merchants, monks, and mariners (and combinations of all three) ventured forth on expeditions. The nomadic military powers of the Vikings and the Mongols as well as the eight expansive military expeditions of the Crusades were prime examples of the fundamental need to discover and conquer. By the end of the Middle Ages, the political map of the world had been dramatically altered, and the impetus was in place for nation building in Europe and colonization in (as well as exploration to) the far-flung regions of the world.

ANN T. MARSDEN

Hatshepsut's Expedition to Punt

Overview

In the ninth year of her reign, the Egyptian pharaoh Hatshepsut (c. 1478-1457 B.C.) sent a number of ships on a trading expedition to the distant land of Punt, located to the south of Egypt. The Egyptians were fascinated by the exotic people, plants, and animals that they encountered in Punt, and proud of making the difficult journey to this mysterious, remote land. Hatshepsut commemorated the expedition in a series of sculptured reliefs, which decorated the walls of her magnificent mortuary temple at Deir el-Bahri. Hatshepsut's account of the trip and the Deir el-Bahri reliefs provide an unparalleled record of Egyptian trade practices, the type of boats they used for commercial voyages, flora and fauna of foreign lands, and the culture of the Puntites. Not only did the exotic land of Punt capture the Egyptian imagination, but the Punt trade also provided goods that were essential to Egypt's internal economic development and to its other international markets.

Background

Hatshepsut was not the first pharaoh to trade with Punt. The Egyptians had commercial relations with Punt as early as the Fifth Dynasty (c. 2470-2350 B.C.), and maintained trade sporadically for over 1,000 years, until trade lapsed in the Twentieth Dynasty (c. 1180-1060 B.C.). After this time, Punt is rarely mentioned in Egyptian texts, and historians cannot be sure whether, or to what extent, contact was maintained. The exact location of Punt is uncertain, but it was probably located south of Egypt near the coast of the Red Sea in what is now Sudan or Eritrea. Because references to Punt appear in a wide variety of ancient Egyptian texts (from love poems to autobiographies), and these texts cover such a long chronological period (roughly 2,000 years), it is not clear that "Punt" means the same thing at all times. Thus, the location of Punt seems to shift over time, and by the Greco-Roman period (310 B.C.-A.D. 395), it seems to have taken on an almost mythical character. Nevertheless during all periods, the Egyptians considered that they had a special relationship with this exotic foreign land, and they always demonstrated a high regard for the Puntites and their business.

The earliest record of Punt is found on the Palermo Stone, a broken monument on which a list of the kings of the first five dynasties is recorded. According to the Palermo Stone, Sahure, Pharaoh of the Fifth Dynasty (c. 2462-2452 B.C.) imported 80,000 units of 'ntyw (myrrh or frankincense) as well as large quantities of electrum (an alloy of silver and gold) from Punt. Another inscription from the Old Kingdom records a trip to Punt in which a dwarf was brought back and presented to the Egyptian court, and a private autobiography relates how a man accompanied his master to Punt and Byblos. This latter text brings up an important factor of the Punt trade; namely that it was only part of the Egyptian mercantile organization. In fact even at this early date, Egyptians regularly made trips up the Levantine coast to Byblos, in addition to voyages to Punt. The early route to Punt apparently led through Wadi Tumilat and the Bitter Lakes of the Nile Delta overland to a port on the Red Sea where large, seagoing boats were built and outfitted for the voyage south to Punt.

At the end of the Old Kingdom (c. 2190 B.C.), Egypt entered a period of chaos and disunity which made it impossible to maintain contact with such a distant land as Punt. However, when order had been restored in the Middle Kingdom (c. 2055 B.C.), trade with Punt was reestablished. Henenu, the chief steward of the Eleventh-Dynasty pharaoh Mentuhotep III (c. 2014-2001 B.C.), led an expedition of 3,000 men to Punt in order to renew trade. Henenu's autobiography gives a detailed account of the trip, including information about planning and organization. Thus we know that Henenu and his men took a new route, leaving Koptos in Egypt and traveling overland through the Wadi Hammamat to a port on the Red Sea. According to Henenu, he sent a team ahead of the expedition to dig wells at intervals through the 90 miles (145 km) of desert between Koptos and the Red Sea. Each expedition member was issued a staff and a leather canteen, and received a food ration of 2 jars of water and 20 biscuits a day. The baggage train even carried extra sandals in case anyone's wore out on the arduous journey. Once at the Red Sea port, the expedition built "Byblos ships," special, large, seagoing vessels used for the Byblos-Punt voyages. Recent archaeological excavation at Mersa Gawasis has evidently uncovered the remains of the Red Sea port used by the Middle-Kingdom traders.

Egyptian painting showing part of the expedition to the Land of Punt. *(Gianni Dagli Orti/Corbis. Reproduced with permission.)*

From this port, Henenu and his men sailed down the coast, landed, and apparently marched inland some distance before meeting the Puntites. They stayed in Punt for two to three months and then returned up the coast. Once they had landed at the Egyptian port, they had to pack all the trade goods onto donkeys and trek across the desert back to Koptos. Trade goods included myrrh, animal skins (leopard and cheetah), ivory, ebony, gold, and other luxury items. The Punt trade involved a major investment of time and resources on the part of the Egyptians, but they made huge profits from it.

Impact

Hatshepsut was one of the few women ever to rule Egypt. The daughter of Thutmose I (c.1506-1493 B.C.), the third king of the Eighteenth Dynasty, Hatshepsut married her half-brother Thutmose II and became queen of Egypt. When Thutmose II died (c. 1479 B.C.), his heir Thutmose III was too young to rule, so Hatshepsut, the young king's stepmother, became regent. After two years as regent, Hatshepsut decided to flout protocol and make herself pharaoh of Egypt. She ruled alone for over 20 years, but after she died Thutmose III finally became king. Because she had es-

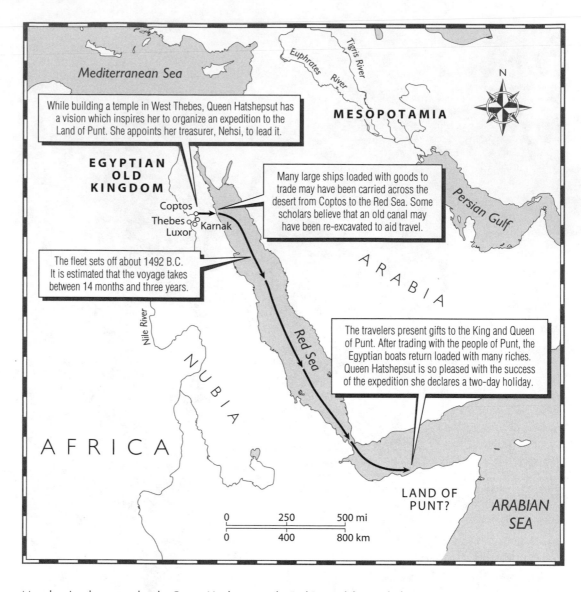

While building a temple in West Thebes, Queen Hatshepsut has a vision which inspires her to organize an expedition to the Land of Punt. She appoints her treasurer, Nehsi, to lead it.

Many large ships loaded with goods to trade may have been carried across the desert from Coptos to the Red Sea. Some scholars believe that an old canal may have been re-excavated to aid travel.

The fleet sets off about 1492 B.C. It is estimated that the voyage takes between 14 months and three years.

The travelers present gifts to the King and Queen of Punt. After trading with the people of Punt, the Egyptian boats return loaded with many riches. Queen Hatshepsut is so pleased with the success of the expedition she declares a two-day holiday.

Map showing the route taken by Queen Hatshepsut to the Red Sea and the Land of Punt.

sentially usurped the throne, Hatshepsut was especially concerned with legitimizing her rule. She spent a great deal of time and effort celebrating all her achievements and presenting them in the best possible light to her subjects. Most of her monuments emphasize her establishment of peace and plenty throughout Egypt. She built a spectacular mortuary temple to herself at Deir el-Bahri, and among the sculptured wall reliefs there, she included the story of one of her greatest accomplishments, the expedition to Punt. For Hatshepsut, who did not accompany the expedition, the Punt trade represented important economic development while also demonstrating the pharaoh's spirit of adventure in exploring exotic, distant lands. Although Hatshepsut claimed incorrectly that her expedition was the first ever to Punt, she did reestablish a trade that had lapsed

during the turbulent years of the Second Intermediate Period (c. 1650-1535 B.C.) The Punt reliefs were an important part of Hatshepsut's message to posterity, illustrating that she was a legitimate pharaoh who ruled effectively and brought peace and prosperity to her people.

Although the Punt reliefs may have functioned essentially as propaganda, they contain a wealth of information about the land of Punt, its people, and the conduct of trade. Egyptian artists, who may have accompanied the expedition, were careful to accurately depict the different characteristics of the Puntites and their surroundings. For example, men from Punt have dark, reddish skin, fine features, long hair and carefully dressed goatees. Men usually wear only a short kilt, but the women wear a type of dress or robe. The most fascinating and controversial figure on the reliefs is the queen of Punt who is depicted with an exaggerated swayback and rolls of fat covering her arms and short legs. The peculiar, but precise, portrayal of the queen of Punt has led many scholars to suggest that she suffered from a variety of serious medical conditions, although it may be that the Egyptians were simply impressed by her enormous size and wanted to record it accurately. In addition to people, the Punt reliefs show villages and surrounding flora and fauna. The typical Puntite village was located on the banks of a river where round, domed, mud huts were built on piers to keep them free of the river's floodwaters and safe from snakes, crocodiles, and hippos. Landscape features, plants, and animals included on the reliefs suggest that Punt was located in the hilly savannah country west of the Red Sea. Because no one has attempted to excavate in this area, many questions remain about the Puntite culture and its trade with Egypt. The Egyptians themselves apparently did not have a permanent camp in Punt, although Hatshepsut built a small shrine there in honor of the god Amun and the queen of Punt.

The Punt expedition provided Egypt with numerous luxury items. Most in demand were aromatic resins, myrrh and frankincense, which the Egyptians used for religious ceremonies. The Egyptians even brought back myrrh trees, their root balls protected in baskets, to be replanted at various temples. Other desirable commodities included panther, leopard, and cheetah skins; ivory; ebony; gold; live animals such as baboons and cattle; semiprecious stones; and spices. Not all these items were native to Punt, but were gathered further inland by the Puntites expressly for the Egyptian trade. In return, the Egyptians traded beer, wine, fruit, meat, jewelry, weapons, and other small items. The Egyptian economy obviously profited hugely from this trade, since much of what they brought back to Egypt was distributed to temples, private individuals, and the Byblos trade, or was given in exchanges of gifts with foreign rulers.

Hatshepsut's Punt expedition reestablished an important and lucrative trade. It is possible that Egypt exercised some form of authority over Punt, although the essential character of their relationship remained mercantile, rather than political. It would have been nearly impossible to rule Punt effectively from such a distance. On the other hand, the fact that Punt was located too far away to pose a threat to Egypt was probably very attractive to the Egyptians, who were always concerned about foreign invasion. Distance also added an element of fascination and adventure to the association. Quite simply, the Egyptians felt that they were traveling "to the ends of Earth" in order to gain untold riches and arcane knowl-

MYTHS REVEALED AS TRUTH: WHEN THE EXPERTS ARE WRONG

A great deal of what is known today about the ancient world represents the triumph of the amateur over the professional, or of what first seemed to be myth over apparent scientific skepticism. An example of the former was the 1952 translation of the Mycenaean script known as Linear B, an effort begun by an architect named Michael Ventris (1922-1956). Though he completed his translation with the help of a professional linguist, John Chadwick, much of the deciphering work had already been completed by Ventris—who, though knowledgeable in areas such as statistical analysis, was far from an expert on Mycenae.

As for the idea of seeming myth triumphing over skepticism, a variety of information related in the Bible has turned out to be accurate historical data. At one point historians, rejecting the biblical Creation story, were ready to throw out the proverbial baby with the bath water, rejecting David and Solomon—not to mention Abraham and Moses—as figures no more historical than Achilles or Heracles. In fact it now appears that virtually all biblical figures after Noah were real human beings.

A particularly interesting case was that of the Hittites of Asia Minor, who, though mentioned in the Bible, were unknown to their next-door neighbors in Greece. In the nineteenth century, archaeologists established the Hittite civilization as historical fact, and discovered the reason why the Greeks had never heard of them: the Hittites disappeared in c. 1200 B.C., about the time Mycenaean Greece was plunged into a dark age following the Dorian invasion.

One story that brings together all the above threads— mythology as fact, the brilliant amateur, and the layers of history in Greece and Asia Minor—is that of Heinrich Schliemann (1822-1890). Among the few joys of his poverty-stricken childhood had been a book of tales about the Trojan War, which everyone at the time assumed to be a myth. But not Schliemann: after amassing a great deal of wealth as a merchant, he set off for Turkey to find the ancient home of the Trojans. Not only did he discover Troy (under present-day Hissarlik), but he eventually uncovered much of Mycenae.

JUDSON KNIGHT

edge. In many ways, the Egyptian expeditions to Punt represent the world's first true explorations.

SARAH C. MELVILLE

Further Reading

David, R. *Handbook to Life in Ancient Egypt*. Oxford: Oxford University Press, 1998.

Grimal, N. *A History of Ancient Egypt*. Oxford: Blackwell, 1992.

Quirke, S. and J. Spencer, eds. *The British Museum Book of Ancient Egypt*. London: Thames and Hudson, 1992.

Trigger, B. G., B. J. Kemp, et al. *Ancient Egypt: A Social History*. Cambridge: Cambridge University Press, 1983.

The Role of the "Sea Peoples" in Transforming History

Overview

At the end of the thirteenth century B.C., the major powers of the eastern Mediterranean, Anatolia and Egypt, entered a period of political turmoil, economic privation, and population shifts that resulted in deep permanent changes in the cultural identity of the ancient world. The Hittite empire in Anatolia (modern-day Turkey) collapsed and disappeared completely; the civilization of Mycenaean Greece was utterly destroyed; cities in Syria and on the coast of the Levant were sacked and abandoned; and Egypt, having lost its territories in Syria and Palestine, just managed to maintain its borders. The ensuing period of disruption lasted for several hundred years. Various circumstances combined to produce this period of collapse, but the migrations and invasions of different population groups throughout the Mediterranean world were a major factor. Egyptian sources call these wandering tribes "peoples of the sea" from which modern scholars adopted the name "Sea Peoples."

Background

The cultures of the Aegean and Near East enjoyed a period of remarkable prosperity and general stability in the fourteenth and thirteenth centuries B.C. The great political powers of the day—the Egyptians, Hittites, Mitanni, and Babylonians—maintained sophisticated diplomatic relations, carried out extensive commercial activity, and struggled with each other to control the economically lucrative areas of Palestine, northern Syria, and the Levantine coast. The commercial centers of the Levant provided access to the Aegean islands and mainland Greece, where the Minoan and Mycenaean cultures prospered. Trade contact and diplomacy led to artistic and cultural exchange on a scale that far surpassed that of any earlier period. For the first time, there was a truly international order.

For most of this period, Egypt was universally acknowledged to be the "leading nation," and as such governed an empire that extended north through the Sinai, up the Palestinian coast into Syria, and south down the Nile into Nubia. The only threats to Egypt's prominence came from the Hittites of Anatolia, rebellious vassals, occasional incursions of Libyan tribes from the west, and sporadic attacks of pirates or nomads. Under the leadership of Ramses II (c. 1290-1224 B.C.), Egypt apparently dealt easily with these threats. In the fourth year of Ramses's reign, the Sherden, pirates from the Aegean islands or the Syrian coast, launched an aggressive attack against the Egyptian delta. Ramses defeated them and solved the problem of any future threat by incorporating the surviving Sherden into his own army. This is the earliest mention of any of the "Sea Peoples," and it is noteworthy that they became an important mercenary contingent in the pharaoh's army.

Inevitably, Ramses came into conflict with the Hittites over control of Syria, and fought them at Kadesh around 1286 B.C. The battle was inconclusive, but both armies included mercenary contingents whose tribal names would later appear among lists of the "Sea Peoples." Thus the Lukka and the Dardanians, both from the south coast of Anatolia, fought for the Hittites, while the Sherden fought for the Egyptians. At this point the Egyptians and Hittites were still strong enough to deal fairly easily with these aggressive groups. That both powers thought it desirable to include these people in their armies attests to the great fighting ability of the tribes, but it also anticipates a dangerous weakness on the part of the great powers; namely their dependence on mercenaries. The Egyptians and Hittites resolved their differences and signed a peace treaty in c. 1268 B.C. and the political situation in the Near East was apparently stable. However, the lengthy reign of Ramses II (c. 1290-1224 B.C.) led to a succession crisis, political confusion, and economic exhaustion which weakened Egypt and left her vulnerable to attack. Toward the end of the thirteenth century B.C., the Hittites also suffered from internal political problems that drained central authority and provided vassals with an excuse to rebel.

The Aegean Islands and Greek mainland also experienced prosperity and economic growth in the thirteenth century B.C. The Mycenaeans (named by scholars after the city of Mycenae) controlled Greece and the Aegean from separate city-states whose power depended on a strong warrior class. Each city was autonomous, ruled by a king, and protected by heavy fortifications. Exactly how much contact the Mycenaeans had with the Hittites, Egyptians, or Levantine trading ports is not known, but later Greek tradition dates the famed Trojan War to the end of the thirteenth century B.C., and there is evidence of trade and possible Mycenaean colonization in Anatolia and the Levant. The Mycenaeans fought each other frequently, and inevitably this constant warring took its toll. Starting in c. 1250 B.C., the Mycenaean economy suffered a period of decline which weakened the city-states and left them susceptible to outside threats.

Eventually, general economic decline and bad environmental conditions (drought) throughout the eastern Mediterranean made it impossible for the great powers to function effectively against increasingly active pirates and land raiders. Just what started the deadly attacks of these raiders remains subject to debate, but the devastation they caused is certain.

Impact

In the fifth year of the reign of the Egyptian pharaoh Merneptah (c. 1224-1214 B.C.), Egypt was attacked by the Libyans and a coalition of "Sea Peoples" including the Ekwesh, Shekelesh, Sherden, Lukka, and Teresh, all apparently originating in coastal Anatolia. This was not intended to be a simple raid to gain booty, but was a concerted effort to invade Egypt for the purpose of settling there. Merneptah managed to fight off the invasion but the worst was yet to come. About 30 years later, the pharaoh Ramses III (c.

1194-1162 B.C.) confronted a large invading army of "Sea Peoples." According to Ramses,

> . . . as for the foreign countries, they made a conspiracy in their lands. All at once the lands were on the move, scattered in war. No country could stand before their arms: Hatti (the Hittites), Kode, Carchemish, Arzawa, and Alashiya (Cyprus). They were cut off. A camp was set up in one place in Amor (Amurru, i.e. northern Syria). They desolated its people, and its land like that which has never come into being. (cf. A. Kuhrt, The Ancient Near East c. 3000-330 B.C., vol. II. London: Routledge, 1995, p. 387)

This inscription provides our only written description of these events. According to the Egyptians, the Hittite Empire, the cities of the Levant, and Cyprus had already succumbed to the invaders who then swept down the coast to invade Egypt. In fact, excavations in the cities of the Hittite Empire, northern Syria, and the Levant have shown massive destruction levels. The Hittite civilization, which had thrived in Anatolia for nearly 1,000 years, was so utterly destroyed that it was completely forgotten until its rediscovery in modern times. The Levantine cities of Emar and Ugarit were devastated and never reoccupied, as were several sites in Palestine. In Egypt, Ramses III and his army fought a desperate battle against the combined forces of the Peleset, the Tjerkru, the Shekelesh, the Da'anu, and the Washosh. The Egyptians prevailed but lost their holdings in Syria-Palestine and much of their land to the south of Egypt in Nubia as a result. Although Egypt managed to repel the invaders, the Twentieth Dynasty effectively represents the end of the New Kingdom and in some ways, the end of pharaonic Egypt. Never again would Egypt attain such high political and cultural levels; never again would Egypt lead the international order.

While the eastern Mediterranean succumbed to the invasions of the "Sea Peoples," Mycenaean Greece suffered total destruction as well. Just who was responsible for destroying the Mycenaeans is debated, but many scholars attribute their extinction to an invasion of new people, the Dorians, from the north. The devastation was so complete that many cities ceased to exist and the society sank into illiteracy, having lost the ability to write in the Linear B script used by the Mycenaeans. Greece had entered a "dark age" so severe it would last for about 400 years.

The end of the Bronze Age is characterized by migrations of different ethnic groups and the collapse of very old, established political bodies such as the Hittites. The destruction wrought by the "Sea Peoples" brought the Bronze Age to a bloody end, but many positive changes occurred as a result. The roving tribes, having no one left to prey upon, finally settled. The Peleset mentioned in the inscription of Ramses III have been identified by scholars as the Philistines, who settled in Palestine at this time. The Sherden and Shekelesh are associated with the islands of Sardinia and Sicily, respectively, while the Teresh may be linked to the Etruscans of Italy. Although these identifications are uncertain, they do underscore some of the key movements that occurred as a result of the invasions described above. Sometime during this period the Israelites, who were not "Sea Peoples" settled in Palestine and made the transition from a nomadic to an urban way of life.

The destruction of Bronze Age cultures left a political vacuum that would eventually be filled by new people and new political concepts. The great Bronze Age powers had all been monarchies in which the economy was controlled by a strong central authority. Most of these cultures functioned by a system of tax and distribution, with little opportunity for independent commerce. The collapse of this type of political system paved the way for innovation. For example, the Greeks, still living in separate city-states, abandoned the old aristocratic warrior society and eventually developed new types of governments, few of which included kings. The decline of the Hittite and Egyptian states allowed other Near-Eastern countries such as Assyria and Babylon to become more powerful. The influx of new ethnic groups throughout the Mediterranean led to technical innovations, such as the invention of the alphabet by the Phoenicians and the development and use of iron. The elusive "Sea Peoples" may have initiated a period of decline in the eastern Mediterranean, but out of this ruins rose the great cultures of the Iron Age—the Assyrians, Babylonians, Greeks, and Romans.

SARAH C. MELVILLE

Further Reading

Barnett, R. D. "The Sea Peoples." In *The Cambridge Ancient History*, third ed., vol. 2, part 2. (1975): 359-378.

Drews, R. *The End of the Bronze Age: Changes in Warfare and the Catastrophe of c. 1200* B.C.. Princeton, NJ: Princeton University Press, 1993.

Kuhrt, A. *The Ancient Near East c. 3000-300* B.C., vol. 2. London: Routledge, 1995.

Redford, D. B. *Egypt, Canaan and Israel in Ancient Times*. Princeton, NJ: Princeton University Press, 1992.

Sandars, N. K. *The Sea Peoples*. London: Thames and Hudson, 1978.

The Phoenicians: Early Lessons in Economics

Overview

The Phoenicians were members an ancient culture located in the region of the modern Middle East. They were renowned for their aggressive pursuit of trade and colonization in the Mediterranean Sea region during the last three millennia B.C. They established important cities and colonies throughout the region, including Sidon, Tyre, Carthage, and Berot. After its establishment, Carthage became the most important city in the western Mediterranean. It was the chief site of commerce and served as an important link in the trail of colonies that Phoenicia had established. There were also several other Phoenician settlements that provided an easy route to Spain. Spain was an important destination because of its wealth of precious metals. Other important imports were papyrus, ivory, ebony, silk, spices, precious metals, and jewels.

The Phoenicians made unique items that were desired all over the world. They were skilled craftsman noted for the fine detail of their work. Because of their wide range of travel, they often took an idea from one culture and improved upon it, or brought materials to parts of the world where they had previously been unavailable. Their most important exports were cedar wood, glass, and Tyrian cloth. Cedar was very important in the ancient Middle East because this natural resource was sorely lacking in many areas. In addition, the nobility desired its

A Phoenician galley. *(Corbis Corporation. Reproduced with permission.)*

fragrance. The Phoenicians developed the technique of glassblowing, which enabled glass products to be available to all strata of society. Perhaps the best-known and most desired export was their famous purple Tyrian cloth, made from the snail-like shellfish *Murex*. In fact, the Tyrian cloth was so identified with this group of people that the name Phoenicia comes from the Greek word for purple. Other significant exports included fine linen, embroideries, wine, and metalwork. Lastly, the Phoenicians conducted an important transit trade that shuttled people from one place to another.

There is little evidence concerning how the Phoenician civilization came about, or even how they referred to themselves in their own language; however, historians believe they used the term *kena'ani*. Interestingly, in Hebrew this same word can also mean "merchant," an apt description for the trade-loving Phoenicians. They probably arrived in the Mediterranean Sea area around 3000 B.C.

In general, Phoenicia was not a sovereign nation for much of its existence. It was constantly threatened and overrun by powerful nations. Egypt initially seized some control over Phoenicia, but once the Phoenicians wrestled free from their grip, they were often controlled by other entities, such as the Syrians and the Persians, until the Romans eventually assimilated the Phoenicians into their society. Despite this, the Phoenicians were able to make a lasting impact on the world.

The Phoenicians were instrumental in disseminating their form of writing, from which our modern alphabet is derived. They encouraged trade with other cultures and through commerce they exposed various civilizations and cultures in the Mediterranean basin to each other. Through their constant travel of their trade routes, the Phoenicians encouraged cultural exchange between various civilizations. This helped to hasten the spread of science, philosophy, and other ideas throughout the ancient world. The Phoenicians have even been incorrectly credited with inventing such important technologies as glass; nevertheless, they certainly were vital in the dissemination and improvement of that technology through the known world.

Background

The Phoenicians had established trade routes that used both land and sea. There is strong evidence that all of western Asia was served by land caravans led by Phoenicians. Phoenicia was involved in trade with most known cultures, and those they could not reach by land, they traveled to by sea.

There is little evidence that remains regarding Phoenician land trade. It is theorized by scholars that the extent of Phoenician trade stretched far beyond what the scarce historical records indicate. It may have extended to such places as central Africa. What is known is that the land commerce of the Phoenicians was carried out almost exclusively by caravans. This was done primarily for safety reasons, since large groups were much less vulnerable than small ones. This allowed the precious cargo to be protected from thieves that were invariably found along the route. There are records dating from 1600 B.C. that indicate Phoenician caravans traveled east with wood and returned with spices. They provided their closest neighbors with grain and other products, while supplying other cultures with goods they needed or desired. The Phoenician trade with Egypt was carried out on a large scale, where they imported such items as linen sails, papyrus, and scarabs, while exporting wine fabric and manufactured items. However, their most important land routes led to Arabia.

The Phoenician trade with Arabia was especially important because they were able to not only trade for desired Arabian goods, but it was the only way they could obtain products from India as well. Arabia was the main source of spices, such as cinnamon, for Phoenicia and in turn for the entire Western world. This area of the world was also known for its production of fine wool. There were other important goods that came from Arabia as well. In turn, it is believed that Phoenicia exported principally manufactured goods to Arabia, such as linen fabrics and glass, where it is believed they would be strongly desired.

As extensive as the Phoenician land trading routes were, the sea routes were much broader. Their voyages either went to trade with their own colonists or the natives of various countries. Most of the colonies that were established by the Phoenicians were trading settlements that were strategically placed near the supply of a particular commodity. The intent of this strategy was to provide the Phoenicians with a monopoly on that item so that they could ask any price. As an example, it is thought that Cyprus was originally colonized for its copper mines, while Lycia was established for access to timber. The colony worked to secure the commodity for Phoenicia, who in turn, provided the colonists with many different types of their own manufactured goods.

One colony, Carthage, was different. Carthage (New Town) was a great city of antiquity founded on the north coast of Africa by the Phoenicians for the purposes of establishing an important commerce center. The site was chosen in a natural bay that provided a safe anchorage and an abundant food supply. It served as a starting point for treks into the interior of Africa or voyages to Spain. Carthage was an extremely important city to Phoenicia until it was completely destroyed by the Romans in 146 B.C. during the Third Punic War.

The sea trade routes carried the Phoenicians to the ends of the known world. They traded throughout the Mediterranean Sea and even left to brave the Atlantic Ocean. There is speculation that these trade routes went as far as circumnavigating Africa, reaching Great Britain, and establishing trade with the Canary Islands.

In trading with other countries, Phoenicia desired to attain three goals. The first was to sell their manufactured goods for profit. The second was to sell goods from other nations at a profit. Third, Phoenicia wanted to obtain commodities from that country that would be desired by other nations. Thus, Phoenicia wished to profit from each country in three different ways. The main thrust of Phoenicia's economic philosophy was to always gain a monopoly. The traders would often come into a country and sell items necessary for living so reasonably as to put the native merchants out of business. Once this goal was achieved, the native market was now dependent solely on Phoenicia for support.

Impact

Some historians have unjustly characterized the Phoenicians as merely being passive peddlers of art and merchandise. Their historical achievements were not only noteworthy, but they were essential to the dissemination of knowledge and ideas in the ancient world. In many ways, their influence can be regarded as intermediary. The fact that they may have not been the originators of certain concepts or technologies should not diminish their contributions. They served as the conduit of thoughts and ideas between various cultures and enhanced the exchange of technology and information between them. Thus, many cultures that would have remained isolated from each other, possibly for extended periods of time, were able to benefit from each other's existence. The route of information can be traced from Mesopotamia and Egypt to Phoenicia, then on to Cyprus, Anatolia, and Syria. Thus, the Phoenicians can be credited with transmitting knowledge to many cultural groups.

The Greeks, in particular, owe a huge debt of gratitude to the Phoenician culture. First and foremost, they adopted the Phoenician alphabet for their own use with little variation or change. The Greeks also implemented the Phoenician standards of weights and measurement. Furthermore, the Greeks adopted the art of Phoenicia, from decorative motifs on pottery to the architectural styles of buildings.

In order to reign supreme in the area of commerce, the Phoenicians had to be skilled in navigational techniques. Their trade routes spanned the known world and beyond. They were characterized as patient yet fearless navigators who were willing to venture into regions where no one else would dare to go. The Phoenicians have been credited with the circumnavigation of Africa, the discovery of islands in the Azores, and they may have even reached Great Britain. They did this with the hope of establishing business monopolies or expanding their existing trade. They closely guarded the secrets of their trade routes and the techniques used to navigate them, but this information slowly leaked out to other societies. For instance, the Phoenicians are credited with utilizing Polaris (Pole Star) as a navigational aid. This

and many other routing techniques were a great help to subsequent seafaring people.

The Phoenician's influence on the world was primarily an economic and cultural one. They had little political influence and steered away from confrontation whenever possible. They made use of well chosen sites with natural harbors to build their cities and colonies. These geographical locations enabled the Phoenicians to build up a large merchant trade where they could provide an exchange of not only goods, but also information and ideas between cultures. Certainly subsequent cultures owe a great debt of gratitude to the Phoenicians.

JAMES J. HOFFMANN

Further Reading

Aubet, Marua E. *The Phoenicians & the West: Politics, Colonies & Trade.* New York: Cambridge University Press, 1996.

Bullitt, Orville, H. *Phoenicia & Carthage: A Thousand Years to Oblivion.* Pittsburgh: Dorrance Publishing, 1978.

Rawlinson, George. *Phoenicia.* North Stratford: Ayer Company Publishers, 1977.

Persia Expands the Boundaries of Empire, Exploration, and Organization

Overview

At the time of its establishment in the sixth century B.C., the Persian Empire was the largest known, and it gave southwestern Asia and adjoining regions an unprecedented degree of organization. The Persians built roads, dug canals, and established the first important postal system in history to maintain communication between the emperor and his satraps, or governors. Known for their religious tolerance, at least in the early days of the empire, the Persians respected the traditions of the people they conquered, for instance allowing the Israelites to rebuild their city of Jerusalem. Through Judaism and later Christianity, their Zoroastrian faith would have a powerful if indirect effect on the spiritual life of the West. Likewise, Persia would exert an enormous political impact through its influence on Greece.

Background

In about 3000 B.C. groups of tribes today known as Indo-Europeans began moving outward from

their homeland in what is now south-central Russia. Little is known about these groups, who ultimately scattered from India to Europe; in fact, the only evidence that they even existed is the strong relationship between the languages of Iran, India, and Europe. One group of Indo-Europeans, the Aryans, began moving into the region of modern-day Afghanistan, and between 2000 and 1500 B.C. they split into two groups. Some migrated eastward, where they conquered the peoples of the Indus Valley and established the Hindu civilization of ancient India, while others moved southward, into what is now Iran—a land to which they gave their name.

Eventually the Iranians further divided into groups, the most notable of whom were the Medes along the Caspian Sea in the north, and the Persians across the mountains to the south. At first the Medes were the dominant group, but they suffered a defeat at the hands of the Scythians, a seminomadic people from what is now the Ukraine, during the mid-seventh century B.C.

Darius I, also known as "the Great." *(Library of Congress. Reproduced with permission.)*

They recovered, however, and reassumed control over the region after 625 B.C., when the Median king Cyaxares (r. 625-585 B.C.) drove out the Scythians and began making war on Assyria. At that time the latter controlled the largest and most powerful empire in the region, but Cyaxares joined forces with another emerging power, Babylonia, to destroy Assyria in 612 B.C. After that, the Medes and Babylonians divided the Near East between them, and for a time Median influence extended all the way to Lydia in Asia Minor (modern Turkey).

In fact, the Medes and their Babylonian allies had paved the way for a new dynasty, led by the Achaemenid ruling house of Persia. This might not have happened, however, without the emergence of a strong leader: Cyrus II, better known as Cyrus the Great (c. 585-529 B.C.; r. 559-529 B.C.). Cyrus united the Persians against the Medes and defeated them in 550 B.C., thus bringing into being the Persian Empire.

Impact

Cyrus next waged war against Lydia, defeating it and capturing its king, Croesus (r. c. 560-546 B.C.), in 546 B.C., before moving on to wage war against the Ionian city-states of Greece. The latter event was significant in several regards. This was the first time a Mesopotamian power had penetrated the edges of Europe: indeed, the empires of the Assyrians, Babylonians, and Medes, though more multinational in character than those of the Egyptians and Hittites before them, had still been largely local in scope, drawing in peoples of relatively similar linguistic background. Thus the Ionian incursion marked the opening salvos in an attempt to forge an intercontinental realm— an effort that, in Greece at least, would fail, resulting in one of antiquity's most important conflicts.

In the meantime, Cyrus turned his attention toward Babylon, which he captured in 539 B.C. With this conquest, an event depicted in the biblical book of Daniel, the Persians controlled the largest empire that had existed up to that time, encompassing much of modern-day Iran, Iraq, Syria, Israel, Lebanon, and part of Turkey. With the later addition of Egypt, this would constitute the third-largest realm in Western antiquity; and the two larger ones—built by Alexander the Great (356-323 B.C.) and later the Romans—owed their existence in part to the example set by the Persians.

But the Persian Empire was more than merely a large political unit. As would be the case with the Mongols, who built the largest empire in all of history 17 centuries later, the Persians had no highly advanced civilization to impose on the world. Instead, they were more than happy to adapt and borrow from others, and they allowed their new subjects to go on with their lives much as before. Thus the Assyrians and Babylonians continued to worship their gods, and Cyrus even restored the Babylonians' temples. He also permitted the Jews to return to Israel and begin rebuilding their temple and their holy city, Jerusalem.

Cyrus met his end in battle in 529 B.C., and was succeeded by his son, Cambyses II (r. 529-522 B.C.), who conquered Egypt in 525 B.C. After Cambyses's death in the midst of an uprising, a general named Darius (550-486 B.C.; r. 522-486 B.C.) took the throne, and promptly set about dealing with the enemies of Cambyses. It took a year to subdue the insurrection, after which Darius marched into northern India and added large areas of land to his territories. This, too, marked an important event in the forging of multinational realm: never before had conquerors from southwestern Asia marched so far east, and here again the Persians set the example for Alexander.

Indeed the Persians, like the Medes and Babylonians before them, literally paved the way for their successors, and in part this occurred because Persia and Greece became embroiled in a long, bitter struggle that left the Greeks eager for retribution. Though Cyrus had been first to prosecute this conflict, it fell to Darius to fan the flames. In 516 B.C. he marched against the Scythians to stop them from supplying the Greeks with grain, and was prepared to attack Greece itself. As it turned out, the affairs of ruling his empire kept Darius busy for many years, but in 499 B.C. the Ionian city-states forced the issue by revolting against Persian rule. Soon the Athenians, Spartans, and others on the mainland joined their neighbors in Ionia against him, and the conflict came to a head in 490 B.C. with the Battle of Marathon, which ended in a Greek victory. Darius retreated, hoping to attack Greece again, but he died four years later without achieving his goal.

During his long reign, however, Darius had done much to transform the life of Persia. Unlike Cyrus, who does not seem to have held a strong religious belief, Darius accepted and sought to propagate the belief system taught by the prophet Zoroaster, sometimes called Zarathustra (c. 628-c. 551). Zoroastrianism proclaimed that the god Ahura-Mazda was supreme above all others, and it depicted his opponent Ahriman as the embodiment of evil: in other words, the Devil. This idea would have an enormous impact on the Israelites, many of whom had stayed in Persia, and all of whom remained under Persian rule in any case. Old Testament passages written prior to the Captivity certainly discussed the nature of evil; but only in the Book of Isaiah and other later works did the figure of Satan (a name derived from the Persian *Shaitan*) appear in the Jewish scriptures.

Nonetheless, the idea of a Devil never fully took hold in Judaism, a faith that generally depicts God as the father of all things, both good and evil. But as Christianity emerged from Judaism many centuries later, the concept of Satan as a distinct being became fixed. So too was the idea of the struggle between good and evil, which (with its implication that the struggle would eventually come to a head at the world's end), fueled Christians with a sense of mission. This in turn influenced the Christian zeal for hard work and productivity, attitudes that would ultimately propel the societies of Western Europe to unparalleled successes in the period after c. 1450 A.D. (Symbolic of the connection between Zoroastrianism and Christianity was the

Xerxes, king of Persia.

appearance, as recorded in the Gospels, of three Magi or Zoroastrian priests who followed a star to find the baby Jesus.)

As Ahura-Mazda provided a heavenly order, Darius sought to ensure the earthly order through what was by far the most efficiently organized empire up to its time. He set out to establish a system of justice that would be uniform throughout the empire, yet would also take into account local customs. Under his legal reforms, the provinces had two types of courts: one to administer law under the Persian legal code and one to deal with local matters according to the local system. A system of some 20 satrapies, or provinces, also allowed a measure of local rule. The satrap, who was usually a member of the royal family, had a free hand in ruling his local area, but of course he was expected to remain loyal to the emperor in Susa, the Persian capital.

In fact the Persians had three capitals. Susa, the winter capital reserved for reception of foreign visitors, lay at the end of the "Royal Road," which ran for 1,500 miles (2,400 kilometers) from the former Lydian capital at Sardis; but Darius built his palace and many other great structures at Persepolis, a springtime capital hidden away to the southeast. In summertime he used Hamadan or Ecbatana in Media. As for the Royal Road, at the time of its construction it was one of the longest in the world, and even compared with the interstate highways of the United States today, it is impressive. Interstate 75, which runs from the Canadian border in Michigan all the way to the southern end of Florida, is barely as long.

The Royal Road made possible one of the world's first postal systems. Along it lay some 80 stations, where one horse-bound mail carrier could pass the mail to another, a system not unlike the Pony Express used in the American West during the 1860s. The Persian messenger system was so efficient that Herodotus (c. 484-c. 420 B.C.) later wrote, "Neither snow, nor rain, nor heat, nor gloom of night stays these couriers from the swift completion of their appointed rounds." Today these lines are inscribed on the front of the central post office building in New York City.

Mail in the Persian Empire was only for the use of the king and satraps, and Darius maintained order by visible displays of military might. Behind the scenes, he employed one of the world's first intelligence networks to keep him informed of goings-on within the empire. Yet the Persian system of taxation was relatively liberal, at least at first. Subjects of the Persian Empire were taxed a flat 10% of their income, a system that would be adopted by the Islamic caliphates a millennium later. By contrast, the ancient Egyptians paid fully one-third of their income to the government (and of course, most Americans today have to give up more than 10%); but later, as taxes rose, the crippling effect on the Persian economy helped bring about the empire's downfall.

Throughout conquered lands, the Persians introduced a method of irrigation that helped to render areas in Egypt and central Asia fertile. Furthermore—and in another foreshadowing of Mongol rule—the stability provided by their empire facilitated hitherto unprecedented trade between India, central Asia, and the Mediterranean. Later, when Darius's son Xerxes (r. 486-465 B.C.) and his armies marched against the Greeks, Herodotus's catalogue of the assembled fighting force testified to the multinational character of the Persians' vast realm: there were Medes, Persians, Assyrians, Indians, Scythians, Thracians, and Africans.

But with Xerxes, the Persian Empire passed its summit. A less tolerant ruler than his predecessors, he ruthlessly suppressed revolts in Babylonia and Egypt, and tried to do the same in Greece when in 480 B.C. he launched the second attack his father had never lived to make. He defeated the Spartans at Thermopylae and burned Athens, but his navy lost the Battle of Salamis, and by 479 B.C. the conflict had fallen to the Greeks. Thereafter Xerxes lost interest in imperial expansion and spent most of his time in his palace, where he met his death by assassination in 465 B.C.

During the Peloponnesian War (431-404 B.C.) and its aftermath, the Persians tried to play Athens and Sparta off against one another. Though Persia in 387 B.C. signed a peace treaty with Sparta respecting Persian control over Asia Minor, Artaxerxes III (r. 359-338 B.C.) became embroiled in another Balkan conflict. This time he faced a challenger more formidable than any Greek: the Macedonian military leader Philip II (382-336 B.C.; r. 359-336 B.C.), who swore he would conquer the Persians' empire. Philip did not live to do so; instead the job fell to his son Alexander.

Thanks to the conquests of Alexander the Great, the Achaemenid empire of Persia came to an end in 330 B.C., yet it lived on through the empires that took its place. The Persian realms formed the backbone of Alexander's empire, and of that established by his general Seleucus (c. 356-281 B.C.) In 129 B.C. the Seleucid Empire fell to the Parthians; meanwhile, the example of Alexander had influenced the creation of India's own Mauryan Empire. By then, however, an even greater realm was on the rise, one whose leaders had also learned from the conquests of Alexander and the Persians before him: Rome.

JUDSON KNIGHT

Further Reading

Books

Neurath, Marie. *They Lived Like This in Ancient Persia.* New York: F. Watts, 1970.

Persians: Masters of Empire. Alexandria, VA: Time-Life Books, 1995.

Internet Sites

"Persian History." http://www.persian.com/aboutiran/history.

Hanno Sails Down the Coast of West Africa—and Perhaps Even Further

Overview

In about 500 B.C. an expedition led by the mariner Hanno sailed westward from Carthage in what is now Tunisia. Commanding 60 vessels on which were some 5,000 men and women, Hanno was charged with establishing trading colonies along the western coast of North Africa. This he did, founding a number of cities in what is now Morocco; but in a feat that would not be repeated until the golden age of Portuguese exploration some 2,000 years later, Hanno went much further. He and his crew sailed down the African coast, perhaps as far as modern-day Senegal or even Liberia—and perhaps, in the view of some scholars, even further.

Background

Some time after 800 B.C., the Semitic Phoenicians established Carthage near the site of modern-day Tunis. At its height, Carthage was home to some 1 million people, making it an almost unbelievably huge city by ancient standards. It expanded, adding colonies throughout North Africa, the Iberian Peninsula, and Sicily, and by the fifth century B.C. Carthage had emerged as the dominant sea power in the western Mediterranean. In 264 B.C. Carthage would find itself in conflict with the Roman Republic in the Punic (the Latin adjective for "Phoenician") Wars, and 118 years later, Rome would completely destroy the city.

But all of that lay far in the future when Hanno undertook his historic voyage. It appears that his was not the first group of Carthaginians sent to sail around the African continent: reportedly the Egyptian pharaoh Necho II (r. 610-595 B.C.) hired a group of Carthaginians in about 600 B.C. to sail around the coast of Africa. Some reports maintain that these earlier journeyers completed the feat, hugging the coastline and rounding the southern tip of Africa before coming back up the coast along the Indian Ocean to Egypt.

It is hard to know how to treat this tale, which seems to have an existence independent of Hanno's but which includes many of the same elements—though in this case the commission to undertake the voyage was from a foreign ruler. This in turn raises the question of what exactly Necho, if indeed he sent out the expedition, intended to achieve. Egypt in 600 B.C., be-leaguered as it was after years of attacks from outside powers, was hardly in a position to send out voyagers simply for the sake of curiosity or even to display Egyptian power. The first of these options was almost inconceivable among premodern states, and the latter most likely beyond the reach of Egyptian resources.

Impact

Virtually all details of Hanno's voyage—and indeed of his entire biography—come from an inscription left by Hanno himself in the form of a stele or pillar honoring the gods for him giving him safety on his journey. Known as the *Periplus,* it consists of 18 (and in some versions 19) numbered paragraphs regarding his exploits, and despite its short length is reputedly the longest known text by a Phoenician writer.

The text that has been passed down is a copy of a copy. Within a century of the time Hanno made the inscription, an unknown scholar prepared a serviceable but far from inspired rendering of the Semitic text into Greek. Over the centuries that followed, Greek, Greco-Roman, and later Byzantine clerks copied the original, and of the two versions known today, one dates back no earlier than the ninth—and the other the fourteenth—century. Some scholars maintain that Hanno himself did not actually make the inscription, but that it was the work of a priest who interviewed two sailors from Hanno's expedition.

In any case, the account begins by stating that "Hanno, king of Carthage" engaged in a voyage "to the Libyan lands beyond the Pillars of Herakles," and that the inscription is intended to honor "Kronos." In fact, the term *king* meant simply that he was a high magistrate, while *Libya* was the Greek name for Africa itself. Elsewhere the text refers to "Libyophoenicians"—in other words, Carthaginians—as well as "Ethiopians," the latter a general term for all the dark-skinned peoples of sub-Saharan Africa. Finally, Kronos was the name of a Greek god (or, more properly, a titan) who was father to Zeus. It is unlikely Hanno or any other Carthaginian would have erected a stele to Kronos; probably the intended deity was Baal Hammon, a variation of the god worshipped by the Carthaginians' Phoenician ancestors.

Remains of the circular Punic Gate at Carthage. *(Corbis Corporation. Reproduced with permission.)*

As is indicated in the first numbered paragraph of the inscription, the people of Carthage sent out Hanno's expeditionary force for the purpose of establishing cities or colonies to expand their trading empire. Hanno went on to note that "He set sail with sixty fifty-oared ships, about thirty thousand men and women, food and other equipment." It would have been impossible to fit 30,000 colonists on just 60 ships, and it is more likely Hanno brought with him 5,000 people—still an impressive number by ancient standards.

After sailing past the Pillars of Hercules, or the Straits of Gibraltar, some 1,000 miles (1,600 kilometers) west of Carthage, the journeyers entered the largely uncharted waters of the Atlantic. They then turned toward the shore of what is now Morocco, where they founded the first of several colonies at Thymiaterium, or present-day Mehdiya near the capital city of Rabat.

At the next spot (which may have been Cape Cantin, Cape Beddouza, or Cape Mazagan), the inscription states that they built an altar to Poseidon. Here again, the name is a Greek one, and probably they honored a Phoenician sea-god whose identity was unknown to the Greeks. According to the inscription, the journeyers then sailed eastward, a questionable detail since land lay to the east. Probably they navigated up the river known as

Oum er Rbia and entered a lake, where according to Hanno they found "elephants and other wild animals."

After another day's sail, the voyagers established cities called Karikon Teichos, Gytte, Akra, Melitta, and Arambys. Each of these has been identified with varying degrees of certainty, and the last is associated with a site where modern archaeologists have found Carthaginian remains—including evidence that inhabitants engaged in a signature Phoenician industry, harvesting shellfish to make purple dye.

At each city, the voyagers left settlers behind as they continued southward. Hanno's account first mentions human life (other than the Carthaginians themselves) in describing an encounter beside a river he called the Lixos, where the nomadic "Lixites" befriended the visitors. Apparently some of locals continued on with the Carthaginians, serving as interpreters. Later, however, in what may have been the Anti-Atlas Mountains, the journeyers encountered "hostile Ethiopians."

After sailing "past desert land," Hanno's party reached a small island five stades (about 900 meters or half a mile) in circumference, where they founded a colony named Cerne. This may have been Herne Island off the coast of the Western Sahara, though it is much larger than the dimensions given by Hanno. Soon they encountered more hostile inhabitants "who sought to stop us from landing by hurling stones at us," and afterward they passed a river— probably the Senegal—that was "infested with crocodiles and hippopotami."

They sailed for 12 days beyond Cerne, during which time the party observed a coastline "peopled all the way with Ethiopians.... [whose] tongue was unintelligible to us and to the Lixites in our company." On the twelfth day, they "came in sight of great, wooded mountains, with varied and fragrant trees." This may have been Cape Verde, or Cape Mesurado near the present-day Liberian capital of Monrovia; in any case, it is noteworthy that Hanno seemed to be pointing out the area's valuable resources with an eye toward commerce.

Soon they were in the Gulf of Guinea, where at night they saw numerous fires along the shore. At a place Hanno called the Western Horn, which is perhaps Cape Three Points in modern Ghana, they "heard the sound of pipes and cymbals and the rumble of drums and mighty cries. We were seized with fear, and our interpreters told us to leave the island." Still further on, Hanno's party saw a volcano he dubbed "Chariot of the Gods," which may have been Mount Cameroun. They sailed on for three days "past streams of fire" to what he called the Southern Horn, located either in Gabon or Sierra Leone.

In his final paragraph, Hanno related a strange incident that took place in the Southern Horn: "In this gulf was an island ... with a lake, within which was another island, full of savages. Most of them were women with hairy bodies, which our interpreters called Gorillas. Although we chased them, we could not catch any males: they all escaped, being good climbers who defended themselves with stones. However, we caught three women, who refused to follow those who carried them off, biting and clawing them. So we killed and flayed them and brought their skins back to Carthage. For we did not sail any further, because our provisions were running short."

This was the first written reference to the gorilla, a term that according to *Merriam Webster's Collegiate Dictionary* comes from the Greek *Gorillai*—"a tribe of hairy women mentioned in an account of a voyage around Africa." The word itself is apparently a Hellenic version of the kiKongo term *ngò diida*, meaning "powerful animal that beats itself violently"—but therein lies an intriguing aspect of the Hanno story. Based on the written account, the voyagers would still have had to travel much further, crossing the Equator, to meet speakers of kiKongo.

Thus is raised the question of whether Hanno actually rounded the southern tip of Africa, but chose to keep his further discoveries a secret. Pliny the Elder (c. A.D. 23-79), who stated that the gorilla furs remained on exhibit at a Carthaginian temple until the city's destruction by the Romans, wrote that "Hanno sailed from Gades [Cadiz] to the extreme part of Arabia," in the process circumnavigating the African continent. Most likely, however, Hanno actually turned back when he said he did: though the matter of the word gorilla's derivation is a compelling one, it seems rather less so in light of the fact that there is no evidence of a Carthaginian presence in southern or eastern Africa. More important, the Cape of Good Hope constitutes a formidable barrier, one that Portuguese mariner Bartholomeu Dias (c. 1450-1500)—possessing far more advanced marine technology than that of the Carthaginians—found impassable.

In any case, Hanno's account influenced numerous other writers, among them Herodotus (c. 484-c. 420 B.C.). According to the Greek historian, Phoenician traders on the coast of Africa, probably in the region of modern Senegal, would land on an island and set a certain amount of goods on a beach, then return to their ships. The Africans would then place an amount of gold, which was plentiful in their area, next to the Phoenicians' goods. If the Phoenicians judged that it was a fair exchange, they would take the gold and depart. If they did not, however, they would leave their goods on the shore until the Africans brought out more gold. Once they had agreed on an exchange, the Phoenicians would take their gold and sail away.

Herodotus's description seems to be drawn from Hanno's, and centuries later, Arab journeyers in the region reported that the Africans still maintained those trade practices. Carthage and its colonies, of course, had long since died out, but Hanno and his voyage remained legendary: even if he did turn around well on the west side of Africa, he still traveled further down the African coast than any sailor prior to the fifteenth century. In later centuries, writers as diverse as Montesquieu and Ralph Waldo Emerson wrote admiringly of Hanno and his exploits.

JUDSON KNIGHT

Further Reading

Books

Cary, M., and E. H. Warmington. *The Ancient Explorers.* London: Methuen, 1929.

Simon, Charnan. *Explorers of the Ancient World.* Chicago: Children's Press, 1990.

Sladen, Douglas. *Carthage and Tunis.* London: Hutchinson, 1906.

Internet Sites

Casson, Lionel. "Setting the Stage for Columbus." http://www.millersv.edu/~columbus/data/art/CASSON01.ART

"Hanno's Periplus on the Web." http://www-personal.umich.edu/~spalding/Hanno.

Lendering, Jona. "Hanno." http://home.wxs.nl/~lende045/Hanno/Hanno.html.

The *History* of Herodotus

Overview

The Greek historian Herodotus is known as the "Father of History" because he wrote the first historical work in prose in western literature. His book *History* explains the events leading up to the Greek and Persian wars of the fifth century B.C. Although he was writing long after the war, Herodotus talked to those who lived through it. He traveled the known world to learn the geography of the Persian Empire and to understand the way of life and political interactions of the people who lived there. Herodotus was one of the first to describe the geography, the culture, and the society of these areas.

Background

Herodotus was born about 485 B.C. in Halicarnassus, a Greek city on the west coast of Asia Minor, across the Aegean Sea from mainland Greece. Halicarnassus was a colony on the fringe of the Greek world. Though far from the center of Greek culture, its residents spoke Greek, considered themselves Greek, and followed Greek customs, religion, and politics. Little is known about the life of Herodotus, and he reveals little of himself in his work. Some sources say that his family was prominent in Halicarnassus. In his youth, Herodotus took part in a revolt against a local tyrant and was exiled to the island of Samos for a time.

The subject that interested Herodotus was the war between Greeks and Persians which began in 492 B.C. and finally ended 449 B.C. He was particularly interested in the period between 490 and 479, during which Persia twice attempted to invade mainland Greece. His interest in this conflict may have been kindled as a young boy, when his mother took him to the docks in Halicarnassus to watch the arrival of the defeated Persian fleet. Eventually, Herodotus came to believe that the successful outcome of this conflict had been essential for the preservation of Greek culture, and that the defeat of the Persians saved not only Greece, but the rest of the world, from being overrun by eastern despotism.

The world that Herodotus knew consisted of the Mediterranean Sea and the countries that surrounded it, for very little was known about the rest of the world at the time. Herodotus began traveling as an adult, although it is not known exactly when or why. It is also unclear how he traveled or how he paid his way; he may have been employed by a trader or worked as a trader himself. His writings show that he traveled to distant lands on trading ships, and that he knew a great deal about boats, weights, systems of measurement, and trade goods. He may also have traveled overland by trading caravan. He journeyed as far south as Aswan in Egypt, east to the Euphrates River, west to Cyrene (Libya today), and into North Africa and the area north of the Black Sea. He traveled through much of the Persian Empire, from the Mediterranean Sea in the west to the Indus River in the east, and from Russia in the north to the Arabian Sea in the south. He writes of Niger and Chad in central Africa, the Caspian Sea in Russia, and of Scythia, which included parts of today's Hungary and Rumania. It is doubtful that Herodotus actually visited these places, but he heard them described by people who had.

In the fifth century B.C., Greece was made up of independent city-states variously ruled by dictators, kings, or citizens. These city-states seldom agreed, but they did share the same culture, language, and customs. The largest city-state, Athens, threw out its kings in 680 B.C. and established a system of rule by elected officials with the consent of its citizens. In 560, dictators took over Athens, but they fell from power in 510, and a new plan of government was established—the world's first democracy. Under this new plan, every qualified citizen helped to run the government. This unique system was admired by many writers, including Herodotus, who emphasized the contrast between the governing systems of Athens and Persia.

Impact

Herodotus is known as the "Father of History" because he was the first to undertake a written and unified explanation of an historic event in prose. Previous histories had been written as narrative poetry, like the *Iliad* or other long sagas of rhyming verse. Herodotus may have intended his work to appear in several phases, although almost all this work survives as a unified book called *History*. The first five sections of the work, traditionally called "books," explore the various parts of the Persian Empire. The books added immensely to the ancients' knowledge of the world and the lands on the shores of the Mediterranean. Because Herodotus's work was read in public to Greek audiences, he did not give details about Greece. He did explain the Greek political situation of 500 B.C. because his contemporaries would not know about or remember this period, which was 50 years before.

Herodotus has also been called the "Father of Anthropology" because he recorded the behavior, beliefs, customs, and culture of the people in the Persian Empire. He was astonished by some of these cultures, repelled by some, but admired others. One of his descriptive phrases is familiar in connection with the U.S. postal system. Praising the system of communication in Persia, he says of its messengers: "Neither snow nor rain nor heat nor gloom of night stays these couriers from the swift completion of their appointed rounds." Herodotus often commented on unusual, astonishing, or remarkable events and people, often reporting hearsay without commenting on whether he believed the stories or not. His works reveal his great enthusiasm for the diversity of life.

The main focus of *History,* however, was the war between Greece and Persia. All Herodotus's travels, inquiries, and insights serve to illuminate the background of this conflict and to explain the origin of the enmity between the two countries. Herodotus saw the conflict as a series of struggles between the independent cities of Greece in the west and the huge sprawling empire of Persia ruled by eastern despotic autocrats. He believed that the gods punished humans who displayed an excess of overweening pride, aptly embodied in the Greek word *hubris*, and that the Persian defeat clearly illustrated his belief. However, Herodotus always emphasized

Herodotus reading his historical narrative to an audience. *(Archive Photos. Reproduced with permission.)*

the role played by the character of men, not the interventions of the gods, in his writings. This rationalistic approach to the writing of history was completely new.

Greek city-states shared a common language, culture, religion and history, and in spite of their long-standing differences, they unified in 480 B.C. to defeat the Persians. Persia was ruled by Darius II, who conquered and controlled a collection of varied cultures, with different languages, customs, and religions. Darius was so powerful he was able to maintain a huge army and navy made up of many different peoples under a single, unified command. Herodotus was astonished at the size of the Persian forces, but the numbers he quotes are so fantastic that historians discount them as gross exaggerations.

Herodotus had a storyteller's enthusiasm for a good yarn, as well as a discriminating eye for detail and a keen sense of geography. He also knew what was historically important, and thus understood the significance of the great naval battle at Salamis, the pivotal engagement in the Persian defeat. The Greco-Persian wars were a crucial turning point in the freedom of the Greek city-states, and the Greek victory ensured the triumph of democracy and the survival of the rule of law over Persian despotism.

Herodotus praised the Athenian leader Themistocles for engineering the victory at Salamis, although he condemned the man and his life. The battles of Marathon and Salamis, fought in 490 and 480 B.C., respectively, were won by the Athenians in spite of the fact that their ally Sparta did not arrive to help. These victories led to an Athenian Empire, but also eventually to internal rivalries, constant strife, and finally to the Peloponnesian War (431-404 B.C.) which destroyed Greek power.

Herodotus traced the events that led to the conflict between Greece and Persia, described the important clashes of the war, and recorded what he knew of the events and the people. He did this admirably, though with many side trips and digressions. He included legends and stories that were so incredible even he doubted them. Why use them? Because they were the only data available to illuminate the culture. All his sources were oral because there were no written records, conventional versions, or official documents for him to consult. Herodotus spoke to eyewitnesses or to people who knew others that had participated in the events. Hearsay was often all he had. He included hearsay even when he was skeptical about it but told the reader to believe if he wished. When he recorded two different accounts of an event, he gave his preference for one version over the other. Herodotus's history was not scientific, but his accounts showed what ancient people believed about their own history.

As an early geographer, the first historian, and the first anthropologist of the western world, Herodotus is unparalleled in ancient literature. His successors, like the Greek historian Thucydides (c. 401 B.C.) and the Greek biographer Plutarch (A.D. 46-after 119), could consult written data and had developed more ordered techniques for writing history. Herodotus had only his wit and purpose to build on. No successor produced a work of history or anthropology as readable, entertaining, or complete as the *History*.

LYNDALL BAKER LANDAUER

Further Reading

Herodotus. *The Histories.* trans. by David Grene. Chicago: University of Chicago Press, 1987.

Myres, Sir John L. *Herodotus, Father of History.* Chicago: Henry Reghery Company, 1971.

Romm, James. *Herodotus* . New Haven, CT: Yale University Press, 1998.

Xenophon and the Ten Thousand

Overview

In 401 B.C. Cyrus the Younger (424?-401 B.C.) marched into the heart of the Persian Empire to take the throne from his brother Artaxerxes II (reigned 404-359/58 B.C.). The core of his army was a contingent of Greek mercenaries, later known as the "Ten Thousand." Their ranks included a junior officer named Xenophon (431?-354? B.C.). After Cyrus was slain on the battlefield of Cunaxa, Xenophon helped lead the harried Greek soldiers north to the Black Sea and then home. Their trek through the Kurdistan mountains and Armenian tableland remained the only exploration of these isolated and inhospitable regions until modern times. Xenophon's *Anabasis,* which recounts these exploits, is one of the very few extant eyewitness accounts of the Persian Empire.

Background

The death of Xerxes the Great (519?-465 B.C.) marked the beginning of the Achaemenid dynasty's downslide. Persia was gradually weakened under the generally impotent rule of his successors. Artaxerxes I's reign (465-425 B.C.) was plagued by Greek incursions in Asia Minor and several rebellions. Xerxes II reigned but 45 days before his assassination. His half-brother, Darius II Ochus (reigned 423-404 B.C.), immediately seized the throne. However, his power was compromised by court intrigue and corruption. Darius was also dominated by eunuchs and his half-sister and wife, Parysatis.

In 413 B.C. Darius attempted to reassert Persian suzerainty over the Greek coastal cities of Ionia. Operations were directed by Pharnabazus, satrap (governor) of Dascylium, and Tissaphernes, satrap of Lydia and Caria. An alliance against Athens was formed with Sparta and much of Ionia was recovered. Tissaphernes's limited support of Sparta hampered further success. Consequently, Parysatis convinced Darius to appoint their son Cyrus to replace Tissaphernes (407 B.C.). Cyrus helped rebuild Sparta's fleet, which then decisively defeated the Athenians at Aegospotami (405 B.C.). This quickly ended the Peloponnesian War.

Cyrus was in attendance at his father's death in 404 B.C. When his older brother was crowned Artaxerxes II, Tissaphernes accused Cyrus of plotting the new king's death. Parysatis interceded on Cyrus's behalf and persuaded Artaxerxes to send him back to Asia Minor. Upon returning to Sardis, in Lydia, Cyrus immediately commenced preparations to seize the throne.

Impact

On the pretext of wishing to subdue the defiant Pisidians, Cyrus assembled an army. Its core was composed of approximately 14,000 Greek mercenaries. Over 10,000 of these where hoplites—heavily armored infantry, equipped with 6- to 10-ft (2- to 3-m) spears. The rest were peltasts—lightly armed support troops. These were recruited from all over Greece. Proxenus of Boeotia alone recruited 1,500 hoplites and 500 peltasts. He also enlisted his friend Xenophon. Cyrus appointed the Spartan exile Clearchus commander-in-chief of the Greeks.

In March of 401 B.C. Cyrus marched from Sardis with a mixed force that included his Greek mercenaries, 2,600 cavalry, and an unspecified number of Asiatics. The army headed southeast toward Pisidia. By June, they had marched well beyond the Pisidia and onto Syria. The Greeks realized they had been deceived and refused to advance further. Clearchus won them over with Cyrus's assurance that they would campaign no further than the Euphrates. However, in late July, when the army arrived at Thapsacus, on the western banks of the Euphrates, Cyrus announced his true intentions. Only the promise of rich rewards convinced the Greeks to follow him across the river and into Babylonia.

Keeping the Euphrates on their right, the army eventually reached the Charboras (Araxes, modern Khabur). Crossing this river, they continued their march along the Euphrates through the desert. Xenophon told of the strange beasts they encountered there, including wild asses, ostriches, bustards, and antelope. They next came upon the Mascas river. In mid-stream was a large deserted city that Xenophon referred to as Corsote.

As the army proceeded to Pylae, they were increasingly harassed by forward elements of Artaxerxes's forces. The king had been forewarned of the invasion by Tissaphernes and had hastily assembled an army of 30,000 foot-soldiers and 6,000 cavalry. The opposing armies finally met on the third of September at Cunaxa, about 100 miles (161 km) north of Babylon.

Clearchus, commanding the Greek center, drew up his forces to take advantage of the Euphrates on his right. Proxenus, with Xenophon at his side, commanded those nearest the river. Cyrus took the field further inland. Tissaphernes led the Persians directly opposite Clearchus while Artaxerxes held the center. The King's right wing stretched menacingly beyond Cyrus's left.

The lightly-armed Persian infantry proved no match for the Greek hoplites. While they crushed Tissaphernes's line, the left wing of Cyrus was in danger of being enveloped. Realizing a decisive blow was needed immediately, Cyrus charged into the enemy center with a squadron of 600 cavalry. He succeeded in reaching Artaxerxes and wounding him, but was himself slain. As Clearchus wheeled on Artaxerxes's center, Cyrus's Asian mercenaries were fleeing the field in disarray. However, the Greeks routed what remained of the Persian hosts.

It was not until the next day that Xenophon and the Hellenic commanders learned of Cyrus's death. Weeks of maneuvering ensued before Clearchus satisfactorily negotiated with Artaxerxes for safe passage back to Ionia. While being "escorted" by Tissaphernes's forces, they saw the remains the Median Wall—the great Opis-Sippar fortifications of Nebuchadnezzar (partially preserved between Sippar and Nuseffiat, Iraq). According to Xenophon, the wall was 20 feet (6 m) thick and a 100 feet (30.5 m) high and built of bituminous sun-dried bricks.

After crossing the Tigris (south of Baghdad) and proceeding north along its eastern bank to the tributary Zapatas (Greater Zab), Clearchus, Proxenus, and staff met with Tissaphernes to negotiate further. During this meeting, the satrap treacherously murdered them. Xenophon was among the new commanders immediately elected to lead the Greeks.

Though the hoplites had proven their worth in pitched battle, they were slow and had limited maneuverability. This made them vulnerable to cavalry while on the march. Thus, Xenophon's suggestion was to cross the Zapatas and withdraw northward as quickly as possible to rougher terrain. This would neutralize Tissaphernes cavalry, which were now openly harassing them. They would then look to cross the Tigris in hopes of making their way west to the Aegean. If unsuccessful, they would head north to the Black Sea. With this course of action agreed upon, Xenophon organized the Rhodian slingers, archers, and other peltasts into various screening units to delay enemy skirmishers while the main body of hoplites retreated in hollow-square formation.

Following the course of the Tigris, the Hellenes came upon a large deserted city known as Larissa. This was likely the southwest corner of ancient Nineveh (near modern Nimrud, southeast of Mosul). Eighteen miles (29 km) further on (just north of Mosul) they passed a long abandoned fortress called Mespila. Scholars believe this was northwest section of Nineveh, whose circuit was said to have been 56 miles (90 km).

The Greeks were frustrated in their attempts to cross the Tigris because of its depth and breadth. Furthermore, just north of Jezirah, the Carducians hills hung sheer over its course, making passage along the banks impossible. Thus, they were forced up into the Kurdistan mountains. This region, and the Armenian tableland to the north, had never really been subdued by Persia. Xenophon's *Anabasis* provides the first Western reference to the independent and warlike inhabitants of the area known as the Carducians or Kurds. Strongly opposed by them, the Greeks fought their way north to the Centrites (Eastern Tigris), reaching it in early December.

Determining the route taken by the Greeks from this point has been problematic. After crossing the Centrites, most scholars believe they pushed on to the northwest before turning northeast and heading up into the Armenian tableland. They then made for the Teleboas by way of Mus. Next, they struck out across the trackless countryside, enduring the cold and snow before reaching the Western Euphrates, which they crossed somewhere near Erzerum. In January of 400 B.C. they marched north to the Harpasus river through the territories of the Taochi and Chalybes. The latter earned Xenophon's praise as the fiercest savages they had encountered.

Following the Harpasus, the Greeks finally reached the city of Gymnias, where they learned they were but days from the port of Trapezus (now Trabzon, Turkey) on the Black Sea. They arrived at this Hellenic colony in early February. Their number now stood at approximately 10,000—thus their name the "Ten Thousand." Making their way west along the southern shores of the Black Sea, the Ten Thousand even-

tually reached Chysopolis on the Bosphorus, whence they departed from Asia Minor and crossed over to Byzantium.

News of the successful retreat of the Ten Thousand through unknown territory under harsh conditions and against hostile natives created a sensation in the Greek world. Xenophon recorded these exploits in the *Anabasis,* and he is generally given the lion's share of credit for the expedition's survival. The campaign and exploration of Kurdistan and Armenia were adventures of the first order, but they also had lasting military and political implications.

The battle of Cunaxa reinforced what the Greeks already knew about Persian infantry— they were no match for hoplites. More importantly, the subsequent five-month, 1,500-mi (2,414-km) trek of the Ten Thousand revealed the essential internal weakness of the Persian Empire. This encouraged renewed Greek incursions into Persian territory. The Spartan king Agesilaus II (444?-360 B.C.), who employed Xenophon and elements of the Ten Thousand, campaigned against Persia in Asia Minor with great success, decisively defeating Tissaphernes at Sardis in 395 B.C. These events also influenced Philip II of Macedon's (381-336 B.C.) decision to invade Persia, which was successfully carried out by his son Alexander the Great (356-323 B.C.). Thus, the exploits of the Ten Thousand were indirectly responsible for the fall of the Persian Empire.

STEPHEN D. NORTON

Further Reading

Books

Cary, Max, and E.H. Warmington. *The Ancient Explorers.* Rev. ed. Baltimore, MD: Penguin, 1963.

Cawkwell, George. *The Persian Expedition.* Trans. of Xenophon's *Anabasis,* by Rex Warner. New York: Penguin, 1949.

Dillery, John. *Xenophon and the History of His Times.* London: Routledge, 1995.

Hirsch, S.W. *The Friendship of the Barbarians: Xenophon and the Persian Empire.* Hanover, NH: University Press of New England, 1985.

Jacks, Leo V. *Xenophon, Soldier of Fortune.* New York: Scribner, 1930.

Warry, John. *Warfare in the Classical World.* Norman, OK: University of Oklahoma Press, 1995.

Periodical Article

Roy, J. "The Mercenaries of Cyrus." *Historia* 16 (July 1967): 287-323.

Ultima Thule, Brettanike, and the
Voyage of Pytheas of Massalia

Overview

For centuries, "ultima Thule" has been synonymous with the ends of the earth. Pytheas of Massalia (fl. c. 325 B.C.) first used "Thule" to refer to the northernmost land he visited during his North Atlantic voyage. Seneca (4 B.C.-A.D. 65) later dubbed it "ultima ('farthest') Thule." Though its precise location remains unknown, it seems certain Pytheas ventured at least as far as 62°N. He was also the first to circumnavigate Britain and to record accurate geographic and ethnographic information about northwestern Europe.

Background

Greek awareness of Brettanike (Britain) and sub-Arctic Europe was reflected in its mythology. The mystical river Eridanus was originally thought to flow north through western Europe and was associated with the production of amber. Homer's (fl. c. 850 B.C.) *Illiad* makes references to the land of the Laestrygones where the paths of day and night lie close together as well as the Cimmerians who lived at the ocean's edge in cold and gloom.

A clearer picture of the North Atlantic lands did not begin to emerge until the seventh century B.C., when Greek colonists penetrated the western Mediterranean and began trading with Tartessos (modern Seville, Spain). Tartessian merchants had long since established trade routes with Brittany and Cornwall for tin and Ireland for gold and copper. The Greeks more fully exploited their connections with Tartessos when the Phocaean port of Massalia (modern Marseilles, France) was founded about 600 B.C. There is also evidence that a Phocaean by the name of Midacritus journeyed as far north as Brittany and returned with a load of tin.

Greek access to the Atlantic was severed around 500 B.C. when the Phoenicians drove them from Spain and destroyed Tartessos. The Carthaginians henceforth controlled the Pillars of Heracles (Straits of Gibraltar) from their colony of Gades (founded c. 1100 B.C.). Massalia still maintained control of the coast as far south as Emporion (Ampurias, 75 miles or 121 km northeast of present-day Barcelona, Spain), but their only means of obtaining Atlantic tin and copper was by caravan through Gaul.

Pytheas was the next Greek to sail the Atlantic. Though the exact date of his voyage remains in doubt, it is possible to make an approximate determination. It appears he used a reference work dating to 350 B.C. Further, Dicaearchus of Messene (fl. 326-296 B.C.) referenced Pytheas's treatise. Thus, the voyage must have occurred sometime between 350 and 290 B.C. In addition, Carthage strictly monitored traffic through the Pillars. Therefore, it is commonly believed Pytheas could only have sailed into the Atlantic while Carthage was distracted by its war with Syracuse during the years 310 to 306 B.C.

This dating assumes Pytheas was the leader, or at least a member, of a Massaliote expedition. An alternative hypothesis suggests that he traveled as a passenger on native vessels engaged in regular shipping runs, possibly having traveled by land to Brittany before securing such passage. Though this obviates the need for explaining how a Greek vessel could have breached the Phoenician blockade, it seems rather implausible and has few supporters.

The expedition was likely conducted under official auspices for the purpose of obtaining information to enhance Massaliote commerce. Tradition has it that Pytheas was an exceptional astronomer and geographer, having accurately established the latitude of Massalia. Thus, he would have been a valuable member on such a venture.

Impact

Pytheas described his voyage of exploration in *On the Ocean (Peri Okeanou)*. This was a general treatise of geography on the "Outer Ocean." Unfortunately, the work is no longer extant. What is known of it has been gleaned from later commentaries. From these scattered sources, his route and discoveries have been reconstructed.

After passing through the Pillars of Heracles, Pytheas sailed northwest past Gades. He then rounded the headland at Cape St. Vincent, Portugal, and steered a northerly course along the coast. Passing the northwestern tip of Iberia, Pytheas followed the coastline east into the Bay of Biscay. When the coast again turned north, he determined his position to be only 400 miles (644 km) from Massalia and at the same latitude. He thus discovered Iberia to be a peninsula.

The expedition eventually put in at the Celtic port of Corbilo (modern Saint-Nazaire, France), situated at the mouth of the Loire River. After obtaining provisions, Pytheas continued his journey in a northwesterly direction along Brittany. He eventually reached Cape Kabaion and the island of Ouexisame (Ushant) at the westernmost point of Gaul.

Instead of continuing on his coastal route, Pytheas struck out across the English Channel in search of the tin mines of Britain. He most likely learned of their location from the Celts of Corbilo or Brittany. The Massaliotes eventually sighted the shores of Britain and laid up their vessel to explore.

The land was wild and uninviting, covered with dense forests and swamps. However, they were cheered to find Celtic inhabitants. These Britons lived in peace, though they had many different chieftains and rulers. Pytheas possibly made excursions inland to gather further information. He noted that their domiciles were crudely built of wattle or logs; that they threshed their grain indoors due to inclement weather; and that they had knowledge of the war chariot. He also mentioned their fermented barley liquor (beer) and honey wine (mead).

Continuing his journey, Pytheas next reached Bolerium (Land's End, Cornwall), a rocky promontory at the southwest corner of Britain. Here he discovered what, for centuries, had been one of the Mediterranean's major sources of tin. The Cornish miners showed him their methods for extracting, smelting, and refining ore. He also described how, at ebb tide, they transported the processed ore over the exposed tidal flats to the small island of Ictis (St. Michael's Mount). From here it was shipped to Gaul and further afield.

Pytheas then traveled to the north of England along the eastern littoral of the Irish Sea. Though the corpus of commentaries makes no mention of Ireland, he must surely have sighted the island as he passed through the North Channel where it is but 13 miles (21 km) from southern Scotland. Tales of cannibalism among the natives may have dissuaded him from making land. Regardless, he eventually reached the northern tip of Scotland, where he reputedly encountered 120-foot (37-m) tides. True tides of this magnitude are impossible. The reference probably refers to the gale-enhanced tidal displacements of Pentland Firth, which separates the mainland from the Orkney Islands.

It was after exploring the Orkneys that Pytheas made his most famous discovery. He learned from the island shepherds of another, much larger island to the north. Known as Thule, it was 6 days further away, on the very edge of the frozen Cronian Sea, which marked the world's end according to the beliefs of the time. Pytheas had already noticed the lengthening of the day as he traveled north, and he stated that Thule had 19 hours of sunlight in summer. He also claimed that further north was a region of semicongealed water that was neither earth properly so-called, nor sea, nor sky, but rather a mixture of all three resembling a "sea lung" (jellyfish) upon which one could neither walk nor sail.

The location of Thule has remained cloaked in controversy. Nevertheless, because of the generally high quality of his observations, one is inclined to believe Pytheas went just beyond 61°N, where 19-hour days occur. This suggests the Shetland Islands or the area of Bergan, Norway, as prime candidates. However, Thule is definitely described as a large island, which Norway is not, and nowhere is it described as a group of islands, which the Shetlands are. Arguments have been advanced for Iceland and the Faeroe Islands as well. However, each of these is more than 6 days journey from England.

In any case, Pytheas finally turned south and completed his circumnavigation of Britain. At this stage, he recrossed the channel and proceeded northward until penetrating the Baltic Sea. Seven hundred miles (1,127 km) further on he located the Amber Isle—the primary source of amber for the Mediterranean—and the river Tanais, which marked the ancient boundary between Asia and Europe. Some scholars believe this accurately describes Prussian Samland near the Vistula River, which is today known as the Amber Coast. Others have rejected this idea, believing he reached only as far as the Elbe River. By the time he returned home, Pytheas, even by conservative estimates, had traveled over 7,500 miles (12,070 km).

Pytheas's astronomical measurements contributed to the development of mathematical geography. Eratosthenes (c. 275-c. 194 B.C.) accepted Dicaearchus's line running from the Pillars of Heracles to Rhodes and expanded it into a series of parallels for measuring latitude. He relied on Pytheas's observations to determine the northern parallels. Hipparchus (c. 190-c. 120 B.C.) also believed the Massaliote's observations sufficiently accurate for establishing his own system of projected parallels. Pytheas also correctly described Britain as triangular, accurately estimated its circumference at 4,000 miles (6,437 km), and approximated the distance from northern Britain to Massalia at 1,050 miles (1,690 km), only slightly less than the actual distance of 1,120 miles (1,802 km).

Most writers of antiquity, though, viewed Pytheas as an imaginative liar. The Greek geographer Strabo (c. 63 B.C.-c. A.D. 24) criticized Pytheas because his reports of the far north conflicted with long-accepted theories about uninhabitable torrid and frigid zones. Strabo further ridiculed Pytheas's description of gigantic tides and the "sea lung." Despite being largely discounted, Pytheas's tale opened Hellenistic eyes to western Europe.

Ironically, many details that gave early critics reason to doubt Pytheas now lend credence to his story. His description of flora and fauna, meteorological conditions, and geographic features all comport well with what is known of these areas today. Perhaps not so surprisingly, the Arctic explorer Fridtjof Nansen (1891-1930)

recognized in his "sea lung" an accurate description of the ice sludge and fog that forms at the edge of drift ice. All this confirms Pytheas's place as one of the world's great explorers.

STEPHEN D. NORTON

Further Reading

Books

Carpenter, Rhys. *Beyond the Pillars of Heracles.* New York: Delacorte Press, 1966.

Cary, Max, and E. H. Warmington. *The Ancient Explorers.* Rev. ed. Baltimore, MD: Penguin Books, 1963.

Romm, J. S. *The Edges of the Earth in Ancient Thought: Geography, Exploration and Fiction.* Princeton, NJ: Princeton University Press, 1992.

Roseman, Christian Horst. *Pytheas of Massalia: On the Ocean.* Chicago: Ares Publishers, Inc., 1994.

Articles

Diller, Aubrey. "Pytheas of Massalia." In C. C. Gillispie, ed., *Dictionary of Scientific Biography.* New York: Charles Scribner's Sons, 1975: 225-26.

Whitaker, Ian. "The Problem of Pytheas' Thule." *Classical Journal* 77 (Dec.-Jan. 1981-1982): 148-64.

Nearchus Discovers a Sea Route from India to the Arabian Peninsula

Overview

In 325 B.C. the Greek military commander Nearchus undertook a naval expedition from the mouth of the Indus River in what is now Pakistan to that of the Euphrates River in Mesopotamia, or modern Iraq. His voyage served a number of purposes, not least of which was to ferry a large portion of Alexander the Great's fighting force from India back to Greece; but his principal mission was to find a sea route between the Indian subcontinent and the Near East. This he did, in the process making possible much greater trade and exchange between India and lands to the west.

Background

The career of Nearchus (360-312 B.C.), who came from Crete, is inexorably tied with that of his friend and leader, Alexander the Great (356-323 B.C.) Alexander's father, Philip II of Macedon (r. 359-336 B.C.), conquered the Greek city-states with the aim of uniting all of Greece and going on to subdue the dying empire of the Per-

sians. But he was assassinated before he could undertake his mission, so it fell to his son to become the greatest military leader the world has ever known.

In 335 B.C. Alexander began moving his vast army into Asia Minor, and soon won an engagement against pro-Persian forces led by a Greek mercenary named Memnon. He then moved into Cilicia, where he scored a decisive victory against the Persian emperor Darius III (d. 330 B.C.) at Issus. As a result, the Greeks gained control of the entire western portion of the Persian Empire, and during the period from 334 to 331 B.C., Alexander's forces secured their hold over southwestern Asia and Egypt. In October 331 B.C. they met a Persian force at Gaugamela in Assyria, this time scoring a complete victory over the enemy.

Alexander moved eastward to claim his empire, but he was not content merely to subdue Persia itself: between 330 and 324 B.C. his armies conquered what is now Afghanistan and Pakistan, and ventured into India. But in July

326 B.C., just after they crossed the Beas River in what is now Pakistan, Alexander's troops refused to go on. The men had been gone from their homeland for nearly a decade and were eager to return to their families, so Alexander agreed to begin heading westward again.

Throughout the long years of battle, Nearchus had fought by the side of his friend and commander. His role as a close associate of Alexander was revealed early, when the leader granted him the role of satrap, or governor, over the provinces of Lydia and Pamphylia in Asia Minor. Now, as he began preparations for the return to Greece, Alexander again granted Nearchus a favored position as admiral. While one group of Greek troops returned via a northerly route and another, led by Alexander himself, pursued a southerly route, Nearchus's fleet would sail along the coast to Mesopotamia. With this honor Alexander included a charge: it was Nearchus's job to find the best possible sea route between India and the Near East.

Impact

In fact Alexander appointed Nearchus to the position of admiral in 327 B.C., before the decision to turn back. This indicates one or both of two possibilities: that the commander knew his troops were growing anxious for a return to their homeland, and that he had already conceived the idea of a seaborne mission of exploration. Certainly historians believe that had he lived, Alexander would have devoted his remaining years not to administration and perhaps not even to conquest—given the fact that his troops were weary, and he was a commander who kept a close watch on the sentiments of his men—but to exploration.

It is not known when, or indeed whether, Nearchus gained earlier experience as a commander of a sea force. However, the historian Arrian (d. A.D. 180), who wrote extensively on Alexander's military campaigns, offered insights both regarding Alexander's decision not to lead the naval force himself, as well as his choice of Nearchus. Loosely quoting Nearchus, Arrian wrote that "Alexander had a vehement desire to sail the sea which stretches from India to Persia; but he disliked the length of the voyage and feared lest ... his whole fleet might be destroyed; and this, being no small blot on his great achievements, might wreck all his happiness; but yet his desire to do something unusual and strange won the day."

Nonetheless, Arrian went on to note, Alexander "was in doubt whom he should choose, as equal to his designs; and also as [to] the right man to encourage the personnel of the fleet." Nearchus, again paraphrased by Arrian, wrote that Alexander shared with him his many deliberations regarding the choice of an admiral: "but as mention was made of one and another, and as Alexander rejected some, as not willing to risk themselves for his sake, others as chicken-hearted, others as consumed by desire for home, and finding some objection to each; then Nearchus himself spoke and pledged himself thus: 'O King, I undertake to lead your fleet! And may God help the [enterprise]! I will bring your ships and men safe to Persia, if this sea is so much as navigable and the undertaking not above human powers.'"

Alexander at first feared to send Nearchus on such a dangerous mission, but was eventually persuaded by Nearchus's highly logical argument that if the men saw their emperor place his close friend in charge of the mission, this would set their minds at ease regarding the danger. Having thus commissioned Nearchus, Alexander granted him all troops with seafaring experience, as well as a brigade of Indian shipwrights. The latter built for the Greeks some 800 vessels, many as large as 300 tons, and Nearchus engaged the services of Indian pilots who would serve as guides. Late in September 325 B.C., the fleet departed from the mouth of the Indus.

The journey had an inauspicious start when they were delayed for 24 days at Crocola (modern Karachi, Pakistan) due to unfavorable winds. The next five days' sailing took them to the mouth of the Hab, today on the border between the Pakistani regions of Sind and Baluchistan. The Hab was the first of many rivers, flowing from the Indian subcontinent into the Arabian Sea, that the voyagers would pass, and here again, the signs did not look good: as they sailed westward from the Hab, a storm capsized three ships. Fortunately, the crew members themselves survived.

At Ras Kachari, the fleet anchored and met up briefly with a portion of Alexander's forces. Then, after taking on supplies, they continued to the Hingol River. There they waged a successful battle against some 600 natives who attacked them, individuals Nearchus described as "hairy over their heads as well as the rest of their persons, [with finger]nails like wild beasts." Nearchus's force took several prisoners, and went on their way.

It took them 20 days to pass the coast along Makran, where they searched in vain for fresh water and again found the inhabitants hostile. At the River Kalami, Nearchus desecrated a place revered by a local sun-worshipping cult, violating taboos by setting foot on their sacred island of Astola. He apparently emerged unscathed from this incident, but the troops were running low on food, and had to hunt for wild goats on the shore.

The expedition took on supplies at the town of Pasni in what is now Baluchistan, and as they continued westward they found that the land was more fertile. Perhaps it was during this time that Nearchus observed sugar cane, which he described as a reed "that produce[s] honey, although there are no bees." It was also in this region that Nearchus appears to have seen whales, reporting that he observed great towers of water blowing into the air. He also noted that the local inhabitants paddled canoes, instead of rowing in the Greek fashion.

As he had been in the incident involving the sun-worshippers, Nearchus proved himself rather foolhardy at the town of Gwadar, where in spite of the locals' willingness to trade, he chose to attack the city. This effort ended in a stalemate, however, and in the end he was content to trade with the townspeople for fishmeal. From there they sailed along the coast, reaching Persia at Cape Jask in the Kerman region. The voyagers glimpsed Cape Musandam, the tip of the Arabian Peninsula that serves as dividing line between the Gulf of Oman and the Persian Gulf, but Nearchus resisted a suggestion from his principal lieutenant that they cross the Strait of Hormuz and explore the Arabian coast.

At the mouth of the Minab River, Nearchus turned inland and eventually met up with Alexander, who with the other troops greeted him as a hero. Alexander himself celebrated the successful voyage with feasts and sacrifices to the gods, but after tarrying awhile, Nearchus continued sailing on past Hormuz Island toward Qeshm Island. There the ships ran aground on sandbanks, an incident that cost them three weeks' repair time. Finally, however, they set sail up the Persian Gulf, finally landing at the city of Diridotis on the mouth of the Euphrates. From there they moved up the Karun River to meet Alexander for the last time at the Persian capital of Susa.

Alexander died soon afterward, at Babylon in June 323 B.C., and Nearchus's fortunes diminished. It is likely that in the ensuing power struggle between Alexander's generals, Nearchus lost his satrapies in Asia Minor, and at this point he faded from the historical record. His accomplishments, however—preserved not only in his writings and those of Arrian, but also in the work of Strabo (c. 64 B.C.-c. A.D. 23)—did not. Thanks to Nearchus, the lands of the Near East and Europe gained contact with India, from whence they would import numerous valuable goods and—more important—ideas. Not least among these were the Hindu-Arabic numerals, which began making their way westward in the early years of the Middle Ages.

JUDSON KNIGHT

Further Reading

Books

Cary, M. and E. H. Warmington. *The Ancient Explorers.* London: Methuen, 1929.

Hyde, Walter Woodburn. *Ancient Greek Mariners.* New York: Oxford University Press, 1947.

Kagan, Donald, ed. *Studies in the Greek Historians: In Memory of Adam Parry.* New York: Cambridge University Press, 1975.

Vincent, William. *The Commerce and Navigation of the Ancients in the Indian Ocean.* New Delhi, India: Asian Educational Services, 1998.

Internet Sites

Arian: *Anabasis Alexandri:* Book VIII (*Indica*), Tr. E. Iliff Robson (1933) *Ancient History Sourcebook.* http://www.fordham.edu/halsall/ancient/arrian-bookVIII-India.html.

"Names of Rivers of NW India According to Greek Sources." http://sarasvati.simplenet.com/aryan/alexander. html.

Strabo: *Geography:* Book XV: *On India Ancient History Sourcebook.* http://www.fordham.edu/halsall/ancient/strabo-geog-book15-india.html.

Alexander the Great

Overview

Alexander, a Macedonian king whose fourth-century B.C. military conquests earned him the name "Alexander the Great," united much of the known world of late antiquity into an empire. Alexander crushed Greek resistance, ended the Persian Empire, conquered Egypt, and invaded northern India. Thus, Alexander became the first to unite Europe and Asia Minor. Alexander's persona was both forceful and politically astute. His intelligent military strategy and worldly philosophies, along with striking physical features and both brutal and benevolent treatment of the vanquished, elevated Alexander to the status of a deity during his lifetime. During the reign of Alexander, the culture and language of Greece spread far throughout the empire, becoming the idealized standard for a new, united world. After Alexander's death, the common culture, referred to as Hellenistic, endured, even as Greece fought civil unrest and Alexander's generals battled for pieces of the empire.

Background

Alexander was the son of Philip II, king of Macedonia, who united the city-states of Greece for the purpose of making war with the Persians. The Macedonians spoke a Greek dialect, but were regarded by the Greeks as barbarians from the north with only a veneer of Hellenic sophistication. Philip II, however, was a strong ruler who desired that his people be accepted as Greek by the Greeks—with one major difference. Macedonians regarded themselves as one united people (Macedonia was one of the first nations in the history of Europe). Greeks saw themselves as citizens of separate city-states. In 338 B.C. Philip II, through a complex and efficient military system, along with a knack for diplomacy, organized the Greek city-states into a union, the Corinthian League. The league's states were to be independent and self-governing with Philip as their commander-in-chief. Philip's concept of conquering through unity would later be perfected by his son and heir, Alexander.

In midsummer, Philip II, fresh from his latest victory at Potidaea, received three messages simultaneously: his racehorse had won at the Olympic games, The Illyrians (Albanians) had been conquered in a great battle, and his son Alexander had been born. Thus, the legends of Alexander began with his birth. Sages predicted to Philip that the son who was born among three victories would eventually prove victorious himself. Alexander was a robust youth with a keen, inquisitive mind. From his mother, Olympias, a princess of Epirus, Alexander inherited a passionate nature. Both of Alexander's parents fostered his sense of ambition. Olympias and Alexander shared the closer parent-child relationship, as father and son were separated during much of Alexander's youth by Philip's military campaigns. Still, both parents reveled in Alexander's precociousness and, by the time Alexander was 12 years old, his mind was judged superior enough to require the best available education. Aristotle (384-322 B.C.), the star pupil of Greek philosopher Plato (427?-347 B.C.), was employed as tutor to Alexander.

During the next three years, Alexander developed an enduring interest in Greek culture, philosophy, and heroic history, especially embracing the works of the poet Homer (fl. c. 800s B.C.). Alexander later took with him to Asia a copy of Homer's *Iliad* that Aristotle had prepared for him. Aristotle often referred to a passage in the *Iliad* which, referring to Agamemnon, states, "Both things is he; both a goodly King and a warrior mighty." Aristotle also tutored Alexander in rhetoric, mathematics, botany, medicine, astronomy, and poetry.

Alexander enjoyed an athletic physical stature (although he was not tall), but was not particularly fond of athletic endeavors, according to the Greek philosopher and historian Plutarch (45-120 A.D.). His complexion was fair, and his face was regarded as handsome with a straight nose and clear, piercing eyes. His hair stood above his forehead, resembling a lion's mane. Alexander as a youth was disciplined with his time and personal habits. For relaxation, he played the lyre and enjoyed listening to music. Hunting was a passion Alexander often pursued, as the Macedonian countryside contained an abundance of deer and foxes, with bears and lions still roaming the highlands. The young Alexander was energetic and even restless with his ambition, according to Plutarch, once worrying that his father would conquer all the world, leaving Alexander no opportunity for greatness.

At 16, while his father was laying siege to Byzantium, Alexander was left as regent in Macedonia. During this time, Alexander thwart-

Hellenistic ruins in Turkey. *(Roger Wood/Corbis Corporation. Reproduced with permission.)*

ed a Thracian rebellion. He overtook a city in Thrace and renamed it Alexandropolis, after himself. Future cities named in honor of Alexander would dot his empire. In 336 B.C., at the age of 20, Alexander ascended to the Macedonian throne after Philip II was murdered by Pausanias, a king's guard and friend to Alexander. Some historians assert that Pausanias was encouraged to murder Philip by Olympias, in retribution for Philip divorcing Olympias to marry a woman of full Macedonian heritage.

After the death of his father, Alexander quickly won the allegiance of the Macedonian army and its generals. Once in power, Alexander ordered the execution of all conspirators and domestic enemies, and restored Olympias as Queen of Macedonia. The Greek cities of Athens and Thebes were reluctant, however, to pledge allegiance to the 20-year-old Alexander, and additionally considered the Macedonian culture to be inferior to their own.

A revolt in the two cities soon erupted, conveniently while Alexander was away once again securing Macedonia's northern frontier in Thrace. The revolt was fueled by rumors reaching Athens and Thebes that Alexander was killed

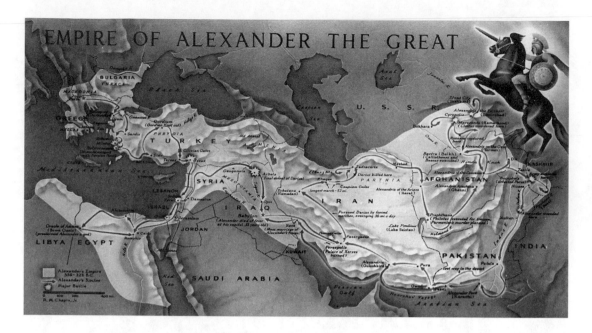

The empire of Alexander the Great. *(Corbis Corporation. Reproduced with permission.)*

in the northern campaign. After successfully securing the northern border in Thrace, Alexander marched 250 miles (402 km) south in a fortnight. The Greek historian Arrian (second century A.D.) wrote of Alexander marching up to the gates of Thebes in 335 B.C. and giving the Thebans a chance to change their minds. The Thebans responded with soldiers, and Alexander stormed the city, burning Thebes (sparing only the temples and the home of the poet Pindar) and killing many of its citizens (others were sold into slavery), as an example to the Athenians and the rest of Greece. Athens quickly capitulated, and remained under Alexander's rule. All other Greek resistance evaporated as well, ensuring Alexander's ascendancy over all of Greece and allowing him to concentrate on the impending campaign against the Persians.

In 334 B.C. Alexander began his eastward march toward Asia Minor to confront the Persian Empire. His army numbered nearly 40,000 men, with 12,000 reinforcements available within the alliance. The Companion cavalry, a 2,000-strong legion of highly trained, heavily armed cavalrymen formed the key to the Macedonian military machine. The Companion cavalry was joined by light troops and the flexible phalanx, a group of highly effective armored foot soldiers who carried short swords along with long spears projecting several feet from the line. The phalanx advanced as a group before the enemy could engage with their short swords. Alexander crossed the Hellespont (modern Dardanelles)

with his generals Antigonus, Ptolemy, and Seleucus, and soon encountered the Persian army at the crossing of the Granicus river. The Persian army was led by their king, Darius III. Alexander led a charge across the Granicus and engaged in a pitched battle of hand-to-hand combat in which most of Darius's generals died.

In 333 B.C. Alexander engaged Darius in battle for the second time in the Syrian coastal city of Issus. Alexander was greatly outnumbered at Issus, but prevailed because of the terrain and his ability to outmaneuver Darius's army. Darius escaped Issus, but left behind his mother, wife, and two daughters, all of whom Alexander treated with courtesy. From Issus, Alexander moved southward along the Mediterranean seacoast, vanquishing several small coastal cities. When Alexander reached the Phoenician naval home seaport of Tyre, a seven-month siege was necessary to obtain the city's surrender. Alexander used tension catapults (his army was the first to use them) and rams, as well as swords during the siege. Alexander executed many who refused to surrender at Tyre, and sold the city's women and children into slavery. The naval power of Persia was broken with the loss of Tyre, and Alexander's navy reestablished control over the Aegean Sea.

With the destruction of Tyre, it was necessary to find another center to host the growing trade of the eastern Mediterranean. Egypt was the logical location, and Alexander arrived there in late 332 B.C. In Egypt, on the westernmost mouth of the Nile river, Alexander founded the great city, his namesake, Alexandria. Alexander was welcomed into Egypt as a liberator by Egyptians weary of Persian rule. In truth, the disorganized army of Egypt would have offered little resistance to Alexander. While in Egypt, Alexander made pilgrimages to the temple and oracle (priest) of Amon-Ra, the Egyptian sun god who was comparable to the Greek Zeus, and to the temple and oracle of Zeus Ammon. During the dangerous trip through the desert to Siwah, where the oracle of Zeus Ammon was located, Alexander was declared to be the son of Zeus who was destined to rule the world. Alexander was pleased with this declaration, but was more concerned with confirming that the great Libyan desert was indeed a barrier to discourage an invasion of Egypt from the west. During his time in Egypt, Alexander exchanged letters with Darius. Darius offered Alexander a truce and some selected Persian provinces, but Alexander declined the offer. In mid-331 B.C. Alexander returned to Persia in search of Darius.

Alexander and his armies marched toward Babylon, an ancient capital in Persia where Darius was mounting an opposition, and toward one of the greatest battles in antiquity. The two armies met at Gaugamela (in modern Iraq). The Macedonians, upon spotting the vast number of Persian army campfires at night, surmised the Persian forces too numerous to defeat. Alexander's men encouraged him to attack at night, as the cover of darkness would provide a surprise advantage. Alexander, realizing the hazards of night battle, insisted he defeat Darius in an equally matched battlefield. The two armies clashed the next morning on a wide plain. Darius's men greatly outnumbered Alexander's by at least 100,000 men; some historians claim Darius's forces to be almost a million strong. In front of Darius's royal squadron were scythe-bearing chariots and 15 elephants. Darius commanded his army from inside his chariot, and was slow to amend his battle plans as conditions changed on the battlefield. Alexander seized upon this misjudgment. Darius failed to bestow sufficient command to his subordinates, so that his army lacked the ability to take advantage of Alexander's weaknesses. Alexander routed the Persian army, while Darius again deserted the men who where fighting for him by fleeing into the mountains.

Alexander marched to Babylon, occupied the city, and was declared King of Persia. Alexander benevolently appointed Persians as provincial governors, but also burned the royal palace

to the ground. Alexander continued his pursuit of Darius, eventually learning that the former king had been arrested by a Bactrian prince. By the time Alexander made his way to Darius, Alexander found him in his coach murdered by his own men. Alexander had the assassins executed and gave Darius a funeral fit for royalty.

Alexander continued eastward towards India, recrossing the Hindu-Kush mountains in 327 B.C. The route to India, roughly down the Cabul valley and through the Khyber pass contained difficult terrain and Alexander encountered much local resistance. When he reached India in 326, Alexander was met by Taxiles, a local reigning prince of Punjab. Taxiles honored Alexander with elephants and other gifts, and convinced Alexander to help him settle a dispute with an enemy, Porus, who was reportedly almost 7 feet (2.1 m) tall. Porus was just east of the Hydaspes river, which Alexander intended to cross before the Himalayan snowmelt and seasonal tropical rains made it impossible. The battle with Porus, fought at the Hydaspes river, taxed Alexander as none had before. Porus had a well-equipped army with more infantry than Alexander, chariots, and over 200 elephants to terrify Alexander's cavalry horses. Alexander defeated Porus and the Indians after a fierce battle using surprise tactics to essentially corner Porus along the rising river during a storm. Alexander had hoped to reach India's Ganges river or a natural ocean border to his eastern empire, but shortly after the battle with Porus, Alexander's Macedonian armies had had enough. After almost nine years and 11,000 miles (17,703 km), they did not mutiny, but persuaded Alexander in 325 B.C. to turn westward towards Macedonia and home.

In 323 B.C. Alexander returned to Babylon, which he intended to make the capitol of his new empire. In early June of the same year, Alexander fell ill with a fever and lingered for almost 10 days before he died. The exact cause of Alexander's death remains a mystery, but malaria or typhoid fever remain the most probable explanations. The wear of Alexander's decade-long campaigns and multiple wounds probably rendered his body unable to tolerate either of these common, but serious, diseases. Alexander attended a large banquet more than a week before he died, and the possibility of poisoning was suspected at the time of his death. The herbalists of Babylon, however, possessed no lethal concoctions with an action sophisticated enough to poison its victim over a 10-day time period. Poisons of the day acted quickly, even for those presumed to be demigods.

As for Alexander's successor, some histories claim Alexander, in his words, willed his empire "to the strongest." Other accounts name Alexander's general Perdiccas as his heir, with Alexander shortly before death joining the hands of Perdiccas and Alexander's pregnant wife Roxanne. Most likely, Alexander died without naming a successor.

Impact

The turbulent times that followed Alexander's death saw his generals and others vying for pieces of Alexander's empire for almost 300 years, eventually leaving it vulnerable to a new power rising in the west, that of Rome. Alexander's son, Alexander IV, was born shortly after his father's death, and for a time it was agreed that the infant would share kingship with Alexander's younger half-brother Arrhidaeus. Olympias soon arranged for the murder of Arrhidaeus, and Alexander IV was slain at age 12. With Alexander's bloodline out of the picture, the military leaders shared an empire split into three large kingdoms and several smaller states. The three generals with the largest prizes included Ptolemy, founder of the Ptolemic dynasty in Egypt, Seleucus, who founded the Seleucid dynasty that ruled over Persia, and Antigonus, who ruled over Asia Minor, and whose successors would eventually win control over Macedonia.

Ptolemy made himself a monarch in the Macedonian tradition of Alexander, and founded a dynasty that lasted until the Caesars conquered Egypt. Ptolemy, like Alexander, claimed he was a descendent of the gods. Ptolemy attempted to forge a religious bond with the Egyptians by introducing a new religion that, like the religion of Hellenism, drew from mythology. One of the new gods, Isis, gained some popularity, but for the most part, Egyptians rejected the new religion. Ptolemy was somewhat more successful appealing to the glory of Egypt's ancient past and declaring himself a new Pharaoh. By staffing his administration with Greeks rather than Egyptians, many continued to view Ptolemy's rule as foreign. Some of his successors would be embraced by the Egyptians. The last descendent of Ptolemy was a princess who was also pharaoh, Cleopatra. Antigonus won Greece and the league of Greek cities that were subject to Alexander. The Antogonids also ruled until the coming of Rome. Seleucus won a reduced portion of the Persian empire, and claimed lineage that extended back to the god Apollo. Seleucus further claimed that Zeus resided at his capital, Antioch. Despite these divine affiliations, Seleucus's Persian prize was the least successful during the Hellenistic Age.

Greek culture became widespread throughout the known world during the Hellenistic Age. The Hellenistic Age encompassed Alexander's lifetime until approximately two centuries after his death. From southern coastal France to India, Greek became the language of the intellectual elite. As a result of Alexander, Greek athletic events would be held in the heat of the Persian gulf, and tales of the Trojan horse or the love story of Cupid and Psyche would be told in Punjab. Homer would be translated into Indian, and distant religions such as Buddhism would become familiar to Greeks. No longer contained to the "frog pond," (Alexander's characterization of the Mediterranean) Greek influence spread across the entire Near East, and blended Greek and Oriental cultural currents. Even as the Roman legions arrived, much of the Near East continued to regard Greek culture as supreme and the one most worthy of imitation.

Alexander hoped to create a new broadened cosmopolitan tradition by merging the cultures of antiquity. Intermarriage was encouraged and the cultural snobbery of the classical Greeks was dismissed. In 324 B.C. Alexander married a Persian noblewoman, Roxana, who bore his son after his death. In all, over 10,000 of Alexander's men married local women, and Alexander often rewarded them dowries or orders to demobilize and return home. Many historians assert that Alexander sought a long-term solution to the prejudices of East versus West, and a new class of nobility with mixed blood would cement a permanent bond between the two cultures.

In an additional marriage assumed to enhance his legitimacy as King of Persia, Alexander married Stateira, one of the daughters of Darius. Nevertheless, after Alexander's death, many of the marriages of officers and Persian noblewomen were dissolved. Alexander also adopted a modified eastern style of dress, and provided instruction on Greek and Macedonian culture to the Persians. When Macedonian troops were demobilized, Alexander replaced them with Persians, creating an integrated force. The Macedonians with pikes were placed in the front rank; the Persians bearing swords and javelins followed in rows behind them. Alexander did not tolerate the actions of his men who showed prejudice toward the vanquished.

Inspired by Alexander, classical Greek art was no longer the concern of a few cities, but reflected the pictorial language of almost half of the world. Classical Greek art, therefore, changed its character in the Hellenistic era, especially as new capitals in Alexandria, Pergamon, and Antioch acquired wealth and influence. In architecture, a new form of column was preferred, based upon one designed earlier in the fourth century B.C. and named after the wealthy merchant city of Corinth. In the new Corinthian column, ornamentation, usually foliage, was added to the Ionic design. This luxurious design was suited to the vast scale of buildings in the newly founded cities of the East (the Corinthian capital of Epidaurus, for example, built around 300 B.C.). Sculpture de-emphasized the harmony and refinement of classical Greece, and instead aimed for dramatic impact. The Altar of Zeus in Pergamon (ca. 170 B.C.) contains sculpture representing the struggle of the gods and the giants portraying wild motion and frenzied expression, all on a grand scale. As Alexander's conquests, the altar was meant to be impressive.

During this age, wealthy families began collecting art. Nature, especially flora, was often portrayed in painting and design. The art of portraiture also began, perhaps as a reaction to a perceived loss of personal identity with the expanding world and its blending cultures or, alternately, due to the abundance of impressive political, military, and scholarly figures of the Hellenistic Era. Early Roman art in Pompeii es-

pecially embraces the Hellenistic era, where a mosaic of Alexander himself at the Battle of Issus was found.

Though it may be interesting to speculate what Alexander might have accomplished had he lived to an old age and come to grips with the problems of governing his expanded empire, it is certain that Alexander's career was sufficient to change the course of human history. Gone was the small democratic city-state, along with the homogenous civilization concentrated along the Aegean Sea. The new culture of unity sought to bind the empire together as a social, political, and economic whole. It was this new culture of the Hellenistic Age that civilized Rome and led to Rome's creation as a world state.

BRENDA WILMOTH LERNER

Further Reading

Arrian. *The Campaigns of Alexander.* Trans. by Aubrey de Sélincourt. Harmondsworth, England, and Baltimore: Penguin Books, 1971.

Fox, Robin Lane. *Alexander the Great.* New York: Dial Press, 1974.

Green, Peter. *Alexander to Actium: The Historical Evolution of the Hellenistic Age.* Berkeley: University of California Press, 1990.

Plutarch. *Plutarch: Lives of Noble Grecians and Romans.* Edited by A.H. Clough, trans. by John Dryden. 2 vols. New York: Modern Library, 1992.

The Silk Road Bridges East and West

Overview

The opening of the Silk Road, which ultimately linked China with Europe, was one of the most important undertakings in the history of exploration prior to the period from A.D. 1400 to 1600. Indeed, the creation of the Silk Road was a phenomenon in many ways mirrored by the great Age of Exploration some 1,500 years later. Both were enormous ventures that involved numerous individuals, yet in both cases, it was possible to trace the impetus to one or two people. In the more recent example, that would be Prince Henry the Navigator, whereas in the case of the Silk Road, the honor belongs to the second-century B.C. Chinese traveler Chang Ch'ien, or perhaps his emperor, Han Wu ti.

The Silk Road was not, strictly speaking, a "road": rather, it constituted a set of overland routes from the Chinese capital at Chang'an (Xian) all the way to Antioch, Damascus, and other cities of the Levant. Nor was it typical for a single journeyer to travel the entire route: instead, merchants would cover a certain distance to a trading town and exchange their wares, which continued to move westward or eastward. At several points, mountains and other obstacles created forks at which it became necessary to take either a northerly or southerly route, for instance either north to Samarkand (in modern Uzbekistan) or south into Bactria, or present-day Afghanistan; but the ultimate destination was the same. At the height of the Silk Road's early history in A.D. 100, it was theoretically possible to travel on established routes from China to Spain, some 8,000 miles (12,800 kilometers) away.

Background

The trunk of the Silk Road ran across central Asia, the enormous landlocked expanse between the lush river valleys of eastern China, the Siberian forests to the north, the Indian subcontinent to the south, and the Iranian plateau to the west. Here the horse had been domesticated as early as 3200 B.C., and became the ideal form of transportation for the nomadic peoples crossing the seemingly endless expanses of desert and grassland. It would be many centuries before the nomads developed the stirrup, one of their greatest legacies to humankind, but they were such accomplished horsemen that a later Roman historian suggested it was as though horse and rider were one being.

To the east and south of these nomads was China, a land rich not only in civilization—something for which the nomads cared little—but also in material wealth (which *did* interest the nomads). Thus from the beginnings of Chinese history, the peoples of central Asia and Mongolia harassed China's borders. The Chinese, for their part, dismissed all these tribes as "barbarians," and indeed from the viewpoint of ancient China there were only two kinds of people: barbarians and Chinese. This resulted in part from the same instinct that led the Greeks to label all non-Greeks as *barbaroi;* but it was also a product of the fact that the Chinese simply knew of no other civilizations. The world's highest mountains separated them from India, and when the Chinese finally did make contact with another civilization, it was not that of India but rather Greece, long since in decline.

The invasion of western Asia by Alexander the Great (356-323 B.C.) had left in its wake the Seleucid Empire, but in about 200 B.C. the Bactrians threw off Seleucid rule in their region. The Hellenistic—that is, heavily influenced by Greek civilization—Bactrians became known to history as the Greco-Bactrians, and over the next century assumed control over much of India. Also important in the region were the Sakas, sometimes known as the Indo-Scythians, who had roots both in India and among the peoples of what is now the Ukraine. Clearly a polyglot cultural mixture was forming, and in the second century B.C. it became even more varied with the arrival of yet another group from the frontiers of northern China.

Of the many nomadic forces menacing China's border, none was more dangerous than the Hsiung-Nu or Xiongnu—better known to the West as Huns. Not only did the Hsiung-Nu threaten the Chinese, they even attacked other nomadic tribes such as the Yüeh-chih. Hsiung-Nu aggression forced the Yüeh-chih westward, and by 165 B.C. they had arrived in Bactria.

Meanwhile, the Chinese set out to do something about the Hsiung-Nu. Ch'in Dynasty ruler Shih-huang-ti (r. 221-210 B.C.) united much of the country for the first time, established China as an empire, and began the building of the Great Wall to force out the barbarians. Though the Wall failed in its purpose, it did help scatter various Hsiung-Nu tribes westward, where they spread toward India and—ultimately—Europe.

The Han Dynasty emperor Wu-ti (r. 140-87 B.C.) adopted a different strategy for dealing with the Hsiung-Nu. Under his rule, China became geographically larger than ever before, including within its control what is now Vietnam and Burma in the south, as well as Korea and Mongolia in the north. China prospered, and he was not about to allow the Hsiung-Nu to roll back the gains he had made. Therefore he devised a plan to send an emissary westward for the purpose of locating the Yüeh-chih and establishing an alliance with them against the Hsiung-Nu.

To travel into the forbidding lands of the unexplored western reaches required a man of both courage and cunning; perhaps for that reason the normally autocratic Wu-ti did not simply order one of his ministers to undertake the job, but asked for a volunteer. One man stepped forward, an imperial bodyguard named Chang Ch'ien or Zhang Qian (fl. 138-114 B.C.) A later historian described Chang Ch'ien as "a man of strong physique and of considerable generosity; he inspired the trust of others, and the barbarians loved him." In 138 B.C. he set out with 100 men for a journey of more than 2,000 miles (3,200 kilometers) across some of the most treacherous terrain on Earth.

Impact

Nature itself did not pose the most severe threat on the journey: much more fearsome were the Hsiung-Nu, who captured Chang Ch'ien and his party in western China. The Hsiung-Nu leader clearly resented what he regarded as an intrusion on his territory: "If I wished to send envoys to [Vietnam]," he asked Chang Ch'ien, "would the Han be willing to let me?" Surprisingly, how-

Attila the Hun (seated on throne) was a descendent of the Hsiung-Nu. *(Bettmann/Corbis. Reproduced with permission.)*

ever, his captors treated Chang Ch'ien well, and he was even allowed to take a Hsiung-Nu wife, with whom he had a son.

For a decade Chang Ch'ien bided his time, never having forgotten his mission; then finally he escaped with his family and at least part of his expeditionary force. In time they arrived at the town of Kokand or Quqon in Ferghana (now part of Uzbekistan), where a friendly local ruler informed him that the Yüeh-chih had passed by on their way to Bactria. It is easy to imagine Chang Ch'ien's disappointment when he finally reached the Yüeh-chih, only to discover that they had set aside the ways of nomads, settling down and becoming civilized; thus they had no interest in returning to China to fight old enemies.

Chang Ch'ien remained among the Yüeh-chih for a year, then headed for home. In the Tarim Basin, an arid region of extreme western China, the Hsiung-Nu again captured him, and he remained with them for a year; but after the Hsiung-Nu ruler died in 126 B.C., he took advantage of the resulting confusion to make a break for Chang'an. He returned to the capital accompanied by his wife, son, and just one of the 100 men who had gone with him: the others had been killed, captured, or forced to turn back.

After his 13 years' journey and the incredible hardships it had entailed, it seemed Chang Ch'ien had achieved little; but he had accom-plished far more than he knew. Thanks to Chang Ch'ien, the Chinese became aware for the first time of lands to the west: Anxi (Persia), Tiaozhi (Arabia), and even Da Ch'in (Rome). Aside from a wealth of political, diplomatic, and economic knowledge with which he returned, there were also new delights to be gotten from a pack of seeds he had managed to smuggle. These yielded grapes, which, in addition to all their goodness in their natural state, introduced the Chinese to wine.

Han Wu-ti sent Chang Ch'ien on a second mission to central Asia (119-115 B.C.), this time with a larger and better-equipped expeditionary force. Along the way, Chang Ch'ien sent out envoys to the court of the Parthian ruler in Persia, who in turn dispatched an ambassador to China. The Parthians also had diplomatic relations with Rome, establishing a concrete link between East and West—and, for the Parthians, an advantageous position as middleman.

In Ferghana again, Chang Ch'ien took notice of the region's exquisite horses, steeds that stood 16 hands (64 inches or 163 centimeters) at the shoulder. Powerfully built, and with fine-looking features, these "Heavenly Horses" acquired legendary status, in part because they apparently sweated blood. (Only in the twentieth century did scientists come to understand the cause of this condition: parasites that burrowed under the horses's skin, causing the surface to swell and bleed.)

Wu-ti became intrigued by the "Celestial Horses," as they were also known, and attempted to purchase some of them from a king in Ferghana. The latter, presumably not the friend of Chang Ch'ien from earlier years, responded by murdering the Chinese envoy and stripping the body as a sign of disrespect. At this point, the emperor sent an army 60,000 strong to Ferghana, where they successfully besieged the capital and brought back several thousand steeds. Thenceforth "Celestial Horses," much like certain makes of automobile today, became status symbols of the wealthy and powerful in China.

In the two centuries following the death of Chang Ch'ien—still celebrated as a hero among the Chinese today—a combination of Chinese and Parthian power made it possible for merchant caravans to travel in relative safety as far west as Mesopotamia. There the road split, with a northerly route terminating at Syrian Antioch, and the southerly road passing through Babylon to Damascus. The latter areas fell under the jurisdiction of a third empire, Rome, which in time would establish direct diplomatic links with China.

Han Wu-ti's expeditions into central Asia had driven back the Hsiung-Nu, but in later years the power of the Han Dynasty fluctuated, resulting in the resumption of Hsiung-Nu control over the Tarim Basin. In the period from A.D. 73 to 102, however, a general named Pan Ch'ao won back the entire region, including Kashgar, today the chief city of Chinese Turkistan. His victories extended China's borders far west of their present-day limits, all the way to the shores of the Caspian Sea, and he sent ambassadors to the courts of lands around the Persian Gulf.

In A.D. 97 Pan Ch'ao commissioned Kan Ying to act as his emissary to Rome. This would have been the first direct Chinese-Roman contact, but in fact Kan Ying never got further west than what is now the city of Nedjef in Iraq, where he was told that a journey to Rome itself would take an additional two years. In reality it probably would not have taken two months, but his Parthian hosts did not want to risk losing their profitable role as middlemen, and therefore greatly exaggerated the distance. Believing the Parthian disinformation, Kan Ying returned without having achieved his goal.

Not until A.D. 166 did Rome, having defeated the Parthians and won control of the Persian Gulf region, establish a direct link with China. Apparently the Roman diplomatic envoy, sent by Marcus Aurelius (r. 161-180), traveled by sea via what is now Vietnam. By that point, Ptolemy (c. 100-170) had attempted to map the Silk Road, and Rome and China had long been enjoying extensive trading relations, thanks to the Pan Ch'ao's reopening of areas that had been threatened.

Since the time of Christ, wealthy Romans had valued silk, unknown in the West until the establishment of the Silk Road. Recognizing the wealth to be gained from silk export, the Chinese government took measures to ensure its monopoly by forbidding the transport of silk-worms beyond China's borders. (In the early Middle Ages, however, smuggled silkworms made their way west.) By contrast, neither Rome nor its vassal kingdoms seem to have placed any such restrictions on the exports of the Near East, but this was hardly because of generosity: Western exports tended to be food items that could not be transported without spoilage. Thus instead numerous varieties of seeds and plants traveled eastward, introducing the Eastern palate to the olive, cucumber, pomegranate, and alfalfa, as well as walnut and sesame. (The oil from the latter is one of the ingredients that today gives Chinese cuisine its distinctive flavor.)

In addition to the product for which the Silk Road was named, Chinese exports along the route included other textiles, bamboo, and iron. The Chinese in turn imported woolen goods and Hellenistic artwork, which in parts of Asia influenced artistic styles. Indeed, the export and import of ideas and beliefs would prove one of the most important aspects of the Silk Road. Thus, for instance, gnosticism, a variety of belief originating among Greeks in Asia Minor and the Hellenic mainland, began moving eastward in the centuries after Jesus of Nazareth.

Mentioned and criticized in the New Testament, gnosticism came in a number of varieties, but all emphasized *gnosis*, the belief in special knowledge available only to initiates in a given sect. Gnosticism had a great influence on the Manichaean splinter religion, which developed in Persia in the third century A.D. Later, Manichaean ideas spread westward, thus reintroducing gnosticism to the lands where it had been born.

Also influential was Nestorianism, a form of Christianity that, after the Church declared it heretical in 431, began to move east along the Silk Road. Nestorian communities in India and China would provide a vital East–West link, and as with Manichaeism, Nestorianism was destined to return to the West in changed form. This time, however, the carriers were not primarily priests but conquerors: Mongol warriors, many of whom adopted the Nestorian belief system.

But if the Silk Road had a religion, it was Buddhism, a legacy that can be attributed in large part to Kanishka (fl. c. A.D. 78-103) The latter came from the Kushans, greatest of the five Yüeh-chih tribes, who in the first century A.D. won control over a vast territory that stretched from the southern end of modern-day Russia to the Ganges valley in southeastern India. This placed them at the crossroads of the world, astride the Silk Road at a time when Pan Ch'ao's victories had ensured the viability of East-West commerce as never before.

Under Kanishka, the Kushan empire became, along with the realms of the Chinese, Parthians, and Romans, one of the world's four great multinational states, but the configuration of influences in Kushan lands was particularly eclectic. Symbolic of this quality were the names for various administrative positions within the empire: the Persian *satrap* for provincial rulers; the Hindi *meridarek* for district officers; and the Greek *strategoi* for military governors. Similarly, Kanishka maintained a variety of titles from lands he controlled, or from ones with which he maintained contact via the Silk Road: king of kings (Bactria and Persia), great king (India), son of heaven (China), and emperor (Greece and Rome).

Kanishka's empire also helped transmit Greek ideas about art to India, where sculpture began to take on a distinctly Hellenistic cast. He built a number of monuments, including a stupa or temple tower in his capital at Peshawar (now in Pakistan on the Afghan border). If its reported height of 638 feet (194 meters) is correct, that would make this the tallest structure of the ancient world—indeed, the tallest prior to the nineteenth century. Certainly the two statues carved into the rock at Bamiyan, 150 miles (240 kilometers) northwest of present-day Kabul, Afghanistan, constitute a powerful legacy of the Kushan empire: one of these figures stands 174 feet (53 meters) high, considerably taller than the Statue of Liberty.

Most important, however, was the fact that the statues represented the Buddha (Siddhartha Gautama; 563-483 B.C.), whose faith Kanishka had adopted. Some historians have questioned the sincerity off Kanishka's Buddhist beliefs simply because adoption of the religion had its practical advantages. Among these was the fact

that the Kushans, as outsiders, could never expect to hold anything better than a mediocre position in the Hindu caste system of India. Furthermore, the gentle ways of Buddhism promoted harmony in an empire composed of diverse ethnicities.

True believer or not, Kanishka did much the same thing for the Buddhist faith that the Roman emperor Constantine (A.D. 285-337) did for Christianity. By embracing Mahayana or "Great Wheel" Buddhism, Kanishka gave the imperial seal of approval to a particular interpretation of the faith, which remains the dominant version of Buddhism today. And just as Constantine later convened the Council of Nicaea, which helped define the heart of Christian doctrine, Kanishka brought together a group of Buddhist monks who formalized the doctrines of Mahayana.

Through his control of the Silk Road, Kanishka ensured that the faith spread throughout not only northern India, but in what is now Afghanistan; Pakistan; Kazakhstan and other nations of central Asia; and China. The Kushan empire came to an end not long after Kanishka, but its influence continued through its impact on the spread of Buddhism. There was an irony in this: by associating the faith with foreigners, Kanishka actually hurt the Buddhist cause in India, where a nativist resurgence of Hindu fervor ensured that religion the dominant position it retains today.

Meanwhile, the faith began to take hold in China, and this led to a number of notable journeys by Buddhist pilgrims to India, including Fa-hsien (c. 334-c. 422) and Hsüan-tsang (c. 602-664). Though the route from China to India would seem to follow a straight southward path, in fact the impenetrable Himalayas forced the travelers to work their way westward, then south, and in both cases the Silk Road made it possible to do so with a degree of safety. So too did the fact that virtually all the lands they passed through belonged to Buddhist kingdoms; yet in Hsüan-tsang's time another force—Islam—was on the rise. In time Islam would sweep over Afghanistan, western India, and central Asia, driving back the influence of Buddhism; but the latter faith had meanwhile spread further east, to Korea, Japan, and Southeast Asia.

The period between Kanishka and the rise of Islam saw the downfall of all four empires that had guarded the Silk Road. Interestingly, the force that put in motion Rome's destruction was the very same nation, the Hsiung-Nu, driven out by the Chinese. Having arrived in Europe in the fourth century, they had a new name: Huns. Behind them came other central Asian nomads, including the Avars, who introduced Europeans to the stirrup—a device that, by making it possible for men to fight on horseback while wearing heavy armor, virtually paved the way for knighthood and thus the feudal system.

In later centuries, two other notable groups emerged from central Asia, pouring into the West along the route of the old Silk Road. First were the Turks, who cut off the last links between Europe and the East and ensured that no European traveler would dare venture far past Syria. Then in the early thirteenth century there were the Mongols, who by subjugating virtually the entire known world facilitated the resumption of trade links that had not been open for centuries. This in turn made it possible for travelers such as Marco Polo (1254-1324) and Ibn Battuta (1304-1368?) to journey along the Silk Road once more, bringing about a renaissance in communication between West and East.

JUDSON KNIGHT

Further Reading

Books

Frye, Richard Nelson. *The Heritage of Central Asia from Antiquity to the Turkish Expansion.* Princeton, NJ: Markus Wiener Publishers, 1996.

Griffiths, Philip Jones. *Great Journeys.* New York: Simon & Schuster, 1990.

Grousset, René. *The Empire of the Steppes: A History of Central Asia.* New Brunswick, NJ: Rutgers University Press, 1970.

Hopkirk, Peter. *Foreign Devils on the Silk Road: The Search for the Lost Cities and Treasures of Chinese Central Asia.* London: Murray, 1980.

Hulsewé, A. F. P. *China in Central Asia: The Early Stage, 125 B.C.-A.D. 23: An Annotated Translation of the History of the Former Han Dynasty.* Leiden, Netherlands: Brill, 1979.

Journey into China. Washington, D.C.: National Geographic Society, 1982.

Internet Sites

Center for the Study of Eurasian Nomads. http://www.csen.org.

"Lost Cities of the Silk Road." http://www.alumni.caltech.edu/~pamlogan/silkroad/index.html.

Silk Road Foundation. http://www.silk-road.com.

Rome's Quest for Empire and Its Impact on Exploration

Overview

From its humble beginnings in c. 753 B.C., Rome emerged to conquer most known portions of Europe, Southwest Asia, and North Africa in the nine centuries that followed. Ruthlessly subduing subject peoples and bringing them into alignment with their imperial vision, the Romans forged an empire breathtaking in its scope, a realm that made possible trade, communication, and travel as never before. At the empire's height in A.D. 116, it became possible to travel from Scotland to the Red Sea without leaving a Roman road except to cross bodies of water. It was no wonder, then, that geographers of Roman lands, perhaps for the first time in history, began to conceive the idea of mapping the known world. Likewise it was fitting that, centuries after Rome crumbled, the vision of a great European-based multinational empire remained alive.

Background

Numerous myths glorified the Romans' origins, none more so than the great *Aeneid* of Virgil (70-19 B.C.), which depicted them as descending from a great prince who led a group of escapees from defeated Troy. A particularly telling aspect of the myth is the unfavorable light in which it casts Dido, the queen regarded by the people of Carthage—Rome's historic rival—as the founder of their own civilization: in Virgil's tale, Dido falls deeply in love with Aeneas, and commits suicide when he leaves her.

Much older were legends of the twins Romulus and Remus, raised by a she-wolf near the site where Romulus, after killing his brother, established the city itself. This tale of betrayal, along with the image of the wolf—which later became Rome's symbol—says much about what was to come when the Romans cast their eyes on the spoils of surrounding nations. So, too, does the story of the Sabine women: eager to get wives, the men of early Rome were said to have tricked the Sabine men into leaving their town; then they charged in, raping and kidnapping the women.

Rome itself consisted of several native groups, including Latins, Etruscans, and Sabines, and the tale of the Sabine women served as an explanation regarding how Rome came to be ruled alternately by kings from each of the three groups. This arrangement was said to have continued throughout the reigns of six legendary kings who ruled from the time of Romulus's death to the founding of the Roman Republic—the first genuinely historic date from Roman history—in 507 B.C.

Supposedly the son of the seventh king, the Etruscan Tarquinius Superbus, raped a virtuous Roman woman named Lucretia, who was so overcome by shame that she killed herself, and this act led to a revolt that overthrew Tarquinius. Thereafter the Romans refused to be ruled by kings, and all citizens of Rome—whether Latin, Etruscan, or Sabine—considered themselves Romans first. Hence they dedicated themselves to establishing dominance over the other peoples (even their own cousins) in the surrounding regions.

Impact

In its first century of existence, the Roman Republic expanded slowly within the Italian peninsula, winning victories by a combination of military might and diplomatic subterfuge—that is, by breaking treaties. Then came an attack on the city itself by the barbarian Gauls or Celts in 390 B.C., an event that proved crucial in fortifying Roman determination to establish their power throughout the Italian Peninsula. This Rome largely achieved in the fourth century B.C., and by 275 B.C. Rome had conquered the part of Sicily formerly controlled by the Hellenic colonists of Magna Grecia.

Not only was eastern Sicily the first Roman territory outside the Italian mainland, but its conquest provided Rome with a staging area for the act that would catapult it into a true empire: the defeat of Carthage. The latter controlled much of North Africa, Spain, and the islands off Italy's western coast; but unlike the Romans, the Carthaginians were a relatively peaceable people. Thus it was at Rome's prodding that the two powers entered into the First Punic War (264-241 B.C.), which resulted in the Roman conquest of Corsica, Sardinia, and all of Sicily.

In the Second Punic War (218-201 B.C.), by contrast, the angered Carthaginians acted as the aggressor in a brilliant 15-year campaign led by Hannibal (247-183 B.C.). The latter opened up a new chapter in the history of exploration with his crossing of the Alps, which proved that the dangerous mountain passes could be traversed

The Roman soldier, as pictured here, was a formidable instrument of Roman imperialism. *(Bettmann/Corbis. Reproduced with permission.)*

by a large army. No doubt his success in doing so opened Roman eyes to the possibilities that lay northward.

Despite the genius of Hannibal's advances, the war resulted in the defeat of Carthage, and in the aftermath Rome added numerous lands. By the beginning of the first century B.C., Spain and southern Gaul (modern France), as well as all of the Italian Peninsula, were under direct Roman rule, as was the former Carthaginian homeland. The conquest of lands in Eastern Europe, undertaken between 230 and 133 B.C., added the entire Adriatic coastline, as well as the greatest prize yet: Greece. Further east, the Romans subdued Asia Minor, thus making theirs a power on three continents.

By this point, Roman leaders had come to appreciate the spoils that went with victory, and each further acquisition added riches to the empire. In turn the Romans brought their new subjects the stability that came with the rulership of a powerful empire: on the coast of Asia Minor, for instance, Julius Caesar (102-44 B.C.) and others broke the power of Cilician pirates who had long threatened trading vessels. In conquering these lands, Rome granted subject peoples rights of citi-

Roman soldiers battling with Germanic tribesmen. *(Bettmann/Corbis. Reproduced with permission.)*

zenship, which carried prestige and political benefits. The Mediterranean world became increasingly Romanized, meaning that the peoples accepted Roman civilization—in many regards it was an extension of its Greek counterpart.

Between 133 and 31 B.C., when Octavian (63 B.C.-A.D. 14) brought an end to the Republic and became Augustus, first Roman emperor, Roman influence spread in all directions. Julius Caesar had conquered much of what is now France and Belgium, and conducted the first campaign to subdue a region previously beyond the reach of Mediterranean civilization: Britain, which Rome annexed in A.D. 43. Meanwhile, Syria and later Judea became part of the empire, and this last fact would have an unintended consequence.

It so happened that the Roman conquest of Judea coincided with the birth of Christianity, and over the years that followed, Christian missionaries would extend their faith to the Roman cities of Asia Minor and Greece. After centuries in which Christians suffered persecutions at the hands of pagan emperors, Constantine (A.D. 285-337) would adopt and legalize their faith, which became the dominant religion of Rome. This in turn spawned a close connection between Christianity and the seat of Roman power, a tie—symbolized in the office of the pope, or bishop of Rome—that would hold even after the Roman Empire itself fell.

The fall of Rome seemed far away indeed during the early days of the empire, but there were already signs that it was reaching its limits. A disastrous campaign by Aelius Gallus in Arabia in 25 B.C., and (to a much greater extent) the defeat of Roman forces in the Teutoburg Forest of Germany in A.D. 9, convinced Augustus that the time for expansion was over. By that point Rome controlled all of the North African coast (including another great prize, Egypt); all of Western Europe except for Britain, Scandinavia, and the regions above the Rhine; all of Eastern Europe south of the Danube; Asia Minor; and the Levant. Rivers, deserts, oceans, and mountains now formed the boundaries of the empire, and Augustus cautioned his stepson and successor Tiberius (r. A.D. 14-37) to be wary of further expansion.

In fact the century after Augustus's death would see a few conquests, most notably in Britain and Western Asia, and the empire reached its greatest extent under Trajan (r. 99-117). Though revolts, uprisings, plagues, and other disasters continued to occur, the first two centuries A.D. would prove a period of unprecedented stability, an era characterized by the *Pax Romana* or Roman peace.

At a time when Rome controlled much of the known world, attempts to map that world reached a degree of maturity unequalled before, and indeed for a long time afterward. As with much else, this early form of scientific geography constituted a legacy from the Greeks, but it was the Roman era that spawned the greatest geographer of antiquity, Strabo (c. 63 B.C.–c. A.D. 23) of Asia Minor. His *Geographica,* despite its flaws, would not be surpassed as a guide to the Western world until late in the medieval period. The early centuries of empire likewise produced a number of other figures, almost all of them Greeks, whose work represents the highest Western achievements in their disciplines for the next 1,000 years: the astronomer Ptolemy (c. A.D. 100-170), for instance, or the physician Galen (130-200), who served as personal attendant to Emperor Marcus Aurelius (121-180).

During the age of *Pax Romana*, the deployment of Roman legions in border lands ensured the stability of the empire, and the construction of roads opened its many lands to trade. Eventually the most impressive highway system known to that time covered Roman territories, and the protection of Roman soldiers ensured that travel was safe. Nor were these roads mere paths: most were 12 feet (3.7 meters) wide or more, built of stone, clay, and gravel 3 feet (0.9 meters) deep. Drainage ditches lined either side, and there were stone markers indicating the distance to and from Rome—hence the famous saying, "All roads lead to Rome."

It says a great deal about these roads, and about the dismal conditions that prevailed in Europe after the fall of the Roman Empire, that at the beginning of the Renaissance the best roads on the continent were still those built by the Romans more than 1,000 years before. Indeed, the next historic parallel to the Roman road system would probably be the German Autobahn, inspiration for the U.S. interstate highway system, which was built by a twentieth-century ruler intent on creating a nightmarish version of the Roman Empire: Adolf Hitler.

Throughout Roman lands, trade flourished, and each region had its specialties. Egypt, where the fertile Nile Delta had long yielded excellent crops, was the empire's breadbasket, but even less civilized countries had their products of value: tin, iron, and wool from Britain, for instance, or grain, honey, and hemp from the Scythian frontier. The Silk Road and the existence of a similarly powerful empire, the Han Dynasty, in China made possible contact with the most distant of known civilizations. Roman protection of shipping lanes facilitated trade with the spice-producing lands of southern Arabia and India, while Rome's control of the Nile allowed the importation of ivory from Central Africa.

It is perhaps no wonder, then, that the historian Edward Gibbon (1734-1794), in his *Decline and Fall of the Roman Empire,* described the era from A.D. 96 to 180 as "the period of the history of the world during which the condition of the human race was most happy and prosperous." But with the death of Marcus Aurelius, Rome entered a long, slow period of decline. Many causes have been offered for the collapse of the Western Roman Empire, a downfall that began in earnest during the late fourth century A.D.: Gibbon, for instance, blamed Christianity. It is more likely, however, that Roman civilization was simply exhausted. The Romans had grown their wealth almost purely through military conquest, rather than through trade or invention, and once there were no more worlds to conquer, there was no more growth: the realm simply collapsed on itself.

Yet the dream of Rome remained alive in the empire's eastern half, ruled from the city of Constantinople or Byzantium beginning in A.D. 330. The Byzantine Empire would continue to exist until 1453, by which time it had transmitted a form of Roman culture to Russia, whose rulers called themselves *czar* or "caesar." Meanwhile, in Western Europe the dream of empire had never died, and with the crowning of Charlemagne (742-814) as "Emperor of the Romans" in A.D. 800, it gained new form in what became known as the Holy Roman Empire. The latter would never prove as magnificent as its name, yet it became the basis for the nation whose people had once helped bring down the original Roman Empire: Germany. In the twentieth century, the idea of Rome would lend itself to the ghastly imperial visions of Hitler, Benito Mussolini in Italy, and Josef Stalin in the Soviet Union; but by the early twenty-first century it would reappear in a much more benign form with the gradual economic unification of Europe.

JUDSON KNIGHT

Further Reading

Books

Bardi, Piero. *The Atlas of the Classical World: Ancient Greece and Ancient Rome.* Illustrations by Matteo Chesi, et al. New York: Peter Bedrick Books, 1997.

Richardson, John. *Roman Provincial Administration, 227 B.C. to A.D. 117.* Basingstroke, England: Macmillan, 1976.

Starr, Chester. *The Ancient Romans.* New York: Oxford University Press, 1971.

Internet Sites

The Interactive Ancient Mediterranean Project. http://iam.classics.unc.edu/.

Maps of the Roman Empire. http://www.dalton.org/groups/Rome/RMaps.html.

The Roman Empire. http://www.roman-empire.net.

Rome and Romania, 27 B.C.-A.D. 1453 http://www.friesian.com/romania.htm.

Caesar and the Gauls

Overview

Julius Caesar's military expeditions to Gaul in the first century B.C. marked a dramatic turning point in the history of continental Europe. After Caesar's successful intervention in the migration of the Helvetii people westward into Gaul, Rome quickly dominated all land between the Mediterranean and the English Channel, and the Atlantic and the Rhine. After Caesar's initial military victories, the process of Romanization instituted a more subtle series of cultural battles.

Julius Caesar campaigned through Gaul from 58 to 50 B.C. The best account of the Gallic War is Caesar's own *Commentaries on the Gallic War,* which provides a clear and detailed account of the campaign. The seven books that make up the *Commentaries* were probably written during the winters between tactical maneuvers. These installments were issued to a Roman public highly sympathetic to Caesar's campaign. While the *Commentaries* appear to be objective and nonbiased, they operated to promote Caesar's own political interests: conquering Rome through the conquest of Gaul. He de-emphasized Roman defeats and played on the Roman stereotype of the Gauls as untrustworthy savages in a text clearly intended to help consolidate his political power in Rome. However, despite the political objective of *Commentaries,* the text remains the most influential source for information on the cultural and tactical dimensions of the Gallic War.

Background

By 58 B.C., Transalpine Gaul, or the Provincia, had been under Roman control for nearly a century. Roman involvement in the region stemmed from the long alliance between the Roman Empire and Massalia, an independent Greek city-state. Massalia, which is now the French city of Marseille, was a successful trading center on the Mediterranean. Established about 600 B.C. by colonists from Phocaea, an Ionian city, Massalia encouraged the development of a Mediterranean lifestyle along the southern coast of Transalpine Gaul. However, Massalia's success was not the result of military strength. Despite Massalia's dominant position in Mediterranean commerce, the city-state faced numerous difficulties with the neighboring Celto-Ligurian tribes.

Following Roman aid to Massalia in 125 B.C., the Romans established a military base at what is today Aix-en-Provence in France. From this tactical position, Roman forces were able to secure the region. By 121, the hostile forces of the Allobroges and Arvenians had been squelched. Rome annexed a huge area that extended from the Pyrenees to the lower Rhône, and from the Rhône valley to Lake Geneva. Rome allowed Massalia to retain its lands, but Massalia became a small enclave of Greek influence in a large Roman territory. Likewise, the Via Domitia, built in 118 B.C., helped to establish Gaul as Roman territory. This extensive highway effectively linked Gaul to Italy to the east and Roman possessions in Spain to the west. Because Spain also provided numerous problems for the Romans, the Via Domitia ensured a steady flow of Roman troops through Gaul.

The process of Romanization in Gaul was aided by Massalia's years of commercial influence. While Massalia sought aid from hostile neighbors, the city-state was engaged in extensive trade with numerous Gallic tribes. Archaeologists have found great numbers of *amphorae,* jugs used for Mediterranean wine, dispersed widely through ancient Gaul. Massalia exposed southern Gaul to Mediterranean influence to such an extent that Justinus remarked that "it seemed as though Gaul had become part of Greece, rather than that Greece had colonized Gaul." However, the process of Romanization was more problematic. This process, which assimilated the conquered in order to transform them into loyal subjects, was particularly difficult in Gaul because of Roman and Gallic attitudes. The Gauls were ready consumers of Mediterranean goods, but they were wary of

Roman social practices and unaccustomed to Roman notions of urban development. Likewise, the Romans viewed the Gauls as boisterous and unstable, more inclined to resort to violence than to rhetoric in order to settle their disputes. Furthermore, the Romans believed the Gauls to be too attracted to the pleasures of the vine.

Impact

The movement of the Helvetii westward across Gaul instigated the Gallic Wars. This mass migration threatened the tenuous stability in Gaul. Prior to this migration, in 60 B.C., Cicero spoke of peace in Gaul. While the management of the area had always been problematic for the Romans, a tolerable equilibrium seemed to envelop the region. Indeed, Caesar's initial desire to control the province did not stem from the need to accomplish any immediate military objectives in Gaul. Instead, Caesar was eager to assume control of Gaul because this position allowed him to amass military forces. He desired these forces so that he could assert his dominance in Rome.

By 60 B.C., the old Roman Republic had degenerated. The Empire was ruled by the impulses of power hungry warlords. Pompey, Caesar, and Crassus competed for ultimate authority. The agreement which the three charted in order to rule has been referred to by historians as the First Triumvirate. However, this unconstitutional ruling agreement did not dispel the political ambitions of the members of the triumvirate. Caesar sought to gain sufficient wealth and military prestige to overtake Pompey and Crassus. The migration of the Helvetii provided him with a series of opportunities to secure his reputation as an effective and dynamic military leader.

While the *Commentaries on the Gallic War* initially functioned as a public-relations tool for Caesar, the work is significant today because of its depiction of Gallic and Germanic culture. There are very few written accounts of pre-Roman Gaul. At best, the Gauls appear as wild barbarians in earlier travel narratives. While the Romans regarded all outsiders as barbarians, Caesar's account cites the similarities and differences between these barbarians and the people of Rome. The *Commentaries* discuss the political and social organization of both Gaul and Germany. However, his *Commentaries* do not belittle the culture of the Gallic and Germanic tribes. Instead, in many places, Caesar praises these barbarians for avoiding the vices and weaknesses characteristic of Roman culture.

Caesar mentions that, in many ways, the Gauls and Romans were nearly identical culturally, and points out that the hierarchical nature of Gallic society paralleled Roman society. Likewise, Caesar was able to compare the roles of the Gallic gods to the roles of the gods of the Roman pantheon. However, he also validates the superiority of the Roman religion. In the *Commentaries,* he plays on the Roman distaste for human sacrifice, indicating the extent to which "the Gallic tribes as a whole are slaves of superstition." Caesar emphasizes this penchant for human sacrifice and details the Gallic practice of stuffing the limbs of gigantic wicker figures full of men and then igniting these structures.

Furthermore, Caesar draws clear distinctions between the Gauls and the Germans. He stresses the extent to which the nomadic and martial lifestyle of the Germanic tribes actually serves as a means of maintaining social equality. The corrupt and inefficient hierarchies of Gallic and Roman culture are absent among the German barbarians. While Caesar does not find many similarities between Germanic and Roman culture, he seems to respect the martial eminence of the Germanic tribes. For Caesar, the discipline inherent in a military culture engenders other positive qualities. For example, he emphasizes the protected position of guests in Germanic culture, and discusses the extent to which the Germans revered hospitality as sacred.

Despite Caesar's sympathy for the Gallic and Germanic people, his military campaign subjugated many Gallic and Germanic tribes to Roman rule. But Caesar's complex description of the Gauls and Germans hints at the difficulties inherent in the Romanization process. It is too simple to characterize Roman and Gallic cultural conflict through a simple dichotomy of Romanization versus resistance. The similarities between Gallic and Roman culture, as well as the fact that Massalia encouraged the development of a Mediterranean lifestyle, suggest that the Gauls should have easily assimilated into Roman culture. But even after the Gallic wars, the Roman army maintained a strong presence in Gaul. Indeed, Roman efforts in Gaul were more focused toward the prevention of violence than the imposition of Roman cultural practices and customs. As a result, the culture that emerged after the Gallic wars was a unique hybrid of Gallic and Roman influences. Indeed, the Gallo-Roman culture which developed at this time both resisted and absorbed Roman cultural influences.

As indicated by the strong Roman military presence in the province, the Romans were slow to accept the Gauls. Instead of replacing the Gallic villages with strong centralized Roman cities, the Romans kept construction to a minimum and expressed strong reservations about permitting the Gauls to assume the status of Roman citizens. Indeed, the distinctions between Gallic and Roman city planning highlights the primary cultural divisions that inhibited Romanization in Gaul. Caesar discusses the extent to which the exclusively nomadic nature of the Germanic tribes and the nonurban villages of the Gauls affected Roman military operations. While it was difficult to pin down the constantly roving Germans, the Romans were able to overtake Gallic villages with relative ease.

The lack of fortifications in Gallic villages made them easy targets for the Roman army. But these features also reveal different attitudes toward the role of the urban center and the role of an urban culture. The Gallic villages lacked the concentration of buildings that typified Roman towns. These Gallic cities seemed to the Romans to lack a center, and they were reluctant to even refer to these spaces as cities. Even after the Gallic war, Gaul had few urban areas. A sharp distinction existed between Roman cities and the culture of the surrounding countryside. The cities erected by the Romans in Gaul operated as scattered outposts of civilization among a scattered network of rural Gallic villages. As a result, Gallo-Roman civilization was also divided spatially. Romans largely occupied the new Roman cities built throughout Gaul, while the Gauls remained in the countryside.

This spatial distinction created a Roman Gaul and a Gallic Gaul. Paradoxically, the close similarities between the Gauls and the Romans inhibited the rapid creation of a culture that was equally Gallic and Roman. Caesar's depiction of the Gauls in the *Commentaries on the Gallic War* seems to anticipate the difficulties inherent in the cultural conflict that follows military conflict.

DEAN SWINFORD

Further Reading

Caesar, Julius. *The Gallic Wars.* trans. by John Warrington. New York: Heritage Press, 1955.

Drinkwater, J. F. *Roman Gaul: The Three Provinces, 58 B.C.-A.D. 260.* Ithaca: Cornell University Press, 1983.

King, Anthony. *Roman Gaul and Germany.* San Francisco: University of California Press, 1990.

Woolf, Greg. *Becoming Roman: The Origins of Provincial Civilization in Gaul.* Cambridge: Cambridge University Press, 1998.

Aelius Gallus Attempts the Conquest of Arabia— and Reaches the Limits of Roman Power

Overview

In 25 B.C., the Roman emperor Augustus sent Aelius Gallus, prefect of Egypt, on a military expedition to the Arabian Peninsula. His aim was to extend Roman control throughout Arabia, and to gain control of the wealthy spice-producing states at the peninsula's southern tip, but instead the Romans became mired in a miserably failed operation that proved costly in terms of lives, finances, and the empire's reputation. It was the first time Rome truly came face to face with the limits of its imperial ambitions, an early sign of the slow Roman retreat that would commence some two centuries later.

Background

Founded as a republic in 507 B.C., Rome had begun its existence fighting with the Etruscans for supremacy over the Italian Peninsula. In 496 B.C., Rome fought a battle with several of its neighbors and won, in large part thanks to the Romans' adoption of Greek military tactics such as the use of the hoplite (a heavily armed infantry soldier) and the phalanx, or massed column. An attack on Rome itself by the Celts or Gauls in 390 B.C. further intensified the Romans' determination to establish military superiority over all foes, and in the years that followed, the consuls who led the young republic undertook to ensure that its lands would never be so threatened again.

Between 343 and 290 B.C., Rome fought the Samnites for power over much of southern Italy, and by 275 B.C. had defeated the Greek colonists who controlled Sicily. This left them staring across the Mediterranean at the one other great

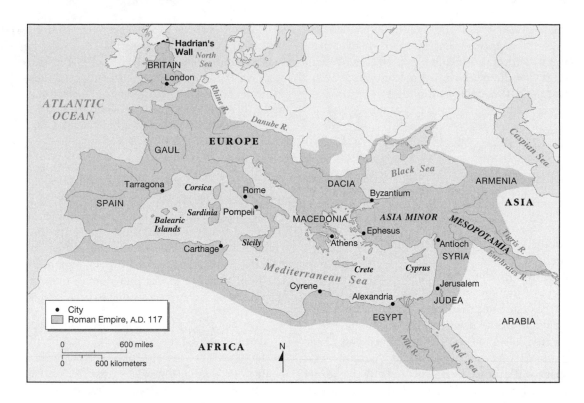

The Roman Empire at its height, circa A.D. 117.

power in the region: Carthage, which Rome defeated in the First Punic War (264-241 B.C.) The latter conflict left Rome in possession not only of Sicily, but of Corsica and Sardinia, and marked the beginnings of Rome's overseas empire. The Second Punic War (218-201 B.C.), despite a heroic series of engagements led by the Carthaginian general Hannibal (247-183 B.C.), resulted in the end of Carthage's power, and in the years that followed, Rome began to build its empire in earnest.

As Rome added Carthage, parts of Spain and Asia Minor, and Greece to its lands, the Roman economy became increasingly dependent on conquest. The ancients were largely ignorant of the idea of economic growth by accumulation, as with investment or business-building; theirs was a zero-sum game of growth through absorption and acquisition. With the conquest of each new land, Rome simply exploited the riches of the vanquished—including the people themselves, many of whom the Romans forced into slavery.

The situation worked well as long as there were lands easily accessible for conquest; meanwhile, time was running out for the republican form of government. A series of civil conflicts led to veritable anarchy, and when Julius Caesar (102-44 B.C.), along with Pompey (106-48 B.C.)

and Crassus (c. 115-53 B.C.), established the First Triumvirate in 60 B.C., many believed that the republic was on its way to recovery. In fact what had happened was the establishment of dictatorship, and the collapse of the First Triumvirate led to the formation of a second, with Caesar's nephew Octavian (63 B.C.-A.D. 14) occupying the apex. With the defeat of his former ally Mark Antony (82?-30 B.C.) at the Battle of Actium in 31 B.C., Octavian became supreme ruler under the name Augustus Caesar. Thereafter Rome would be in theory what it had long been in fact: an empire.

Impact

Augustus himself never used the title "emperor," and in spite of the ruthlessness with which he seized power, he would prove a just ruler once he had it. In foreign affairs, however, he was bound by established practice—and indeed by his own desires—to continue the process of growth by expansion.

In the first years of his reign, Augustus turned southward. Cornelius Gallus (c. 70-26 B.C.), first prefect of Egypt, conducted what appeared to be a successful campaign to extend the Egyptian province into Kush or Nubia (modern Sudan), but when Gaius Petronius attempted to consolidate those gains, he was forced to stop at the First Cataract of the Nile. In 26 B.C., Aelius Gallus had replaced Cornelius (who may or may not have been his relative) as prefect, and soon afterward, he received orders from Augustus to undertake the Arabian expedition.

The Romans conceived of Arabia in three parts, and as it turned out, Arabia Petraea or Petra—the area including the Sinai Peninsula and parts of the Red Sea shore—was the only one they ever conquered. Though Rome would not annex Petra until A.D. 106, it was already under Roman influence, and Augustus hoped to use its aid to win control of Arabia Felix, or southern Arabia. The latter, which included the modern nation of Yemen, was a relatively wealthy area. Not only did it possess riches in the form of frankincense, but spices grew there in such abundance that the Himyarite Arab kingdoms of southern Arabia were nicknamed "the incense states." Between Arabia Petraea and Arabia Felix, however, lay a vast, uncharted region that would prove the downfall of any would-be conqueror, a realm whose forbidding quality is encompassed in its Roman name: Arabia Deserta.

But Aelius had reason to be confident at the outset of his campaign. He had the support of the Nabataean Arabs in Petra, and the help of a Nabataean administrator named Syllaeus, who agreed to serve as guide. He set out from the city of Cleopatris (the modern city of Suez) on the Gulf of Suez with some 10,000 Roman and Egyptian troops, as well as 500 Jewish and 1,000 Nabataean auxiliaries.

The first sign of bad things to come occurred when the Romans' ships ran into ill winds crossing the Gulf of Aqaba, but they finally reached the city of Leuke Come or Haura on the Arabian Peninsula. Soon afterward, however, bad food and water caused an outbreak of sickness that forced them to tarry in Leuke Come throughout the summer and winter. Finally, in the spring of 24 B.C. they set out across the desert with a caravan of water-bearing camels.

The expeditionary force marched for 30 days through lands controlled by the Areta tribe, allies of Rome, and spent 50 more days traveling across uncharted desert. Finally they reached the area of Negrana or Nejran, a region whose fertile lands raised their hopes. The Romans even conquered a few towns and took on supplies, then proceeded to besiege the city of Marsiaba or Marib. Diminishing water supplies, however, forced them to break off the siege, yet Syllaeus insisted that they were just two days' march from the tempting "incense states" of the coast.

Gallus and his troops spent six months wandering through the desert, until finally he realized that Syllaeus and the other Nabataeans were not the trusty allies he had supposed them to be. Eventually he turned his army around, and they took only 60 days to reach the Red Sea coast—further proof that they had been led on a wild goose-chase.

The army that crossed the sea to Myos Hormos, or Abu Scha'ar in Egypt, was much smaller than the one that had departed more than a year before: though they had lost only seven men in battle, disease, hunger, and exposure had claimed many lives. Nor had they gained any valuable knowledge concerning the region: indeed, all they knew was that Roman interests would not be served by a second expedition.

The campaign of Aelius Gallus, about which Gallus's friend Strabo (c. 64 B.C.-c. 23 A.D.) is the principal source, in retrospect seems the first chapter in a gathering saga of Roman retreat. Though Augustus's forces consolidated control over the western reaches of North Africa, as well as Judea and other parts of western Asia, it was becoming clear that certain fringes of the empire could not be won except at unthinkable costs.

Principal among these regions was the area beyond the Rhine, whose blond-haired, blue-eyed inhabitants seemed so alike to the Romans that they had long dubbed them by a Latin word meaning "similar": *germanus*. The Germans, of course, would play a major role in Rome's ultimate downfall many centuries hence, but even in Augustus's time, at the height of Roman power, the German tribes dealt Roman forces a decisive blow at the Teutoburg Forest in A.D. 9.

As a result of the Teutoburg defeat, Augustus—by then getting on in years—withdrew from further attempts at northward expansion, and counseled his stepson Tiberius (r. A.D. 14-37) to be wary of further military adventures. Nonetheless, the empire would expand under Tiberius's rule, and reached its greatest extent under Trajan (r. 99-117) in 116. At that point, Roman authority stretched from the borders of Scotland to the mouths of the Tigris and Euphrates, and the *Pax Romana* or Roman peace established by Augustus prevailed throughout much of the world. But just six years later, Hadrian (r. 117-138) ordered the erection of his famous wall in northern Britain. Intended to keep out the Picts of Scotland, a purpose in which it failed, the wall served as a physical manifestation of the fact that the empire had its limits—and that it would inevitably begin to shrink.

JUDSON KNIGHT

Further Reading

Books

Birley, A. R. *Roman Papers*. New York: Oxford University Press, 1979.

Cary, M., and E. H. Warmington. *The Ancient Explorers*. London: Methuen, 1929.

Internet Sites

"Aelius Gallus' Arabian Expedition Ends in Disaster!" Vox Romana II. http://pages.ancientsites.com/~Donalda_Antonius/VoxRomana2.html.

"Ancient Accounts of Arabia, 430 B.C.-550 C.E." Ancient History Sourcebook. http://www.fordham.edu/halsall/ancient/arabia1.htm.

"A History of Merchant Routes." Sheba Aromatics. http://www.shebaaromatics.i12.com/historyroutes.htm.

Roman Technology, Government, and the Spread of Early Christianity

Overview

Christianity arose in the Roman Empire after the death of Jesus of Nazareth to become one the major world religions. Missionaries such as Paul of Tarsus (10-61 A.D.) would spread this new belief system throughout the Roman world. The success of this spiritual enterprise was the result of two major factors. By A.D. 35 a growing number of people within the empire were searching for an alternative to their outdated and corrupt religious and philosophical systems. Christianity would provide a universal alternative to this Roman malaise. The success of this new value system was also based upon the superiority of Roman technology that had created the greatest system of roads in the classical world. These ancient avenues provided the first Christian missionaries with easy access to the hearts and minds of approximately 55 million people.

Background

By 500 B.C. Rome developed a deliberate policy of expansion. The aristocracy believed the only way to both defend against outside aggression and increase Rome's power was to adopt a policy of imperialistic expansion. From 400 to 360 B.C. the legions of Rome conquered and pacified all of central and southern Italy, and 70 years later the entire peninsula had come under Roman control. This success set the stage for Rome's greatest military challenge to date, when it engaged and defeated the North African city-state of Carthage in what has come to be known as the Punic Wars.

The major economic impact of the Punic Wars was the commercialization of agriculture.

As a result of the Empire's expansionist policy, many middle-class farmers spent more time fighting in the legions than cultivating their land. This placed great economic strain on the families, and eventually they were forced to sell out to wealthy landowners. This new class of commercial farmers combined these individual holdings into great estates known as *latifundia*. As more lands came under Roman control, this agricultural model was extended into the newly conquered provinces. This process led to the specialization in cash crops that were sold to the Empire's ever-growing urban areas.

In time, a new urban empire came into existence that depended upon successful trading networks for its support. The high point of Roman success spanned the years from 20 B.C. to A.D. 180. Historians now refer to this period as *Pax Romana*, the Roman Peace. Trade, manufacturing, and culture flourished during this time of unprecedented stability.

One of the most important reasons for this success was the great expertise of the Roman civil engineers who created a network of roads and bridges that linked the urban centers with the provinces under the protection of the Roman military. Originally, roads were constructed with one purpose in mind, to move the legions as quickly as possible in order to engage and defeat hostile forces. The first major road was constructed in 312 B.C. to support the conquest of the southern part of the Italian Peninsula. By the time of *Pax Romana*, the empire had a network of roads that covered approximately 55,000 miles (88,500 km). In order to protect Rome's

Saint Paul brought Christianity to many Gentiles (non-Jews) in the Roman world. *(Corbis Corporation. Reproduced with permission.)*

extensive trade network, units of both cavalry and infantry were required to move quickly throughout the Empire to keep the peace. The government made the important decision based upon the reality that it would be far too expensive to quarter troops permanently in every part of the Empire. The military opted for the creation of a small mobile strike force of about 180,000 soldiers, which could be quickly moved to any location to enforce and protect Roman law. The Empire's system of roads actually became a launching pad for the army's military action. When word was received that an uprising or invasion was imminent, units of the armed forces would be dispatched to engage the enemy. Depending upon the weather and terrain, a Roman unit could cover up to 24 miles (38.6 km) in a five-hour period. Roman soldiers were so physically fit they could immediately engage the enemy at the conclusion of a long march. This highly effective combination of roads and military tactics allowed Rome to govern 55 million inhabitants.

Impact

The Jews were one of the many minority groups living under Roman rule. They were unique among the religions of the Empire because they were monotheistic, believing in only one God.

By the fourth decade of *Pax Romana*, a young Jewish activist named Jesus was gaining notoriety based upon a message of peace, love, charity, and humility. The Romans had been concerned for many years about the threat of revolutionary activity on the part of the Jewish people. Their devotion to monotheism clashed with the values of Rome and drove a deep wedge between the two societies. The teachings of Jesus of Nazareth not only made the Romans uneasy, but in time they also began to undermine the authority of the Jewish authorities. The suspicions of both groups led to his trial and condemnation, and he was put to death in A.D. 33.

After Jesus's death, his disciples began to spread his message first among their fellow Jews and eventually throughout the entire Empire. The first great missionary to work outside the Jewish community was Paul of Tarsus, who began his work around A.D. 35. He received his education in Jewish law and spent the early part of his career persecuting those who violated its tenets. One day on the road to Damascus he had a conversion experience that changed his life. He believed that he had encountered the risen Christ and that he was directed to bring the teachings of Christianity to the Gentiles. Using the extensive system of Roman roads, Paul made four trips to establish communities of Christians throughout the Empire. He was able to keep these religious groups together and oriented in the new faith by a series of letters that today form a significant part of the New Testament. Scholars agree that at least 13 of the books of the New Testament can be attributed to Paul or his followers. His successful establishment of Christian churches would have been impossible without the excellent system of Roman roads.

Christianity's initial appeal was based upon the widespread spiritual disenchantment of many of Rome's citizens. Its message of universal salvation particularly appealed to the segment of the population that had been economically and politically left behind during the Empire's expansion. Christianity reflected the social grievances of slaves, dispossessed farmers, and the urban unemployed. Many Roman intellectuals were drawn to this message because it filled a spiritual need that could no longer be met by the traditional Roman values and beliefs that had been debased by centuries of corruption.

The spread of the Christian religion also benefited from the stability of *Pax Romana*. The same stability that protected the Empire from invasion and revolt provided safe passage for these

Early Christian art in an Italian parish. *(Elio Ciol/Corbis. Reproduced with permission.)*

early Christian missionaries, which accelerated the spread of this new value system. Christianity received the status of a state religion when the Emperor Constantine (r. 306-337) adopted it as his personal belief system. From that time on, Christian missionaries went out into the Empire under the protection of the emperor.

The greatest advances in spreading the doctrine were made among the Germanic peoples of northern Europe. This was actually an important turning point in the history of the Church. Constantine's conversion necessitated that the new religion have a recorded, chronological history The decision to use December 25 as the birthday of Jesus was also made as an attempt to establish a common spiritual ground with the newly converted Germanic tribes. The pagan re-

ligion of these tribes was oriented to the importance of the changing seasons. In particular, the Winter Solstice was an event of great significance because it celebrated the rebirth of nature. The days begin to last longer, and the life-giving warmth of the Sun begins a regeneration of nature. The early Christian Church adopted this concept of rebirth and renewal to signify the birth of Jesus. The same held true in deciding on a date for the celebration of Christianity's most important feast, the Resurrection. The date of Easter is set according to the spring equinox. Easter Sunday, the ultimate celebration of rebirth, is always scheduled for the first Sunday after the first full moon after the spring equinox. The Easter symbols of rabbits and eggs reflect the importance of fertility and regeneration in the Germanic nature religion.

The work of St. Patrick (457-492 A.D.) among the Irish is also linked to Roman technology and culture. Christian missionaries advanced into the British Isles by the fourth century A.D. Patrick was kidnapped as a youth in one of the many raids that occurred as the result this expansion. He was taken back to Ireland and placed into bondage. He eventually escaped, received Holy Orders and returned to work among the people who were once his captors. His story represents an important break with the cultural heritage of the Empire. Patrick's theology broke from the legalistic, guilt-ridden doctrine of the early Roman Church fathers. He focused on the joy found in God's creation and made his liturgy one of celebration. Patrick also broke with the Roman view of women, which relegated them to the status of second-class citizens. He admired the strength and courage of Irish women and elevated their position within the Church.

From these early connections with Roman culture and technology, Christian doctrine and values would ultimately spread throughout the world. The spread of early Christianity is an important example of the link among technology, government, and society.

RICHARD D. FITZGERALD

Further Reading

Ayers, Robert. *Judaism and Christianity: Developments and Recent Trends.* Lanham, MD: University Press of America, 1983.

Markus, R.A. *Christianity in the Roman World.* New York: Charles Scribners' Sons, 1974.

White, K.D. *Greek and Roman Technology.* Ithaca: Cornell University Press, 1984.

Lindisfarne and Iona: Preserving Western Civilization in the Dark Ages

Overview

When the Roman Empire fell to the barbarians in A.D. 476, much of western civilization's accumulated knowledge began to disappear. Scrolls and manuscripts were lost when the monasteries and libraries that contained them were abandoned. Documents that remained were less and less valuable to people ignorant of their value. As Europe fell into the Dark Ages, only a very few people, most of them Christian monks, had any knowledge of Latin, the language of knowledge and learning. With this decline, the accumulated culture of Rome began to wane; the translated works of Greek antiquity nearly disappeared.

Despite the eclipse of knowledge from Greece and Rome, there was some significant contribution to western scholarship in the Dark Ages. The Roman Church managed not only to survive the fall of the empire, but to thrive in its absence, sending missionaries to the geographical extremes of Europe. Some of these men traveled as far as the British Isles, seeking to Christianize the barbarian tribes (people who were not former citizens of Rome). When they arrived in northern Europe, they established monasteries, educated future generations of priests, and transcribed religious and some secular texts. Monasteries such as Lindisfarne and Iona, both located on islands in the northern British Isles, are known as the "cradle of Christianity" because of their contributions to medieval theology, promotion of ecclesiastical order, dedicated teaching of Latin, and transcription of numerous manuscripts.

Background

The collapse of Rome created tremendous political, economic, and social instability in Europe. The empire had built and managed a network of roads, supported fleets of trading vessels, and set currency standards that facilitated trade. The spread of Latin among the empire's elites made communication easier and laws more uniform. Before the collapse, Roman artists, philosophers, and scholars not only transcribed the notable works of Greek civilization into Latin, they made substantial contributions to those fields in their own right. Roman civilization produced great technological advancements and masterful urban planning. Without an administrative structure, however, most of these endeavors were impossible. During the next 300 years Europe languished in the Dark Ages, an era that earned its name when the scientific, technological, and philosophical advancements of the Roman Empire were largely forgotten.

The island of Lindisfarne, off the coast of Northumberland, England. *(James Murdoch;Cordaiy Library, Ltd./Corbis. Reproduced with permission.)*

One institution did survive the fall of Rome, however. The Roman Church continued to spread Christianity within the borders of the former empire. With the end of Roman rule, Church law and administrative structure became a dominant political force in western Europe. But during the first centuries of the Dark Ages, the church's libraries and scriptoria (places where manuscripts were written and copied by hand) were threatened by barbarian invaders. Many monks fled to larger monasteries or to the fringes of Europe and Asia Minor, often taking Latin texts with them. The body of Roman knowledge was scattered.

The British Isles

Although Julius Caesar invaded Britain around 54 B.C., it was not until A.D. 43 that Roman armies actually conquered Britain. Four centuries later, the Roman presence in Albion, as Britain was known, had nearly disappeared, and the island began to be overrun by Germanic tribes that became known as the Anglo-Saxons. The centuries that followed were rife with conflict. While the Romans had introduced Christianity into the region, the Germanic invasion sparked a widespread abandonment of the new religion. Thus, the isle became a target of later missions by the Church to reclaim their former subjects. The best known of these missionaries was St. Augustine,

who, for a while at least, converted the kingdom of Kent to Christianity in 597.

Other missionaries traveled outside the borders of Roman Britain. Saint Columba established a monastery on Iona, an island off the coast of Scotland, in 563. From there he launched his mission to Christianize Scotland. Other missionaries founded monasteries of their own, each overseeing the immediate surrounding region.

Columba died in 597, but his successors continued their work in the British Isles. In 635, a Scot-Celt monk at Iona, Saint Aidan, established a missionary outpost on Lindisfarne, an island off the coast of northern England in the North Sea. Lindisfarne had the strategic advantage of being an island, but it was also linked to the British mainland by a natural causeway at low tide, thus enabling easier contact with nearby Northumbria. By 637, Saint Aidan had expanded the outpost into a church and monastery that, like Iona, served as the main ecclesiastic see of the region, sending missionaries into the interior of the British mainland. Lindisfarne also became one of the most important theological centers and manuscript production sites of the Dark Ages.

In 685, Saint Cuthbert, bishop of Hexam, exchanged his post for one at Lindisfarne. He gained fame for the island and its monastery through his personal renown as a healer and worker of miracles. When he died, he was revered as a saint and his tomb at Lindisfarne became a destination for pilgrims and a site famous for miracles. This period was perhaps the monastery's "golden age," both in prosperity and influence.

The first major Viking raid on the British Isles occurred at Lindisfarne in the ninth century. In 875, after decades of repeated Viking attacks, the monastery, including Saint Cuthbert's remains, which remained miraculously uncorrupted, was relocated to Durham, on the mainland in Northumbria. By 1082, the relocated monastery was thriving much in the same way it had on Lindisfarne. Durham became one of the most important cathedrals and bishoprics of the Middle Ages, controlling the northernmost third of England.

Impact

The conversion of the British Isles to Christianity also had practical implications. The establishment of new churches and monasteries required monks with theological training and copies of the Scriptures. In addition, clergymen had to be able to read in Latin. Lindisfarne's monastery became a bastion of literacy and scholarly study in an era when even literacy was exceedingly rare. The men of Lindisfarne were not only trained to read, but to write so they could make copies of ecclesiastical texts. Often, a monk would work years to transcribe his personal copy of the Bible (usually only the Gospels, Psalms, Proverbs, and a few other notable books) before being sent out to head his own parish or conduct missionary work. This form of monastic education promoted Latin as the primary language for the Church and other educated Europeans during the Middle Ages.

The most impressive surviving work of the monastery is the Lindisfarne Gospels, an illuminated (ornately illustrated) manuscript of the Gospels (the first four books of the New Testament) that also contains writings by St. Jerome, Eusebius of Nicomedia, and other texts. The Lindisfarne Gospels were most likely created for the island's bishop, Eadfrith, sometime in the late seventh or early eighth century. (A similar surviving work, Book of Kells, also containing the four Gospels, was most likely finished at Iona around the same time.) The Lindisfarne manuscript, written in Latin, is historically important in its own right, but in the ninth century, after the monks of Lindisfarne had fled to nearby Durham the Gospels were revised. Between the lines of the text, someone translated the Latin into the Northumbrian Saxon dialect, making them one of the first Bibles translated into a vernacular language during the Middle Ages.

As missionaries were trained at Iona and Lindisfarne, and then sent out on their respective missions, they established a network of parishes, churches, and monasteries. These often sprang up near villages, integrating Church administration and local society. More importantly, as the rulers of various British kingdoms converted to Christianity, they joined and supported the Church; some of the earliest kings of Scotland and Ireland were buried at the Iona monastery.

The growing number of monasteries and churches required an effective means to govern parishes and parishioners. The use of Latin permitted communication among Church leaders across ethnic and regional boundries, and allowed for more uniform interpretation of doctrine and law. The bishop of Rome, or pope, became the Church's chief political administrator as well as its spiritual leader.

Lindisfarne and Iona, however, had a history of difficult relations with Church hierarchy. By 700, both monasteries had adapted their religious ideals to make Christianity more appealing to people in their region incorporating native holidays, traditions, and mystic views of nature not sanctioned by the Roman Church. The tension was so strong that Iona later declared itself the center of the Celtic Church, rather than a subject of Rome. The two were reconciled in the Middle Ages, but the differences remained. In fact, the Scottish Presbyterians, a Protestant sect established in the seventeenth century, claim Iona as a theological ancestor.

Iona and Lindisfarne helped preserve Latin, promulgate learning, and spread Christianity through northern England. Both most likely produced many copies of the Bible and other theological manuscripts, and may have transcribed secular works as well. In 1997, archaeologists found what may be the remains of the scriptorium at Lindisfarne, giving new insight into the monstery's role in the preservation of Western civilization during the Dark Ages.

ADRIENNE WILMOTH LERNER

Further Reading

Bonner, Gerald, David Rollason, and Clare Stancliffe. *St. Cuthbert, His Cult and His Community to A.D. 1200.* Woodbridge, Suffolk, U.K.: Boydell Press; Wolfeboro, NH: Boydell & Brewer, 1995.

Alexander, Jonathan J. G. *Medieval Illuminators and Their Methods of Work.* New Haven, CT: Yale University Press, 1994.

Saint Brendan's Epic Voyage

Overview

Saint Brendan, a Celtic monk, is sometimes credited as the first European to reach the new world. If the accounts of his travels are accurate, his journey to North America predates Leif Erikson's discovery of Vinland by 400 years and Christopher Columbus's Caribbean voyages by nearly 1,000 years.

We know very little about the life of the historical Brendan. Textual references to him begin in the seventh century. The evidence provided by these texts indicates that he was born around A.D. 489 in county Kerry in the southwest of Ireland. He was a student of Erc, the bishop of Kerry. Brendan was baptized by Erc at Turbid, near Ardfert. He was educated by Saint Ida, the "Brigid of Munster," over a period of five years. Then he completed his studies under Bishop Erc, who ordained him in 512. After he was made an abbot, he began his ministry in Ireland, and established numerous monasteries and churches throughout the country. Many topographic details in Ireland, and even as far as Brittany, bear his name. Between 512 and 530, Brendan built monastic cells at Ardfert and Shanakeel. Shanakeel, located at the bottom of Brandon Hill, became the departure point for Brendan's voyage to the "Promised Land." An influential member of the clergy, Brendan is credited with undertaking other far-ranging pilgrimages to Britain, Brittany, and the Orkney and Shetland Islands. Indeed, prior to his legendary voyage, Brendan seems to have gained a reputation as a traveler. He is credited with establishing the abbey of Llancarfan in Wales and the celebrated monastery on the remote island of Iona, located off of the northwest coast of Scotland.

Background

There are two main sources that recount Brendan's legendary seafaring adventures. The first, *Vita sancti Brendani,* or *Life of Saint Brendan,* was first composed in Ireland in the ninth century. This document follows the fairly standardized pattern of early medieval saints' lives written in Latin. It provides significant biographical information and indicates that Brendan undertook two voyages. According to the *Vita,* the first voyage took five years and was an unsuccessful attempt to find a holy island that appeared to Brendan in a vision. The second successful mission took two years to complete. However, Brendan's voyages are not a focal point in *Vita,* but feature as relatively minor details in the life of a powerful and significant clergyman.

The second source, *Navigatio Brendani,* is concerned entirely with a single seven-year voyage. Scholars have determined that *Navigatio* is also the work of an Irish writer, and existed by the beginning of the tenth century. *Navigatio* does not concern itself with biographical details. Indeed, it presupposes a familiarity with the life and legend of Brendan the Saint as provided by *Vita.* To this extent, its structure differs significantly from *Vita.* The form of *Navigatio* most closely resembles that of the Irish *immram,* or "voyage-tale." The *immram* details a series of unrelated adventures which are linked by a sea journey undertaken by a heroic adventurer. The format of an *immram* is similar to that of Homer's *Odyssey.*

Impact

The social conditions of the early medieval period were not particularly sympathetic to voyages of discovery. However, in at least one of the distant points of European civilization, religious and missionary zeal allowed for considerable exploration. The Irish, converted to Christianity through the course of the fourth and fifth centuries, had developed a complex monastic culture. Numerous isolated convents were scatted across the island, where the religious led an ascetic lifestyle far from the lure of earthly temptations. Irish clerics devoted themselves to the translation and copying of texts and to missionary pilgrimages. These clerics also traveled to the numerous isolated and nearly inaccessible islands scattered around the British Isles in order to construct isolated monasteries or solitary cells (small religious houses).

These clerics achieved daring seafaring feats along a particularly rocky coast in small boats known as *curraghs.* These boats, about 30 feet (9 m) in length, featured wooden frames that were covered with leather hides sewn together into a single covering. Like the longboats of the Vikings, these boats flexed with the ocean, which allowed for fast sailing.

Early medieval monks traveled widely in these ships. The celebrated monastery of Iona, located on a desolate island off the northwest coast of Scotland, was founded in 563 by a group of monks led by Brendan, who had sailed from Ireland in a curragh. Monks used these seemingly unstable ships to undertake pilgrimages throughout the British Isles. By the end of the ninth century, groups of Irish monks had traveled as far as France and Iceland.

The Scandinavians were the most powerful naval force in Europe between the ninth and twelfth centuries. At different times, the Viking empire extended as far as Sicily in the south, as far as Russia in the east, and as far as Greenland and the mysterious Vinland in the west. The Vikings installed a base in Greenland in the tenth century, and began the ambitious exploration of the farthest reaches of the west Atlantic. But documents such as the *Navigatio* indicate that the Scandinavians were not the first to conquer the grim tempests of the north Atlantic.

When the Vikings arrived in Iceland in 870, they encountered a community of Irish monks, who they referred to as *papars*. The word *papar* was the Norse word for "fathers," directly referring to the colonization of Iceland by Irish monks. Indeed, many place names along the southeast coast of Iceland still incorporate the word *papar*. The *Landnamabok*, or *Book of Settlements*, a twelfth-century Icelandic text, describes the result of these early encounters between isolated Irish monks and the Norsemen. Shortly after the Norse arrival, these Irish monks, distraught at the prospect of sharing their land with heathens, fled in their boats, leaving behind books, bells, and crosiers (a staff carried by a bishop as a symbol of office).

The work of Dicuil, a monk writing in the ninth century during the time of Charlemagne, also demonstrates the feasibility of travel between Ireland and Iceland by medieval monks. Dicuil indicates that Irish monks were traveling regularly to an island so far north that, during the summer, the sun never set. This information seems to refer to an island like Iceland located in the far northern latitudes.

We know that Irish monks traveled considerable distances in their curraghs. But did they cross the Atlantic? *Navigatio* indicates that monks were aware of distant lands, and their desire to achieve a distance from worldly temptations serves as a viable motive for cross-Atlantic journeys. Indeed, while *Vita* indicates that Brendan's voyage was the result of a mystical vision, *Navigatio* presents a different motive for the trip. In *Navigatio*, Brendan hears of an earthly paradise which the hermit Barintus encountered on a sea voyage he undertook in order to find Mernoc, his godson. Mernoc, the text recounts, lived with a group of other monks on this distant paradise.

Interpretations of the accuracy of *Navigatio sancti Brendani* vary considerably. To some, the tale is too readily apparent as a Christianized version of an *immram*, an Irish tale of a fantastic ocean voyage. There are numerous parallels that link the Latin account of Brendan's voyage with, among others, the Celtic *Voyage of Mael Duin*, a text written in Gaelic. The Christian emphasis of Brendan's journey is secularized in *Voyage of Mael Duin*, but the islands described by both texts are virtually identical.

Navigatio features an odd mixture of realistic and fantastic details concerning the natural world. Unlike contemporaneous documents which detail the lives of saints, neither *Vita* nor *Navigatio* attribute any miracles to Brendan. Instead, Brendan is portrayed as a shrewd leader and skillful adventurer. In this sense, Brendan is similar to Odysseus. He never displays fear in the face of danger, and is certain that God will lead his ship successfully to its destination.

On one hand, the text of *Navigatio* is clearly allegorical. Many of the characters are flat and appear for a single purpose. For example, Brendan allows three latecomers to accompany his team of monks on the voyage, each of whom perishes because of sin. Likewise, the monks' adventures are unified through their religious and dogmatic purpose. Each adventure serves as an exemplum which demonstrates the crew's progressive development of trust in God. The crew visits three "places" three times: the Island of Sheep, the Island of Birds, and Jasconius, the largest fish of the ocean. At first, the monks are astounded by the flocks of talking birds and frightened when they realize that the island on which they have landed is actually a whale. But as they repeat this cycle, they gain the confidence in God's plan which marks Brendan's temperament. Furthermore, they pass by Hell and witness the punishment of Judas. Brendan commands the demons who administer Judas's punishment, demonstrating the true power of righteousness. The heavily Christian emphasis of these and other repeated motifs echoes that found in Celtic *immrams* as well as in the Latin classics familiar to medieval monks.

However, a recent voyage sheds new light on the realistic accuracy of accounts of Brendan's voyage. In 1976, Tim Severin and his crew recreated Brendan's voyage. Their success indicates that Brendan's voyage, which likely was from Ireland to Newfoundland, was possible. Severin reconstructed a leather ship using techniques described in *Navigatio*. His passage took him along the "stepping stone" route which crosses the North Atlantic. This route is still used by modern ships and planes, and may have also been used by Lief Ericson in his tenth-century journey to Vinland.

Severin found numerous correspondences between the seemingly fantastic details recounted in *Navigatio* and daily occurrences on his transatlantic journey. For example, their leather ship, named, quite fittingly, the *Brendan,* was frequently surrounded by schools of curious whales. Their appearance and reappearance parallels the monks' repeated adventures on the back of Jasco-

nius, the biggest fish in the ocean. Indeed, the fairly simple, natural explanations for the fantastic details that characterize this narrative indicate that *Navigatio* is, more than likely, based on fact.

DEAN SWINFORD

Further Reading

Benedeit. *Anglo-Norman Voyage of St Brendan.* Edited by Ian Short and Brian Merrilees. Manchester: Manchester University Press, 1979.

Bouet, Pierre. *Le Fantastique dans la Littérature Latine du Moyen Age: La Navigation de Saint Brendan.* Caen: Centre de Publications de l'Université de Caen, 1986.

Ohler, Norbert. *The Medieval Traveller.* trans. by Caroline Hillier. Woodbridge: Boydell Press, 1989.

O'Meara, John. *The Voyage of Saint Brendan.* Dublin: Dolmen Press, 1978.

Severin, Tim. *The Brendan Voyage.* New York: McGraw-Hill, 1978.

Hsüan-tsang Forges a Link Between China and India

Overview

Though Fa-hsien in the fifth century was the first Chinese Buddhist pilgrim to visit India, the trip by Hsüan-tsang more than two centuries later was equal if not greater in terms of historical significance. As Fa-hsien spurred Chinese interest in Buddhism by bringing back scriptures from its birthplace in India, Hsüan-tsang helped influence much wider acceptance of the faith among Chinese. He also became the first Chinese visitor to go to all major regions of India, and he is remembered today as the initiator of Sino-Indian relations.

Background

Buddhism had its origins in the sixth century B.C. ministry of an Indian prince named Siddhartha Gautama (563-483 B.C.) After years of spiritual seeking in which he rejected wealth and worldly pleasures, as well as the precepts of both Hinduism and Jainism, he experienced a spiritual transformation, after which he was known as the Buddha, or "the awakened one." A faith grew up around his teachings, which included the idea that desire is the cause of pain. The Buddha also taught that only through reaching nirvana, a state of inner

peace, can the individual transcend the cycles of reincarnation that characterize the Hindu worldview.

Buddhism initially gained adherents in India, but it was destined to enjoy its greatest influence in China. The new faith made its first appearance there during the Later Han period (A.D. 23-220), but initially the Chinese rejected it as a "foreign" religion. Only later, during a period of turmoil between dynasties (220-589), did Mahayana or "Great Vehicle" Buddhism finally begin winning Chinese adherents.

One of the principal agents of this change was the monk and pilgrim Fa-hsien (c. 334-c. 422). Dissatisfied with existing Chinese translations of Sanskrit Buddhist scriptures, Fa-hsien set out for India at age 65, intent on finding originals. What followed was an odyssey of some 10,000 miles (16,000 kilometers) in 15 years, during which he traveled through Buddhist lands in China, central Asia, Indian, Ceylon, and the East Indies. He finally returned to his own country bearing the scriptures he had sought, and as a result of his work, knowledge and acceptance of Buddhism in China increased enormously in the years that followed.

Impact

More than two centuries after Fa-hsien, another pilgrim named Hsüan-tsang (c. 602-664) set out for India with much the same purpose in mind: to increase his understanding of the Buddha's teachings by going to the source—not only the original Buddhist texts, but the geographical homeland of Buddhism.

A child prodigy, Hsüan-tsang had been raised as a Buddhist monk, but under the Sui Dynasty (589-618) and the newly founded T'ang Dynasty (618- 907), he and other monks faced a government suspicious of their influence. Not only did T'ang China's first ruler, Kao Tsu (r. 618-626), embrace the rival faith of Taoism, but he had placed restrictions on travel in western portions of the country—precisely the area through which Hsüan-tsang would have to pass if he wanted to go to India.

And Hsüan-tsang certainly wanted to go. Whereas Fa-hsien's mission started from his dissatisfaction with Chinese translations of the Buddhist scriptures, Hsüan-tsang's longing arose from his preoccupation with difficult theological questions. If he intended to answer these growing quandaries, he needed to consult the *Yogacarabhūmi sastra* (fourth and fifth century A.D.), which could only be found in India. Therefore he resolved to make the arduous, extremely challenging, journey across the mountains.

It should be noted that though civilizations thrived in India and China during ancient times, their two peoples were ignorant of one another for nearly two millennia, a fact that highlights the great barrier posed by the Himalayas and other ranges that separate the two lands. Added to this was the emperor's restriction on travel, which made Hsüan-tsang's trip across China doubly dangerous.

Starting from Ch'ang-an, the T'ang capital in east-central China, Hsüan-tsang followed a route more southerly than that of Fa-hsien. He made his way deep into the west, but he was preceded by messengers from the emperor, bearing news of a monk trying to defy imperial orders against travel in the west. He later wrote, "As I approached China's extreme outpost at the edge of the Desert of Lop, I was caught by the Chinese army. Not having a travel permit, they wanted to send me to Tun-huang to stay at the monastery there. However, I answered, 'If you insist on detaining me I will allow you to take my life, but I will not take a single step backwards in the direction of China.'"

As it turned out, the leading government official in the region was a devout Buddhist, and he chose to look the other way, allowing Hsüan-tsang to pass the military outposts that separated China from the lands of central Asia. Hsüan-tsang continued on, making his way over mountains and across deserts, where he encountered both bandits and marauding tribes, as well as admiring rulers and welcoming groups of sages. Much of what is "known" about his journeys comes from hagiographic accounts that exaggerate many of Hsüan-tsang's accomplishments; in any case, he traveled much further west than Fa-hsien, visiting the cities of Tashkent, Samarkand, and Balkh. (The first two are today in Uzbekistan, and the last in Afghanistan. All three were important trading and cultural centers of the premodern era.)

In about 631 Hsüan-tsang reached India, where he visited numerous sites important to the Buddha's life and ministry. In time he made his way to the monastery at Nalanda, India's largest Buddhist center, where the esteemed master Silabhadra taught him personally for 15 months. Hsüan-tsang would spend a total of five years at Nalanda, off and on, during which time he composed three religious treatises in Sanskrit.

He also traveled from his base at Nalanda to various parts of India, including Bengal in the east, the Deccan Plateau of central India, and both the Coromandel (eastern) and Malabar (western) coasts. In addition, he journeyed through the Indus River Valley by which he had entered the country, and in time he became eager to follow that route back to China.

However, a king named Kumara invited Hsüan-tsang to visit him in Assam, in northeastern India, an offer Hsüan-tsang could not safely refuse. This in turn led Kumara's rival Harsha (c. 590-647)—India's greatest ruler of the early medieval era—to make an invitation of his own. At Harsha's court in 642, Hsüan-tsang greatly impressed a gathering of several thousand kings and wise men, winning arguments with Hindu and Jain theologians. Harsha showered him with gifts, but Hsüan-tsang accepted only a buffalo-skin coat to keep him warm and dry, and an elephant to transport the many books he had brought with him. Finally, in 643, he set off for China.

Given the fact that he had left illegally, Hsüan-tsang undoubtedly returned with much apprehension. From the oasis at Khotan, he sent a letter to the emperor announcing his return, and eight months later he received a welcoming reply. As it turned out, Kao Tsu had been ousted by his son T'ai Tsung (r. 626-649), who was a Buddhist and eager to meet Hsüan-tsang. The latter arrived at Ch'ang-an early in 645, and the crowd that came out to greet him was so large that at first he could not enter the city.

He met with the emperor, who debriefed him on all manner of details concerning the lands he had visited. T'ai Tsung even offered him a position as his personal advisor, and when Hsüan-tsang demurred, the emperor instead set him up at nearby Hung-fu Monastery with a fleet of assistants to help him in his translation work. The only stipulation was that Hsüan-tsang write a record of his travels, *Ta T'ang Hsi-yü-chi,* or "The Great T'ang Record of Travels to the Western Lands," which he completed in 646.

Hsüan-tsang's translation work continued under the reign of Kao Tsung (r. 649-683), and after 19 years yielded 76 books. When the great monk died in 664, it was said that some 1 million people attended his funeral, and in later years he became a legendary figure. Not only did his translations, commentaries, and those of his close followers make up fully one-quarter of the extant Buddhist literature in Chinese, but the contact he had initiated with India led to increased T'ang relations with the southern power.

Nine hundred years after his death, Hsüan-tsang became the subject of a fictional narrative, *Hsi-yü-chi* by Wu Ch'eng-en (c. 1500-c. 1582). Translated in the twentieth century as *The Jour-* *ney to the West* (1977-83), the book is one of the classics of Chinese literature, an enthralling comic adventure in which Hsüan-tsang becomes the quixotic monk Tripitaka, accompanied by the companions Monkey and Pigsy. Much like legends such as that of King Arthur in the West, this fictionalized version of Hsüan-tsang's story has permeated virtually every facet of Chinese cultural life, from opera to comic books and animated cartoons.

JUDSON KNIGHT

Further Reading

Books

Boulting, William. *Four Pilgrims.* New York: E. P. Dutton, 1920.

Grousset, René. *In the Footsteps of the Buddha.* London: G. Routledge & Sons, 1932.

Hwui Li. *The Life of Hiuen-Tsiang.* Translated, introduced, and edited by Samuel Beal. Westport, CT: Hyperion, 1973.

Kherdian, David. *Monkey: A Journey to the West: A Retelling of the Chinese Folk Novel by Wu Ch'eng-en* (fiction based on the life of Hsüan-tsang). Boston: Shambhala, 1992.

Waley, Arthur. *The Real Tripitaka, and Other Pieces.* London: Allen and Unwin, 1952.

Internet Sites

Marx, Irma. "Travels of Hsüan-Tsang—Buddhist Pilgrim of the Seventh Century." *Silk Road.* http://www.silk-road.com/artl/hsuantsang.shtml.

Fa-Hsien Travels Around the Outskirts of China, to India and the East Indies

Overview

When he was 65 years old, the monk Fa-hsien resolved to travel from northern China to India, homeland of the Buddhist faith. The result was a journey of some 10,000 miles, or 16,000 kilometers, across China, central Asia, the Indian subcontinent, and the Malay Archipelago. The expedition, details of which Fa-hsien later recorded, constituted one of China's first notable encounters with India. In accomplishing his mission of retrieving original Buddhist texts, Fa-hsien influenced widespread acceptance of the new faith in his homeland.

Background

The most ancient of Chinese faiths are various folk religions and forms of ancestor-worship, many of them still practiced today. The sixth century B.C., however, saw the birth of new belief systems based on the teachings of three remarkable contemporaries: Siddhartha Gautama or the Buddha (563-483 B.C.), Confucius (551-479 B.C.), and Lao-tzu (fl. sixth century B.C.) The latter two men, both Chinese, did not set out to establish Confucianism and Taoism, respectively; rather, those systems developed around their writings and those of their disciples during the Former Han Dynasty (207 B.C.-A.D. 9).

Buddhism, on the other hand, came from faraway India. As a young man, Siddhartha chose to renounce his life as a prince, and set out from his father's palace to find enlightenment. He rejected Hinduism, with its endless cycle of birth and death through reincarnation, and after studying with the Jain ascetics, he likewise rejected their aggressive self-denial. It was then that he underwent a deep spiritual transformation, emerging as the Buddha, or "the awakened one." Thereafter he devoted his life to his newfound teachings: that desire is the cause of pain, and can be overcome by reaching nirvana, or a state of inner peace; and that only by doing so can the individual transcend the cycles of reincarnation.

Though the Buddhist faith initially gained adherents in India, Hinduism remained the dominant religion, and Buddhism might today be a mere splinter sect had it not caught on in the Far East. This, however, was a long time in coming. During the troubled Later Han period (A.D. 23-220), the influence of Taoism spread, and Confucianists and folk religionists gradually began to accept the idea of coexistence: after all, Taoism, with all its peculiarities, was still a native Chinese religion.

Only in the interregnum between dynasties, a period from 220 to 589, did Buddhism begin to gain influence. This era, of course, coincided with the decline and fall of the Western Roman Empire, and just as pagan Romans feared Christianity's challenge to their ancient religion, many in China were concerned about the destructive effect a new faith would have on traditional values. By urging believers to concentrate on inner peace and enlightenment rather than social concerns, Buddhists seemed to challenge the dominant Confucian system of thought, which emphasized loyalty to family, respect for elders, hard work and study, and obedience to rulers.

Perhaps if China had possessed a more stable system of authority during those centuries, the forces of tradition would have kept out the "foreign" religion. Instead, Mahayana or "Great Vehicle" Buddhism began to gain adherents, and Fahsien (c. 334-c. 422) was one of the principal instruments for this change. Fa-hsien became an ordained monk at about the age of 20, and spent the next four decades studying the Buddhist scriptures. During this time, he grew increasingly dissatisfied with Chinese translations of the Sanskrit, and at the age of 65—hardly a time to begin a great odyssey—he struck out for India to obtain original texts for making fresh translations.

Impact

Taking with him a small group of other monks, Fa-hsien left the great city of Ch'ang-an, which served as capital for numerous dynasties, in 399. The group crossed the Lung Mountains and made a summer retreat at Ch'ien-kuei, then crossed the Yang-lou range. At the city of Chang-yeh, they were joined by more monks also intent on making a pilgrimage to India.

It is interesting today to observe a map of Fa-hsien's route, which in the westward stint of his journey closely paralleled the modern border between China, Mongolia, and (further west) Russia. Likewise as they turned southward, the pilgrims traced what today forms the boundary between China and Kazakhstan. The reason for this parallelism, of course, is that most of China's frontiers are drawn as much with an eye toward natural features as toward politics.

Among those features are the Nan Shan Mountains and the Takla Makan Desert, which together separate China from the heart of central Asia. Later, Fa-hsien would write compellingly of his experience crossing the desert: "The sands are full of evil spirits and burning winds, and anyone who encounters them dies; no one is left unharmed. No birds fly overhead, no animals run across the ground. Squint one's eyes, gaze as one may in the four directions, he can find no place to turn, nothing to guide him. Only the dried bones of the dead serve as markers on the trail."

Eight centuries later, another journeyer would have much the same experience crossing the Takla Makan: Marco Polo (1254-1324). Indeed, it is interesting to note the degree to which Fa-hsien's and Polo's routes intersected, especially considering the difference in their points of origin. Both crossed the Pamir Mountains, for instance, and both later found themselves in the East Indies.

South of the Takla Makan, Fa-hsien entered Khotan, a Buddhist kingdom that he described as a sort of earthly paradise: "The land is exceedingly prosperous and happy, and its people are numerous and thriving." But he kept on going, and two months past Khotan he reached another city later visited by Polo, Kashgar, at the extreme westernmost edge of China. Rather than cross the Himalayas, he chose to move westward to an only slightly less forbidding range, the Pamirs.

As he crossed the Pamir plateau, where the average altitude is more than 12,000 feet (3,658 meters), Fa-hsien went westward into the area where modern-day Afghanistan and Tajikistan intersect. Later he described the Pamirs as "inhabited by poison dragons. If one arouses the ill humor of the dragons, they will at once call forth poisonous winds, cause the snow to fall, or send showers of sand, gravel, and stones flying. Of the persons who have encountered such difficulties, hardly one in ten thousand has escaped without injury. The natives of the region call the poison dragons the Snow Mountain people."

The group had another summer retreat at the town of Mamuk, and passed by the city of Peshawar. The latter, which today is on the Afghan-Pakistani border, was at that time still a Buddhist city. (Indeed, Islam would not come into existence for more than two centuries, and virtually all of Fa-hsien's journey took him through Buddhist lands.) Peshawar had been established by Kanishka (fl. c. 78-103 A.D.), who by opening up the Silk Road between East and West helped spread Buddhist ideas into China.

The group entered India via the treacherous Hanging Passage along the gorges of the upper Indus River. "The trail is precarious," wrote Fahsien regarding this part of the journey, "and the cliffs and escarpments are sheer, the mountains forming stone walls which plunge thousands of feet to the valley below. Peering down, one's eyes grow dizzy, and when one tries to go forward, one can find no spot on which to place one's foot." During this juncture of the trip, the party had to make use of some 700 suspension bridges, ladders, and sets of footholds. Finally, however, they reached India itself, but had to spend the winter waiting to cross the Safed Koh range.

Along the way, some members of the party chose to return to China, and one of Fa-hsien's closest friends died. But finally they reached the more temperate climes of the Indian subcontinent, and began making their way eastward. Along the way, they visited places important in

the life of the Buddha: Kapilavasu, where he was born, and the kingdom of Magadha, where he conducted much of his ministry. Fa-hsien also visited Pataliputra, capital of the great Buddhist empire of the Mauryan dynasty.

After three years collecting and copying texts in Magadha, Fa-hsien continued down the Ganges River to the port city of Tamralipti. There he stayed for another two years, translating the texts, before boarding a ship for Ceylon, or modern Sri Lanka. He spent two more years in Ceylon before heading homeward.

Fa-hsien's ship wrecked off the coast of Sumatra in what is now Indonesia, however, and Fa-hsien wound up in Java, where he had to wait half a year for a boat to China. Finally he caught a ship, but it was blown off course and drifted for more than two months. Only in 414, some 15 years after he had left, did Fa-hsien finally reach his homeland at the Shantung Peninsula.

By now 80 years old, Fa-hsien immediately went to work in the capital city of Chien-k'ang, now Nanjing or Nanking, translating more of the texts he had brought home from India. Later he retired to a monastery in the province of Hupei, where he wrote a record of his travels. Fa-hsien was about 88 years old when he died, and in the years that followed—in large part thanks to his efforts—the influence of Buddhism in China gradually increased.

JUDSON KNIGHT

Further Reading

Books
Legge, James, translator. *A Record of Buddhistic Kingdoms: Being an Account by the Chinese Monk Fa-Hien of His Travels in India and Ceylon (A.D. 399-414) in Search of the Buddhist Books of Discipline.* New York: Paragon, 1965.

Internet Sites
"Fa Hsien." Theosophy Library Online. http://theosophy.org/tlodocs/teachers/FaHsien.htm.

Marx, Irma. "Travels of Fa Hsien—Buddhist Pilgrim of Fifth Century." Silk Road. http://www.silk-road.com/artl/fahsien.html.

Alexander the Great
356 B.C.-323 B.C.
Macedonian Conqueror

Though the Romans would rule more land, no one man has ever subdued as much territory in as short a period as Alexander the Great, or Alexander III of Macedon, who conquered most of the known world before his death at age 32. Yet Alexander did more than win battles: trained in the classic traditions of Greece, he brought an enlightened form of leadership to the regions he conquered. His empire might have been a truly magnificent one if he had lived; as it was, he ensured that the influence of Greece reached far beyond its borders, leaving an indelible mark.

Macedon was a rough, warlike country to the north of Greece, and though the Macedonians considered themselves part of the Greek tradition, the Greeks tended to look down on them as rude and unschooled. But Greece's own day of glory had passed, and Alexander's father, Philip II (382-336 B.C.; r. 359-336 B.C.), subdued all of southwestern Europe between 354 and 339 B.C. At the age of 17, Alexander himself led the Macedonian force that conquered Thebes.

Philip, who revolutionized infantry tactics, might well be remembered as the greatest of Macedonian rulers, had he not been eclipsed by his son. Also remarkable was Alexander's mother, Olympias, who brought up her boy on stories of gods and heroes. Under her influence, he became enamored with the figure of Achilles from Homer's *Iliad,* and came to see his later exploits as fulfillment of the heroic legacy passed down by his mother. A third and significant influence was Aristotle (384-322 B.C.), who tutored Alexander in his teen years. It is an intriguing fact that one of the ancient world's wisest men taught its greatest military leader, and no doubt Alexander gained a wide exposure to the world under Aristotle's instruction.

He was not, however, a thinker but a doer. A natural athlete, Alexander proved his combination of mental and physical agility when at the age of 12 he tamed a wild horse no one else could ride. Alexander named the horse Bucephalus, and the two would be companions almost for life: later, when Bucephalus died during Alexander's campaign in India, he would name a city for his beloved horse.

Soon after Philip took control of Greece, he was assassinated, and in claiming the throne, Alexander had to gain the support of the Macedonian nobility. He did so with a minimum of bloodshed, establishing a policy he would pursue as ruler of all Greece: leaving as much good will as he could behind him, he was thus able to push forward.

Alexander next turned to consolidation of his power in Greece, which he did by a lightning-quick movement in which he captured Thebes and killed some 6,000 of its defenders. After that, he faced no serious opposition from the city-states, and embarked on a mission that had been Philip's dream: conquest of the vast Persian Empire to the east. The latter had once threatened Greece, only to be defeated in the Persian Wars (499-449 B.C.); now Greece, led by Macedon, would take control of the Persians' declining empire.

The main body of Alexander's army, some 40,000 infantry and 5,000 cavalry, moved into Asia Minor while their commander crossed the Hellespont with a smaller contingent so that he could go on a personal pilgrimage to the site of Troy. Eventually, Alexander and his army passed through the ancient Phrygian capital of Gordian. In that city was a chariot tied with a rope so intricately knotted that no one could untie it. According to legend, the fabled King Midas or Mita (fl. 725 B.C.) had tied the Gordian Knot, and whoever could untie it would go on to rule the world. Alexander simply cut the knot.

After a successful military engagement against a Greek mercenary named Memnon, Alexander moved down into Cilicia, the area where Asia Minor meets Asia. The Persian emperor Darius III (d. 330 B.C.) came to meet him with a force of 140,000, and at one point—because Alexander's armies were moving so fast—cut him off from his supply lines. Darius chose to wait it out, letting Alexander's forces come to him, and Alexander, taking this as a sign of weakness, charged on the Persians. Alexander nearly got himself killed, but the Battle of Issus was a decisive victory for the Greeks. Darius fled, leaving Alexander in control of the entire western portion of the Persians' empire.

Instead of raping and pillaging, as any number of other commanders would have allowed their troops to do, Alexander ordered his armies

Alexander the Great.

to make a disciplined movement through conquered territories. In 332 and 331 B.C., Alexander's forces secured their hold over southwestern Asia, and by the latter year he was in Egypt, where he founded the city of Alexandria, destined to become a center of Greek learning for centuries to come. In October 331 B.C., he met a Persian force of some 250,000 troops—five times the size of his own army—at the Assyrian city of Gaugamela. It was an overwhelming victory for the Greeks, and though Darius escaped once again, he would later be assassinated by one of his own people.

Alexander now controlled the vast lands of the Persian Empire, but with the agreement of his men, he kept moving eastward. Over the next six years, from 330 to 324 B.C., his armies subdued what is now Afghanistan and Pakistan, and ventured into India. Alexander secured his position in Afghanistan by marrying the Bactrian princess Roxana (d. c. 310 B.C.), but as he became aware that some of his troops were growing weary, he sent the oldest of them home.

He wanted to keep going east as far as he could, simply to see what was there, and if possible, add it to his empire. But in July of 326 B.C., just after they crossed the Beas River in India, his troops refused to go on. There might have been a rebellion if Alexander had tried to force the issue, but he did not. He sent one group, led by Nearchus (d. 312? B.C.), back by

sea to explore the coastline as they went, and another by a northerly route. He took a third group through southern Iran, on a journey through the desert in which the entire army very nearly lost its way.

In the spring of 323 B.C., they reached Babylon, and Alexander began plotting the conquest of Arabia. But he was unraveling both physically and emotionally, and he had taken to heavy drinking. He caught a fever, and was soon unable to move or speak. During the last days of his life, Alexander—the man of action—was forced to lie on his bed while all his commanders filed by in solemn tribute to the great man who had led them where no conqueror had ever gone. On June 13, 323 B.C., he died.

Alexander was no ordinary conqueror: his empire seemed to promise a newer, brighter age when the nations of the world could join together as equals. Though some of his commanders did not agree with him on this issue, Alexander made little distinction between racial and ethnic groups: instead, he promoted men on the basis of their ability. From the beginning, his armies had recruited local troops, but with the full conquest of Persia, they stepped up this policy. It was his goal to leave Persia in the control of Persians trained in the Greek language and Greek culture, and he left behind some 70 new towns named Alexandria. Thus began the spread of Hellenistic culture throughout western Asia.

But Alexander's empire did not hold. The generals who succeeded him lacked his vision, and they spent the remainder of their careers fighting over the spoils of his conquests. Seleucus (c. 356-281 B.C.) gained control over Persia, Mesopotamia, and Syria, where an empire under his name would rule for many years, and Ptolemy (c. 365-c. 283 B.C.) established a dynasty of even longer standing in Egypt. His descendants ruled until 30 B.C., when the last of his line, Cleopatra (69-30 B.C.)—also the last Egyptian pharaoh—was defeated by a new and even bigger empire, Rome.

JUDSON KNIGHT

Asoka

c. 302 B.C.-c. 232 B.C.

Indian Emperor

Asoka ruled India's Mauryan Empire at its height, and brought most of the subcontinent under his control, but he was more than a mere conqueror. At the end of a bloody conquest in the eighth year of his reign, he experienced a religious conversion that led him to dedicate the remainder of his life to doing good for his people. Thereafter he devoted himself to making life better for his subjects, establishing rest stations and other public works projects throughout his realm, and he commanded that his principles of morality be carved onto rocks where his words can still be viewed today.

In 326 B.C., armies under Alexander the Great (356-323 B.C.) invaded India. They soon left, but the impressive conquests of Alexander inspired a dream in a young Indian monarch named Chandragupta Maurya (r. 324-301 B.C.). Chandragupta ruled over the kingdom of Magadha in eastern India, a place where the Buddha (Siddhartha Gautama; 563-483 B.C.) conducted much of his ministry. From his capital at Pataliputra, northwest of modern-day Calcutta, Chandragupta launched his armies, and soon they swept over most of India.

The Mauryan Empire was a splendid and well-organized realm, and Pataliputra was said to be the greatest city of that time. Like most ancient emperors, however, Chandragupta ruled with a fist of iron, maintaining a network of spies and punishing even a hint of rebellion. Yet in an eerie foreshadowing of his grandson's career, Chandragupta stepped down from the throne in 301 B.C. to become a member of the Jain sect, and later died of starvation.

Chandragupta was succeeded by his son Bindusara (d. 270 B.C.), whose wife Subhadrangi bore her husband a son she named Asoka, meaning "I am without sorrow" in Sanskrit. According to legend, Bindusara did not care for his son, yet he apparently trusted him enough to give him the job of suppressing a revolt in the city of Taxila; after that, Asoka became prince over the city of Ujjain, later a great center of scientific study, in west-central India.

It appears that Asoka did not assume the throne until some time after his father's death, and though there was a power struggle, the story that Asoka murdered 99 of his brothers is probably a legend. There are a number of such tales—a tradition known as "Black Asoka"—all intended to convey the fact that Asoka was a ruthless leader prior to his conversion. It appears that Asoka maintained a prison, with an extensive network of torture chambers, for dealing with his enemies. Certainly it is clear that he fought a number of wars, and spilled plenty of blood, in the course of securing his empire. Then, in 262 B.C., the bloodshed became too much for him.

The occasion for this abrupt about-face was his victory over the Kalinga people of southeastern India. In the course of the campaign, his troops captured more than 150,000 people, and killed many times more, either directly or as a result of the general havoc created by the war. "Just after the taking of Kalinga," according to one of the many inscriptions he left behind, "His Sacred Majesty began to follow Righteousness, to love Righteousness, to give instruction in Righteousness."

His statement regarding his conversion comes from one of the 35 inscriptions that Asoka put in various places around India beginning in 260 B.C. Some of these edicts, or statements, were carved into rocks in high places; others on man-made pillars. Hence they are referred to either as rock edicts or pillar edicts. Though he never identified the "righteousness" by name, it appears reasonably certain that Asoka adopted Buddhism. However, the inscriptions speak of morality in broad terms, and make no mention of certain key Buddhist concepts. It is possible, therefore, that what Asoka embraced was a mixture of Buddhism and a "universal religion," or a belief that all people who worship with sincerity worship the same god.

Whatever the faith, Asoka set out to conquer the world with it, as he had once conquered with the sword. He sent missionaries to bring his message to far-flung places, including Egypt and Greece. Though Buddhism never took hold in those countries, it did spread to the island of Ceylon (modern Sri Lanka), where it replaced Hinduism as the dominant religion.

In 257 B.C. Asoka appointed officials to a position whose name is translated as "inspectors of morality," charging them with distribution of gifts to the poor, and with ensuring that his people treated each other with kindness. He devoted much of his time to traveling around the country, and even oversaw the planting of trees to provide travelers with a shady place torest.

At some point in the 230s B.C., officials in his court—perhaps taking advantage of the fact that Asoka was consumed with humanitarian concerns—managed to gain influence over his grandson Samprati, who Asoka had picked as his successor. Apparently Samprati forced his grandfather into virtual exile within the palace, and in his latter days Asoka lived on a meager ration of food. It was an unfortunate end to a reign that had brought an unprecedented degree of justice and humanitarian concern to one of the ancient world's greatest empires.

JUDSON KNIGHT

Saint Brendan. *(The Granger Collection, Ltd. Reproduced with permission.)*

St. Brendan
484-577
Irish Missionary

One of the most puzzling figures from the history of medieval exploration was the Irish monk St. Brendan. Some claim that Brendan, who in about 522 sailed with a group of some 60 men for what he called the "Land of Promise," actually reached North America. Others have maintained that he at least visited the Canary Islands or Madeira, and still others hold that the entire story of Brendan's journey was a legend.

Known variously as St. Brendan of Ardfert and Clonfert, or as Brendan the Voyager, he was born near the present-day city of Tralee in Ireland's County Kerry. Educated under several clerics who like Brendan himself were destined for canonization or sainthood, he received his ordination as a priest in 512, and promptly began building monastic centers in a number of towns around Ireland.

About 10 years after his ordination, Brendan allegedly sailed westward or southward, depending on which of several written accounts—most were composed between the ninth and eleventh centuries—one consults. So extensive did this body of literature become, in fact, that writings about Brendan's voyage were known

collectively as the *Navigatio Brendani,* or "The Voyage of St. Brendan."

Most versions of the *Navigatio Brendani* hold that Brendan and a group of other monks (estimates of the number range from 18 to 150) set out for an earthly paradise, and that after seven years' journey they finally found such a land, known as *Terra repromissionis.* Maps made in the fourteenth century identified this area with the "Fortunate Isles," the premodern name for the Canaries; or with Madeira. But as knowledge of those areas increased, the hypothesized destination of Brendan's voyage edged further westward. By the late sixteenth century, geographers were placing the "Land of Promise" to the west of Ireland, but by the early nineteenth century the entire story had been dismissed as a mere fabrication.

In later times, however, there arose a new theory concerning Brendan's voyage, one that placed his destination in America. Proponents of this idea have often tended to be motivated by something other than a pure desire for scientific knowledge—usually Irish nationalism and ethnic pride—yet a number of findings do at least make this an interesting position.

In 1976 and 1977, British navigational scholar Tim Severin and others undertook a voyage in a craft similar to the one Brendan would have used, and successfully traveled from the

Irish coast to Newfoundland. There is also some basis for the claim that legends of the Norse colonies established by Leif Eriksson (c. 970-c. 1020) and others include some mention of an Irish colony, though it is not clear that this has any direct connection with Brendan himself. The same goes for Shawnee Indian legends regarding a white tribe with iron tools that inhabited Florida long before the Europeans arrived, as well as modern claims that medieval tablets inscribed in Ogham, the Old Irish alphabet, have been found in West Virginia.

Certainly it is likely Brendan undertook some sort of voyage; as to his exact destination, that remains a matter of conjecture. In any case, he lived nearly half a century after his return in about 529, during which time he founded a number of monasteries throughout the British Isles. He also traveled extensively in Wales, and visited the monastery on the island of Iona, an important center for the preservation of learning during the early years of the medieval period.

JUDSON KNIGHT

Gaius Julius Caesar
102 B.C.-44 B.C.
Roman General and Statesman

Without a doubt the most significant figure in the history of Rome, Julius Caesar paved the way both for the end of the republic and the creation of the empire under his nephew Octavian, or Augustus Caesar. As a general he led military operations in Britain and elsewhere, and as dictator of Rome, he put through valuable reforms. But his actions, including his celebrated affair with Cleopatra, earned the distrust even of his closest friends, who conspired in his assassination.

Caesar was born on July 13, 102 B.C., to an aristocratic but not wealthy family. During his childhood, Rome was caught in a struggle between the aristocratic party, led by Sulla (137-78 B.C.), and the popular party, which—though its members were aristocrats as well—favored a greater distribution of power. With his aunt Julia's marriage to the popular party leader, Marius (c. 157-86 B.C.), Caesar became linked with that faction, and he further increased his standing by marrying Cornelia (d. 67 B.C.), daughter of Marius's ally Cinna (d. 84 B.C.) They later had a daughter, Julia.

Sulla established a dangerous example, one Caesar himself would later imitate, when in 88 B.C. he marched his troops into Rome. This ulti-mately meant the end of Marius's power, and Sulla demanded that Caesar divorce Cornelia. Caesar refused, and had to go into hiding. His mother's family convinced Sulla to relent, and Caesar returned to Rome and entered the army.

While serving in Asia Minor in 80 B.C., Caesar earned a high military decoration for bravery in the battle to take Mitylene on the isle of Lesbos, and went on to take part in a war against pirates from Cilicia. Following the death of Sulla, Caesar went back to Rome in 77 B.C., but soon after earning a name for himself as prosecutor in an important legal case, he traveled to Rhodes for further training in rhetoric. On the way, he was captured by Cilician pirates and held for ransom. After his release, he led a force to victory against the pirates, then—without being commanded to do so—led a successful attack against Mithradates of Pontus (r. 120-63 B.C.). Soon after this, he gained his first elected office as military tribune.

Cornelia died in 67 B.C., and within a year, Caesar remarried. Again the marriage had a political angle: Pompeia was the granddaughter of Sulla, and Caesar wanted to establish closer ties with Crassus (c. 115-53 B.C.), a leading figure in the aristocratic party.

To get ahead in Rome, a politician had to spend money on bribes and lavish entertainments for fellow politicians and the Roman citizens. As *aedile,* a type of magistrate responsible for all manner of public affairs, Caesar heavily outspent his colleague Bibulus (d. 48 B.C.), sponsoring the most magnificent set of gladiatorial games Rome had ever seen. Later, he obtained a position as governor in Spain, where he made back all the money he had spent—probably by means that were less than honorable.

Meanwhile Pompeia became involved in scandal when a character named Pulcher got into an all-female party at her house disguised as a woman. Caesar promptly divorced her. He was soon elected consul with Bibulus, but by then it was clear that only three men in Rome really mattered: Caesar, Crassus, and Pompey (106-48 B.C.)

The latter had just returned from his defeat of Mithradates, and together the three formed the First Triumvirate in 60 B.C. Pompey even married Caesar's daughter Julia to solidify their bond, but despite their mutual claims of loyalty, the alliance was an uneasy one.

The conflict would be delayed for many years, however, while Caesar went to Gaul. Anxious to gain military glory, he went looking for a

At the same time, Mithradates's teenaged son Pharnaces (r. 63-47 B.C.), taking advantage of Caesar's distraction in Egypt, had attempted to regain his father's kingdom in Pontus. Caesar went to Asia Minor, and destroyed Pharnaces's army in just five days. Afterward, he made his famous report of his victory: "I came, I saw, I conquered"; or in Latin, "*Veni, vidi, vici.*"

In 47 B.C. Caesar returned to Rome, where he assumed the powers of a dictator, and quickly pushed through a series of reforms. Most notable among these was his effort to reduce unemployment by requiring that every landowner hire one free man for every two slaves working in his fields. He increased the membership in the senate from 300 to 900, and included Celtic chieftains from Gaul in Rome's legislative body.

Caesar managed to combine the authority of numerous political offices, giving himself more power than any Roman leader had ever enjoyed; yet he seemed to want more. He placed his portrait on coins, an honor previously reserved only for the gods, and declared that the month of his birth would no longer be called Quintilis but "Julius" or July. Instead of standing before the senate when he spoke to them, as rulers had always done before, he sat—more like a monarch than a citizen. His junior colleague Mark Antony (82?-30 B.C.) even tried to convince him to wear a crown.

Had Caesar accepted the crown, it would have been such an offense to the Romans' views on government that he would have been an instant target for assassination. As it was, assassination was not long in coming. Just before leaving for a campaign in Persia, Caesar planned to address the senate on the Ides of March, or March 15, 44 B.C. Unbeknownst to him, however, a group of some 60 influential Romans—led by his supposed friends Brutus (85-42 B.C.) and Cassius (d. 42 B.C.)—had joined forces to assassinate him. As he entered the senate chamber, the assassins jumped at him with daggers, stabbing him 23 times. It would fall to Caesar's nephew Octavian (63 B.C.-A.D. 14) to avenge the murder, which he did on his way to assuming power and establishing what would become the Roman Empire.

JUDSON KNIGHT

Julius Caesar. *(Corbis Corporation. Reproduced with permission.)*

war, and he soon had one when the Helvetii, from what is now Switzerland, tried to cross Gaul without permission. He drove them back, then dealt with the Suebi from Germany. Next, as a means of preventing the Celts of Britain from aiding their cousins on the mainland, he led the first Roman invasions of that island in 55 and 54 B.C. In the course of these campaigns, Caesar killed perhaps a million people, but eventually placed all of Gaul firmly under Roman control.

Julia died in 54 B.C., breaking the bond between Caesar and Pompey, and in the following year, Crassus was killed in Asia. Soon Pompey ordered Caesar back to Rome, and Caesar, knowing he would be killed if he went back alone, brought his army with him. By crossing the River Rubicon, a shallow stream which formed the boundary between Cisalpine Gaul and Italy, he passed a point of no return, making conflict with Pompey inescapable.

Pompey moved his forces to Greece in order to regroup, while Caesar defeated Pompey's legions in Spain. The two met in battle at Pharsalus in Greece in 48 B.C., and though Caesar's armies won, Pompey managed to escape. He fled to Egypt, where he was killed, and Caesar, hot in pursuit, soon had his attention diverted by Cleopatra (69-30 B.C.). They began a romance, and Caesar aided her in war against her brother, Ptolemy XII (r. 51-47 B.C.).

Ezana

r. 325-360

Aksumite King

During the Middle Ages, Europe was not the only stronghold of Christianity: far to the south, separated from Rome and Constantinople by hostile Muslim forces, was a Christian stronghold in Africa. The influence of Christianity in Ethiopia, where the religion remains a powerful force today, can be traced to a single man, King Ezana of Aksum. Under his leadership, Aksum grew to become an enormously influential regional power, and extended its influence to neighboring lands through conquest, trade, and exploration.

Ezana of Aksum (sometimes spelled Axum), in what is now Eritrea, grew up around the Red Sea port of Adulis after about 500 B.C. As a seafaring power, it enjoyed contact with the Greek culture of Ptolemaic Egypt, and by the first century A.D. was engaged in trade with regions as far away as China and India.

During the reign of Ezana's father Ella Amida, a philosopher named Meropius, from Tyre in what is now Lebanon, undertook a voyage to India. He took with him two boys, Frumentius and Aedesius, but when their ship put in at Adulis to take on supplies, they were attacked by a crowd hostile to them: the visitors were Roman citizens, and Ella Amida had long been in conflict with Rome. All of the foreigners were put to death except for the two boys; not only did Ella Amida spare their lives, but he took them into his household.

The fourth century historian Rufinus, principal source for the tale (which has been corroborated by Aksumite records), reported that though Aedesius had a good heart, he was a simple lad. Frumentius, on the other hand, proved himself wise beyond his years, and in time the king appointed him as treasurer and secretary. After Ella Amida died, his wife Sofya—who served as regent for her young son Ezana—begged the two foreigners, by then young men, to stay and assist her.

Both Frumentius and Aedesius were devout Christians, and in time, both grew desirous of an opportunity to leave Aksum and pursue careers in the Church. As king, Ezana allowed them to go: Aedesius to Tyre, where he became a priest, and Frumentius to Alexandria—where that city's patriarch consecrated him as a bishop and sent him back to Aksum as a missionary. It was most likely after his return, and under his influence, that Ezana converted to Christianity.

The fact of Ezana's conversion is preserved in the coins of Ezana, which had previously borne pagan symbols but thereafter bore the cross, as well as in numerous inscriptions throughout his empire. These inscriptions are in Geez or Old Ethiopic, Sabean (the language of southern Arabia), and Greek, thus attesting to the cosmopolitan nature of Ezana's realm. They preserve, among other things, a record of conquests: according to one inscription, Ezana deputized his brothers Shaiazana and Hadefan to subdue the Beja people, who had been raiding trading caravans to the north. Other inscriptions indicate journeys to regions that have not been identified. It is possible to conclude from these records that Ezana began the conquest of southern Arabia, which Aksum's rulers would dominate by the sixth century.

Though Egypt had once been the dominant power in the region, it had long since been overtaken by Rome—and by Ezana's time, Roman power, too, was fading. Another great power of previous centuries was Kush, in what is now Sudan, which had controlled Egypt during the latter's Twenty-Fifth Dynasty (712- 667 B.C.) But the influence of Kush, too, had waned, and Ezana dealt that empire its final blow when he conquered the ancient Kushite capital of Meroë.

Thanks to Ezana, Aksum would remain a force for stability in sub-Saharan Africa and southern Arabia for centuries to come. It also remained Christian even after the Muslim conquest of Egypt neutralized the influence of Alexandria's bishop. As a result, Ethiopia developed its own Coptic version of Christianity, which remains highly distinct from the Roman Catholic and Eastern Orthodox churches.

JUDSON KNIGHT

Fa-hsien
c. 334-c. 422
Chinese Pilgrim

At the beginning of the fifth century A.D., the Buddhist monk Fa-hsien undertook a heroic journey around China's western periphery to India, where he gathered texts sacred to his religion. These writings would later play an important role in the development of Chinese Buddhism, and helped spur acceptance of the faith in a land far from the place of its birth.

Born in about 334, Fa-hsien came from the village of Wu-yang in Shansi province. After all three of his older brothers died, Fa-hsien's father dedicated him to a religious society in hopes that the Buddha would protect his son's life. Thus at the age of three, Fa-hsien became a monk. His father died when he was 10, and he resited an invitation from his uncle to return to the family home. Except for a short visit after his mother died a few years later, Fa-hsien never returned to the place of his birth.

He became an ordained monk at about the age of 20, but the next four decades of his life are a mystery. Only one thing is clear about Fa-hsien's years as a monk in China: over time, he became increasingly convinced that existing Chinese translations of the original Buddhist texts, written in Sanskrit, were inadequate. Eventually he resolved to travel to India himself and retrieve original manuscripts so that he and other monks could make fresh translations.

Obviously, then, the decision to undertake the arduous journey to India was not the impetuous act of a boy: Fa-hsien was approximately 65 years old when he—along with a small group of other monks—set out in the year 399. The party traveled far across northern China, following a route very close to the line that today forms the border between China and first Mongolia, then Russia. When he finally turned southward at the extreme northwestern corner of China, he again followed a line similar to the present-day boundaries between China and Kazakhstan.

The reason for the similarity between Fa-hsien's path and modern political divisions, of course, is that boundaries are typically drawn along natural barriers, of which Fa-hsien encountered many, including the Nan Shan Mountains and the Takla Makan Desert. After veering deep into central Asia in the region of modern Afghanistan, he crossed the high Pamir range and began moving southeastward into India. Gradually he worked his way across the country to Magadha, where the Buddha (Siddhartha Gautama; 563-483 B.C.) had once lived and taught, and where Fa-hsien spent three years collecting and copying texts. He then continued down the Ganges River to the port city of Tamralipti, where he stayed for another two years.

From Tamralipti, Fa-hsien continued to Ceylon (modern Sri Lanka), where he spent two years. Then he began the long trip home, setting sail for China. His ship was wrecked off the coast of Sumatra in what is now Indonesia, however, and he wound up in Java, where he spent half a year waiting for passage to China. Even when he did catch a ship, that one was blown off course and drifted for more than two months. Finally in 414, fully 15 years after he had left, Fa-hsien made it back to his homeland.

Though he was now about 80 years old, Fa-hsien must have been in good condition to endure all that he had. Nor was his work finished when he returned to China: at the capital city of Nanjing, he busied himself preparing translations of the texts he had brought back. Then he retired to a monastery in the province of Hupei, where he wrote a record of his travels. Fa-hsien was about 88 years old when he died.

JUDSON KNIGHT

Aelius Gallus
fl. 25 B.C.
Roman Soldier

In 25 B.C., Aelius Gallus led a military expedition to the region the Romans called "Arabia Felix," or modern-day Yemen. His mission was to extend Roman control throughout the Arabian Peninsula and gain for his emperor the wealthy spice-producing states at the peninsula's southern tip, but things did not turn out as planned. What he gained was not a new province, but a lesson for the seemingly invincible empire: that even Rome, with all its power and irresistible influence, had its limits.

The details of Aelius's life prior to his expedition are a mystery. He served as prefect of Egypt, a position in which he succeeded Gaius Cornelius Gallus (c. 70-26 B.C.), though his familial relation to that other Gallus—if any—is likewise unknown. In any case, soon after taking his post, Gallus received orders from Augustus (63 B.C.-A.D. 14), Rome's first emperor, to undertake an expedition to Arabia.

Roman knowledge of Arabia was sketchy at best. The Romans knew that trading caravans plied the peninsula, and that the southern kingdoms possessed great wealth, but as to what lay in between, they possessed only vague knowledge of a wide desert expanse. Departing from the city of Cleopatris on the Gulf of Suez, Gallus took with him a force composed of Egyptians, Jews, and Nabataeans, the latter an Arab people with whom the Romans had an uncertain alliance. The expedition got off to a bad start when their ships ran into trouble crossing the Gulf of Aqaba, and when they reached the city of Leuke Come or Haura on the Arabian Peninsula, an outbreak of sickness forced them to delay the expedition for half a year. Finally, however, they set out across the desert with a force of camels bearing water.

Traveling for some 80 days, the force conquered towns in the region of Negrana, or modern Nejran, then went on to a failed siege against Marsiaba or Marib in what is now Yemen. By then water supplies were running low, but Gallus's Arab guides insisted that they were only two days' march from the wealthy "incense states" of the coast. Two days eventually turned into six months, and finally Gallus retreated in disgust. His suspicions of his putative allies were confirmed when his return trip to the Red Sea took only 60 days instead of six months.

After reaching the sea, the army crossed to Myos Hormos, or Abu Scha'ar in Egypt. They had lost only seven men in actual battle, but their numbers had been greatly reduced by disease, hunger, and exposure—and not only had they failed to conquer Arabia, but they had gained little knowledge concerning the region. The principal intelligence gained from the Roman expedition, in fact, was much the same information gathered by the Soviets in their failed invasion of Afghanistan two millennia later: that some countries are so well protected by natural features and cultural barriers that the cost of conquest is simply too high for any imperial power.

JUDSON KNIGHT

Gregory the Great
c. 540-604
Italian Pope

When Gregory I, or Gregory the Great, became pope in 590, the Church and indeed all of western Europe was still reeling from the destruction of the Western Roman Empire. After failing to win assistance from the Eastern Roman (or Byzantine) emperor, Gregory gave up hope of reestablishing Roman power in the West, and set about building the Church as a powerful, self-sufficient political entity. In so doing, he established a locus for Western civilization during the most frightening days of the Dark Ages, and began to consolidate all of the West under Catholic leadership.

Gregory was born a member of a wealthy and powerful Roman family, but the Rome he knew was a mere shadow of its former glory: two centuries of destruction by barbarian tribes such as the Lombards had left it in ruins. His was one of the last generations to have access to the old-fashioned learning that had produced so many well educated Romans in the past, and

after receiving his education, Gregory assumed the ancient Roman office of prefect in 573.

In his thirties, Gregory became interested in the growing monastic movement. At first he was content to establish monasteries for others—he set up one in Rome and six more on lands owned by his family in Sicily—but in 574 or 575, he decided to leave the outside world and become a monk himself.

Six years later, Gregory received a commission from Pope Pelagius II (r. 579-590) to serve as emissary to the imperial court in Constantinople. Though he was unsuccessful in his mission to gain Byzantine aid in defending Rome against another wave of Lombard invasions, his work attracted the attention of church leaders, who chose Gregory as Pelagius's successor.

In the year he became pope, there was a famine and flood in Rome, as well as a plague in various parts of Europe. Gregory and other Christians took such events as evidence that the world was coming to an end as predicted in the New Testament, and this gave him a sense of urgency to bring as many people as possible into the fold before Christ returned.

It was ironic that Gregory would find himself in a position of great spiritual and political responsibility, because he stated on many occasions that he would have preferred to live his life quietly in a monastery. Indeed, throughout his years as pope, he maintained a monkish lifestyle, including little food and sleep, that often weakened him for his duties. Yet he pressed on with seemingly boundless energy.

At that time, the bishops of Constantinople, Antioch, and Alexandria vied with their counterpart in Rome for leadership of Christendom. Maintaining that a biblical basis existed for Roman supremacy, Gregory once again attempted unsuccessfully to gain the support of the Byzantine emperor, who naturally favored Constantinople. At that point Gregory determined to lead the Christians of Italy with or without the help of the eastern emperor, a decision that assumed vital importance when Jerusalem and Antioch fell to Muslim invaders a few years after his death. (Constantinople went on to become the seat of the Eastern Orthodox Church, which formally separated from Rome in 1054.)

Gregory's decision formed the cornerstone of medieval politics: instead of relying on a king to protect against the Lombards, or to distribute grain in times of famine, Gregory took on the job himself. Thus the papacy became a political

tones of Gregorian chants—a type of singing performed by Benedictine monks, committed to writing for future generations under his orders—to the many tales of miracles associated both with his life and death in 604. Yet his greatest legacy was in the formation of the church as a political power, and of Western Christendom as a political and spiritual alliance united under the leadership of a strong pope.

JUDSON KNIGHT

Hanno
fl. c. 500 B.C.
Carthaginian Mariner

Some time around 500 B.C., the Carthaginian mariner Hanno sailed westward from his hometown in what is now Tunisia. He passed the Pillars of Hercules and continued along the African coast, perhaps as far as modern-day Senegal or even Liberia. Though the full extent of his journeys is not known, it is certain that he sailed farther south than any navigator in the Atlantic Ocean would do until the golden age of Portuguese exploration some 2,000 years later.

The only part of Hanno's life known to historians is the saga of his voyage, and virtually all the details of that chapter in his biography come from a sometimes bewildering inscription left by Hanno himself. Having completed his expedition, he dedicated a stele or pillar to the Carthaginian gods, providing 18 (in some versions 19) paragraphs regarding his exploits.

Most notable of the Phoenician trading colonies along the Mediterranean, Carthage was a great seafaring power. Apparently the people of Carthage commissioned Hanno—the inscription refers to him as a "king," but in Carthaginian parlance this meant merely that he was a high official—to undertake his voyage for the purpose of establishing cities. Hanno's inscription records that his fleet consisted of 60 Phoenician ships, and that he brought with him 30,000 men and women, but 5,000 is a far more likely (and certainly still impressive) number.

Simply reaching the Pillars of Hercules, or the Straits of Gibraltar, was a feat in itself. Not only did this already place Hanno's ships some 1,000 miles (1,600 kilometers) west of Carthage, but the Pillars served as the gateway to the forbidding, unexplored Atlantic. Having passed through them, Hanno's party sailed southward along the coast of modern-day Mo-

Pope Gregory the Great. *(Corbis Corporation. Reproduced with permission.)*

as well as a spiritual office. The power of the popes would rise to staggering heights in centuries to come, and it could all be traced back to Gregory—yet Gregory did what he did not because he wanted power, but because he felt he had no choice.

During his extraordinarily active career, Gregory ransomed prisoners and constantly welcomed refugees from war-torn Italy into the relatively safe confines of Rome. He took power over corrupt bishops guilty of simony (buying and selling church offices), adultery, and other acts inappropriate to their roles as spiritual leaders. He dealt harshly with heresy, yet often negotiated with Lombard leaders who still embraced Arianism, a belief declared heretical in 325.

Among Gregory's achievements, in fact, was the conversion of the Lombards from the Arian heresy. He was also responsible for the conversion of the Anglo-Saxons from paganism under Augustine of Canterbury (d. 604), whom he had sent as a missionary to Britain. Augustine established the archbishopric of Canterbury, a position that retains religious leadership over English Christians today. Gregory also showed his ability to negotiate with the Franks in France and the Visigoths in Spain, even though neither of thesee groups formally accepted th leadership of the pope.

The effects of Gregory's work could be observed throughout Europe, from the haunting

rocco, where they founded the city of Thymiaterium, or present-day Mehdia.

At Cape Cantin, they built an altar to the gods, and then—according to the inscription, at least—sailed eastward. The latter detail has led to some questions regarding Hanno's account, since it is not possible to sail eastward at that point on the African coast without running into land. Probably they navigated up the river Oum er Rbia and entered a lake, where according to Hanno they found "elephants and other wild animals."

The voyagers established a number of cities, leaving settlers to maintain them as they continued southward on their way to a river Hanno called Lixos. The exact identity of this river is a matter of dispute; in any case, it was at this point in the narrative that Hanno made his first mention of human life, indicating that the journeyers befriended the nomads of the area. (Apparently, to judge from a later part of his narrative, some of the "Lixites" sailed as guides with the Carthaginians.) Further on, however, in what may have been the Anti-Atlas Mountains, the journeyers encountered "hostile Ethiopians," the latter a general term for all sub-Saharan Africans.

After sailing "past desert land," Hanno's party reached a small island where they founded a colony named Cerne, perhaps Herne Island off the coast of the Western Sahara. At this point Hanno noted that they had traveled as far beyond the Pillars as the latter were distant from Carthage. Soon they encountered more hostile inhabitants "who sought to stop us from landing by hurling stones at us," and afterward they passed a river—probably the Senegal—that was "infested with crocodiles and hippopotami."

They sailed for 12 days beyond Cerne, during which time the party observed a coastline "peopled all the way with Ethiopians.... [whose] tongue was unintelligible to us and to the Lixites in our company." On the twelfth day, they "came in sight of great, wooded mountains, with varied and fragrant trees"—a place variously identified as Cape Verde, or perhaps what is now Liberia. Soon they were in the Gulf of Guinea, where at night they saw numerous fires along the shore. At a place Hanno called the Western Horn, which is perhaps Cape Three Points in modern Ghana, they "heard the sound of pipes and cymbals and the rumble of drums and mighty cries. We were seized with fear, and our interpreters told us to leave the island."

Still further on, Hanno's party saw a volcano he dubbed "Chariot of the Gods," which may

have been Mount Cameroun. They sailed on for three days "past streams of fire" to what he called the Southern Horn, located either in Gabon or Sierra Leone. There they found what Hanno identified as gorillas, three of whom they captured, killed, skinned, and brought back with them to Carthage.

Hanno's narrative ends abruptly with the words "We did not journey any farther than this because our supplies were running low." Some scholars posit that in fact the Carthaginians continued sailing, circumnavigating the entire African continent but choosing not to divulge the fact to trade competitors. More likely, however, the expedition turned around at this point and returned home. Even so, Hanno's party had sailed further south than any expedition would until Portugal's Prince Henry the Navigator (1394-1460) sent out his sailors to chart the West African coast.

JUDSON KNIGHT

Hatshepsut
c. 1510 B.C.-1458 B.C.
Egyptian Pharaoh

Hatshepsut was not the only woman ruler of ancient times; however, as Egyptian pharaoh during the Eighteenth Dynasty, she was the only one to rule a world power at the height of its greatness. She did so by usurping the throne, and therefore did not engage in risky military exploits. Instead she devoted herself to building monuments, one of which records a historic visit by her emissaries to the land of Punt, a region usually identified with modern Somalia.

The elder of two daughters born to Thutmose I (r. 1493- 1481 B.C.) and his wife Ahmose was named Hatshepsut (pronounced hah-CHEPsut), which means "foremost of the royal ladies." When she was still a young girl, her parents married her to her half-brother Thutmose II (r. 1481-1479 B.C.), son of Thutmose I by a different wife. Marriage between relatives was common in Egypt, and in the case of Thutmose II, it strengthened his claim on the throne since he, unlike Hatshepsut, did not have two royal parents.

Some historians have suggested that Hatshepsut had Thutmose II poisoned in hopes of becoming pharaoh herself, but if so—and it is unlikely she did—she was in for a disappointment when he died. Egypt was not about to crown a woman, and if Hatshepsut had a son, the child would have become king. But she had no son, so a boy born to her husband by one of his concubines became Thutmose III (r. 1479-1425 B.C.)

Initially Hatshepsut served merely as regent for the young king, who was only 10 years old when he assumed the throne. She sent an expedition southward to Upper Egypt in order to acquire granite for a pair of obelisks, which were then covered in gold and placed in the Temple of Karnak, an important ceremonial center. Leading the expedition was her chief advisor, Senenmut.

In the sixth year of Thutmose's reign, however, Hatshepsut claimed that the god Amon had spoken to her personally and told her to take the throne. She therefore proclaimed herself "king" of Egypt, and thereafter presented herself as a man, complete with a ceremonial beard. Powerful officials, most notably Senenmut, helped her secure her position, though she maintained the fiction that she ruled jointly with Thutmose III.

In about 1472 B.C. Hatshepsut sent five boats from the Red Sea port of Kosseir to the land of Punt. Though she claimed hers was the first expedition, Egypt had actually been trading with this country, which historians have located anywhere from modern-day Eritrea to Somalia, for as much as seven centuries. A relief sculpture in her tomb records the voyage, and depicts the people of Punt as physically similar to the Egyptians themselves: slender, delicate of features, with long hair. The queen of Punt, however, was apparently an odd-looking woman—short, fat, long-armed, and with a prominent behind. Inscriptions also record that the primary purpose of the expedition was trade: "the ships were laden with the costly products of the Land of Punt and with its many valuable woods, with much sweet-smelling resin and frankincense, with quantities of ebony and ivory...." Also recorded were precise observations of flora and fauna.

Later, Hatshepsut sent another expedition to Upper Egypt, as well as several to the Sinai Peninsula, to acquire rock for monuments. Her inscriptions on these monuments portray her as a queen chosen by the gods for the special mission of resurrecting Egypt from the damage left by the Hyksos when they invaded about two centuries before. This exaggerated the situation considerably, no doubt in an attempt to further justify her legitimacy as ruler.

Some scholars have suggested that Hatshepsut and Senenmut were lovers; whatever the case, the bond between them was strong. In the nineteenth year of her reign, he disappeared from

Hatshepsut. *(Corbis Corportation. Reproduced with permission.)*

the official record, perhaps because he fell from favor, and three years later, the name of Hatshepsut too disappeared from the list of pharaohs. Presumably she died, though it is possible she was removed from power by Thutmose III.

During his long reign, Thutmose, who went on to conquests that expanded Egyptian territory into Palestine and Kush—thus bringing Egypt to its greatest territorial extent—sought to remove all evidence that Hatshepsut had ever ruled. He had her statues destroyed and her name removed from a number of monuments, and later king lists would include a fictional pharaoh, "Amensis," in her place. Despite all these efforts to erase her memory, however, the name of Hatshepsut has endured, and along with it the knowledge that a woman once ruled the most powerful nation on earth.

JUDSON KNIGHT

Hecataeus of Miletus
fl. 400s B.C.
Greek Geographer and Historian

In his *Genealogia* or *Historiai,* Hecataeus of Miletus presented an overview of Greek mythologies and traditions. The work may—to some extent at least—be considered a forerunner of modern anthropological writing; but it is primarily for a second book, *Ges periodos* or *Periegesis,* that Hecataeus is remembered. With the *Periegesis* (Tour round the world), Hecataeus introduced a systematic approach to geography, using a style that would later be employed by other noted geographical writers of the classical world.

Hecataeus came from Ionia, the region of islands off the western coast of Asia Minor (now Turkey) where the long set of wars between Greece and Persia (499-449 B.C.) had their beginnings. He opposed the Ionian revolt against Darius I (r. 522-486 B.C.) in 500 B.C., and after Darius's forces defeated the Ionians six years later, Hecataeus served as one of his region's ambassadors to the court of the Persian satrap, or governor, Artaphernes. As such he was instrumental in persuading Artaphernes to restore the constitutions of the Ionian city-states; but soon afterward, the war spread to the Greek mainland, embroiling Athens, Sparta, and other powers in a famous series of battles with Persia at Marathon (490 B.C.) and other cities.

The events involving the Persian Wars are the only part of Hecataeus's career that can be dated with any certainty. As for his two known texts, the *Historiai* has exerted little influence, mostly because little of it remains except for scattered fragments. Nonetheless, it is known that in his overview of gods and mythologies Hecataeus attempted to provide rational explanations for various tales; in other words, he was applying science and not superstition or prejudice to the study of a culture—his own, at that—and thus he may rightly be called one of the earliest social scientists.

The *Periegesis,* too, is preserved only in pieces, but there are many more of them: some 300 fragments survive. In his *Tour Round the World,* Hecataeus examined the lands of Europe and what he called "Asia"—which actually included Egypt and other parts of North Africa. Apparently he had personally visited these regions, and his travels also included forays into Persia itself, as well as the lands surrounding the Black Sea.

Much of Hecataeus's writing is preserved in that of another noted historian and geographer from Asia Minor, Herodotus (484-425 B.C.) The latter referred extensively to the *Periegesis* in his own *History,* though some scholars maintain that Herodotus credited Hecataeus only when he found fault with his predecessor. In any case,

Hecataeus's systematic treatment of geography and the histories of peoples, laid the groundwork for the encyclopedic approach applied by Herodotus and many others. Included among the latter was yet another noted scholar from the Asia Minor region, one of the last important geographical writers of the ancient world, Strabo (c. 63-23 B.C.)

JUDSON KNIGHT

Herodotus
484-420 (or earlier) B.C.
Greek Historian

Hailed as the "Father of History," Herodotus wrote the first known prose history in Western literature. Writing at a time when history was still a dubious mix of myth and verse, he gathered information on the geography and people of the eastern Mediterranean and Persia as background for his detailed account of the war between Greece and Persia from 499 to 479 B.C. His interest in the people and culture in his account is also recognized as an early foray into anthropology.

Though Herodotus is one of the most famous ancient writers, little is known of his life. He was born in Halicarnassus, a Greek city on the west coast of Asia Minor, now Turkey. He reports seeing the defeated Persian fleet arrive in the harbor at Halicarnassus when he was five years old. This occurred in 479 B.C., so it seems evident that he was born in 484. This event may have led to his interest in the war between the Greeks and Persians.

His education was probably good, though he never mentions it. His family was well known in the city and prosperous. Though he was Greek, some of his relatives may have been Persian, which also may have fueled his curiosity about the huge Persian empire. As a young man, he was involved in an attempt to overthrow a local leader and was exiled to the island of Samos for several years.

Herodotus had no known occupation, but he traveled extensively. He may have worked for a trader, or was a trader himself. He knew about boats and modes of travel as well as lands as far as Mesopotamia, the Black Sea, Egypt and North Africa. When not traveling, he lived in Athens, and in 443 emigrated to Thurii, a Greek colony in southern Italy. During this time he wrote his master work, *History* of the Greco-Persian Wars. He read parts of his *History* aloud from unwieldy

Persian Wars, it is a major and generally reliable reference about this important conflict. Because his work is readable, full of fascinating stories and anecdotes as well as historically significant facts, it was circulated, copied, and recopied through the centuries. He was successful in conveying the idea that the Greeks fought for their independence and the rule of law. His assertion that their victory over the Persians saved Western democracy and prevented the spread of Eastern despotism and decadent living in the Western world was long undisputed.

LYNDALL BAKER LANDAUER

I-ching
634-712
Chinese Pilgrim

As a Buddhist monk who traveled to India from China, I-ching followed in the footsteps of earlier pilgrims Fa-hsien (c. 334-c. 422) and Hsüan-tsang (602-664). He did not, however, follow them in a literal sense: instead of crossing the mountains that divided China and India in the west, I-ching took an easterly route, via the waters between the Malay Peninsula and Archipelago.

After hearing of Hsüan-tsang's exploits, I-ching, along with 37 other monks, resolved to visit the homeland of the Buddha (Siddhartha Gautama; 563-483 B.C.) himself. The group traveled to Canton in order to board a ship for India, but at the last minute, the other monks got cold feet; therefore I-ching was alone when in 671 he boarded a Persian ship bound for what is now Indonesia.

Sailing past Poulo Condore off the southern coast of the Malay Peninsula in what is now Vietnam, the ship went on to Palembang, a trading center on the southeastern coast of Sumatra. I-ching tarried there for half a year before taking a Sumatran vessel to the Nicobar Islands in the Bay of Bengal. (Today the Nicobars, just west of Sumatra and south of the Andaman Islands, constitute the extreme southeastern fringe of India's territory.) From the Nicobars he sailed to Tamralipti, a port on the Ganges delta where Fa-hsien had spent part of his Indian sojourn nearly three centuries before.

I-ching devoted his first three years in India to learning Sanskrit at the Buddhist temple of Varaha in Tamralipti. Then he sailed up the Ganges to the Buddhist center of Nalanda, where Hsüan-tsang had gone before, and spent a

Herodotus. *(The Granger Collection, Ltd. Reproduced with permission.)*

and expensive scrolls to paying audiences. At the time, this was a customary method of presenting literature to the public.

He returned to Athens around 430, near the beginning of the Peloponnesian War between Athens and Sparta, and died shortly afterward, perhaps of the same plague that struck down the powerful Greek leader, Pericles. The date of Herodotus's death is obscure, but occurred sometime between 430 and 420 B.C.

Herodotus's great achievement in his *History* involved his focusing on a single historical event—the war between the Greeks and Persians—providing a historical context and explanation for the conflict, and documenting it in prose writing. He investigated the reasons for animosity between these peoples and described the great deeds of the Greeks and Persians alike. Though he often recorded the stories he gathered without comment or judgment, he was also uncritical and often exaggerated in his writing. However, writing long before formal techniques for gathering and evaluating history had been established, he had no guidelines or no written records to emulate, and was further hindered by the lack of accurate maps with which to work.

His *History* remains an indispensable source of information about the ancient Mediterranean world and western Asia in the fifth century B.C.. Written only 50 years after the end of the Greco-

decade studying the Buddhist scriptures there. He also compiled a library of as many as 10,000 sacred texts, which he brought back to China and with which he greatly increased the scope of Buddhist learning in his homeland.

In all, I-ching visited some 30 different principalities and kingdoms throughout India before concluding his time in that country. He then traveled back to Sumatra, where he lived for many years while translating the texts he had brought with him. It is possible that he returned briefly to Canton in 689 to recruit monks who could help him with his translation work. In any case, he finally returned to his native Honan province in 693 or 694, nearly three decades after he left China.

During the remainder of his life, I-ching stayed busy translating some 56 Buddhist texts into Chinese. He also wrote a book on the travels of numerous Buddhist pilgrims to India, as well as one discussing various religious practices of the Sumatran and Indian peoples.

JUDSON KNIGHT

Kanishka
fl. A.D. 78?-103?
Chinese Ruler

Kanishka is considered to be one of the most significant and important rulers of the Kushan Kingdom, an area that included the northern part of the Indian subcontinent, Afghanistan, and possibly regions north of Kashmir in central Asia. During his reign he expanded the territory of his kingdom through a series of successful military campaigns, accumulated great wealth for his empire, built a considerable number of monasteries and temples, and encouraged the spread and adoption of Buddhism. Art, particularly Gandharan art, and culture also flourished under his rule.

It is not known when Kanishka was born. The dates of his reign are also subject of debate, but the most reliable dates seem to be from A.D. 78 to 103. Just how Kanishka came to power is also not known, but he succeeded Wema Kadphises II and was the fourth king of the Kushan. The Kushan Empire came into existence under Kajula Kadipheses in the north of India in Bactria, where Kadipheses unified the five tribes of the Yeuh-chi, who had fled the warring Hsiungnu from the Chinese frontier in 125 B.C. When Kanishka came to the throne, the Kushan Empire was already powerful, but he took it to new

heights and made it one of the greatest nations of its time.

Under Kanishka, the Kushan spread their boundaries to include the area of present-day Afghanistan, northern India, and parts of middle Asia. The capital city of this kingdom was Puruhapura, or modern Peshawar, Pakistan. The empire flourished economically, as the Kushan were in contact with many of the world's most important civilizations, including Rome, China, and Parthia (a part of what today is Iran). Trade and commerce were encouraged and cities and urbanization followed. The ruins of a city established by Kanishka have been discovered under present-day Taxila, India. The Kushan also had a great deal of control over a major trade route known as the Silk Road.

Although much of the prosperity of the Kushan was gained through conquest, it is important to note the tolerance of Kanishka towards other cultures. Though Kanishka was a Buddhist, coins from that era, the Saka era, depict religious deities of Greek, Sumerian, Zoroastrian, and Indian origin. These cross-cultural influences were seen in the art of the Kushan, with images of Buddha depicted in Greco-Roman style in what is referred to as Gandharan art. These were the first visual images of Buddha. The school of Mathuran art also flourished under Kanishka, and was more indigenous to the area in its style. Kanishka was also a great builder, and monasteries and temples were built in great numbers, with most located in the capital. One of the monasteries was called by Kanishka *Mahavihara* (great monastery), and the ruins of a great temple commissioned by him can be found near modern Peshawar, Pakistan. This temple was 400 feet (121 m) high, and rested on a five-stage base that was 150 feet (45 m) in height. It was excavated in 1908 and 1909, and a casket known as the "Kanishka Casket" was discovered. It contained holy Buddhist relics reportedly placed there by Kanishka himself.

One of the most important contributions made during Kanishka's reign was the spread of Buddhism to central Asia and China. Kanishka probably adopted Buddhist faith later in life. Confused by conflicting doctrines within the Buddhist faith, Kanishka convened a special council under the Buddhist poet Asvaghosa. The council was held in Kashmir and became known as the fourth Buddhist council. The use of Sanskrit in manuscripts produced from this council was significant because it was the first time it had been used and was later to become the tradition. Kanishka adopted a form of Buddhism known as Mahayana, and was probably responsible for the spread of Buddhism into China, first noted in the second century A.D.

Military conquests marked Kanishka's rule, but the achievements of his empire were not limited to these exploits. Under Kanishka, the Kushan reached a new zenith in wealth and culture that highlighted art, architecture, literature, and science. The tolerance of the kingdom's ruler toward different cultures and religions placed allowed Kushan to benefit from the knowledge and wealth of some of the greatest civilizations of the time. As a result of such cross-cultural connections, the Buddhist religion owes much of its development and spread throughout Asia to Kanishka and the Kushan.

KYLA MASLANIEC

Megasthenes
c. 340-c. 282 B.C.
Greek Historian and Diplomat

Megasthenes was neither the first European to travel to India nor the first to write of it. Yet his work attracted considerable attention in ancient times, and exerted an impact on the portrait of India in the writings of Strabo, Arrian, and others.

Some 70 years before Megasthenes, Ctesias of Cnidus (416 B.C.), served as physician in the courts of the Persian emperors Darius II Ochus (r. 432-404 B.C.) and Artaxerxes II Mnemon (r. 404-358 B.C.). Ctesias wrote the *Persicha,* which covers the histories of Assyria, Babylonia, and Persia in 23 books. Though modern scholars regard the work as unreliable in many particulars, for many years it was the principal Greek source on the area from Mesopotamia to India.

Something crucial happened in the period between Ctesias and Megasthenes: the conquest of the Persian Empire by Alexander the Great (356-323 B.C.), who by 326 B.C. controlled most of the known world from Sicily to western India, which is now Pakistan. After Alexander's death, his general Seleucus (c. 356-281 B.C.) took power over Persia, Mesopotamia, and Syria. Meanwhile, the Greco-Macedonian conquerors lost control of India, where the conquests of Alexander inspired a young monarch named Chandragupta Maurya (r. 324-301 B.C.) to build an empire of his own.

Probably in the last decade of the fourth century B.C., Seleucus sent Megasthenes as his diplomatic representative to the court of Chandragupta Maurya. All that is known of Megasthenes himself is that he came from Ionia, a group of city-states on the western coast of Asia Minor, or modern Turkey. As for his most notable achievement, this was not necessarily his work as Seleucid ambassador to the Mauryan Empire, but the four books of his *Indica,* which for centuries provided Greeks with their most important eyewitness account of India's history and geography.

The manuscript of the *Indica* has been lost, but echoes of it live on in the work of other writers, particularly the geographer Strabo (c. 64 B.C.-c. 23 A.D.) and the historian Arrian (d. A.D. 180). For instance, the former, discussing the size of India in his *Geography,* favored "a more moderate" (and in fact more accurate) estimate made by Megasthenes over observations made by Ctesias or other authorities, including officers in Alexander's army. Strabo went on to reference Megasthenes's reports concerning animal life and social organization among the Hindus.

Much of what Strabo quoted seems valid, but he also included enough fantastic stories— for instance, of gold-mining ants—to suggest the less reliable side of Megasthenes's writing. Arrian, as a military historian, was particularly interested in what Megasthenes had to say about the size, population, and ethnic makeup of India and the Caucasus, a region about which the Seleucid ambassador also wrote. Nonetheless Arrian, who quoted Megasthenes as saying that there were 118 different "tribes" or ethnicities in India, also relayed a number of strange tales— including the tale of the gold-mining ants—that came from Megasthenes.

Though the manuscript of the *Indica* has long ago disappeared, the many references in Strabo, Arrian, and others make it possible to gain some idea of the original. Thus in the twentieth century the German scholar E. A. Schwanbeck reconstructed a portion of the book from the many surviving fragments.

JUDSON KNIGHT

Nearchus
360 B.C.-312 B.C.
Greek Military Commander

A commander under Alexander the Great, Nearchus served as admiral over a fleet that sailed from the coast of the Indian subcontinent to the Euphrates river in distant Mesopotamia. In so doing, he proved that a sea route between India and the Near East existed, and thus made possible later Greek and Roman trade with India.

Nearchus apparently came from Crete, and at some point in his career gained naval experience. It is likely that if he did serve on a ship, this must have happened before the time he joined the forces of Alexander the Great (356-323 B.C.): in conquering his vast Greco-Macedonian empire, Alexander relied on land troops, using his navy almost solely for transport.

Nearchus's role as admiral would come much, much later; but from the outset it was clear that as a longtime friend of Alexander, he was one of his commander's most trusted aides. Early in his conquests—and thus long before he acquired huge territories that he could distribute lavishly—Alexander made Nearchus satrap, or governor, over two choice provinces in Asia Minor, Lydia and Pamphylia.

The grant occurred in 333 B.C., when Alexander was in the process of moving from Asia Minor into the Levant, and thence to Egypt. In the years that followed, he defeated the Persians, absorbing their empire and eventually placing under Greek dominion a realm that stretched from Sicily to India. Eventually, however, his troops grew weary of conquest and longed to see their families, from whom they had been separated for a decade.

In July 326 B.C. Alexander began the long westward movement, separating his force into three groups. One returned via a northerly route, while a second—led by Alexander himself—would take a southerly path through Iran. The third would sail under Nearchus, whom he had appointed admiral. That appointment, in fact, may have occurred prior to the decision to head for home; in any case, Alexander directed Nearchus to find a sea route, if such existed, that would connect the Indus and Euphrates rivers.

Alexander also placed at Nearchus's disposal all troops with seafaring experience. Under Nearchus's leadership, Indian shipwrights built for the Alexandrine navy some 800 vessels, many as large as 300 tons, and Nearchus procured the services of Indian pilots who would serve as guides. Finally, in September of 325 B.C., the fleet departed from the mouth of the Indus.

The first month of journey was a difficult one, beginning with a long period of delays and concluding with a storm in which three ships were capsized. At Ras Kachari in what is now Pakistan, the fleet anchored and met up briefly with a portion of Alexander's forces. Then they continued on to the Hingol River, where they waged a successful battle against hostile inhabitants, and thence to Makran in the area of the present-day border between Pakistan and Iran.

At Makran, the expedition underwent hardships as they searched for fresh water supplies, and later they ran short of food, a situation that forced them to hunt for wild goats on the shore. At Pasni, a town today in the Pakistani province of Baluchistan, they were able to take on fresh supplies, and as they continued westward they found that the land was more fertile.

During his voyage, Nearchus appears to have seen whales, which must have been an unusual sight that far south, and provided the first known written reference to sugar cane. He also fought an apparently pointless battle at the town of Gwadar, where in spite of the locals' willingness to trade, he chose to attack the city. After this effort ended in a stalemate, however, he was content to trade with the townspeople for fishmeal.

In time his fleet met up with Alexander in Persia, then continued through the Strait of Hormuz and up the Persian Gulf. They finally landed at the mouth of the Euphrates, having proven that it was possible to sail from India to Mesopotamia. Later Nearchus and his fleet rejoined Alexander in the capital city of Susa, but the commander did not have long to live: he died in June 323 B.C., leaving behind a massive power vacuum. In the ensuing struggle between generals, Nearchus most likely lost his satrapies in Asia Minor.

JUDSON KNIGHT

Saint Patrick
fl. fifth century
British/Irish Clergyman

One of the best-known Catholic saints, Saint Patrick was sent by the Church to Ireland in the mid-fifth century to bring Christianity to the Irish. For over a half century, he traveled through Ireland, spreading the Christian religion among the Druids. He is largely credited with almost single-handedly converting much of Ireland to Christianity in a single lifetime. In so doing, he became the Irish patron saint and national apostle.

Patrick was born in Britain in the late fourth century or early fifth century. His parents were

Saint Patrick brought Christianity to Ireland. *(Archive Photos. Reproduced with permission.)*

Romanized Britons, and his father, Calpurnius, was an official in the local government and a deacon in the British Christian faith. Patrick was taken from his parents and sold into slavery by Irish raiders at the age of sixteen. Isolated by language, religion, and distance from everything familiar, Patrick turned to prayer and faith to see him through. After six years of servitude, Patrick escaped his master, begging his way aboard a ship to Britain. There, he was reunited with his family after nearly starving to death and being briefly recaptured. Around this time, too, he decided to devote himself to God, becoming a priest.

In Patrick's *Confessio*, his spiritual autobiography, he mentions having a vision in which he received a letter from the Irish begging him to return to them. He impressed his superiors sufficiently that they agreed to send him to Ireland, hoping he could succeed where his predecessor, Palladius, had failed. Although he was filled with doubts about his ability to adequately serve in Ireland, Patrick came into his own almost as soon as he stepped ashore, and rarely doubted again that he could succeed.

Patrick concentrated his efforts in the north and west of Ireland, never claiming to have converted all of Ireland, although his words and fame certainly spread throughout the island quickly. To achieve his aims, however, Patrick was forced to be as much politician as saint, be-cause his goals were entirely contrary to those of Ireland's rulers.

In particular, Patrick was careful to visit local kings, bringing small gifts to them while refusing to take gifts himself. He was similarly careful with others whose influence was important in swaying the opinions of others. Like many other Christian evangelists, he realized that converting a king often led to converting the king's subjects, because many kings simply decreed the religion of their subjects. However, Patrick was willing to speak with, baptize, and convert all who came to him, king and commoner alike. It was this fundamental fairness that earned him the respect and admiration of so many of his followers.

In spite of this, Patrick was often arrested and was several times threatened with death during his work. Each time he was eventually released, and he returned to his preaching as soon as he could. Throughout, he never lost his faith or his confidence in his ability to complete his mission.

There are only two writings by St. Patrick that are known to exist. The most famous is his *Confessio*. Written as a response to British church officials, the *Confessio* details Patrick's spiritual growth and development through much of his career. Although his Latin is often rough and unpolished, many scholars have been impressed with the moral and spiritual greatness that is evident, as well as the simple strength of his faith. His other work, the *Epistola*, is a strong indictment of British treatment of the Irish Christians, even at this early stage in the countries' relationship.

By the end of his life, Patrick had traveled widely through Ireland, becoming a beloved clerical fixture in the process. It is said that his death shroud was made for him by St. Brigid, and St. Tassach administered his last sacraments. These facts, however, are not as important as what he accomplished during his career and his many travels through Ireland.

P. ANDREW KARAM

Pytheas of Massalia

fl. c. 350 B.C.

Greek Astronomer, Geographer, and Explorer

Pytheas of Massalia was the first Greek to undertake an extended voyage in the North Atlantic, possibly visiting Iceland. He was certainly the first to circumnavigate Britain, first to document the midnight sun, and first to write on British ethnography. He was also the first to record the connection between the tides and the Moon.

Almost nothing of Pytheas's life or motivation for his voyage are known. He was from the Greek colony of Massalia (modern Marseilles, France), which was an important trading center in the western Mediterranean. Tradition has it he was an exceptional astronomer and geographer, having accurately established the latitude of Massalia. He would thus have been a valuable member on any voyage of discovery, especially an expedition conducted under official auspices for the purpose of obtaining information to enhance Massaliote commerce.

Pytheas presented the results of his voyage in a general treatise of geography titled *On the Ocean (Peri Okeanou)*. Though no longer extant, 18 ancient writers (from 300 B.C. to A.D. 550) directly drew upon or referred to it. From these scattered sources, Pytheas's route and discoveries have been reconstructed.

The voyage is widely thought to have occurred sometime between 350 and 290 B.C. However, the Carthaginians controlled the Pillars of Heracles (Straits of Gibraltar) during this period and strictly monitored traffic through the straits. Therefore, it is commonly believed Pytheas could only have sailed into the Atlantic while Carthage was distracted by its war with Syracuse during the years 310 to 306 B.C.

Having passed through the Pillars into the Atlantic, Pytheas followed the coast of Iberia to the mouth of the Adour River. Here he determined he was at the same latitude as Massalia but only 400 miles (644 km) away. He thus discovered Iberia is a peninsula. Continuing northward and crossing the English Channel, he located the famous Cornish tin mines at Bolerium (Land's End, Cornwall). Pytheas then traveled along the eastern littoral of the Irish Sea until he reached the northern tip of Scotland. Here, he encountered the gale-enhanced tidal displacements of Pentland Firth, which separate the mainland from the Orkney Islands.

Pytheas made his most famous discovery next. He learned of a large island to the north. Known as Thule, it was 6 days away, situated on the edge of the frozen sea. Pytheas also stated that Thule had 19 hours of sunlight in summer and claimed that further north was a region of semi-congealed water—possibly referring to the combination of ice sludge and fog found near drift ice.

Thule has variously been identified as Iceland, the Shetlands, the Faeroe Islands, and Norway, but its location remains cloaked in controversy. Regardless, it seems certain Pytheas ventured at least as far as 62°N, where 19-hour days occur.

Pytheas finally turned south and completed his circumnavigation of Britain. He may also have penetrated the Baltic Sea and traveled as far as Prussian Samland near the Vistula River, which was the primary source of amber for the Mediterranean. By the time he returned home, Pytheas, even by conservative estimates, had traveled over 7,500 miles (12,070 km).

Pytheas's astronomical observations contributed significantly to the development of mathematical geography. Eratosthenes (c. 275-c. 194 B.C.) accepted Dicaearchus's (fl. 326-296 B.C.) line running from the Pillars of Heracles to Rhodes and expanded it into a series of parallels for measuring latitude. He relied on Pytheas's measurements to determine the northern parallels. Hipparchus (c. 190-c. 120 B.C.) similarly relied on the Massaliote's observations in establishing his own system of projected parallels. Pytheas also correctly described Britain as triangular, accurately estimated its circumference at 4,000 miles (6,400 km), and approximated the distance from northern Britain to Massalia at 1,050 miles (1,690 km), slightly less than the actual distance of 1,120 miles (1,800 km).

STEPHEN D. NORTON

Strabo

c. 63 B.C.-c. A.D. 23

Greco-Roman Geographer

Strabo's *Geographica*, with its 17 books discussing the world known to the Greeks and Romans of his day, is perhaps the greatest geographical text of antiquity. Vast in scope, the work is filled with the author's observations on philosophy, history, and science, all of which give the modern reader enormous insight regarding the ancient mind.

Raised in Pontus, a kingdom of Asia Minor conquered by Rome shortly before his birth, Strabo came from a wealthy family and received the finest education in Greek and Roman traditions. In his youth he received instruction from tutors who had also taught the sons of several great Romans, among them Tyrannio, tutor of Cicero's (106-43 B.C.) sons and an expert on geography. Most likely Strabo acquired his interest in that subject from Tyrannio, who taught him in Rome.

Also in Rome, the young Strabo came under the influence of the Stoics, whose philosophy later found expression in many passages from the *Geographica*. Another powerful influence was Homer (fl. ninth or eighth century B.C.), whose epics Strabo regarded—and would later refer to in his *Geographica*—as historical works. He also became acquainted with the writings of various historians and scientists of the ancient world, and traveled widely throughout Asia Minor and the Mediterranean.

By 25 B.C. he was in Alexandria, Egypt, where he would remain for the next six years. Some historians maintain that at the beginning of his time in Egypt, he accompanied Aelius Gallus (fl. c. 25 B.C.) on a voyage down the Nile. However, Aelius's own biography does not record such a voyage, which he would have taken just before beginning his failed mission to conquer Arabia—a mission of which Strabo became one of the principal, if second-hand, chroniclers. Perhaps Strabo traveled instead with Gaius Petronius, a Roman general in Egypt who undertook a Nile expedition around this time. In any case, he did follow the Nile to the borders of Kush or modern-day Sudan.

After his time in Alexandria, Strabo returned to Rome, where in about 20 B.C. he composed his *Historical Sketches,* a work of which only fragments survive. Consisting of 43 books, Strabo's history covers the period from the Roman destruction of Corinth and Carthage in 146 B.C. to around the time of the Battle of Actium in 31 B.C. Thus it would have been a continuation of the historical writings of Polybius (c. 200-c. 118 B.C.), whom Strabo greatly admired.

At some point Strabo wrote what became his magnum opus, the *Geographica,* the world's first text to attempt a summary of all existing geographical knowledge. In the first two of the 17 volumes, Strabo examined the history of the discipline, praising those he admired (e.g., Homer, who he called the first geographer) and attacking those he did not, including Herodotus (c. 484-c. 420 B.C.) and Eratosthenes (c. 276-c. 194 B.C.).

Book 3 discusses Spain, and Book 4 Britain, France, and the Alps. In the fifth and sixth books, Strabo presented the geography of Italy and surrounding islands. Somewhere in this section he noted that Mt. Vesuvius seemed to be a volcano, a statement that turned out to be quite accurate when the mountain erupted in A.D. 79, destroying the nearby city of Pompeii. Book 7, which examines northern Europe, suffers from the fact that neither Strabo nor most other Romans knew much about the world beyond the Alps.

The next three books, on Greece and surrounding islands, also lack a great deal of useful information, but for the opposite reason: Strabo assumed that his readers would already know plenty about the area, and therefore he devoted much of his attention to such preoccupations as Homer—and to further discussions of volcanic activity, in this case around Thera and Methone. Five more books (11 through 16) concern Asia, and the part that addresses Strabo's homeland in Asia Minor is fairly accurate. Also of interest are his discussions of India, derived from lost writings of soldiers in the armies of Alexander the Great (356-323 B.C.). Finally, in Book 17 Strabo examined Africa. Again, in the sections involving areas he had seen (Egypt, the Nile, and Ethiopia), his observations are useful; but concerning the remainder of the continent, the book falls back on conjecture and supposition.

The *Geographica* is rife with mistakes, from the author's placement of the Earth at the center of the universe to his statement that the Caspian Sea is connected to the great northern sea, which he and most ancients believed was in turn part of a larger ocean surrounding all the Earth's lands. Despite its shortcomings, however, the *Geographica* proved a work as influential as it was ambitious. Not only did it provide highly useful summations of the author's firsthand knowledge, but it included references to, and insights on, writings that have long since been lost.

JUDSON KNIGHT

T'ai Tsung
600-649
Chinese Emperor

T'ai-Tsung was one of China's greatest emperors. During his reign, China achieved perhaps her highest cultural peaks and regained much of her national identity. T'ai-Tsung and his father helped to expel foreign forces and reunify China, setting the stage for a return to former power.

T'ai-Tsung was born as Li Shih-min around 600 in a China alive with the memory of past glories. Little is known of his early life except that he had at least two older brothers. His father, a general in the imperial army, tired of China's status as a former great power and rose against the government, taking power himself in about 617. This uprising was the final blow for the Sui dynasty and established the T'ang dynasty with

Tsung's father as the first T'ang emperor. Although many accounts claim that T'ai-Tsung played a major role in his father's revolt, more recent material suggests his role was relatively minor. However, there is little doubt that as a commander in his father's army he was instrumental in taking the Eastern capital Lo-yang and much of the strategic eastern plain. When the uprising was successfully concluded, T'ai-Tsung was named the military and civil leader of the eastern provinces, and he surrounded himself with a government of talented administrators.

T'ai-Tsung's relationship with his older brother took a dramatic downward turn at about this time when his brother raided the government for talented personnel, led an attempted coup against his father, and made at least one attempt on T'ai-Tsung's life. A final assassination attempt by his older and younger brothers was defeated when T'ai-Tsung, who had been informed in advance, ambushed his brothers and had them killed. After hearing about the ambush, his father abdicated the throne, allowing T'ai-Tsung to ascend as emperor.

The revolt that led to the formation of the T'ang dynasty was caused in part by the perception that China was no longer a great power, a perception that was largely true. At that time, China had lost territories to outside nations, and outsiders (primarily Turks) were influential in internal Chinese affairs. T'ai-Tsung successfully in remedied these affronts to Chinese national pride, leading campaigns to expel the Turks and Turkish influences from China and reuniting those lands with the rest of China. This restoration of Chinese territory and influence helped spark a revival of Chinese culture and resulted in the blossoming of Chinese arts and sciences. In fact, many scholars consider the final years of T'ai-Tsung's reign and the reigns of his immediate successors as representing the pinnacle of Chinese culture, during which time Chinese achievements in these areas far exceeded those of the West.

As emperor, T'ai-Tsung realized that he could not continue taxing his people at the rate of previous dynasties, primarily because the people had nothing left to tax. Comparing heavy taxes to a man eating his own flesh, he tried to transfer land from the wealthy landowners to the peasants. Although this helped the peasants and actually resulted in a net increase in cash flow, it also angered the powerful.

Another of T'ai-Tsung's innovations—the first civil service examinations—led to the regu-

larization of the Chinese bureaucracy. This made working for the government more a matter of education and intelligence than of family connections and resulted in a government made up of the best and the brightest, rather than the best connected. This and other improvements led to better government.

Finally, T'ai-Tsung encouraged contact with other cultures. Unlike most earlier dynasties, the T'ang dynasty enjoyed a period of relative tranquility and the absence of large-scale famine. These advantages gave them the opportunity to contact other civilizations, as well as the willingness to do so. Also, unlike other eras in Chinese history, the T'ang did not view China as the sole outpost of civilization in a sea of barbarians. Instead, T'ang emperors eagerly sought knowledge of the Roman, Islamic, Hindu, and other non-Chinese powers, seeking to expand their horizons.

T'ai-Tsung died in 649 following a spectacularly successful reign as emperor. He left behind a thriving Chinese empire and a dynasty that would continue for another 250 years.

P. ANDREW KARAM

Wu-ti
156-87? B.C.
Chinese Emperor

During Wu-ti's reign, he expanded China's borders and influence through most of the world known to China. Because of this, he went down in history as one of the greatest emperors of China's Han Dynasty. After his death he was given the name Wu-ti, which means "martial emperor," in honor of his victories.

Wu-ti was born Liu Ch'e about 156 B.C., the son of Emperor Ching-ti. He was definitely not the eldest son of the emperor, in fact he is believed to have been son number 11, making him far from first in line to ascend to the throne. However, by his seventh birthday relatives assured his status as heir apparent, and he succeeded the throne in 140 or 141 about the age of 15.

For the first several years of his reign, Wu-ti was heavily influenced by his relatives, who urged a more moderate, defensive approach to national security. By about 133, he launched the first of many attacks against threatening neighbors, attempting to both secure and expand China's borders. In this first attack, he decided to secure China's northern border by attacking the Hsing-nu, a nomadic tribe that constituted

China's primary threat from that direction. Successful in this endeavor, Wu-ti set about an era of national expansion.

Wu-ti next looked east, conquering northern and central Korea, which had slipped from Chinese control before his reign. He continued southward, bringing Vietnam and what is now southern China under his control by 128. Wu-ti continued this series of conquests, driving his armies with ruthless determination and little tolerance for error.

At its greatest extent, Wu-ti's empire reached into Fergana, today's Uzbekistan, and controlled most of the world known to China. His exploits resembled those of Alexander the Great (356-323 B.C.), except that Alexander personally led his troops while Wu-ti remained in his capital. However, with a greater population at his command, Wu-ti controlled larger armies of up to 100,000 men plus the supply and support need to keep such a large army in the field. Of course paying for such a huge army was difficult, and Wu-ti reorganized the Chinese bureaucracy and raised taxes to more effectively administer his realm and pay for his government.

In addition to his military and administrative exploits, Wu-ti dispatched missions of exploration and discovery, sent Chinese settlers into newly acquired territories, and helped foster trade with neighboring states. He also attempted to make political and military alliances with western nations against the Huns, another group that threatened China at the time.

Although his attempts to ally China with western nations were unsuccessful, Wu-ti was instrumental in establishing the Silk Road, a major source of transcontinental trade during subsequent centuries. Other initiatives were somewhat less successful. Obsessed with immortality, Wu-ti sent expeditions in search of a purported island of the immortals, but to no avail, of course. Another expedition was sent to bring "blood sweating" horses from central Asia, feeling that their presence would signify heaven's grace for his empire. However, Wu-ti was almost entirely responsible for making Confucianism China's official religion, opening Confucian universities and other centers for Confucian teachings.

Wu-ti's expeditions and military campaigns cost more than even his new taxes could raise. During the final years of his reign, China's expenses far outstripped her income, and Wu-ti was forced to retrench, giving up some of his conquered territories. In spite of this, China reached her greatest territorial extent during Wu-ti's reign, and he is considered to have been one of China's greatest leaders. Wu-ti died in 87 or 86 B.C. at about the age of 70.

P. ANDREW KARAM

Xenophon
431?-354? B.C.
Greek Soldier and Historian

Xenophon is best known for writing the *Anabasis*. It recounts the details of Cyrus the Younger's (423?-401 B.C.) Persian campaign and the role Xenophon played in leading his Greek mercenaries back to the Mediterranean after Cyrus's death. Xenophon wrote on a wide range of topics, and his prose was greatly admired in antiquity and strongly influenced Latin literature.

Born around 431 B.C. to the wealthy Athenian aristocrat Gryllus, Xenophon came of age during the Peloponnesian War (431-404 B.C.). He joined the intellectual circle that gathered about Socrates (470?-399 B.C.) and sympathized with their critical attitude towards Athenian democracy. He supported the short-lived oligarchic regimes of the Council of 400 (411 B.C.) and the Thirty Tyrants (404-403 B.C.). When Thracybulus reestablished democracy in Athens, Xenophon became disillusioned and chose to seek his destiny elsewhere.

In 401 B.C. Xenophon joined the Greek mercenaries of prince Cyrus at Sardis, in Lydia. Cyrus's pretext for assembling this force was to rid his satrapy of the Pisidian hillmen. He marched inland with a mixed force that included 14,000 Greeks. As the army headed into Syria, the Greeks realized Cyrus's true goal was to overthrow his brother Artaxerxes II, king of Persia. Promise of rich rewards convinced the Greeks to follow Cyrus down the Euphrates towards Babylon.

The opposing armies met at Cunaxa in early September. The lightly armed troops of the Persian Empire proved no match for the Greek hoplites. However, with victory in his grasp, Cyrus was slain and his Asian mercenaries fled the field in disarray. The Greeks remained undefeated and refused to surrender. During negotiations, their generals were treacherously murdered, leaving them leaderless. Xenophon was among the new commanders immediately elected. He played an important role as the "Ten Thousand," as they were later known, fought their way northward through Kurdistan and Armenia to the Greek port of Trapezus on the Black Sea. Their five-month, 1,500-mi (2,414-km) trek revealed the essential internal weakness of the Persian Empire.

Xenophon continued leading the Ten Thousand. They fought in Bulgaria for the Thracian prince Seuthes before returning to Asia Minor in 399 B.C. with a force of Spartans. There, they defended the Ionian cities against renewed Persian aggression. It was during this time that Xenophon met the Spartan king Agesilaus II (444?-360 B.C.), whom he admired greatly. He returned to Sparta with Agesilaus and served on his staff. He was present at the Battle of Coronea (394 B.C.) when the Spartan king defeated a coalition of Greek cities that included Athens. Shortly thereafter, Xenophon was banished from Athens as a traitor.

In gratitude for his services, Sparta rewarded Xenophon with an estate at Scillus, near Olympia. He married, had two sons, and devoted his leisure time to hunting, entertaining friends, and writing. After Thebes defeated Sparta at the Battle of Leuctra (371 B.C.), Xenophon sought refuge in Corinth. When his banishment was revoked (365? B.C.), he may have returned to Athens. In any case, both his sons were educated there, and the oldest died at Mantinea (362 B.C.) fighting for Athens.

Aside from *Anabasis,* Xenophon's most important work was *Hellenica,* which was intended as a continuation of Thucydides's *History of the Peloponnesian War,* covering the period from 411 to 363 B.C. He wrote various Socratic dialogues and compiled his recollections of Socrates in *Memorabilia.* His other works include *Cyropaedia,* a hagiography of Cyrus the Great; *Peri hippikes* and *Hipparchicus,* manuals on horsemanship and cavalry tactics; *Cynegeticus,* a treatise on hunting; and *Agesilaos,* an encomium of Agesilaus. *Peri poron,* a pamphlet dealing with Athen's financial difficulties, was written shortly before his death around 355 B.C.

STEPHEN D. NORTON

Biographical Mentions

Amunirdis I
fl. 710 B.C.

Kushite princess who participated in the establishment of Egypt's Twenty-Fifth (Nubian) Dynasty (712-672 B.C.) Most likely the first notable woman of sub-Saharan African descent, Amunirdis was sister to Piankhi (769-716 B.C.), who first invaded Egypt in the 720s B.C., and Shabaka (r. 716-695 B.C.), who established the dynasty four years after Piankhi's death. Amunirdis later held an important position as "god's wife of Amon" in the temple at Thebes.

Antiochus the Great
242-187 B.C.

Seleucid Syrian king who briefly reunited the Alexandrine empire in western Asia, from Palestine to India. Antiochus, who ruled from 223 B.C. as Antiochus III, spent most of his reign at war. He sealed an alliance with Egypt by marrying his daughter Cleopatra to Ptolemy V (their descendant was the famous Cleopatra VII); but unwisely allied himself with Hannibal of Carthage (247-183 B.C.) and Philip V of Macedon (238-179 B.C.) against Rome. Antiochus gave Hannibal refuge when the latter was on the run from the Romans, but eventually Rome defeated them both, as well as Philip.

Saint Augustine of Canterbury
d. 604

Christian missionary and churchman who journeyed to England across Europe from Italy to reestablish the English Christian Church in 597. Born in Rome, he became a Benedictine monk at an early age. Sent to England by Pope Gregory I, he was welcomed by Bertha, the Christian wife of the Saxon king Aethelberht I and began his evangelistic work at Canterbury. Augustine converted the king and thousands of others, extending his mission throughout southeastern England.

Augustus Caesar
63 B.C.-A.D. 14

Roman emperor who opened up new areas to exploration and trade. He was born Octavius (Octavian), the nephew and adopted son of Julius Caesar, and came to power after the latter's murder. Ruling initially as a member of a triumvirate, he became the first emperor of the Roman Empire by eliminating the other two (Mark Antony and Marcus Lepidus). He extended Rome's boundaries to the Danube, the Rhine, and the Euphrates Rivers.

Chandragupta Maurya
r. 324-301 B.C.

Indian emperor and founder of the Mauryan Empire (324-184 B.C.). Born a peasant, Chandragupta rose to overthrow the Nanda dynasty in the Magadha state of eastern India. Inspired by Alexander the Great's (356-323 B.C.) conquests in western India, he raised a vast army, and created an empire that extended to include almost the entire subcontinent. His capital city at Pataliputra was said to be the greatest city of its time. Chandragupta later converted to Jainism and, in sorrow over a famine, starved himself to death.

Chang Ch'ien (Zhang Qian)
fl. 138-114 B.C.

Chinese official who journeyed to central Asia, establishing the first contact between China and other civilizations. Prior to Chang Ch'ien, the Chinese had known only of barbarians beyond their frontiers. It was to deal with just such a group of nomads, the Hsiung-Nu or Huns, that the Han emperor Wu-ti (156-87 B.C.) sent Chang Ch'ien westward to form an alliance with the formerly nomadic Yüeh-chih people. The latter, however, had settled down and become civilized, adopting aspects of Hellenistic civilization that lingered in Bactria (modern Afghanistan) from the invasions of Alexander (356-323 B.C.) centuries before. They had no interest in returning to China, and thus Chang Ch'ien, who was imprisoned a total of 11 years by the Hsiung-Nu, failed in the immediate purpose of his 13-year mission. However, his efforts led to the opening of the trade routes known as the Silk Road, which greatly heightened contact between East and West in the centuries that followed.

Cleopatra VII
69-31 B.C.

Egyptian queen who attempted to increase the power of the Ptolemaic dynasty in Egypt by becoming romantically involved with Julius Caesar and Mark Antony. She married Antony in 37 B.C., even though he had yet to divorce his wife Octavia, the sister of his Roman ruler Octavian (Augustus Caesar). Both Antony and Cleopatra committed suicide after Octavian defeated their army in 31 B.C. Cleopatra's efforts to enhance the power of Egypt thus led to its total control by Rome.

Clovis
c. 466-511

Frankish king who was the first ruler to succeed in building a stable nation after the fall of the Roman Empire. A Germanic barbarian who converted to Christianity, he treated his conquering Franks and the conquered Gallo-Romans as equals, providing a civilizing influence unknown in other areas of the former Empire. He was the first king of the Merovingian dynasty; from his nation have arisen the modern states of France, Belgium, the Netherlands, and Switzerland.

Constantine the Great
285-337

Roman emperor who united and strengthened a fragmented Empire. Proclaimed emperor in 306, he finally defeated other contenders for the throne in 325. He established Christianity as the state religion, thus creating the foundation of medieval Europe. He founded Constantinople (Istanbul) as his capital, thereby strengthening the eastern part of the Empire, which became a separate entity that survived the collapse of the Empire in the West, leading to the distinction between western and eastern Europe.

Cyrus the Great
c. 600-529 B.C.

Persian king and emperor who established the Persian Empire. He succeeded his father as prince of Persis, a part of the Median Empire. He rebelled against and defeated the Medes and went on to conquer other neighboring nations including Iran and the Babylonian Empire. The resulting Persian Empire remained powerful until conquered by Alexander the Great. Cyrus freed the Israelites from captivity in Babylon, allowing them to return to Jerusalem.

Darius I
550-486 B.C.

King of Persia who reorganized the empire into provinces called satrapies, which were ruled by governors. Under his regime an efficient postal system was devised and royal roads were constructed. Darius is also credited with initiating the development of a cuneiform script to represent the Persian language. The Persian army, sent by Darius to punish a Greek revolt, was defeated by the Greek states at Marathon in 490 B.C., one of the most famous ancient battles.

David
d. c. 960 B.C.

Israelite king who brought his nation close to the zenith of its power. Virtually all that is known about David comes from the Old Testament, and for years historians regarded the facts of his existence with skepticism, but today he is widely accepted as a historical figure. It was under his leadership that the Israelites took Jerusalem (thenceforth the "City of David"), and during his reign the kingdom expanded to more than three times the size of modern Israel.

Publius Herennius Dexippus
c. 210-c. 270

Greco-Roman general and historian who wrote a chronicle of world history, among other works. In c. 267 Dexippus led the defense of Greece against the nomadic Heruli. In addition to his world history, his writings—of which only fragments exist today— include a history of Rome's third century wars with the Goths, as well as a chronicle of Alexander the Great's (356-323 B.C.) successors.

Dio Cassius
c. 150-235

Greco-Roman administrator and historian whose *Romanika,* written in Greek, provides a valuable source regarding the latter years of the republic and the early empire. Dio served in various positions throughout the Roman world, from Africa to Asia Minor, and this aided his research in compiling *Romanika,* a work in 80 books. Beginning with the Roman people's mythical beginnings as descendants of Aeneas, the work chronicled Roman history through the reign of Alexander Severus (222-235), and served as a model for writings by Byzantine historians of a later era.

Gundicar
c. 385-437

Founder of the kingdom of Burgundy and an important figure in the transition from ancient to medieval times. In 413 Gundicar's Burgundians became one of the first nomadic tribes to settle down and establish a kingdom, which would maintain Burgundian lands in eastern France until 1477. An ally of Rome, Gundicar died in a battle against Attila the Hun (c. 400-453), and was later celebrated in numerous medieval legends.

Hannibal •
247-183 B.C.

Carthaginian general who is regarded as one of the greatest military commanders of the ancient world. Hannibal, whose father was the general of the Carthaginian army and who waged the First Punic War (264-241) against the Romans, succeeded his father and led the offensive against Rome in the Second Punic War (218-201). Hannibal led the army of Carthage, a city-state in northern Africa, to a series of remarkable victories against the Romans in Spain and Italy, but was eventually defeated. He is especially remembered for leading his army, with elephants, across the Alps.

Harsha
c. 590-647

Indian ruler who created the only stable kingdom in the densely populated northern part of the subcontinent between the fall of the Gupta Empire in 540 and the first Arab invasions nearly two centuries later. Though he was primarily a conqueror, Harsha established a realm in which the arts flourished. He fostered the Buddhist faith, serving as host to the Chinese pilgrim Hsüan-tsang (602-664), and wrote several plays. His empire, however, did not long outlast him.

Homer
fl. 900s-800s B.C.

Greek poet who is widely credited with writing the epic poems the *Iliad* and the *Odyssey.* Homer, who was said to be blind, is thought to have lived in Smyrna or Chios and earned a living by telling stories. The *Iliad* recounts the story of Achilles and the Trojan War, and the *Odyssey* relates the mythic wanderings of Odysseus.

Hsüan-tsang (Xuan Zang)
602-664

Chinese Buddhist monk who made an important pilgrimage to India. Eager to settle certain theological questions in the land where Buddhism had its origins, Hsüan-tsang set out for India in the 620s. Travel into China's western regions was forbidden under the T'ang Dynasty, so he slipped across the border, making a perilous journey across what is now southern Russia, Afghanistan, and Pakistan. He was the first Chinese traveler to see all the major regions of India, and he visited the courts of Harsha (c. 590-647) and other kings. Hsüan-tsang brought back Buddhist scriptures that helped lead to the expansion of the religion in China, and he remains a celebrated figure whose deeds are recorded in Chinese operas, paintings, films, and even comic books.

Flavius Josephus
37?-100?

Jewish historian who wrote extensively on the Roman occupation of Israel and the Middle East. Originally named Joseph ben Matthias, Josephus was a member of the Pharisees, a Jewish sect that practiced strict observance of the law, and later became a Roman citizen. He wrote not only on the current events of his time, but also, in his *Jewish Antiquities,* documented the history of the Jewish people beginning with Creation.

Juba II
c. 50 B.C.-A.D. 24

North African king of Numidia and Mauretania (modern Algeria and Morocco respectively) who sent an expedition to the Canary Islands. Son of Juba I (c. 85-46 B.C.), as a child he had been paraded through the streets of Rome after the defeat of his father. However, Octavian (the future Augustus Caesar; 63 B.C.- A.D. 14) befriended him and in 29 B.C. established him as ruler of Numidia, by then a Roman province. Four years later, Juba additionally became ruler over Mauretania, also a province of Rome. Husband of Cleopatra Selene—daughter of the famous Cleopatra VII (69-30 B.C.)—he wrote a number of scholarly works on history, geography, grammar, and drama. His expedition to the Canaries made the Romans aware of those islands, which had once been known to the Greeks.

Justinian
482-565

Byzantine emperor who sought to restore the Roman empire by reclaiming territories in Western Europe that had been taken by Germanic tribes. With his armies, under the command of General Belisarius, Justinian was able to reclaim North Africa as well as regions across southern Europe and Spain. Justinian also supervised the codification of Roman law in a new unified system, gathered as *Corpus juris civilis,* which became a cornerstone of European law.

Livy (Titus Livius)
59 B.C.-A.D. 17

Roman historian who wrote the most highly regarded history of Rome in the ancient period. His *History of Rome* chronicles events of the city-state from 753 B.C. to 9 B.C. It was published in 142 books, but only 35 of these survive in their entirety, while the others are available in fragments or summaries. Livy is known for his literary style rather than his accuracy: he incorporated the reports of others without critical examination.

Manetho
fl. c. 300 B.C.

Egyptian priest whose chronicle of his nation's history established the framework for the study of Egypt. His history, now lost, was in Greek, and probably had been commissioned by Ptolemy I (r. 305-282 B.C.). Elements of Manetho's work, including his tables of kings and their reigns, have survived in the writings of other historians. Manetho was the source for the division of Egyptian history into 30 dynastic periods, a system still used by Egyptologists today.

Marcus Aurelius
121-180

Roman emperor and philosopher who extended the knowledge of the Middle East and the area along Rome's northern frontier through defensive and offensive military campaigns in these regions. He was adopted by his uncle, Emperor Antoninus Pius, and succeeded him as emperor in 161. His reign brought to an end Rome's extended period of peace and prosperity known as *Pax Romana.* His writings, known collectively as *Meditations,* are significant expressions of Stoic philosophy.

Moses
fl. 1300s-1200s B.C.

Hebrew leader who liberated the Hebrews from slavery in Egypt, leading them on an epic journey through the Middle East to Canaan. The story of these exploits is contained in the biblical books of Exodus and Deuteronomy. He is regarded as the founder of the nation of Israel and the recorder of its first laws. A section of these laws, the Ten Commandments, was fundamental to the development of western civilization.

Nebuchadnezzar II
r. 605-562 B.C.

King of Babylonia who founded the Babylonian Empire and made it the primary military power in the Middle East. When his father, who was governor of Babylon, revolted against Assyria, Nebuchadnezzar became commander of the army. His father soon died and Nebuchadnezzar became king. Babylonia prospered under his rule. Extensive territory was added, and he completed numerous construction projects. Nebuchadnezzar's forces destroyed Jerusalem in 586 B.C. and took many Jews back to Babylon as slaves.

Necho II
r. 610-595 B.C.

Egyptian pharaoh who reportedly began construction of a canal between the Nile and the Red Sea, and who allegedly sponsored a voyage that circumnavigated the African continent. The latter, if indeed it happened, would have preceded the famous voyage of Hanno (fl. c. 500 B.C.) and his Carthaginian colonists by nearly a century. As for the canal, Herodotus (c. 484-c. 420 B.C.) wrote that Necho desisted from the project after an oracle warned him against continuing. He very nearly lost power by forming a disastrous alliance with the Assyrians against the Babylonians under Nebuchadnezzar II (c. 630-

562 B.C.), who defeated him at the Battle of Carchemish in 605 B.C. Nonetheless, Egypt kept the Babylonians out for a time, but eventually succumbed. Later, when the Babylonians themselves lost power to the Persians, Darius II (550-486 B.C.) completed Necho's canal.

Saint Paul
c. A.D. 2-62

Christian missionary and theologian whose accounts of his personal missionary travels and whose theology encouraged exploration during medieval and early modern eras. The first Christian theologian, his writings and journeys were especially directed to Gentiles (non-Jews) and were responsible for the spread of Christianity to Rome and further west into Europe. His emphasis on the importance of evangelical activity among non-Christians provided one of the most important rationales for subsequent exploration and conquest.

Piankhi
769-716 B.C.

Kushite king whose invasion of Egypt in the 720s B.C. led to the establishment of the Twenty-Fifth (Nubian) Dynasty of pharaohs. Perhaps the first well-known figure from sub-Saharan Africa, Piankhi was a devoted worshipper of the Egyptian god Amon, and invaded Egypt to prevent the Libyan Tefnakhte from seizing control of the country. He conducted brilliant military campaigns against the Libyans, but seemed uninterested in occupying the country. Four years after his death, his brother Shabaka (r. 716- 695 B.C.) established the Twenty-Fifth Dynasty (712-672 B.C.)

Plutarch
c. 46-119

Greek biographer and historian whose extensive writings on history, politics, religion, ethics, philosophy, and literature contributed significantly to the linking of Greek and Roman culture. He traveled in Egypt, Italy, and Greece, lecturing in Rome and visiting Athens often. Although a great champion of Greek civilization, he supported Rome's rule over Greece, believing it to have positive effects. His works are divided into two parts: *Parallel Lives* (biographies) and *Moralia* (essays on a wide variety of subjects.)

Procopius
d. 565?

Byzantine historian, born in what is now Israel, whose writing is one of the principal sources on the era of Justinian (r. 527- 565). Procopius served as advisor to Justinian's brilliant general

Belisarius (c. 500-565) in military campaigns from 527 to 540 against the Persians in what is now Iran; the Goths in Italy; and the Vandals in Spain and North Africa. He chronicled these events in *Procopius's History of Our Time,* and also wrote *On Buildings,* a six-volume work concerning structures erected under the reign of Justinian. Behind the scenes, however, Procopius was writing what came to be known as the *Secret History,* in which he portrayed Justinian and the empress Theodora (c. 500-548) as literal fiends in human form, while Belisarius appeared as a cuckold and a fool.

Ruth
fl. c. 700s B.C.

Moabite woman who married a Judaean living in Moab. After her husband died, Ruth traveled to Judea with her mother-in-law. She subsequently married another Judaean and became the great-grandmother of King David, and thus, a progenitor of Jesus. Her story in the Old Testament portrays early migration between nations and teaches the importance of the acceptance of foreigners, who may make valuable contributions to one's own nation and culture.

Sargon of Akkad
c. 2334-2279 B.C.

Mesopotamian ruler who created the first empire in history. Sargon came from the lower classes, and like Moses a thousand years later was said to have been rescued from a boat of reeds and pitch. He eventually seized power, becoming apparently the first Semitic ruler in his region. After taking most of the lands between what is now southern Turkey and the Persian Gulf, he symbolically washed his sword in the waters, a gesture that would be replicated by conquerors throughout ancient times. Sargon may have also established trading contacts with lands as far away as Oman, the Indus Valley, and Crete, but by 2150 B.C. the nomadic Gutians from the north had destroyed the Akkadian empire.

Suetonius
69-c. 122

Roman historian whose *De viris illustribus* (Concerning illustrious men) played a pivotal role in forming later generations' views of classical Rome. The work, which examines rulers from Julius Caesar (102-44 B.C.) through Domitian (r. A.D. 81-96), is full of compelling detail on the corruption and intrigue that characterized life at the top of Roman society. Suetonius, who wrote a number of other works—including, apparently, an encyclopedia—went to great pains to portray such specialized topics as games and pastimes, famous courtesans, and the growth of Rome's civil service.

Tacitus
c. 56-120

Roman historian and official whose works include *Germania* (A.D. 98), one of the few contemporary accounts of the Germans and Britons. Also important were his *Histories* (c. 109) and *Annals* (c. 116), which together comprise a chronicle of the empire during the 82 years from the death of Augustus (63 B.C.-A.D. 14) to the assassination of Domitian (r. 81-96). Though Tacitus is often regarded as the greatest of Roman historians, he sometimes let his conclusions guide his narrative, and is admired more as a commentator on morals than as a recorder of historical fact.

Saint Thomas
d. A.D. 53

Christian evangelist who was one of Jesus' original disciples and who traveled extensively as a missionary. He was called "Doubting Thomas" because he insisted on personally verifying Jesus' identity after the Resurrection. He traveled to Parthia and to India. Most of what is known about his later life is contained in apocryphal literature. It is said that he was martyred on St. Thomas Mount near Madras, India, after founding a church, which still exists there.

Vercingetorix
d. 46 B.C.

Gallic chieftain who, because of his famous conflict with Julius Caesar (102-44 B.C.), became one of the first "barbarian" figures known to history. In 52 B.C. Caesar had nearly completed the conquest of Gaul when Vercingetorix, chief of the tribe called the Arverni, led an uprising in what is now east-central France. Through skillful guerilla tactics, Vercingetorix lured Caesar into fighting in a position disadvantageous to the Romans. However, Caesar destroyed Vercingetorix's reserve army and successfully besieged the fortress, capturing the chieftain and bringing him back to Rome. Six years after he was marched through the city as an exhibit in Caesar's triumphal procession, Vercingetorix was executed.

Zenobia of Palmyra
c. 231-after 271

Syrian queen who built a short-lived empire that briefly challenged Roman hegemony in the East. Born Bat Zabbai, Zenobia was destined to become one of the most significant female rulers of antiquity. Her town of Palmyra served as an important trading post for caravans crossing the Syrian desert, and today its ruins reflect the powerful impact of Hellenism in the region. Following the death of her husband Odenathus, Zenobia ruled Palmyra as regent for her son Vaballath, and eventually began building a realm that included much of Egypt. The situation became so serious that the Roman emperor Aurelian (r. 270-275) personally came to Syria, defeated Zenobia, and brought her back to Rome in chains. Legend holds that the wily Zenobia was not executed, however, but married a Roman senator and lived out her days in comfort.

Bibliography of Primary Sources

Dio Cassius. *Romanika*. c. third century A.D. A valuable record of the latter years of the Roman republic and the early empire, beginning with the Roman people's mythical beginnings as descendants of Aeneas and chronicling their history through the reign of Alexander Severus. Drawn from Dio Cassius's experiences in various positions throughout the Roman world, from Africa to Asia Minor, the work, comprised of 80 books written in Greek, served as a model for later Byzantine historians.

Hecataeus. *Periegesis* (Tour round the world, sixth-fifth century B.C.). In this work, of which only fragments survive, Hecataeus examined the lands of Europe and what he called "Asia"—which actually included Egypt and other parts of North Africa. Apparently he had personally visited these regions, and his travels also included forays into Persia itself, as well as the lands surrounding the Black Sea. Hecataeus's systematic treatment of geography and the histories of peoples laid the groundwork for the encyclopedic approach applied by Herodotus and many others.

Herodotus. *History* [of the Greco-Persian Wars]. Fifth century B.C. An indispensable source of information about the ancient Mediterranean world and western Asia in the fifth century B.C. Written only 50 years after the end of the Greco-Persian Wars, and regarded as the first unified explanation of an historic event in prose, it is a major and generally reliable reference about this important conflict. Because his work is readable, full of fascinating stories and anecdotes as well as historically significant facts, it was circulated, copied, and recopied through the centuries.

Homer. *Iliad* and *Odyssey*. ninth-eighth century? B.C. These companion works, epic poems derived from ancient oral tradition, form the foundation of much of Western mythology and literature and provide a wealth of information about the customs, religion, warfare, geography, and technology of the ancient Greeks. The *Iliad* relates events surrounding the siege and fall of Troy, and the *Odyssey* involves the perilous travels of Odysseus as he attempts to return to his native land after the Trojan War.

Josephus, Flavius. *Jewish Antiquities*. First century A.D. Documents the history of the Jewish people beginning with Creation.

Lindisfarne Gospels. Late 600s. An illuminated manuscript of the Gospels (the first four books of the New Testament) that also contains writings by St. Jerome, Eusebius of Nicomedia, in addition to other texts. The Lindisfarne Gospels were most likely created for the island's bishop, Eadfrith. The manuscript, written in Latin, is historically important in its own right, but in the ninth century, after the monks of Lindisfarne had fled to nearby Durham, the Gospels were revised. Between the lines of the text, someone translated the Latin into the Northumbrian Saxon dialect, making them one of the first Bibles translated into a vernacular language during the Middle Ages.

Livy. *History of Rome* c. first century B.C.-1st century A.D. One of the most significant histories of ancient Rome. Livy chronicles events of the city-state from 753 B.C. to 9 B.C. The work was published in 142 books, of which only 35 survive in their entirety, while the others are available in fragments or summaries. Livy is known for his literary style rather than his accuracy, and he incorporated the reports of others without critical examination.

Megasthenes. *Indika*. c. 300 B.C. A four-volume work that for centuries provided Greeks with their most important eyewitness account of India's history and geography. Though the manuscript of the *Indika* has long ago disappeared, the many references to it in Strabo, Arrian, and others make it possible to gain some idea of the original.

Procopius. *Polemon* (Wars, sixth century A.D.). One of the principal sources on the era of Justinian (reigned 527-565). Procopius served as advisor to Belisarius, Justinian's general, in military campaigns from 527-40 against the Persians in what is now Iran; the Goths in Italy; and the Vandals in Spain and North Africa. Procopius also wrote the *Anecdota* (Secret history), a behind the scenes account of Justinian, the empress Theodora, and Belisarius.

Ptolemy. *Geographike hyphegesis* (Guide to geography, second century A.D.). An eight-volume work that provided a summary of all of the knowledge of geography at that time, including discussion of the basic principles of map and globe construction, locations of various cities, theories of mathematical geography, and instructions for preparing maps of the world. The maps and directions are often crude approximations derived from discussions with travelers, but they were accurate enough to show relative locations and direction. Though the work had little initial influence and was largely forgotten, it was rediscovered some 1,400 years later.

Pytheas of Massalia. *Peri Okeanou* (On the ocean, c. 300 B.C.). Contains Pytheas's account of his voyage to the North Atlantic, during which he circumnavigated Britain and may have visited Iceland. In addition to being the first to write on British ethnography, he was the first to document the midnight sun and to record the connection between the tides and the Moon. Though no longer extant, many ancient writers drew directly upon Pytheas's work or referred to it. From these scattered sources his route and discoveries have been reconstructed.

Suetonius. *De vita Caesarum* (Lives of the Caesars, second century A.D.). Provides compelling, often sensational, accounts of the first 11 Roman emperors, from Julius Caesar through Domitian. Seutonius's biographical sketches document the corruption and intrigue that characterized life at the top of Roman society, and played a pivotal role in forming later generations' views of classical Rome. The work also includes discussion of games and pastimes, famous courtesans, and the growth of Rome's civil service.

Strabo. *Geographica*. c. 7 B.C. Perhaps the greatest geographical text of antiquity, Strabo's work, comprised of 17 books, provides an overview of the world as it was known to the Greeks and Romans of his day. Vast in scope, the work is filled with the author's observations on philosophy, history, and science, all of which give the modern reader enormous insight into the ancient mind.

Tacitus. *Germania*. A.D. 98 One of the few accounts of the customs, manners, and organization of the ancient Germanic people in Europe during the early Roman Empire.

Tacitus. *Historiae* (Histories) and *Annals*. c. A.D. 100. Together these works comprise a chronicle of the Roman Empire during the 82 years from the death of Augustus (A.D. 14) to the assassination of Domitian (A.D. 96). Though Tacitus is often regarded as the greatest of Roman historians, he sometimes let his conclusions guide his narrative, and is admired more as a commentator on morals than as a recorder of historical fact.

Xenophon. *Anabasis*. c. 400 B.C. Recounts Cyrus the Younger's Persian campaign and Xenophon's role, after Cyrus's death, as leader of the "Ten Thousand" Greek mercenaries that retreated through territories in the Near East and Asia Minor to the Black Sea. This text, which represents one of the very few extant eyewitness accounts of the Persian Empire, also provides the first Western reference to the Carducians or Kurds.

JOSH LAUER

Life Sciences and Medicine

Chronology

c. 2650 B.C. Imhotep, multitalented Egyptian administrator and first historical figure of note, attempts to seek medical—rather than spiritual—causes for disease.

c. 1500 B.C. Medical practices based on the Vedic religion, a predecessor to Hinduism, make their appearance in India.

c. 700 B.C. Writing of the *Artharva-Veda,* a Hindu text containing quasi-scientific information on healing.

c. 500 B.C. The first known dissection of a human body is performed.

c. 400 B.C. Greek physician Hippocrates and his disciples establish a medical code of ethics, attribute disease to natural causes, and use diet and medication to restore the body.

c. 350 B.C. Aristotle establishes the disciplines of biology and comparative anatomy, and makes the first serious attempt to classify animals.

c. 300 B.C. Greek physician Herophilus identifies the brain as the primary organ of the nervous system, which he correctly associates with movement and sensation; and establishes a medical school in Alexandria, where the first accurate anatomical observations using dissection are made.

c. 250 B.C. Erasistratus, a Greek physician, correctly describes the division of the brain into a larger and smaller part; discovers the sinuses of the dura mater; and first notes cirrhosis of the liver.

c. A.D. 100 Aretaeus of Cappadocia, a Greek physician, distinguishes between mental and nervous diseases, and describes aspects of epilepsy.

c. A.D. 160-c. 200 Career of Greek physician Galen, last great doctor and scientist of antiquity, who founded experimental physiology; discovered the pineal gland and many other key ideas; and wrote numerous texts that remained authoritative for centuries.

c. A.D. 400 Christian noblewoman Fabiola establishes the first hospital in Western Europe, in Rome.

c. A.D. 650 Byzantine physician Paul of Aegina becomes the first to practice obstetrics.

Overview:
Life Sciences and Medicine 2000 B.C.-A.D. 699

Previous Period

The achievements of ancient peoples are difficult to document because few written records have survived from the period before 2000 B.C. Most of the available evidence comes from archaeological investigations of the physical remains of various cultures. For example, paintings found on cave walls, dating back some 30,000 years, reveal that prehistoric artists were close observers of animals and knew a great deal about animal anatomy. Also, the discovery of punctured human skulls that display regrown bone seems to indicate that medical practice was sophisticated to the extent that surgery, perhaps to relieve pressure on the brain after an injury, was successful; at least the patient survived long enough for the wound to heal and for bone regrowth to occur. But evidence of such achievements is admittedly quite meager.

Surgery

Even after 2000 B.C. the written record remains sparse and very incomplete. The few documents that have been preserved are usually fragmentary, so it is difficult to measure the full extent of the biological knowledge and medical accomplishments of this period. Nevertheless, historians have been able to construct a picture of how ancient civilizations viewed the living world and how these civilizations dealt with issues of health and disease. For example, knowledge of Egyptian medicine comes from both physical evidence and written sources. The way mummies were prepared, with the removal of the internal organs (where putrefaction usually originates), suggests that there was some understanding of internal human anatomy. A papyrus manuscript from about 1550 B.C. indicates that the Egyptians had rather sophisticated surgical practices, and the same seems true of the Babylonians, some of whose clay tablets contain instructions on surgery. Ancient surgical instruments have also been found at a number of sites excavated by archaeologists.

Hippocratic Medicine

The most important physician of the ancient world is thought to have been the Greek doctor Hippocrates (460?-377? B.C.). While there is evidence that such an individual existed, it is less

clear that he was the author of all the writings attributed to him, including the Hippocratic Oath, a vow to practice medicine responsibly, that is still taken by doctors when they begin their careers. Some historians point to indications that the basis of some Hippocratic writings, including the oath, lies in the earlier work of Pythagoras (582?-500? B.C.) and his followers. In any case, these documents record one of the most complete descriptions of medical practice in the ancient world, including the concept of four humors or fluids in the body that control health. They were blood, phlegm, yellow bile, and black bile, which were said to originate in the heart, brain, liver, and spleen, respectively. Good health resulted when there was a balance among the four, and disease was thought to be due to an excess or deficiency in one of these humors. This concept of the controlling role of the body's humors was one of the most influential in ancient medicine and was still the dominant view of how the body functioned well into the Middle Ages.

A major focus of Hippocratic medicine was on diet, as it was recognized that there is a clear relationship between diet and health, both in maintaining good health and in curing illness. Other cultures also evolved medical practices based on rules about diet. Among the most thorough of these were the Hebrew dietary laws, which were set down in the first books of the Old Testament. Prohibitions against eating such foods as pork and shellfish were combined with careful instructions on how food was to be prepared. The basis of these regulations was good sanitation and the avoidance of foods that often carried parasites or other infectious organisms.

After the Hippocratic writings, there was little added to medical practice until the time of Galen (129-199?) in the first century A.D. Once the influence of the dogmatists (the immediate followers of Hippocrates) waned, the center of Greek civilization and medicine shifted to the Egyptian city of Alexandria. There several schools of medicine succeeded one another, from the empiricists, who argued for direct observation as the focus of medical practice, to the methodists of the last century B.C., who reduced medicine to a few simple methods. One of the most influential of the later Greek anatomists and physicians was Galen, whose writings were preserved and dogmatically accepted until the Renaissance. His descriptions of human anatomy, many of which were based on dissections of animals and therefore were incorrect, led anatomists astray for centuries.

The Romans

One area in which Roman physicians did excel and made advances was in military medicine. The far-flung Roman Empire was supported by its large armies, whose physicians developed sophisticated techniques for the treatment of wounds. The surgical practices they originated were passed on through the centuries, and were not improved upon until the Renaissance.

The Romans were also the first to develop institutions similar to hospitals. These arose first as places to harbor sick slaves, but eventually grew into facilities where Roman citizens also sought medical care. Called infirmaries or *valetudinaria,* they became sites where medical care could be provided on a continuing basis. They were also set up during military operations to service sick and wounded soldiers.

Medicine in the East

In the East, the Chinese had developed acupuncture techniques at least as early as 2700 B.C. The use of acupuncture continued throughout this period and, in fact, is still practiced up to the present day. The Chinese also perfected a sophisticated use of plants in medical practice. In India, Brahmanic or Ayurvedic medicine was the high point of ancient Hindu medical practice. Charaka, Susruta, and Vagbhata are considered the three great Hindu healers of this tradition, which included advice on the use of herbal medicines as well as recommendations on diet, surgical procedures, and other therapies. Later, during the time of the Gupta Empire (A.D. 320-540), Indian physicians used inoculation as a way to prevent against the spread of infectious disease.

The Beginnings of Biology

In the ancient world, as this summary indicates, the major focus of the life sciences was on medical issues. It was primarily in Greece that interest went beyond this practical concern, as philosophers began asking questions about the functions of organisms and their relationships to each other. The most noted figure in this regard was Aristotle (384?-323 B.C.). As with so many of the great minds of the ancient world, only a small fraction of his writings has survived. Almost a quarter of these deal with biology, including a book on the classification of animals, which was the first serious attempt to organize species with related characteristics. In all, Aristotle identified over 500 animal species, while also developing theories on reproduction and the inheritance of various traits. His work was based on close observation and was so masterful that it dominated biological writing until the Renaissance.

Although there is evidence that Aristotle also wrote about plants, these writings have been lost. However two botanical works by his student Theophrastus (371?-287? B.C.) have been preserved. Theophrastus was the most noted botanist of the ancient world; his writings contain his observations on the structure and classification of plants as well as on their use in medicine. His knowledge was passed on to the Romans, and it was through Roman texts, called herbals, that this botanical information was handed down through the Middle Ages.

The Future

The period covered in this volume ends in what has been termed the Dark Ages, the early part of the Middle Ages during which there was little scientific or medical progress, and when the traditions and achievements of the past were only dimly understood and followed. As the Dark Ages waned at the end of the millennium, there was a rediscovery of the knowledge of the ancient world. Even with all its inaccuracies and limitations, this knowledge eventually served to spark a renewed interest in the natural world and led ultimately to the dawn of modern science and medicine in the Renaissance.

ROBERT HENDRICK

Doctors, Drugs, and Death in Ancient Egypt

Overview

Ancient Egyptian civilization flourished in the fertile valley of the Nile. Physicians flourished there as well, especially in the centuries between 2000 and 1200 B.C. Much of Egyptian medicine was based on little more than superstition. However, doctors gained some skill at treating wounds, and documents from the time refer to hundreds of different drugs that were then in use. Many of these were likely useless or even harmful, but some have proven to be effective at preventing bacterial infection. The Egyptians also developed skill at fighting bacteria beyond the grave (even though they had no knowledge of these microorganisms). Their mummification process helped to protect corpses from decay—a ritual that not only involved biology but also religion and politics as well. Some of the concepts of Egyptian medicine would live on in later civilizations, such as those of ancient Greece and Rome.

Background

One of the ways scientists and historians have learned about ancient Egyptian medicine is by studying the writings their physicians left behind. The Egyptians were one of the first civilizations to develop a system of writing. Using the fibers of a plant called papyrus, they made a parchment-like paper. A scroll of this ancient paper is also called a papyrus (plural, papyri). Six papyri have been discovered that focus on medicine.

The most important of these is called the Edwin Smith Papyrus. Edwin Smith (1822-1906) was an American Egyptologist who purchased the ancient papyrus in 1862. He made an attempt to translate it, but its importance was not fully realized until 1930 when it was fully translated by James Breasted (1865-1935) at the University of Chicago. The papyrus itself dates from about 1550 B.C., but it is presumed to be a copy of an even earlier work.

The central focus of the Smith papyrus is surgery, specifically in regards to wounds and fractures (broken bones). Part of this papyrus is divided into sections. Each section is organized into four parts in a surprisingly scientific fashion: title, examination, diagnosis, and treatment. The title states a specific type of wound or condition, such as a gash in a patient's forehead. The examination lists the type of symptoms the physician should look for regarding this type of wound. The diagnosis is the writer's opinion of the outcome; this might be favorable or uncertain (in which case the writer thinks the physician should attempt to help the patient) or unfavorable (in which case the writer considers the situation to be hopeless). The treatment portion gives specific instructions regarding surgical techniques and the type of medicines to be used in healing the patient.

Many of the cases discussed in the Smith Papyrus involve the closing of wounds. One method of doing so might have involved the use of stitches. However, the earliest known stitches found in a body have been dated to about 500 years after the papyrus was probably composed. In addition, these stitches had been placed in the body not by a physician, but by an embalmer after the removal of the body's internal organs. There is also evidence in the papyrus to suggest that Egyptian physicians used adhesive strips to close wounds. Doctors applied gum obtained from acacia trees to thin bands of linen. One end of these sticky tapes would be placed on either side of the wound, holding it together—a very early form of Band-Aid. Such tapes would have helped to prevent infection. Pus could leak freely from the wound, and no foreign body (such as thread used for stitches) would have to be inserted in the patient's flesh.

Before about 1500 B.C., iron blades were very rare (and very expensive) in Egypt. Thus, knives, when used by physicians, were usually made of stone. However, Egyptian surgeons did very little cutting, since they were normally repairing a cut or tear rather than making one of their own. Physicians also used blades made of sharpened bamboo. Bamboo cutting tools were inexpensive and could be thrown away after a single use: as such, they were perhaps some of the world's first disposable blades.

A second ancient Egyptian medical document is the Ebers Papyrus. Edwin Smith also purchased it in 1869, but he sold it to George Ebers in 1872. Like the Smith papyrus, it is dated to about 1500 B.C., but may be much older. It consists of a collection of medical texts arranged in a somewhat random order. A large portion is devoted to the wide variety of drugs

Egyptian physicians used to treat their patients. The sources of these drugs included plants, animals, and minerals.

One of the treatments described in the Ebers Papyrus is using a heated knife to treat a "vessel-tumor"—perhaps a swelling in an artery or vein. As the physician cut away the swelling, the heat from the knife would burn and seal the blood vessel, keeping it from bleeding. This may represent one of the world's first surgical attempts to stop bleeding.

Clues to the Egyptians' understanding of biology can also be found by studying their burial practices. The heat and dryness of the Egyptian climate can naturally preserve bodies when they are buried in sand. However, once the Egyptians began laying bodies in tombs, the reduction in temperature and increase in humidity allowed the natural process of decay to proceed. Since the Egyptians believed that life after death was possible only if the body could be preserved, such decay was a serious problem. To combat it, they developed an elaborate process of mummification.

A major obstacle faced by the Egyptian embalmers was stopping the action of bacteria. One way to do so is to remove water from the body, which bacteria need to survive, or to use a chemical called a fixative agent, which prevents the growth of living things. Egyptian embalmers possibly could have used vinegar—a readily available chemical—as a fixative, but the idea of facing eternity as a human pickle probably wasn't overly attractive. For similar reasons, they probably also chose not to use salt as a drying agent.

Instead, they used a combination of drying and fixing. In the first step of the most expensive type of mummification, embalmers removed the internal organs (except for the heart) and then pulled the brain out through the nose with iron hooks. These organs were placed in jars that would be sealed with the body inside the tomb. The heart was left in place because it was considered to be the center of intelligence (as opposed to the brain) and would therefore be useful in the afterlife.

Next, the embalmers treated the body with natron, a drying agent composed of a mixture of sodium bicarbonate (or baking soda) and sodium carbonate. The Egyptians collected this chemical when shallow ponds near the Nile evaporated after annual floods, leaving behind a crust of natron. The embalmers would pack the body's internal cavities with bags of the drying agent, cover the body with more natron, and

leave it for a period of 40 days. This method rather inefficiently preserved the body, since the natron would only have been effective to a depth of a few millimeters.

Next, the embalmers packed mud, sand, or linen under the skin to fill it out, and then wrapped the body in layers of linen. Resins, or tree saps, such as frankincense and myrrh were

TAILS FROM THE CRYPT

The ancient Egyptians not only mummified humans; they also mummified animals. Many of these animal mummies were intended as religious offerings. Each Egyptian god and goddess was represented by or associated with a particular animal. For example, Anubis, the god of the dead, was depicted with the head of a jackal. One way the Egyptians worshipped their gods was to offer a sacrifice of the animals connected with them. Once the animal was killed (usually by a priest), the worshipper would pay to have the animal mummified. The same embalming procedures used for humans were used for animals. The mummy would then be left at a temple, and it would later be buried at a special animal cemetery.

Because of the wide variety of Egyptian gods, the Egyptians mummified a wide variety of animals: everything from cats to gazelles, from wild dogs to crocodiles. The practice of animal mummification grew quite widespread. For example, scientists uncovered one animal cemetery that contained more than a million bird mummies. At the city of Bubastis, where the cat goddess Bastet was worshipped, priests killed hundreds of kittens each day to be used as mummy offerings.

Not all animal mummies were religious sacrifices, however. Wealthy Egyptians would also pay embalmers to mummify beloved pets that died of natural causes. It was believed that pets would experience the afterlife if (and only if) they were mummified.

STACEY R. MURRAY

heated and poured over each layer. The resins acted as a fixative agent and helped to preserve the outer layer of skin, much like amber can preserve the remains of insects for millions of years. This process was quite costly since none of the resins came from Egypt itself.

The Egyptians had a great fear of decay, not only from death and disease, but also from normal processes within the body. For instance, they

Female druggists making herbal medicine in ancient Egypt. *(Bettmann/Corbis. Reproduced with permission.)*

considered the breakdown of food in the intestines to be a form of "internal decay" that could become a source of disease. The Egyptians had only a basic understanding of the body's vascular system—the system of blood vessels that connect all parts of the body. They correctly realized that blood vessels were connected to the heart. However, they mistakenly thought that the anus was also a center of the vascular system. Thus, they believed that if the intestines became overfull, excrement would leak into the other vessels, spreading throughout the body and resulting in illness. To keep this from happening, the Egyptians made frequent use of enemas—injecting liquid into the intestines via the anus.

Impact

By the time of what are known as the Middle and New Kingdoms of Egypt (2060-1070 B.C.),

cities were large enough that doctors were able to compare similar cases. Physicians not only made observations of various diseases and discussed them with one another, but also recorded information about them. Eventually, once enough data had been accumulated, they were able to see patterns and make diagnoses, something that would not have been possible in small, widely separated villages. One example of the importance of doctors in the cities of ancient Egypt is that in addition to general physicians, there were also specialists, especially in the royal court. For instance, there were special physicians for the eyes, the teeth, and the stomach. There was even a doctor given the title of "Shepherd of the Anus." The growth of cities also gave physicians a much larger network for obtaining drugs through trade, often with nations outside of Egypt.

Ancient Egyptians embalming a corpse as part of the process of mummification. *(Bettmann/Corbis. Reproduced with permission.)*

The great variety of Egyptian drugs indicates the degree to which Egyptian civilization was involved in trade. They imported medicines from civilizations on the Mediterranean, from other regions of Africa, and from Asia (perhaps even from as far away as China). In all, about 700 drugs are referred to in ancient Egyptian writings. Some of the more exotic drugs listed in the medical papyri include hippopotamus fat, fried mice, pigs' brains, and tortoise gall. However, it is not known if some of these names are meant literally or not. (For example, an insect that we call a caterpillar has nothing to do with cats or with pillars. Likewise, an Egyptian drug called tortoise gall may in fact be a type of herb and have nothing at all to do with tortoises.) The identities of most Egyptian drugs remain a mystery. In many cases, it can be determined that a drug came from a plant, but the exact species of plant cannot be identified. Even in those instances where the identity of a drug is known, few have been tested in modern experiments, so it is not certain how effective they were.

Egyptian medicine was closely related to Egyptian religion. In some cases, they are so intertwined that it is impossible to know which came first—the religious or the medical uses of a chemical. Take, for example, Egyptian eye paint. For centuries, the Egyptians colored their eyelids with either black or green pigments. The instruments used to prepare these pigments were called by the Egyptian word that means "to protect." The Egyptians likely believed that the paint would protect them from eye diseases that were common in Egypt's dry and dusty climate. It was so valued that it was offered to the gods in sacrifice, and even the eyes of religious statues were lined with it. In addition, physicians applied the pigments directly to wounds to aid in healing.

It turns out that the belief in the protective powers of eye shadow actually had a medical basis. The Egyptians obtained their green pigments from copper-containing compounds. In large enough doses, copper can be poisonous to people, and it is also toxic to bacteria although in much smaller quantities. Modern experiments have shown that the copper-compounds used by the Egyptians would have had at least some effectiveness in preventing infections caused by bacteria.

Religion also played an important role in the Egyptian concept of sickness. They believed that most illnesses, especially those with unknown causes, were the results of evil spirits that had been sent by their gods as a punishment for bad behavior. Therefore, many of the methods described for curing disease in the Egyptian medical papyri relied on the supernatural. There were two basic parts to these magical spells: a chant and a ritual. The words of the chant called on the evil spirit to leave the patient, while the ritual might involve burning patients or stabbing them with needles in order to drive the demon out. That the ritual would harm the patient as well as the disease-causing spirit was seen as an unfortunate but necessary side effect. Appeals to magic rather than medicine were especially common when the outcome was likely to be death, such as in the case of poisonous snake bites or scorpion stings.

A contrasting view, however, is seen in the Edwin Smith Papyrus. In the large majority of the surgical cases it describes, there is no mention of magic or evil spirits. Instead, the treatments directly deal with observable symptoms. One reason for this may have been that the focus of the papyrus is on visible wounds. Therefore, the source of the wound would clearly be natural (a crocodile bite, for example) rather than supernatural (the influence of an evil spirit). This may mark a small, but early first step in the separation of medicine and magic. (Based on the Ebers Papyrus, however, physicians would recite chants even while applying or removing bandages on fairly minor wounds.)

The mummification ceremony also had powerful religious significance for the Egyptian people. The burial rituals for a king, for example, represented his transformation into a god. This in turn established the king's son—the next ruler—as the son of a god, helping to ensure a smooth transfer of power. Thus, it was of great importance that the embalmers prevent the king's body from decaying—an event that had the potential to upset the entire religious and political order.

Embalming and healing were two distinctly different occupations practiced in different locations in Egyptian cities, and it is not believed that embalmers' knowledge of anatomy was passed on to doctors. In fact, any knowledge of anatomy that the embalmers did possess was probably not great, since they removed the internal organs of dead bodies through small slits. Egyptian hieroglyphs show human figures with the internal organs of animals, suggesting that physicians studied anatomy by examining the animals sacrificed in temples rather than by dissecting human corpses. However, since the Egyptians were not forbidden to touch a dead body the way people in some other cultures of the time were, their practice of embalming may have influenced later civilizations to pursue the study of anatomy through dissection.

Egyptian medicine and biology influenced other cultures in other ways as well. A few of their practices would even reappear centuries later. For instance, European doctors began using adhesive tapes to close wounds during the nineteenth century. At the time, the rates of infection from using stitches were so high that surgery of any type had become quite dangerous. Adhesive tapes are still used today to close certain types of wounds.

Another practice that was to reappear hundreds of years later was the use of honey as a medicine for wounds. Honey was the most popular Egyptian drug; there are more references to it in the medical papyri than to any other. One reason the Egyptians may have chosen to use honey in the first place is that, despite its stickiness, it keeps wounds from adhering to bandages.

Honey helps to prevent the growth of bacteria—a main cause of infection. A 2,500-year-old sample of honey unearthed from an Egyptian tomb, for example, showed almost no signs of decay. Honey (like natron) acts as a drying agent, drawing water from cells and causing them to die. In addition, certain chemicals in honey have antibacterial effects. The use of honey as a medicine spread to other ancient cultures such as those of Greece and Rome. Centuries later, honey was used in China during World War II to treat minor wounds when other medicines were not available. Today, scientists are investigating the effectiveness of honey in fighting certain types of bacteria that are resistant to standard antibiotic drugs.

Not all of the Egyptians' practices that lasted were beneficial, however. One such practice involved placing substances on wounds that would cause evil spirits to flee from the patient in revulsion. Such treatment included rubbing excrement (especially from crocodiles) onto wounds, a practice that could not have been very helpful medically. However, during outbreaks of the black plague in Europe during the Middle Ages, similar behavior was common. Many associated this deadly sickness with a force of supernatural evil. Some believed that by surrounding themselves with foul odors (such as by rubbing excrement on themselves or by keeping dead animals in their homes), they could keep the evil spirit from entering their bodies and making them ill.

The Egyptians' idea of an imbalance in the blood resulting in disease (based in part on their mistaken notion of the anatomy of the vascular system) would also last for hundreds of years, eventually leading to the common practice of bloodletting in many different cultures. The ancient Greek and Romans, for example, believed that health was determined by the balance of four humors, or fluids, in the body. It was thought that an excess of a particular humor,

such as blood, would lead to disease. To remove the excess fluid, physicians placed leeches on patients or cut open one of their veins, often letting them bleed until they went into shock from blood loss. The practice was continued until the end of the Middle Ages and was revived again in the eighteenth century, hastening the deaths of countless patients.

Despite their failures, Egyptian physicians made remarkable progress given their knowledge and technology. Many of their medical practices were passed along through trade routes, eventually reaching Assyria and Babylonia in the Middle East and Greece and Rome in the Mediterranean. Their work helped to form a base from which future medical advancements could be made.

STACEY R. MURRAY

Further Reading

Majno, Guido. *The Healing Hand: Man and Wound in the Ancient World*. Cambridge, MA: Harvard University Press, 1975.

McGrew, Roderick E. *Encyclopedia of Medical History*. New York: McGraw-Hill Book Company, 1985.

Mertz, Barbara. *Temples, Tombs, and Hieroglyphs: A Popular History of Ancient Egypt*. Revised ed. New York: Dodd, Mead & Company, 1978.

Parker, Roy, ed. *The Cambridge Illustrated History of Medicine*. Cambridge: Cambridge University Press, 1996.

Acupuncture in China

Overview

Acupuncture, the insertion of thin needles into specific points of the body in order to relieve pain or treat illness, is an ancient Chinese art first developed more than 4,000 years ago. Acupuncture was used to help restore a healthy balance of energy throughout the body and, along with herbal remedies, was the cornerstone treatment offered by traditional Chinese medicine. Although the origin of acupuncture predates written history, acupuncture's evolution through the ages remains grounded in the ancient Chinese naturalistic philosophies of Taoism.

Background

Evidence of practices similar to acupuncture are found in Chinese relics dating back to the New

Stone Age (8000-2000 B.C.). During this time the first acupuncture needles were crudely shaped and made from stone. Known as bian stones, they were used to apply pressure to the body, treat wounds, and lance infectious lesions. Later, needles replaced the bian stones. The first written reference to acupuncture is found in *The Yellow Emperor's Inner Cannon of Medicine* (*Nei Ching Sue Wen*) written in approximately 200 B.C., but assumed to include writings from much earlier times. There are numerous sections in *Nei Ching Sue Wen*, addressing core subjects of medicine such as the physiological constitution of the body, diagnosis, fevers, and treatments (including acupuncture) based on the wisdom and experience of earlier sages. The work contains a dialogue between the ancient "yellow emperor" Huang-yi (c. 2650 B.C.) and his chief minister that sets down the philosophical basis of traditional Chinese medicine embraced by acupuncture.

Traditional Chinese medicine holds that the body contains an essential life energy known as *qi* (pronounced "chee"). Chinese language scholars maintain that the concept of *qi* is greater than any type of energy described by Western science. *Qi* was described in the *Nei Ching Sue Wen* as being responsible for change, movement, and life itself. All mental and physical aspects of life were made possible by the flow of *qi* throughout the body. *Qi* flowed through the body in channels or meridians which were arranged in a mostly symmetrical pattern. There were 14 main meridians running up and down the length of the body. Early references portray the meridians as nonphysical, unique energy pathways in the body. When the pathways became obstructed or inefficient, the flow of *qi* was interrupted and an imbalance in the body could occur, resulting in illness. Traditional Chinese physicians practiced acupuncture to restore that balance.

Qi was composed of two parts, *yin* and *yang*. The concepts of *yin* and *yang*, both fundamental to Taoist philosophy, can be used to describe all that is in the universe, including every part and function of the body. *Yin* and *yang* are complimentary opposites that, when balanced, work together to form a whole. *Yin* and *yang* are interdependent, and their relationship was used to describe the fluctuating balance in nature. Man, for example, (masculine attributes are synonymous with *yang*) is dependent upon woman for his birth, while woman (female attributes are associated with *yin*) is dependent upon man to conceive. In the body, the *yang* element represents the capacity for action and transformation. The *yin* element represents the capacity for circulation, nourishment, and growth. The Taoist concept of health emphasizes the aspiration to attain perfect harmony between the opposing forces of the natural world, between *yin* and *yang*. Any imbalance of the *yin* or *yang* elements of the body could interrupt the flow of *qi*, bringing pain or disease. With his acupuncture needles, the Chinese physician restored the balance of *yin* and *yang*, and the flow of *qi*.

When considering acupuncture treatment for a patient, the Chinese physician first took time to ask questions and record specific physical observations of the patient, much like the patient history of modern times. The physician used his five senses to gather much of the information relevant to his diagnosis. Most afflictions were considered internal ones (even skin disorders and some injuries), thereby lending themselves to acupuncture treatment. Needles were inserted at acupuncture points along the meridians. Acupuncture points are specific locations where meridians come to the surface of the skin and are easily accessible with acupuncture needles. The needles were inserted at various angles and depths, depending upon the desired effect. Often the needles were twisted, tapped, or warmed with burning wool placed at the tip. A paste made of the Chinese mugwort plant was heated and placed at the insertion site in the process of moxibustion (the burning of small cones of dried leaves on certain points of the body). Frequently, acupuncture, moxibustion, and herbal remedies were prescribed to be carried out at the same time.

Impact

Events during the "Warring States" period of Chinese history (475-221 B.C.) helped to secure the practice of acupuncture as a mainstay of Chinese medicine. Taoism and Confucianism both became mainstream ideologies and exerted great influence on Chinese thought. Confucianism held that the body was sacred as a whole, and that it was vital to present oneself to one's ancestors intact upon death. Amputation and execution by decapitation, therefore, held an even greater fear than death. Acupuncture offered a logical method of treating internal disease while maintaining an intact body. Taoism was a more passive philosophy, and its balance of the *yin* and *yang* coincided with the medical concept of acupuncture for maintaining the balance and flow of *qi* for health. Throughout the periodic wars of the next 700 years which established and strengthened the feudal system in China, acupuncture proved an easily portable method of medical treatment during turmoil.

As the ancient practice of acupuncture found an enduring position in Chinese traditional medicine, the tools to administer it improved as technological skills developed. Crude bian stones gave way to more skillfully shaped needles made of pottery. With the development of metallurgical techniques, the metal needle superseded the pottery needles and bian stones. Metal needles (mostly iron) originated in about 100 B.C., and greatly enhanced the practice of acupuncture. Metal needles could be refined for a specific purpose, and several distinct designs were created. In the excavated tomb of the Prince of Chungshan, dating from about 115 B.C., nine distinctly designed needles were unearthed, some made of silver, others made of gold. Needles that followed eventually took one of the nine classic forms found in the tomb: the arrowhead needle for pricking just beneath the skin; the sharp, round needle for rapid or repetitive pricking; a needle shaped specifically for draining abscesses; the multi-edged needle for piercing a vein; the longer needle for deep muscles; the large needle for the joints; a blunt-edged needle for applying small pressure points; the round needle for massaging; and the filliform needle. New acupuncture points and channels were found to accommodate the improved needles. By the end of the third century A.D., nearly all of the major channels and almost 400 insertion points had been identified. The basics of acupuncture therapy were solidified, and (with the exception of further needle technology and insertion points) acupuncture remained essentially unchanged for 1,500 years.

From approximately 220 B.C., traditional Chinese medicine offered a nature-based explanation to illness, rather than contributing illness to evil spirits or other supernatural causes. *Qi* was perceived as a natural substance, and acupuncture was a rational method of maintaining and restoring its balance. Personal responsibility for keeping the body's *qi* in balance came into favor, as the Chinese adopted a long-lasting culture of disciplined moderation in the everyday ways of life. Through acupuncture and moderation in diet, physical activity, sexual practices,

and spiritual meditation, the Chinese practiced some of the world's first preventive medicine. The Chinese concepts of *yin* and *yang* made Chinese medicine the first holistic medicine, as no ailing function or part of the body was considered separate from the whole person.

Ultimately, traditional Chinese medicine took a different branch in its evolutionary process than Western medicine. While both Eastern and Western medicine embarked upon a system that sought its answers in nature at approximately the same time (approximately 200 B.C.), Western medicine eventually sought facts brought about by the scientific method, while Eastern medicine continued with a holistic approach. The upholding of traditional Chinese values and ancient texts of medicine and philosophy ensured acupuncture's place in Chinese medicine through modern times.

BRENDA WILMOTH LERNER

Further Reading

Beinfield, Harriet, and Efrem Korngold. *Between Heaven and Earth: A Guide to Chinese Medicine.* New York: Ballantine Books, 1991.

Eckman, Peter. *In the Footsteps of the Yellow Emperor: Tracing the History of Traditional Acupuncture.* San Francisco: Cypress Books, 1996.

Ni, Maoshing, trans. *The Emperor's Classic of Internal Medicine.* Boston: Shambala Press, 1995.

Unschuld, Paul. *Medicine in China: A History of Ideas.* Berkeley: University of California Press, 1985.

Herbal Medicine

Overview

Herbal or botanical medicine employs roots, leaves, and barks as drugs for the treatment of disease. The medicinal use of herbs and other botanical products is probably as old as medical treatment itself. A common belief throughout history is that nature provides plants in each region that are appropriate for the cure of local diseases. Drug collectors and healers in many cultures used "herbals"—manuals that provide guidance in the identification of medicinal plants and recipes for preparing remedies. Healers in ancient Asia, India, Mesopotamia, Greece, and Rome employed hundreds of medicinal plants. With the invention of the printing press, ancient herbals and their descendants became

widely available. Even in the ancient world, the search for new medicinal herbs played an important role in exploration.

Background

In very early times, in virtually every part of the world, the most important uses of herbs and spices were medicinal, either for internal use or in ointments, balms, and poultices. Ancient herbals, including those of China, India, Sumer, Assyria, Egypt, Greece, and Rome, testify to the widespread use of spices and herbs in the treatment of disease. Herbs and spices retained their medicinal reputation throughout history; their curative virtues are still highly respected, especially in Asia and India.

Traditionally, herbal medicine employed herbs and spices, often in combination with animal parts and products and minerals. Many medicinal herbs and spices are cultivated for their aromatic, pungent, or otherwise desirable qualities. Often the important parts are dried for storage and to concentrate the valuable components. Spices and herbs consist of rhizomes, bulbs, barks, flower buds, stigmas, fruits, seeds, and leaves. Herbs are the fragrant leaves of plants such as marjoram, mint, rosemary, and thyme.

More than any other culture, China has maintained its traditional medicine, especially its rich drug lore, largely based on herbal remedies. The herbalists of China may have studied and employed as many as 5,000 plants. When Li Shih Chen (1518-1593), China's "prince of pharmacists," published his great *Materia Medica (Pen ts'ao kang mu)* in 1578, his scholarly compilation of Chinese herbal lore contained 1,892 drugs from the vegetable, animal, and mineral kingdoms, and more than 8,000 prescriptions. Today, Chinese scientists are attempting to isolate specific active ingredients from traditional remedies.

Shen Nung, one of the Three Celestial Emperors revered as the founders of Chinese civilization, is also known as the "Divine Peasant." Shen Nung is said to have personally tasted "the hundred herbs" so that he could teach the people which were therapeutic. His findings were allegedly recorded in the first pharmacopoeia, the *Pen-ts'ao,* or *Great Herbal.* Huang-Ti, the last of the Celestial Emperors, is considered the author of the *Nei Ching,* or *The Yellow Emperor's Classic of Internal Medicine*, a text that has inspired and guided Chinese medical thought for over 2,500 years. According to the *Nei Ching,* the first remedies were found among the herbs, trees, plants, and animals that served as foods. The use of tea, a beverage made from the leaves of the tea shrub, illustrates the overlap between "foods" and "drugs." Tea contains small amounts of nutrients, but it is rich in physiologically active alkaloids (caffeine, theobromine, and theophylline). The three classes of drugs—vegetable, animal, and mineral—were said to correspond to heaven, man, and earth. Animal parts and products were regarded as sources of remarkable "vital principles." Traditional remedies included seaweed and sea horse powder (good sources of iodine and iron) for goiter and chronic fatigue, and ephedra for lung diseases and asthma. Ginseng, the "queen of medicinal herbs," was credited with almost miraculous powers.

Medical therapy can take two general forms: doctors can try to strengthen the body so that it can heal and defend itself, or they can attack the agents of disease directly. The primary goal of Chinese herbal medicine was to strengthen and protect the body, restore its normal balance of energy, and promote longevity. Ginseng exemplifies the classical Chinese approach to healing. It has been used as a tonic, a rejuvenator, and an aphrodisiac. Modern researchers have called it an "adaptogen," a substance that increases resistance to all forms of stress, from disease to misfortune. Many other cultures have adopted similar ideas about herbal remedies.

Impact

Like Chinese medicine, Indian medicine attempted to prolong life, preserve health, and prevent disease. Ayurveda, the learned system that forms the basis of the traditional medicine that is still widely practiced in India today, is known as "the science of life." Herbal medicines and dietary regulations are of special importance in Ayurvedic medicine.

The richly diverse flora and fauna of India provided a wealth of medicinal substances. Almost 1,000 medicinal herbs are referred to in the major medical classics of ancient Indian civilization, but many are unidentifiable materials or "divine drugs" such as "soma." Although plants provided the majority of medicinal substances, minerals and animal products, such as honey, milk, snake skin, and excrements, are also well represented. Because of the importance of using ingredients that were pure and unadulterated, and herbs that were harvested at auspicious times, the wise physician compounded his own drugs from ingredients that he had gathered himself.

Unlike the learned medical systems of China and India, those of ancient Mesopotamia are no longer extant, but ancient drug lore was not entirely lost. Many civilizations, including those known as Sumerian, Chaldean, Assyrian, and Babylonian, once flourished in Mesopotamia, the land between the Tigris and Euphrates Rivers in modern Iraq. One of the oldest known pharmaceutical documents is a clay tablet probably inscribed about 4,000 years ago by a Sumerian scholar. The tablet contains a series of drug formulas and suggests considerable knowledge of many medicinal herbs and minerals. Plants and herbs were so important to ancient medicine that the terms for "medicine" and "herbs" were essentially equivalent. The ancient civilizations of Mesopotamia developed a very

ΔΙΟCΚΟΥΡΙΔΗC

ΕΥΡΕCΙ

Dioscorides receiving a mandrake root from the goddess of discovery. *(Library of Congress. Reproduced with permission.)*

comprehensive materia medica. Scholars who studied clay tablets from ancient Assyria have identified about 250 vegetable drugs and 120 mineral drugs, as well as alcoholic beverages, fats and oils, parts of animals, honey, wax, and various milks thought to have special medicinal virtues. Botanical drugs included asafetida, cannabis, crocus, hellebore, mandragora, myrrh, opium, pine turpentine, and so forth. Drugs made from seeds, bark, and other parts of plants were dissolved in beer or milk and administered by mouth, or mixed with wine, honey, and fats and applied externally. The sources used by the herbalist were essentially those typical of folk medicine, but Mesopotamian pharmaceutical texts reflect familiarity with fairly elaborate chemical operations for the purification of basic ingredients. Purgative remedies are very prominent in the medical tradition of Mesopotamian civilizations because illness was regarded as a divine punishment for sins committed by the patient. Healing, therefore, required both physical and spiritual catharsis, or purification. The status of the herbalist seems to have deteriorated as Mesopotamian civilizations became more interested in the magical approach to healing.

Greek writers like Homer (ninth to eighth century? B.C.), Herodotus (484-430/420? B.C.), and Theophrastus (372?-287? B.C.) praised the physicians of Egypt for their wisdom and skill and took note of the valuable medicinal plants they prescribed. Unfortunately, only a few fragmentary medical papyri, which were probably composed between about 1900 and 1100 B.C., have survived. The medical papyri provide information about ancient Egyptian ideas about health and disease, anatomy and physiology, magic and medicine, in the form of case histories, remedies, drug formularies, recipes, and incantations. The most complete and famous is known as the Ebers papyrus. Written about 1500 B.C., the Ebers papyrus includes an extensive collection of prescriptions, as well as incantations and extracts of medical texts on diseases and surgery. About 700 drugs, made up into more than 800 formulas, are found in the Ebers papyrus. Many recipes call for incomprehensible, exotic, or seemingly impossible ingredients, which may have been secret or picturesque names for various plants. Physicians apparently relied on specialized assistants and drug collectors, but sometimes they prepared their own remedies. In contrast to Mesopotamian custom, Egyptian prescriptions were precise about quantities. Although the Egyptians were familiar with the sedative effects of opium and henbane, there is no direct evidence that they were used as surgical anesthetics.

In 332 B.C. Alexander the Great (356-323 B.C.) conquered Egypt and brought it into the sphere of Hellenistic culture. How much the Greeks learned from the Egyptians and how much they taught them is difficult to determine, but Greek physicians adopted many of Egyptian drugs. Hippocrates (460?-377? B.C.), Dioscorides (40?-90?), Galen (129-199?), and many other Greek physicians prescribed herbal remedies.

Theophrastus of Eresus (390-286 B.C.), a Greek philosopher who studied with Plato (427?-347 B.C.) and Aristotle (384-322 B.C.), is credited with founding botany. Theophrastus became one of Aristotle's favorite disciples and inherited his library. Thus the writings of Theophrastus provide important insights into the botanical teachings of Aristotle, who left no botanical works of his own. *De historia plantarum,* the most important surviving work by Theophrastus, includes important information about plant lore and gathering herbs for medicinal purposes. Theophrastus collected and organized the existing botanical knowledge of his time and described about 500 plants. Theophras-

tus classified plants as trees, shrubs, and herbs. His basic concepts of morphology, classification, and the natural history of plants were accepted without question for many centuries.

The Greek physician Crateuas (first century B.C.) composed the earliest known illustrated herbal. The text classified plants and discussed their medicinal uses. Unfortunately, only fragments of the works of Crateuas have survived, generally as extracts in the works of other writers.

From the second century B.C. to the first century A.D., a succession of Roman writers prepared Latin treatises on farming, gardening, and fruit growing. Although these Roman collections were not scientific, they do provide information about the plants that practical Romans considered most valuable. Pliny the Elder (23-79) compiled an encyclopedia known as the *Historia naturalis* (Natural history). Sixteen of its 37 volumes were devoted to plants. Pliny had high praise for the healing powers of many herbs and spices. The *Historia naturalis* is valuable as a compilation of some 2,000 works representing 146 Roman and 327 Greek authors. Although Pliny was rather uncritical in assembling his materials, his encyclopedia preserved excepts of many texts that would otherwise be totally lost. A work composed in A.D. 47 by Scribonius Largus (fl. A.D. 40) provides a compilation of drugs and prescriptions, as well as the first accurate description of the preparation of true opium.

The Greek physician and pharmacologist Dioscorides wrote a more systematic treatment of herbal medicine. Dioscorides is considered the originator of European materia medica. His *De materia medica* was the most important source of herbal lore and pharmacology for almost 16 centuries. Many of the drugs described by Dioscorides were still commonly used in medical practice. Dioscorides's travels as a surgeon with the armies of the Roman emperor Nero provided him an opportunity to study the features, distribution, and medicinal properties of many plants and minerals. The text refers to approximately 1,000 simple drugs. *De materia medica* includes descriptions of nearly 600 plants, including cannabis, colchicum, water hemlock, and peppermint and refers to sleeping potions prepared from opium and mandragora as surgical anesthetics. It is interesting to note that Dioscorides called attention to the Egyptian origins of about 80 of the vegetable drugs described in his herbal. Dioscorides grouped his plants under three headings: aromatic, culinary, and medicinal. He also discusses the medicinal and dietetic value of various animal parts and products, such as milk and honey, and the medicinal use of minerals, including mercury, arsenic, lead acetate, calcium hydrate, and copper oxide. An illustrated Byzantine version of Dioscorides's famous herbal, known as the Constantinopolitan Codex was prepared in the sixth century A.D. Some of its illustrations and plant names are probably derived from Crateuas.

Many manuscript herbals, drawing largely from Dioscorides and Pliny, were reproduced in medieval Europe. The printing press revolutionized the availability of all types of literature, including that of medicine and pharmacology. Many herbals were published in the fifteenth and sixteenth centuries. Written by physicians and botanists, the earliest printed herbals were generally derived from the work of Dioscorides and Theophrastus. Traditional herbal medicine is practiced in much of the world today, especially in Asia and India, and herbs still play a role in Western medicine. Present-day herbalists extol the efficacy of herbs, spices, and spice seeds in the treatment of certain ailments and argue that herbs are far less likely to cause dangerous side-effects than prescription drugs.

LOIS N. MAGNER

Further Reading

Anderson, Frank J. *An Illustrated History of the Herbals.* New York: Columbia University Press, 1997.

Arber, Agnes Robertson. *Herbals: Their Origin and Evolution. A Chapter in the History of Botany, 1470-1670.* 3rd ed. Cambridge: Cambridge University Press, 1986.

Bensky, Dan, Andrew Gamble, and Ted J. Kaptchuk. *Chinese Herbal Medicine: Materia Medica.* Seattle, WA: Eastland Press, 1986.

Chatterjee, Asima, and Pakrashi, Satyesh Chandra, eds. *The Treatise on Indian Medicinal Plants.* New Delhi: Publications & Information Directorate, 1991.

Grieve, Maud. *A Modern Herbal; the Medicinal, Culinary, Cosmetic and Economic Properties, Cultivation and Folklore of Herbs, Grasses, Fungi, Shrubs & Trees, with All Their Modern Scientific Uses.* New York: Hafner, 1959.

Huang, Kee Chang. *The Pharmacology of Chinese Herbs.* Boca Raton, FL: CRC Press, 1993.

Kapoor, L.D. *CRC Handbook of Ayurvedic Medicinal Plants.* Boca Raton, FL: CRC Press, 1990.

Leung, Albert Y. *Chinese Herbal Remedies.* New York: Universe Books, 1984.

Sivarajan, V.V., and Indira Balachandran. *Ayurvedic Drugs and Their Plant Sources.* New Delhi: Oxford & IBH, 1994.

The Hebrew Dietary Laws

Overview

The Hebrew dietary laws, or laws of *kashrut*, were first set forth in the biblical books of Leviticus and Deuteronomy about 3,000 years ago. As such, they were among the earliest such restrictions ever promulgated, although they do have parallels in other ancient civilizations. By slowing the assimilation of the dispersed Jewish people into other cultures, the dietary laws helped to maintain the Jews' uniqueness and thus facilitated their role in history.

Background

The Hebrew way of life evolved in the context of the ancient Middle East. In many cases, Hebrew customs were similar to those of other Semitic tribes, or adopted from other peoples in the region such as the Egyptians or Sumerians. However, in some instances new rules or habits were intentionally put in place by the monotheistic Hebrews to distinguish themselves from their pagan neighbors.

Like other customs, dietary taboos and laws developed over time. It was their codification in the books of Leviticus and Deuteronomy in about 1000 B.C., their subsequent elaboration in the Talmud, and the central place of the Hebrew scriptures in the culture of the Jewish people that crystallized the dietary laws and preserved them over the millennia.

A central part of the dietary laws are the lists of permitted and taboo animals. These consist of general rules based on broad classifications. Among quadrupeds, only animals that both have a cloven hoof and chew their cud may be eaten. Examples of animals that fail this test are specifically mentioned in the Bible: the camel, rock-badger, hare, and pig. Food animal taboos are quite common among ancient peoples around the world, and may derive from a fear of absorbing the undesirable characteristics of a particular creature. Pigs, for example, were forbidden to many of the peoples of the Middle East (pork is proscribed for Muslims as well as for Jews, and comparable restrictions existed among the ancient Egyptians and Babylonians). They were also banned by the ancient Hindus, traditional Navajo, the native peoples of Guyana, the Lapps, and a tribe in Borneo.

For edible aquatic animals, the Hebrew law mandates both fins and scales, which eliminates all shellfish as well as eels, squid, and octopus. All insects are forbidden, with the exception of those that "have jointed legs above their feet, wherewith to leap upon the ground," that is, locusts, crickets, and grasshoppers. Reptiles and amphibians, carnivorous and aquatic birds, bats, and animals with paws such as rodents and members of the weasel family are also banned.

In addition, laws apply to the conditions under which an animal may be eaten. Animals that died naturally, as well as carrion, are forbidden. Detailed procedures for slaughtering food animals require cutting the throat in a single slash with a razor-sharp knife. This ensures that death is immediate, thereby minimizing the animal's pain. The laws also mandate that neither the blood nor the fat around the internal organs is consumed. Scholars believe that such laws derive from ancient notions that the vital essence or soul of an animal resided in these substances. Similarly, eating the sciatic nerve of the thigh, associated in Biblical times with the function of procreation, is banned.

Meat and milk may not be eaten, cooked or stored together, based on a Biblical proscription against boiling a kid goat in its mother's milk. In fact the law may have been intended to be more specific than later interpretations held. It has been suggested that it was meant to ban a particular Canaanite fertility ritual known from a 3,400-year-old text, in which a kid was boiled in milk. The first five and most ancient books of the Hebrew Bible, called the Torah, repeatedly command the people to avoid such pagan rites and customs.

Impact

In the Hebrew dietary laws, the words for permitted or forbidden foods (familiar to many in the Yiddish *kosher* and *treif*) are generally translated "clean" and "unclean" respectively. However, the dietary laws are couched in terms not of physical cleanliness but of religious or ritual purity. Some people believe that they were in fact a way to institute health practices under the guise of divine authority. Others believe that the laws exist purely for religious reasons, to improve human nature and enhance mindfulness of God

by investing everyday activities with spiritual significance. In this view, health benefits were either incidental or a natural consequence of following God's law. In any case, some of the dietary laws did turn out to result in a healthier food supply.

For example, an animal that died naturally might be diseased, and carrion might have started to rot or be infested by vermin. Prohibitions against eating carnivorous birds and animals force the people to use animals a bit lower on the food chain, where any toxins in the environment are less concentrated. "Bottom-feeders," scavengers, and filter feeders such as shellfish, which tend to carry parasites, are also prohibited.

The strictly monitored slaughtering procedures elaborated in the Talmudic writings of about A.D. 500 and later rabbinical texts require extremely careful handling of meat. All animal carcasses are inspected, and any evidence of disease renders the meat not kosher. To remove any traces of blood, the meat is soaked in cold water, salted, and washed three times. Concern about the forbidden sciatic nerve has in practice generally meant that the entire hindquarters of the animal, the most difficult part to clean effectively, are not used. Meanwhile, all fruits and vegetables, which are nutritious and less prone to carrying disease than meat, are permitted. They may be eaten with any foods, as can grain, fish, and eggs.

Despite the health benefits of the dietary laws, the result was not always salutary for the Jewish people during their long diaspora—the period after they began migrating from their traditional home in the Middle East to other regions around the world. When waves of disease hit Europe in the Middle Ages, Jews often were, or at least were perceived as, somewhat less susceptible. This may have been due to a combination of their dietary laws, additional religious edicts requiring more bathing and hand washing than was common in medieval Europe, and their relative social isolation from the non-Jewish community. Whatever the reason, the result was that they were sometimes suspected of actually causing the epidemic. This often brought on murderous attacks against a group for whom religious and racial persecution was common in any case.

Another important effect of the Hebrew dietary laws was to help preserve the unity of the Jewish people for thousands of years, despite the pressures of dispersion. Strict regulations about what one could eat served as a powerful tie binding Jews to their own community. Social relationships with the outside world were of ne-

Jews celebrate Seder, the Passover meal, March 22, 1989. *(Corbis/Roger Ressmeyer. Reproduced by permission.)*

cessity limited, and Jews tended to congregate in groups large enough to ensure a reliable source of kosher food.

The requirement of ritual slaughter meant that the tradition of hunting never took hold among Jews, and they were not likely to be found alone on a frontier. In addition, since they could never be sure that food sold by non-Jews had been processed in accordance with the dietary laws, many Jews became involved in the food trades. They were butchers and bakers, dairy and egg farmers, and producers and merchants of grains, oils, and wines for Jews and non-Jews alike. However, beginning in the late Middle Ages, European Jews were excluded from most trades and forbidden to own land. As a result, many communities became desperately poor.

There have been only a few times and places in which Jewish adherence to the dietary laws was seriously questioned. These included the Greek-influenced period of the early Christian era, and again since the nineteenth century, with the rise of Reform Judaism and the mass emigration of European Jews to the United States. In both cases, large segments of the Jewish people assimilated into the general population and lost their identities as Jews.

The cohesiveness of this small Middle Eastern tribe over thousands of years is of significance to the rest of the world because of the im-

pact it has had on history. Jews, particularly through the influence of the Hebrew scriptures, helped build the worldview of the ancient Fertile Crescent into the bedrock of Western civilization. Christianity began as a Jewish sect, later spreading around the globe. The Jewish tradition of scholarship has resulted in many important contributions, particularly in science and medicine. Today, Jewish leaders, engaged in the modern world but concerned about assimilation, continue to wrestle with the place of the dietary laws in their community as it encounters the twenty-first century.

SHERRI CHASIN CALVO

Further Reading

Cahill, Thomas. *The Gifts of the Jews: How a Tribe of Desert Nomads Changed the Way Everyone Thinks and Feels.* New York: Nan A. Talese, 1998.

Douglas, Mary. *Purity and Danger: An Analysis of the Concepts of Pollution and Taboo.* London: Routledge and Kegan Paul, 1994.

Gaster, Theodore H. *The Holy and the Profane: The Evolution of Jewish Folkways.* New York: William Sloane Associates, 1955.

Goody, Jack. *Cooking, Cuisine and Class: A Study in Comparative Sociology.* Cambridge: Cambridge University Press, 1994.

Harris, Marvin. *Cows, Pigs, Wars and Witches.* New York: Vintage Books, 1974.

Life Sciences & Medicine

2000 B.C.
to A.D. 699

MacClancy, Jeremy. *Consuming Culture: Why You Eat What You Eat*. New York: Henry Holt and Company, 1992.

Roden, Claudia. *The Book of Jewish Food: An Odyssey from Samarkand to New York*. New York: Alfred A. Knopf, 1996.

Simoons, Frederick J. *Eat Not This Flesh: Food Avoidances from Prehistory to the Present*. Madison: University of Wisconsin Press, 1994.

Smith, W. Robertson. *Lectures on the Religion of the Semites*. London: Routledge/Thoemmes Press, reprinted 1997.

Hippocrates and His Legacy

Overview

Hippocrates, known as the "Father of Medicine," was a central figure in the medical and philosophical world of Greek antiquity. By rejecting the widely held belief of the time that disease resulted from the disfavor or intervention of the gods, Hippocrates set the stage for Western learned, written medicine. Hippocrates compiled the first written medical library in the fifth century B.C., and is associated with the Hippocratic Oath, a pledge outlining the ethical responsibilities of the physician. As a practicing physician, Hippocrates gained notoriety with his approach to illness. While considering disease as an entity affecting the whole person, Hippocrates replaced superstition with diagnostic observation and treatment. The impact of Hippocratic teaching was widespread in the Western world for over a millennium—first in the students Hippocrates himself trained, and eventually in the work of the famed Roman physician Galen (A.D. 129-c. 216), who sought to perfect the legacy of Hippocrates.

Background

Little is known of Greek medicine before the appearance of texts written in the fifth century B.C. Folk healing was often used, and the healer in turn often summoned help from the gods. Some herbal drug therapies existed. The narratives of Homer give accounts of battle wounds treated with salves. Apollo, the god of healing, also appears in Homer's works, both causing plagues and providing healing relief from them. Various other gods were identified with disease and injury. Prometheus, for example, was said to have had his liver torn from him by an eagle. Asclepius, the son of Apollo and a mortal mother, was the most celebrated god of medicine. Homer portrayed Asclepius as a skilled healer and the father of sons who also became physicians, calling themselves Asclepiads. Asclepius was usually portrayed holding a staff intertwined with snakes, serving as the origin of the caduceus sign of the modern physician. The shedding of the snake's skin was a symbol for the renewal of life. Hygeia, goddess of health or hygiene, and Panacea, goddess of cures, were the daughters of Asclepius. By the time of Hippocrates's birth, approximately 460-450 B.C., most Greek towns had erected a temple to the god Asclepius. The sick made pilgrimages to the temples, sleeping within them to await a cure brought about by a dream or performed directly by Asclepius himself.

Priests and other religious leaders often helped interpret healing dreams, so religious philosophy and medicine were intertwined. Intellectual curiosity of the workings of man, along with its medical implications, found its way into the discussions of the fifth-century Greek scientists and philosophers who were part of the culmination of the Classical period. For the Greek mathematician Pythagoras (c. 530 B.C.), humans' place in the universe lay in the order and symmetry that geometry provided. The symmetry of the body and the sensations it experienced—hot and cold, sleep and alertness, wet and dry—provided Pythagoras with the conception of health and disease as part of this ordered symmetry.

Philosophers such as Democritus (c. 460 B.C.) and Empedocles (c. 540 B.C.) each proclaimed his own explanation of the nature of man. Some scholars find Empedocles was the first to advance into Greek medicine key physiological doctrines, namely the body's innate heat as a source of living processes and the cooling function of breathing. Nevertheless, Hippocrates defied the conjecture of the philosophers, declaring that only through the careful observation of nature could man as a natural being be understood.

Hippocrates was born in a time (fifth century B.C.) when there was not yet a distinction between science and philosophy. His childhood home was the Greek island of Cos, site of the best known Asclepion, or temple honoring the god of healing. Because there is little written history of Hippocrates, trustworthy information about his life is scanty. As with the Asclepiads, Hippocrates was the son of a physician. In the *Protagoras,* the philosopher Plato (427-347 B.C.), a younger contemporary of Hippocrates, described Hippocrates as "the Asclepiad of Cos," who made his living teaching medical students and enjoyed a respected and well known status throughout his long life. Hippocrates died in 377 B.C. in Larissa, Thessaly. Among other contemporaries of Hippocrates was Alcmaeon of Croton (c. 470 B.C.), whose doctrine of health as a balance among the powers of the body greatly influenced Hippocratic medicine.

Impact

Hippocrates defined disease with physical and rational explanations, thus separating religion from the realm of medicine. Hippocratic medicine shunned the priests, shamans, exorcists, diviners, and other traditional healers. Followers of Hippocratic medicine were advocates of the patient, no longer an intermediary between the patient and the gods. The medical writings of Hippocrates contained no mention of divinity. The rational system of medicine relied heavily upon careful observation of the patient and symptoms. Although grounded in nature, the teachings of Hippocrates first gave the physician an independent status. Not only was the physician separated from the divine or traditional healer, he was also separated from the philosophers of nature. Hippocrates helped confine the physician to medicine, and medicine evolved into an art that was practiced based upon the reason of science.

Hippocratic physicians believed the body was to be considered and treated as a whole, with each part contributing to the overall picture of health or illness. Hippocrates was the first to accurately describe the symptoms of several diseases, including pneumonia and epilepsy in children. Hippocrates recorded the symptoms as facts, then referred to his additional recorded works, as well as his experience, to assist in diagnosing disease. Teaching that the body had a powerful ability to heal itself naturally, Hippocrates often prescribed treatments of rest, fresh air, and cleanliness instead of the usual mystical cures of the day. Many of his therapies involved dietary regimes. Only after these natural remedies failed did Hippocrates advocate the use of bloodletting or herbal drugs, and still less frequently, surgery. Hippocrates noted that the ability to cope with disease varied among individuals, as did the severity of their symptoms

Hippocrates examining skulls. *(National Institutes of Health/Corbis. Reproduced by permission.)*

and their response to treatment. The concept of a prognosis was also developed by Hippocrates, whereby a physician can predict the course and outcome of a disease based upon previous observation of similar cases. The skill of forming a prognosis had a function in society. By foretelling the future of a disease process, physicians garnered confidence and trust from their patients and were elevated in status above traditional healers. The prognosis remains an integral part of modern medical reasoning.

Hippocrates assembled a collection of medical literary works into the first known medical library. The contents of the collection, the Hippocratic corpus, were highly varied. Some works held case histories, others were teaching texts or philosophies. The thread that united the works was the conviction that health and disease were explained by reasoning about nature and its universal laws, independent of supernatural influence. The development literary form used in the corpus, the prose treatise, was an important literary milestone. Previous Greek writings such as the histories of Homer were told in the form of poetry. The development of written prose, as embraced by Hippocrates, rapidly gained acceptance in expressing rational thought.

The Hippocratic corpus became the foundation for the work of the Roman physician Galen almost 500 years later. Galen's medicine enhanced the rationality of Hippocrates, depending upon logic, physics, ethics, philosophy, and limited experimentation. It is mostly through Galen's prolific writings that knowledge of Hippocratic medicine survived (references can also be found in the works of Plato, Aristotle, and Menon). Galen's influence lasted a millennium, and was rediscovered during the Renaissance together with its reference to classical Hippocratic medicine.

The principal concept in the Hippocratic corpus was that health was perceived as equilibrium, and illness was an upset of that equilibrium. The body was regarded as stable until an imbalance in the concentration of body fluids occurred in a particular body zone.

Hippocrates borrowed from the Greek philosopher Empedocles (490-430 B.C.) the principle of the four elements in nature: earth, air, fire, water, and added to it the four fluids (humors) of man: phlegm, blood, yellow bile, and black bile, to produce a pattern that was believed to hold an explanation for disease. For example, winter colds were due to excess in phlegm, and mania resulted from excess bile in the brain. A diagram was drawn showing the four humors and their coinciding elements. The four seasons or four ages of man were also often added to the diagram to help explain disease. The diagrams could be made to fit almost any set of observations and symptoms and became a cornerstone of Hippocratic medicine.

Scholars wrestle with the question of authorship of the Hippocratic corpus. In antiquity, the 60 to 70 treatises included in the collection were attributed to Hippocrates himself. Most scholars agree that the corpus was probably an assembly of medical works written by different authors over a period of almost a century. Among the most influential works of the corpus attributed to Hippocrates was the Hippocratic Oath. The oath foreshadowed the notion of a professional (as one who professes an oath) with obligations to both the client and the profession it serves. Setting the stage for medical ethics, the oath attested to the physician's concern for his patient and his art above fortune or self-promotion.

There were two parts to the oath. In the first, the physician's responsibilities to his teachers and students are stated. The second part outlines rules for personal and professional conduct.

A physician of antiquity who accepted the rules contained in the oath was not in agreement with many of his fellow colleagues. Far from being universally accepted, the oath held physicians to standards far more restrictive than the practice of the time. In particular, the oath prohibited the prescription of a "destructive pessary" to cause abortion at a time when infanticide and abortion were familiar practices. The moral requirements of the oath more closely represent those of the fourth-century Pythagoreans, whose doctrines favor the transmigration of souls. As a result, many scholars believe the oath was written by a follower of Pythagoras or even the elderly Pythagoras himself. Nevertheless, the Hippocratic Oath has endured as a standard-setting example of medical ethics and the ideal medical practitioner.

Hippocrates and the followers of Hippocratic medicine did not offer all solutions. The widespread impact of the humoral theory on European medical thought lasted more than a millennium (Galen embraced it), and its simple explanations put a brake on the speed of advancing scientific thought. Little was known of the physiology of the body, because in the classical period reverence to the human body extended to a prohibition of human dissection. Firsthand observation of physiological processes was limited to what could be gleaned from observation of surface anatomy or wounds. Theory often was untested and bowed to experience. The enduring contribution of Hippocrates was in the separation of medicine from divinity, laying a foundation for a medical profession based upon science and ethics.

BRENDA WILMOTH LERNER

Further Reading

Jouanna, J. *Hippocrates*. Baltimore: Johns Hopkins University Press, 1999.

Longrigg, James, ed. *Greek Medicine: From the Heroic to the Hellenistic Age: A Sourcebook*. New York: Routledge, 1998.

Porter, Roy. *The Greatest Benefit to Mankind: A Medical History of Humanity*. New York: W. W. Norton, 1998.

Smith, Wesley. *The Hippocratic Tradition*. Ithaca, New York: Cornell University Press, 1979.

The Philosophy of Greek Medicine

Life Sciences & Medicine

2000 B.C. to A.D. 699

Overview

The evolution of medicine in ancient Greece was directed by a combination of knowledge, beliefs, and rituals handed down from previous civilizations and cultures, and newly developed and ever changing philosophical principles. The Greeks adopted the beliefs and practices of many other civilizations. They borrowed the idea of a godlike figure that dispensed all medical wisdom from the Egyptians. This eventually developed into the concept of their own medical knowledge god, Asclepius. Their symbol of healing, a serpent wrapped around a staff, and much of their hygienic practices were taken from the Minoan culture. Ancient Mesopotamia provided the belief that evil spirits invaded the body to cause sickness, which in turn supported the notion that only gods could then cure that sickness. However, the Greeks also made their own significant contributions to medical knowledge and thought.

The ancient Greeks are thought of as a philosophical and enlightened society. Their pursuit of the true underlying causes of the world that surrounded them helped to provide many improvements to the medical practices that they had adopted from other societies. This is considered the beginning of the Golden Age of Greece, during which new ideas came from observation and thought, rather than superstition and ritual. Like much of the ancient world, most of the advances their society made in medicine were lost or ignored during the Middle Ages, only to be newly rediscovered at the beginning of the Renaissance. However, there are certainly exceptions to that generalization, with the great anatomist Galen (129-199?) and the father of doctors, Hippocrates (460?-377? B.C.), being two of the most noteworthy examples.

Thus, there were initially two distinct medical factions in ancient Greece. One group was a cult closely intertwined with religious rites and temple rituals that worshipped Asclepius. The other group was ironically known as the Asclepiads, which was a secular collection of early physicians that used reason and observation of the symptoms to treat disease and illness. This group would eventually consist of many subgroups, as differences in philosophy would lead to newly formed factions.

Background

The worship of Asclepius, the god of medical knowledge, was widespread throughout Greece. The physicians of this cult were the priests and their assistants who believed that all illness was caused evil spirits. In order to be cured, patients must spend at least one night in the temple where the god would reveal something about the illness in a dream that the priest would interpret. The course of treatment would be based on the dream interpretation, largely using folklore derived from Minoan civilization. These treatments also consisted of a ritual cleansing and the presentation of sacrificial gifts to the priests from the patients as a form of payment. These temples became great centers of healing, much like a modern day health spa. However, the use of this type of medical practice began to wane in the face of the growing body of medical knowledge based on observation and diagnosis, rather than folklore and rituals.

Secular physicians gained increasing popularity in Greece due to the fact that their treatments were more beneficial because their theory was more firmly rooted in scientific principles. Certainly there were many mistakes and assumptions that Greek physicians made, but many of their treatments were successful. For example, laxatives of today bear little difference between the laxatives used by early Greeks. One problem that existed was that physicians at those times were considered to be artisans, so just about any male desiring this as a career could take on an apprenticeship. There was no other formal training. Because of this, one early doctor, Hippocrates, put forth ideals that could be held by all physicians.

Hippocrates had a tremendous influence on physicians in both modern and ancient times. He believed strongly in medical ethics and his teachings were so advanced and far-reaching that they are still widely used today. For instance, all new doctors take the Hippocratic Oath (a pledge that they will do no harm to their patients) upon finishing their training. While Hippocrates had considered many areas, his teachings focused on three main principles. First, he believed that all illness was due to some malfunction of the body. Next he thought that the patient's environment needed to be studied to arrive at a satisfactory diagnosis. Lastly, he supposed that doctors could only help the

The Hippocratic Oath. *(Corbis Corporation. Reproduced with permission.)*

symptoms of a disease and that the patient will eventually get rid of the sickness on their own.

Although Hippocratic medicine rested on the belief that disease was a result of natural causes, there were other important factors to consider as well. One theory that was well accepted was the humoral theory of disease. This idea became so indoctrinated into medicine that it was unquestionably accepted for the nest 15 centuries. This theory states that there are four humors in the body: blood, phlegm, yellow bile, and black bile. A healthy person has these four humors mixed in exactly the right proportions, whereas someone who is ill has a deviation in the correct mixture. The goal of the physician was to restore the humors to the natural equilibrium and thereby restore health. While this idea was not disputed, the steps needed to understand and treat the disease were disputed to such an extent that various competing medical sects were established.

The dogmatists were a medical sect whose members believed that all medical knowledge could be gained through authorities rather than through clinical experience. They felt they needed to have knowledge of hidden causes of the disease in order to understand it. The dogmatists sought to supplement clinical practice with reasoning and conjecture. To this end, they performed dissections, vivisections, and other experiments in order to better understand the underlying principles of the workings of the body. The dogmatists firmly believed that in order to reestablish the harmony of the four humors, a prescription containing plants and animals of the missing humor type needed to be administered.

The empiricists disagreed with the dogmatists. Empiricism is often defined as the doctrine that all knowledge comes from experience, thus they believed that doctors must deal with patients on a case-by-case basis. Thus, there could be no knowledge of disease processes. Rather, the empiricists believed that experience was the most important factor in medical learning. Therefore, treatment of the imbalanced humors was not based on known scientific principles, but rather on the individual case.

The great Greek physician Galen helped to tie these opposing viewpoints together. He believed that background learning was important and authorities should be studied. However, he cautioned against blindly following them. He reasoned that the authorities needed to be proven correct in clinical experience. Galen had a relatively modern view of medicine. He envisioned a system where one studied such established medical thought and appended it through clinical experience. This idea had significant impact for many centuries to come.

There were other quacks, charlatans, and sects that practiced at that time as well. One noteworthy group was known as the gymnasts. These men firmly believed that the road to health was through fitness, and they engaged young people in physical training and sport.

From their observations and practice of massage, they gained a tremendous amount of knowledge regarding the human body. While they saw the value of exercise for health, they relied too heavily on this concept by purporting that every illness could be cured solely on the basis of exercise and diet. It is interesting that similar claims are now being made over 2,000 years later.

Impact

Like much of the ancient world, most of the Greek advances in medicine were lost or ignored

THE MYTH OF ASCLEPIUS

Asclepius was the Greek god of medicine. A son of Apollo, he was raised by the wise centaur Chiron, who taught him the art of medicine. In addition, the goddess Athena gave him a vial of blood from the Gorgon Medusa, a substance that could cure any illness. So effective were Asclepius' cures that he could even raise the dead. This angered Zeus, who feared that Asclepius' skill might make all humans immortal. Zeus struck the physician dead with a thunderbolt, but granted him immortality afterward.

In reality, Asclepius was probably a chieftain from Thessaly who had amassed a great deal of medical knowledge and developed an ability to cure people. As time went on, his legend was embellished. His sons appear in the *Iliad* as skilled healers. His daughters became the goddesses Hygeia, who safeguarded health, and Panacea, who could cure any illness. Human physicians who followed in Asclepius' footsteps were called Asclepiads. By the third century B.C. every town had an Asclepion, a temple of healing, where the sick went to be evaluated and healed.

AMY LOERCH STRUMOLO

during the Middle Ages, only to be newly rediscovered at the beginning of the Renaissance. However, there are certainly exceptions. For instance, the integration of dogmatism and empiricism by Galen was largely ignored in medieval medicine. Instead, the belief that disease was caused by evil spirits pervaded. This is somewhat puzzling since many of Galen's medical beliefs and ideas were never questioned until after the Renaissance had begun. That mixture of ideas is heavily relied upon today and forms the basis of our medical education system, as there is both a large theoretical base and ample clinical experience for the student.

No discussion on the impact of Greek medicine would be complete without an entry on Hippocrates. Like Galen, his contributions were most significant for those immediately after his death and then again centuries later. His most important contribution was his separation of the supernatural from the natural, thereby establishing the idea that disease is a result of natural causes and should be treated accordingly. The effect of such an approach can hardly be overestimated. This theory leads to such modern-day practices as patient examination and history. The importance of these things are so inherently obvious that it is almost incomprehensible that they should have fallen into disuse shortly after the time of Hippocrates, and not brought into general use again until almost 2,000 years later.

These early Greek medical sects helped to form the basis of entire philosophical systems and ideas. In most cases their philosophy was only limited by the technology and knowledge at that time. It is interesting to observe, however, that such ancient ideas have important ramifications in present-day society. Even some of the less mainstream sects, such as the gymnasts, were prophetic in their contention of the extreme importance of exercise and fitness to health.

JAMES J. HOFFMANN

Further Reading

Lloyd, G. E. R. *Science & Morality in Graeco-Roman Antiquity.* New York: Cambridge University Press, 1985.

Longrigg, James. *Greek Medicine From the Heroic to the Hellenistic Age: A Source Book.* New York: Routledge, 1998.

Longrigg, James. *Greek Rational Medicine Philosophy & Medicine from Alcmaeon to the Alexandrians.* New York: Routledge, 1993.

The Doctrine of the Four Humors

Overview

The "Doctrine of the Four Humors" dominated the theory of health, illness, and personality from the time of Empedocles (490-430 B.C.) until the eighteenth century, when bloodletting was finally ended. The doctrine taught that four basic elements comprised all matter: fire, earth, water, and air. Each element had two qualities. For example, fire was hot and dry, earth was dry and cold, water was cold and wet, and air was wet and hot. Based on that structure, the human body was believed to have four humors corresponding to these natural elements: blood with air, black bile with earth, yellow bile with fire, and phlegm with water. The Greek physician Galen later added personality types to go with each humor.

Background

The doctrine's origin can be traced to a number of unrelated but similar theories from ancient Greece and Rome. The Greek philosopher Thales (640-546 B.C.) believed that the basic element in all life was water, and from water developed earth and air. Anaximander (611-547 B.C.), another Greek philosopher, added to that

the idea that the universe was constituted of opposite forces in balance, governed by universal laws. His pupil Anaximes (fl. 546 B.C.), considered the third of the three great Milesian thinkers along with Thales and Anaximander, thought air, rather than water, the primary element and wrote that it could be transformed to other substances through condensation and rarefaction.

Heraclitus (fl. 500 B.C.), a Greek philosopher from Ephesus, differed with the three Milesians by believing fire to be the primary element rather than water or air, but he shared their belief that tensions between opposing forces were essential to life. Since these opposites were constantly at battle with one another, all things were in perpetual change. According to Heraclites, these changes were governed by a principle of order.

By the time of Pythagoras (fl. 530 B.C.), who emigrated to southern Italy from Samos, Greece, a different focus and theory about the makeup of the universe was presented. This philosopher and mathematician emphasized the spiritual rather than material, and a science of numbers rather than elements. Although the larger part of his belief system concerned the transmigration of souls into successive bodies, another part of his philosophy that is relevant to the development of the humoral theory is the importance of the number four. Pythagoras taught that all nature consisted of four elements. These were air, earth, water, and fire.

Empedocles (c. 490-430 B.C.), the Sicilian philosopher, wrote both about opposites and four elements. In his "Poem on Nature," he stated, "From all of these—sun, earth, sky and sea—are united in all their parts" and "Out of water and earth, and air and fire mingled together arose the forms and colours of all mortal things." This is the concept of the macrocosm that formed the germ of the doctrine of the four humors. Empedocles believed that the elements were stable and did not change.

Aristotle (384-322 B.C.) believed that the simplest occurrences of matter existed in the form of four elements. These elements could be analyzed in terms of two pairs of primary contrary qualities (for example, hot and cold or wet and dry) or elementary principles acting as forms. They differed in taste, smell, and color but were all either hot or cold, wet or dry. Each two elementary principles made up the elements, therefore hot-dry = fire, cold-wet = water, hot-wet =air, and cold-dry=earth. Aristotle differed from Empedocles in that he believed that the elements were changeable and elements could be transformed into others. There were two kinds of changes: one where one or both qualities changed, the other where two elements came together and interchanged their qualities to produce the other two. An example of this would be that water and fire could make earth and air. The idea of transformation of metals led to the quest for the "Philosopher's Stone," a mythic stone that could be turned into gold with the proper catalyst.

By the time of Hippocrates of Cos (460-370 B.C.) and his followers in the theory and practice of medicine, the four humors referred to the composition of the human body. This was the corresponding microcosm to macrocosm. The qualities of the elements—heat, cold, humidity, and dryness—produced blood, phlegm, yellow bile, and black bile. Hippocrates taught that when all the elements were balanced and mingled correctly, people felt in perfect health, but if one were out of proportion, a person would become ill. As a corollary to this, Hippocrates believed that there were fundamental differences between the sexes based on "heat" and "cold" qualities. Women were "cold and wet," men "hot and dry."

Although not entirely supportive of Hippocratic humorology, Erasistratus (310-250 B.C.) accepted the part of the doctrine that involved the conversion of food to blood. In his scheme of physiology, he believed that disease was due to an excess of blood from undigested food. If this blood accumulated in the veins, it could seep from the veins to the arteries and the body's tissues could be harmed. As treatment he recommended emetics (drugs to induce vomiting), diuretics (drugs to stimulate urinary output), hot baths, and starvation.

Galen of Pergamon (270-200 B.C.) modified Hippocratic theory by stating that in blood the four elements were mixed in equal quantities, but the other bodily items contained only one element; thus, phlegm consisted mainly of water, yellow bile was mainly fire, and black bile was mainly earth. From this grew the belief that individuals could be grouped into four main physiological/psychological categories: sanguine, phlegmatic, choleric, and melancholic. Galen taught that nutritional intake influenced the amount of humors in the blood.

By now, a theory of pathology was being developed from the basic structure of the four humors. Any imbalance in the amount of a particular humor in the body would result in illness. The treatment was through removal of the various fluids, hence bloodletting was born. For other excesses, there were equally drastic treatments. Purging (induced vomiting), enemas, various applications of heat and cold, or the forced consumption of liquid were used. Starvation, another treatment during this time period, was one of the early treatments for diabetes mellitus before insulin was discovered.

Interestingly, Asclepiades of Bithynia (120-70 B.C.), completely rejected the doctrine of the four humors and instead developed his own explanation. In his system, the body was composed of an infinite number of atoms in motion. Between these atoms flowed bodily fluids. According to Asclepiades, when the motion of the atoms was interrupted, sickness developed. His pupil, Themison, elaborated on the atomistic theory, founding the school of philosophy known as methodism. He described illness as the result of constriction or relaxation (oversecretion) of the pores. However, for treatment he depended on the theory of opposites, defining his treatments as relaxing the overconstricted or constricting the excessively relaxed pores. According to Themison, there were seven naturals and nine non-naturals. The naturals could be likened to what we consider basic internal, anatomical, and physiological structures and functions: elements, humors, temperaments, body parts, faculties, operations, and spirits. The non-naturals were external necessities such as food, water, movement, rest, sleep, and wakefulness as well as excretion, retention, and exercise of passion.

Impact

The commonalties among these various theories were a belief that body humors resulted from food intake, a definition of elements and innate qualities, and the use of opposites to treat imbalances. Despite the many differences such as what was carried in arteries (pneuma versus blood), the doctrine of four humors dominated medical treatment and belief about how the human body functioned. It explained psychological differences in terms of a predominance of one humor over another—hence the terms "choleric," "melancholic," "phlegmatic," and "sanguine," can be traced to ancient times. A choleric individual had an excess of yellow bile. Since the humor of yellow bile was associated with hot and dry qualities, this individual was said to be a thin, energetic, intelligent, fussy eater with a strong rapid pulse, a proclivity to sexual pleasure, and good blood vessels. Choleric people were easily angered. A phlegmatic individual was cold and moist, fat with flaccid skin tone, pasty white skin, and thin hair. He or she was slow in thought and movement and did not have well developed appetites in either food or sex. The sanguine person (hot and moist) had an excess of blood. They were also heavy in body build but had active lifestyles. They were more muscular and robust with strong drives in both sex and gastric indulgences. The melancholic had a preponderance of black bile. This individual (cold and dry) was thin, dark, hairy with a slow pulse, narrow blood vessels, and large appetites.

Although the doctrine of four humors is no longer the paradigm for either physiologic or psychological etiologies (causes and background) of illness, the words "choleric," "phlegmatic," "melancholic," and "sanguine" are still used to describe people. A hot-headed person is choleric, whereas an unemotional, lethargic, cool individual is phlegmatic. A depressed person is described as melancholic, and the ruddy, energetic, stout person is said to have a sanguine character.

LANA THOMPSON

Further Reading

Clendening, Logan. *Source Book of Medical History.* New York: Dover, 1960.

Crombie, A.C. *Medieval and Early Modern Science.* Vol. 1. New York: Doubleday Anchor, 1959.

Crystal, David. *The Cambridge Biographical Encyclopedia.* New York: Cambridge University Press, 1995.

Magner, Lois. *A History of Medicine.* New York: Marcel Dekker, 1992.

Skinner, Henry. *The Origin of Medical Terms.* Williams & Wilkins, 1961.

Aristotle and the Founding of Biology

Overview

In addition to his great reputation as a philosopher, Aristotle (384-322 B.C.) is also regarded as the father of biology and the first ecologist. Despite that fact that his observations on biology, astronomy, and physics comprise the vast majority of his writings, Aristotle gets much more respect for his contributions to ethics, politics, and moral philosophy.

Aristotle offered an all-encompassing system for finding the meaning of reality through the senses. His curiosity about the world of living things and his attempts to fit these into a comprehensive system is unrivaled. Indeed, Aristotle laid the groundwork for the formulation of the scientific method.

Background

Aristotle's influence on the founding of biology was closely tied to the conjunction of several factors. First, as the son of a physician of the medical guild of Asclepius, he had the privilege of being in the court at Pella in Macedonia. His father encouraged him to study biology and to develop investigative techniques, which would prove useful to science. Second, at age 17 he went to Athens and enrolled at Plato's Academy to study the philosophy of idealism. Although

Aristotle observing the natural world. *(Corbis Corporation. Reproduced with permission.)*

Aristotle held Plato (427?-347 B.C.) in high esteem, he did break with him in later life to form his own school based on realism and observation through the senses. Third, Aristotle's relationship as teacher of Alexander the Great (356-323 B.C.) offered him the privileges of the royal palace. Alexander later endowed Aristotle's museum in Athens and contributed to his collection of flora and fauna with rare specimens from the eastern Mediterranean. Aristotle also inspired Alexander to initiate research projects, including a study of the flooding of the Nile river.

Aristotle's logic is very individual. While Plato had focused on reality as being in ideas, Aristotle departed from this mystical idealism and relied on observable phenomena. He encouraged the cool and objective eye of the empiricist—one who looks at the world in a logical, rational, manner. The structure of modern science owes much to his untiring exploration of truth. He is thought to have written over 400 scholarly works on education and scientific observation, of which 50 survive.

Impact

If Aristotle were to attend a seminar on ethics, philosophy, or moral theology today, he would be right at home. If he were to go into a seminar on DNA or neural stem cell analysis, he would be lost like most people. Yet, it was his approach

to the world that offered an all-encompassing system of scientific thought.

Some modern biologists discount Aristotelian biology because he was incorrect about a number of things. According the Aristotle, the cosmos was an order striving for perfection, imitating the "Unmoved Mover" who set the heavens in an enclosed system revolving around the Earth. His vision of the cosmos was central to Western thought until Nicolaus Copernicus (1473-1543) challenged in the 1500s.

Using knowledge from around him, Aristotle viewed the world's substance in four observable elements: earth, wind, fire, and water. Different mixtures of these elements were deemed to be responsible for different humors and temperaments. These ideas persisted for almost 2,000 years and formed the basis of the alchemist's view of the elements and the physician's adherence to the humoral theories. However, only by examining Aristotle's errors were later scientists able to use them as a springboard for subsequent discoveries.

Scientific reasoning as Aristotle saw it should not be confused with the scientific method established today through experimentation and testing of hypotheses. He used demonstration to determine a logical framework to validate ideas. He sought premises to gain scientific knowledge and formal conditions to make an argument valid. This ideal of science as a deductive system based on axioms had considerable impact on the development of scientific theory.

Aristotle was a pioneer of the method that was summarized by Francis Bacon (1561-1626): observe, measure, explain, verify. These four processes represent Aristotle's principles and are still the methodology used by field biologists. Even Charles Darwin (1809-1882) regarded Aristotle as a preeminent biologist, a tribute to his great influence.

Most of Aristotle's work was in the field of biology. His zoological treatises made up about one-fourth of his entire work. The chief collection of data is contained in the *Historia animalium,* a textbook designed to be studied and kept at hand. The nearly 500 animals discussed are not identified for their own sake, but for how they fit into the world, reflecting Aristotle's attempt to establish a rational cosmology. This idea of interrelatedness is also the basis of ecology, the study of the interrelationship of all living things.

The *Historia animalium,* perhaps better called "Zoological Researches," is an amazing book in that one individual amassed such a great amount of knowledge, much of which his predecessors had not previously studied. Shortly after Plato's death, Aristotle spent two years on the island of Lesbos and developed many of his generalizations there.

Aristotle identified 495 species—one more than Pliny (23-79), another biological observer, did 400 years later. The opening of the book begins, "Differences between animals relate to their ways of life, their activities, their habits, and their parts." He noted differences between animals with blood and those without blood. He also distinguished between oviparous, or egg-laying, animals and viviparous, those that give live birth. Man stands separate as a simple species with no differentiation.

The first book deals with the physiology of man, for which there are many limited observations. For example, he proposed a three-chambered heart and suggested that there were no differences between arteries and veins. Knowledge of man was extended through the dissection of animals.

The second book deals exclusively with animals. Aristotle's knowledge was imperfect, but his observations were masterful. He described the four chambers of the stomachs of ruminants, such as the cow, perhaps from careful observation. At other times he would "double up" when one animal seemed to fit into more than one class. For example, apes double up between humans and quadrupeds, and sea anemones double between plants and animals.

The third book contained a labeled diagram and a detail of the testicles of mammals based on animal dissection. On the heart and blood vessels, however, he completely missed the mark. Other topics discussed include bones, cartilage, hair, skin, and some excellent observations of seasonal changes in the plumage of birds.

The next book analyzed animals without blood, such as the cephalopods, crustacea, testacea, and insects, which he described in rich detail. The book also gathered information on the sounds of animals. For example, he described the cause of the humming noise of some insects, made by the rapid movement of wings, and how the grasshopper produced its strange sound by rubbing its legs together. He distinguished the different calls of birds as being related to courtship and aggression. The book concluded with a discussion of sleep and mating among various animals.

Books five, six, and seven targeted the theme of reproduction. The work on insects is very good, showing his skills of observation. The theme of the other books included diet, habitat, migration, hibernation, locomotion, and animal disease and health.

Historia animalium is not a systematic treatment, and certainly not a thorough classification. But Aristotle was a systematist and his studies of morphology and structure may be described as preclassification. Aristotle's goal was to understand harmony and he was one of the first to discern some order in the seeming chaos of nature. Based on his work, subsequent natural philosophers tried to organize nature, basing classification on comparative anatomy, which changed from one period to another. Aristotle inspired Carl Linnaeus (1707-1778), the Danish biologist, to set up his binomial (genus and species) system and his logical, methodical system of classifying animals. Although Linnaeus also erred in some of his details, the order and adaptability of his system ensured survival.

When Aristotle is considered in the context of his day, one can appreciate his great intellect and innovative ideas. The world that surrounded him was based on myth and superstition. His teacher and mentor, Plato, believed that reality was found only in ideas and that man only understood though "shadows." Aristotle was courageous to break with his revered Plato and to insist that man experiences knowledge through the senses and everything we know comes through the five senses.

In his attempts to understand the harmony and interrelationship of animals and plants, Aristotle can truly be called the first ecologist. Aristotle taught us to look at the universe biologically—as a living, exciting, and beautiful world. He encouraged us to look at the world of patterns. Those who do not give Aristotle the credit he deserves for his biological observations, claiming that they are rife with errors, do not recognize how revolutionary these were for the time. Some criticize the successors of Aristotle for holding on to the traditions and thwarting scientific investigations. However, in the spirit of inquiry and scientific pursuit, Aristotle would likely have supported new discoveries that effectively disproved his own. Indeed, if he lived during the Renaissance, it is easy to imagine that he would have sided with Copernicus, Galileo (1564-1642), and Bacon against the conservative Aristotelians.

EVELYN B. KELLY

Life Sciences & Medicine
2000 B.C.
to A.D. 699

Further Reading

Clendening, Logan. *Source Book of Medical History.* New York: Dover, 1942.

Ferguson, John. *Aristotle.* New York: Twayne Publishers, 1972.

Gribbin, John. *A Brief History of Science.* New York: Barnes & Noble, 1998.

Porter, Roy. *A Medical History of Humanity: The Greatest Benefit to Mankind.* New York: W.W. Norton. 1997.

The Origins of Botany

Overview

Botany, the study of plants, developed as a science in ancient Greece, with Theophrastus (c. 371-287 B.C.) considered the father of botanical science. But a practical interest in plants extended far back before recorded times, because plants were not only a source of food but also of medicines. With the dawn of agriculture, interest in plant growth became more focused, as better methods for cultivating crops and protecting them from weather and pest damage were developed. Also, in many ancient civilizations, including those in China, Egypt, Babylonia, and Greece, sophisticated medical practices developed in which plant materials were important medicinals. What Theophrastus brought to such practical interests in plants was an inquiry that was more theoretical. After Theophrastus, botanical science progressed very little until the rediscovery of his writings in the fifteenth century, just as the Renaissance was beginning.

Background

Close observation is an important key to learning about nature for both practical and theoretical reasons, though most early interest in plant growth was clearly practical. It was related to finding and encouraging the growth of plants that were good sources of foods, medicines, building materials, and other products. Agriculture, which is the systematic growth of plants as opposed to just harvesting plant materials that happen to be available, is thought to have originated in western Asia at least 10,000 years ago.

By the time written documents came to be important in ancient times, several sophisticated civilizations had developed in different parts of the world, all with advanced agricultural practices based on close observation of geographic, climatic, and biological factors. The Egyptians had learned to harness the flood waters of the Nile River for irrigation, and with this came an interest in plants not only as food but for their use as medicine, as well as for decorative purposes in gardens. The Assyrians in Babylonia discovered sexual reproduction in date palms that allowed them to cultivate these trees, and from remote times the Chinese cultivated citrus fruits such as orange and lemons. Each of these civilizations also recorded information on the use of plant materials in the treatment of diseases. Plants were the primary sources of medicines in ancient times, and while each culture refined its own medical practices, there is also evidence that information from Egyptian and Babylonian medicine was used later by the Greeks.

The writings that are ascribed to Hippocrates (c. 460-377 B.C.), the greatest of the ancient Greek physicians, include descriptions of over 250 plants that were used either to promote health or cure disease. These documents also contain a great deal on proper diet as well as information on food plants. So it is clear that the Greeks studied plants for the very practical reasons of human health and nutrition. But in the sixth century B.C., an intellectual movement was developing in Greece, a movement leading to the development of Western philosophy, which asked basic questions less for their practical importance than for building a systematic exploration of the natural world.

The first Greek philosophers came from the island of Ionia. They investigated questions about such things as the origin of the world and its composition, and thus began a tradition that led to the development of scientific ways of thinking. The decline of the Ionian school was followed by such great Greek philosophers as Socrates (469-399 B.C.), Plato (c. 427-347 B.C.), and Aristotle (c. 384-322 B.C.). Of these, it was Aristotle who was most interested in the study of the living world, including plants. But as with many ancient writers, a great deal of what Aristotle wrote has been lost, including most of his botanical works. We can get some idea of his thoughts on plants from the writings of his student, Theophrastus. Theophrastus himself wrote more than 200 works, most of which have been lost. But two of his long works on botany have survived. *Inquiry into Plants* deals with the description and classification of about 550 plant species, and *Causes of Plants* discusses plant physiology and reproduction.

Impact

Because of the wealth of information and analysis found in these works, Theophrastus is called the father of botany. In *Inquiry into Plants* he describes not only those plants that are native to Greece, but also species that are found on the Atlantic coast, around the Mediterranean Sea, and even as far east as India. In both this work and *Causes of Plants*, he draws on the writings of earlier philosophers and scientists, including the Greeks Empedocles (c. 495-435 B.C.), Menestor, and Democritus (c. 460-370 B.C.), who had all written about plants. Theophrastus also includes reports from farmers, physicians, and others who had intimate knowledge of plants, so his writings contain a great deal of practical information and accurate observations.

In attempting to classify plants, Theophrastus argued that this task differed from the classi-

fication of animals because plants had less in common with each other than animals did. For example, he noted that while all animals he observed had in common a mouth and stomach, not all plants had leaves, stems, or even roots. For this reason, he thought that while the classification of animals could be based on generalizations, for plants it was specific features that were important. This meant that the essential thing was not reasoning to universals, but direct observation. Theophrastus used generalizations only when he had specific examples with which to support them. So even this early in the history of botany, there was a reliance on careful attention to detail.

It is not surprising that some of Theophrastus's observations are incomplete or inaccurate, considering that he didn't even have a hand lens with which to observe small plant structures. Still, he made a number of important and lasting contributions to botany. He was the first to distinguish between two basic categories of flowering plants, the monocots (those having a set of characteristics that include leaves with parallel veins) and dicots (those having leaves with branching veins). He also differentiated the flowering plants or angiosperms from the more primitive gymnosperms. In *Causes of Plants* he presented a great many observations on plant reproduction, including an accurate description of the germination of seeds, a description that was not improved upon until the seventeenth century.

Some historians of science consider Theophrastus to be the founder of the fields of plant geography and ecology because he didn't just describe the structural features of plants, but also wrote on the relationship between plant structure and habitat. He showed an awareness of plant communities, writing that certain groups of species were often found growing together in the same geographic areas or in the same types of environments. He also wrote of the adaptation of plants to particular environmental conditions, showing, for example, that some species were better adapted to dry conditions while others grew more vigorously in damp areas. In writing of plant pathology, he made a distinction between those diseases due to climate or soil conditions and those that were the result of pests.

The contributions of Theophrastus are particularly outstanding because they were not followed by work of comparable quality. Very little of scientific value was added to botanical knowledge until the Renaissance, which began in the fifteenth century, almost 2,000 years after the time of Theophrastus. Greek science, in general, declined after his time. Athens, which had been the seat of learning, faced political turmoil, and the center of Greek learning shifted to the city of Alexandria in Egypt. While there were a number of writers who produced works on botany, there was little new information in these works, and as the writings of Theophrastus continued to be copied, inaccuracies crept in. So botanical knowledge deteriorated rather than improved.

There were, however, a few figures who made lasting contributions to botany, though to a much lesser degree than Theophrastus. Two of the most noteworthy lived in the first century A.D. The Roman Pliny the Elder (A.D. 23-79) wrote *Natural History*, in which 16 of its 37 books were devoted to plants. Much of Pliny's information is based on the work of the Greek Crateuas (c. 120-60 B.C.) who had produced an herbal, a book on plants used in medicine, in the first century B.C. Pliny's botanical books contain a great deal of information on plants, though this material is not as well organized as that of Theophrastus because Pliny had little interest in classification. Pliny's writings are particularly important because they were highly valued during the Middle Ages, when they were copied many times, leading to the introduction of many errors. The works of Crateuas and Theophrastus, on the other hand, were lost, though Theophrastus's work was rediscovered in the fifteenth century.

The other first-century figure of note in botanical history is Pedanius Dioscorides (c. A.D. 40-90), a Greek born in Sicily. He wrote an herbal that was recopied many times in succeeding centuries, and became the main source of information on the medicinal uses of plants through the Middle Ages. At some point early in its history, Dioscorides's text was combined with illustrations from the herbal of Crateuas, whom Pliny described as the first to use botanical illustrations among the Greeks. It was the illustrated version of Dioscorides's work that was the subject of so much recopying, and this led to errors not only in the text but in the illustrations, which eventually became so simplified that it was impossible to identify the species from the illustrations.

It was in part because of the reliance of the scholars of the Middle Ages on such flawed documents that botanical science failed to progress very far beyond the contributions of Theophrastus. It was only in the late Middle Ages, with the work of scholars such as Albertus Magnus (1193-1280), that botanical science began to progress again. With the introduction of the printing press in the fifteenth century, accurate reproduction of texts and illustrations greatly spurred further development.

MAURA C. FLANNERY

Further Reading

Blunt, Wilfred, and Sandra Raphael. *The Illustrated Herbal*. New York: Thames and Hudson, 1994.

Iseley, Duane. *One Hundred and One Botanists*. Ames, IA: Iowa State University Press, 1994.

Magner, Lois. *A History of the Life Sciences*. 2nd ed. New York: Marcel Dekker, 1994.

McDiarmid, J.B. "Theophrastus." In *Dictionary of Scientific Biography*. Vol. 13. Ed. by Charles Gillispie. New York: Scribner's, 1976: 328-334.

Morton, A.G. *History of Botanical Science*. New York: Academic Press, 1981.

Serafini, Anthony. *The Epic History of Biology*. New York: Plenum, 1993.

Singer, Charles, and E. Ashworth Underwood. *A Short History of Medicine*. 2nd ed. New York: Oxford University Press, 1962.

Stannard, Jerry. *Pristina Medicamenta: Ancient and Medieval Medical Botany*. Aldershot, Great Britain: Ashgate, 1999.

Ayurvedic Medicine

Overview

Ayurvedic medicine, the traditional medical system widely practiced in India today, probably originated some 3,000 years ago. The origins of Ayurveda, "the science of life," are uncertain, but the remedies used may have evolved from prehistoric drug lore. The theoretical concepts and methodologies of Ayurveda are set forth in the classic texts attributed to Charaka, Susruta, and Vagbhata, semilegendary physicians, authors, and founders of the Ayurvedic system of healing.

Background

Mythic tales of ancient India survive in the form of four collections known as the Vedas, sacred books of divinely inspired knowledge. The Vedas are accompanied by later commentaries known as the Brahmanas and the Upanishads, which explain the texts and speculate about the nature of the universe. Further insights into the ancient origins of Indian civilization were made possible in the 1920s when archeologists began to explore the ruins of Mohenjo-daro and Harappa, cities that were part of the forgotten Indus River civilization that had flourished from about 2700 to 1500 B.C.

Ayurveda, the learned system that forms the basis of traditional medicine in India, is called "the science of life." The precise origin of Ayurvedic theory is uncertain, but its materia medica may have evolved from Vedic or even prehistoric drug lore. Vedic medicine was inseparable from Vedic religion, which incorporated many gods and demons. Healing required confession, incantations, and exorcism, because illness was attributed to sin or to demons. Herbal remedies and charms were employed against disease-causing demons. Surgeons treated wounds and snake bites, removed injured eyes, extracted arrows, amputated limbs, and fitted patients with artificial legs. Skulls discovered at ancient Harappan sites suggest that Indian surgeons also performed trepanations. However, appropriate magical or religious rituals had to accompany medication and surgery.

Only fragments of the original *Ayurveda,* the most ancient account of Hindu medicine, have survived. According to Hindu beliefs, the *Ayurveda* was composed by the god Brahma and is thus of divine origin. Scholars believe that the text was probably composed between 1400 and 1200 B.C. Charaka, Susruta, and Vagbhata, the semilegendary authors of the classical Sanskrit texts of Ayurvedic medicine, are honored as the "Triad of Ancients." Their respective treatises are known as the *Susruta Samhita, Charaka Samhita,* and *Astangahrdaya Samhita.*

Impact

The primary objective of Indian medicine was the maintenance of health, rather than the treatment of disease. Health was seen as a state that could only be attained by following an elaborate, individualized program prescribed by the Ayurvedic doctor. According to Ayurvedic principles, the attainment and maintenance of health and happiness was a meritorious pursuit that benefited human beings in their present and future lives.

The *Charaka Samhita,* a treatise on general medicine, is the most complete of the classics of Hindu medicine, but the uncertainty about Charaka's existence is such that some scholars say he lived about 800 B.C. and other claim that he lived about A.D. 100 The *Charaka Samhita,* which is sometimes called the *Handbook of the Physician,* might have been composed anywhere between 1000 and 500 B.C. The text is a guide to the eight branches of Ayurveda: 1) general principles of medicine, 2) pathology, 3) diagnostics, 4) physiology and anatomy, 5) prognosis, 6) therapeutics, 7) pharmaceutics, 8) methods of assuring successful treatment. Other texts use somewhat different lists and include surgery and toxicology. Charaka explains the three forms of medicine: 1) mantras and religious acts; 2) diet and drugs; and 3) psychic therapy, or subjection of the mind. The text provides lists of the plant, mineral and animal substances required for the preparation of medicines, as well as methods for the diagnosis and treatment of hundreds of diseases.

Susruta's existence is similarly uncertain. Some scholars say that "Susruta" does not refer to a real person, but is a kind of title that might be translated as "famous." In any case, like the *Charaka Samhita,* the *Susruta Samhita,* is a guide to the eight branches of Ayurveda. However, the special emphasis of the *Susruta Samhita* is the art of surgery. The text attributed to Vaghbata, who probably lived in the seventh century, is largely based on the *Charaka Samhita* and the *Susruta Samhita.* It was translated into Tibetan from Sanskrit and became a fundamental text of traditional Tibetan medicine. Vaghbata's treatise is sometimes referred to as *The Collection of the Essence of the Eight Limbs of Ayurveda.*

The classic Ayurvedic texts devote considerable attention to the characteristics that distinguished the true physician from quacks and charlatans. The good physician exhibited four primary qualifications: theoretical knowledge, clarity of reasoning, wide practical experience, and personal skill. The surgeon must have courage, steady hands, sharp instruments, a calm demeanor, unshakable self-confidence, and the services of strong-nerved assistants. As members of a professional group, physicians escaped some of the restrictions associated with the Hindu caste system. Physicians did not have to restrict their contact to members of their own caste, but were free to accept students from the three upper castes. Students were expected to live with and serve their teacher until the master was satisfied that their training in medicine and surgery was complete.

According to Ayurvedic physiology, bodily functions could be explained in terms of the three *dosas,* the primary fluids, humors, or principles—*vata, pitta,* and *kapha*—which are usually translated as wind, bile, and phlegm. The three dosas have also been associated with the three gods of Vedic wisdom: Vishnu, Shiva, and Brahma. Although the basic principles of Indian humoral pathology are similar to these of Hippocratic medicine, the Ayurvedic system included additional factors. Ayurvedic physicians believed that the body was composed of a combination of the five elements (earth, water, fire, wind, empty space) and the seven basic tissues. Bodily functions were dependent on the three humors in combination with blood, five separate "winds," the "vital soul," and the "inmost soul." The physician must determine which of the dosas or elements predominates in each individual in order to establish harmony and balance and guide the patient to health.

The delicate balance among the humors and elements could be disturbed by many factors, such as stress, wounds, accidents, and demonic possession. Imbalance among the primary humors was the fundamental cause of disease. Discord among the three humors produced a disturbance in the blood. Depending upon the degree of imbalance, the disease that resulted could be minor, major, or incurable. Almost all people were born with some degree of disharmony among the three humors and this resulted in a predisposition to particular diseases. Knowledge of the natural balance of each patient was essential to prescribing appropriate diet and herbal remedies. In order to restore balance, the physician might also have to remove "bad blood" by venesection or leeching.

Fever was regarded as the king of diseases, but the classical texts listed more than 1,000 different diseases. Accurate diagnosis was the key to selecting the proper treatment for curable disease. The art of diagnosis involved attention to the patient's own narrative of the illness and the physician's observations of the patient's general appearance, abnormalities, internal noises, blood, body fluids, and excretions. The most famous diagnostic test was for the "honey urine disease" (diabetes).

Almost 1,000 drugs derived from plant sources are referred to in the major medical clas- sics, but many are unidentifiable materials or "divine drugs" such as soma. Vedic myths say that Brahma created soma to prevent old age and death, but the identity of this "king of plants" was a mystery to later sages. Although plants provided the majority of medicinal substances, minerals, and animal products, such as honey, milk, snakeskin, and excrement, are also well represented. For diseases involving corruption of the bodily humors, the proper remedies included internal cleansing, external cleansing, and surgical remedies.

Indian surgeons seem to have carried out remarkable operations, despite the lack of a tradition of systematic human and animal dissection. Ancient surgeons are said to have performed cesarean sections, cataract surgery, tonsillectomy, amputations, and plastic surgery. The *Susruta Samhita* does suggest that physicians and surgeons should study human dissection in order to become familiar with all the parts of the body. In order to avoid the religious prohibitions against coming in contact with a corpse or using a knife on the dead, Susruta proposed an unusual form of anatomical investigation. The anatomist must secure a cadaver that was suitable for dissection. The basic method involved preparing the body, covering it with grasses, placing it in a cage of fine mesh, and letting it soak in a quiet pond. About a week later the anatomist would be able to remove the upper layers of skin and muscle by gently rubbing the body with soft brushes so that the inner parts of the body could be observed.

Indian practitioners believed that the body incorporated a complex system of "vital points," or *marmas*. The marmas appear to be sites where major veins, arteries, ligaments, joints, and muscles unite and where injuries are likely to be incapacitating or even fatal. Each point had a specific name and the classical system included 107 points. When the physician examined an injured patient, his first task was to determine whether a wound corresponded to one of the marmas. If he discovered that an injury would be fatal because of the *marma* involved, the surgeon could amputate the limb at an appropriate site above the *marma*. In performing bloodletting or other surgical interventions, the surgeon had to avoid damage to the *marmas*.

All operations were described as variations on the basic techniques of excising, incising, probing, scarifying, suturing, puncturing, extracting solid bodies, and evacuating fluids. When preparing for an operation the surgeon had to pay special attention to the patient, the operating room, and the "one hundred and one" surgical instruments. One reason for the large number of surgical tools was the preference for instruments resembling various animals, such as the lion-mouth forceps. Specific operations required tables of different shapes and sizes. A special "fracture-bed" was used to stretch fractured or dislocated limbs. The cleanliness of the room used for surgery was particularly important. Some historians believed that the ancients had discovered important anesthetic agents, but these claims are probably exaggerated, because the *Susruta Samhita* specifically recommended wine before and after surgery to prevent fainting and deaden pain. The fumes of burning hemp may have been used to create a narcotic effect, but the text also recognized the need for securely tying down the patient.

Surgeons commonly performed bloodletting and cauterization. The *Susruta Samhita* states that the healing properties of the actual cautery (red-hot irons) were far superior to those of the potential cautery (chemically induced burns). Cauterization was used to treat hemorrhages and when diseases resisted herbal remedies. Bloodletting was regarded as a valuable remedy for many ailments, but the operation was dangerous because blood was the source of strength, vitality, and longevity. Leeching was recommended as a gentle form of bleeding in which the leech instinctively discriminated between healthy and unhealthy blood.

Among the operations described in the *Susruta Samhita* are cataract surgery, lithotomy (removal of bladder stones), opening the chest to drain pus, and the repair of torn bellies and intestines. Different kinds of needles and thread were used for closing wounds. If the intestines had been torn, the ancient surgeon could use large black ants as "wound clips." Plastic surgery featured the use of the "sensible skin flap" technique for the reconstruction of damaged noses, ears, and lips. Using a leaf as a template for the new nose, the surgeon would create a flap of "living flesh" (now called a pedicle flap) from the cheek or forehead. The flap was then sewn into place and the wounds were carefully bandaged. The flap could not be cut free until the graft had healed into the new site.

Although Western science and medicine have won a place in modern India, Ayurvedic remedies are still used by millions of people. Indian physicians and scientists continue to find valuable insights and inspiration in the ancient writings. The Indian Medical Council, which was established in 1971, recognizes various forms of traditional medical practice, and sponsors attempts to integrate Indian and Western forms of medicine.

LOIS N. MAGNER

Further Reading

Chatterjee, Asima, and Satyesh Chandra Pakrashi, eds. *The Treatise on Indian Medicinal Plants*. New Delhi: Publications & Information Directorate, 1991.

Gupta, N.N.S. *Ayurveda: The Ayurvedic System of Medicine*. 3 vols. Calcutta: K.R. Chatterjee, 1901-07.

Heyn, Birgit. *Ayurveda: The Ancient Indian Art of Natural Medicine & Life Extension*. Rochester, VT: Healing Arts Press, 1990.

Kapoor, L.D. *CRC Handbook of Ayurvedic Medicinal Plants*. Boca Raton, FL: CRC Press, 1990.

Lad, Vasant, and David Frawley. *The Yoga of Herbs: An Ayurvedic Guide to Herbal Medicine*. Santa Fe, NM: Lotus Press, 1986.

Leslie, Charles M., ed. *Asian Medical Systems: A Comparative Study*. Berkeley, CA: University of California Press, 1976.

Sivarajan, V.V., and Indira Balachandran. *Ayurvedic Drugs and Their Plant Sources*. New Delhi: Oxford & IBH, 1994.

Zimmermann, Francis. *The Jungle and the Aroma of Meats: An Ecological Theme in Hindu Medicine*. Delhi: Motilal Banarsidass, 1999.

Zimmer, Heinrich Robert and Ludwig Edelstein. *Hindu Medicine*. Baltimore, MD: Johns Hopkins Press, 1948.

Zysk, Kenneth G. *Asceticism and Healing in Ancient India: Medicine in the Buddhist Monastery*. New York: Oxford University Press, 1991.

The Science of Physiology: Galen's Influence

Overview

Physiology—the study of how the body works—had its start nearly 2,000 years ago in the studies of Galen, a Greek physician. Galen (also known as Claudius Galenus) studied the writings of the ancient scholars and learned from some of the major scientists of the day, but also brought along his own ideas together with an enthusiastic search for new knowledge. He gained much of his understanding of body organs and systems through the many dissections he performed, particularly during the earlier part of his career. Over the years his work led him to see the benefit of the pulse as a diagnostic tool, to examine the blood and determine that it flowed through the veins and arteries, and to develop hypotheses about the roles of the heart, liver and brain. Galen is now widely considered to be the father of experimental physiology.

Background

As a child, Galen (129-216) was tutored by his father, a mathematician, astronomer, and architect. As a teen, Galen continued his studies with his father, but also learned from many philosophers in his home of Pergamum, Asia Minor, in what is Turkey today. Pergamum was the site of a shrine to the Roman god of medicine, and was visited by many distinguished Romans who were seeking a cure. The teenager soaked up knowledge, but took his father's advice and refrained from joining any particular philosophical sect. Instead he selected portions of the teachings of current philosophers and the writings of the ancient scholars, and combined them to form his own views.

At his father's urging, Galen took up medicine and spent the next nine years studying at Pergamum, at Smyrna, at Corinth on the Greek Peninsula, and at Alexandria in Egypt. He grew well acquainted with animal dissection under his anatomy teacher Satyrus in Pergamum. During this time, Galen became convinced that experimentation, and in particular anatomical dissection, was a requirement for understanding the human body. That understanding, in turn, was necessary for determining the proper treatment for illnesses or injuries. Galen remained true to this conviction and performed numerous animal dissections, sometimes in public, as a way of learning more and teaching others.

While following his studies, Galen began what would become a prolific writing career. His first works included dictionaries designed to help fellow students gain a greater understanding of medicine and philosophy, and a publication titled *On the Movements of the Heart and Lung*.

Galen returned to Pergamum after his schooling and spent four years as chief physician to a troop of gladiators maintained by the high priest of Asia, treating their often massive injuries and gaining new insights into the workings of the human body. Afterward, he went to Rome where he eventually became court physician to three emperors. He continued to write about medicine and philosophy, penning some 700 books and treatises over the years. Some of the most influential were his 17-volume set called *On the Usefulness of the Parts of the Body* and another publication titled *On the Natural Faculties*.

Dissection was key to most of Galen's accomplishments. His many dissections, especially those performed early in his career, helped him gain an outstanding comprehension of the anatomy and physiology of the human body. Through dissection, Galen identified new structures, began to distinguish the functions of known and unknown organs, surmised the relationships between organs, and developed ideas about entire internal systems. Through dissection, he was able to provide nearly incontrovertible evidence for his statements about anatomy.

The dissection of human bodies was frowned upon during Galen's time, so he mostly dissected animals and inferred that humans had the same or similar organs. He also assumed that human organs had the same structure and function as that of animals. While this reasoning was accurate in most cases, Galen wrongly concluded that the human liver had five lobes, that the uterus was horned (based on the shape of the dog's uterus), and that the femur (thigh bone) was curved.

In addition to learning through dissection, Galen turned to the works of earlier scientists to help him gain a more thorough understanding of life processes. Although many of his contemporaries preferred to accept as truth the writings of the ancient scholars and spend their time interpreting their wisdom, Galen believed that science should consider the valid ideas of the ancients as starting points and build upon them.

However, Galen held in high regard many hypotheses of the ancients, including Hippocrates's belief that good health relied on the balance of the four bodily fluids, or humors, which were defined as blood, phlegm, choler (yellow bile), and melancholy (black bile). To maintain the balance between these bodily fluids, Galen maintained a well stocked pharmacy and meticulously documented the effects that different doses of various medications had on a

THE TRUTH BEHIND CRUCIFIXION

The goal of crucifixion was to produce a very slow, painful death. While the Persians probably invented this method of execution, the Romans perfected it. Preparation involved stripping the prisoners and whipping them with a short leather whip called the *flagrum*. Thongs of the *flagrum* held tiny lead balls with sharp bits of sheep bone to bite into the flesh. The 39 strokes caused so much blood loss, pain, and shock that the person would be in a very weak state. By custom, the prisoner had to carry the *patibulum* or crossbar to the place of execution.

Long nails driven into the median nerve in the wrist caused intense pain and secured placement on the cross bar. Then four soldiers lifted the crossbar onto an upright beam. The feet might have been nailed to a wooden footrest. Nerve injury produced fiery pain that shot through the arms producing muscle cramps. Insects bored into the wounds and vultures ripped at eyes, ears, and nose.

A major effect was respiratory failure. As the body hung, the muscles were taxed. The person could not exhale and carbon dioxide was not fully expelled. The person gasped for oxygen. If the prisoner proved especially hardy, the soldiers would break each leg below the knees. Without the ability to push upright to aid in breathing, the victim would eventually suffocate.

EVELYN B. KELLY

series of illnesses. His studies generated considerable insight into the medical benefits of plant, animal, mineral, and combined remedies.

Galen also expanded upon Hippocrates's idea by contending that a person's temperament was directly related to an abundance of one of the four humors; in other words, physiology affects psychology.

Galen was intrigued by the "pneumatic theory," which focused on the importance of air

(pneuma) to the function of the human body and defined three types of "pneuma,"—the natural, vital, and psychic spirits found in the blood and heart, nerves and brain, and liver, respectively. Galen believed that the three forms of pneuma combine to permit life. He also held that a disproportion of one of these spirits could throw off the humoral balance and cause illness. Although this pneumatic theory fell out of favor with the medical community many years later, Galen's investigations in this area helped reveal the heart as a pumping organ and allowed him to see how a changing pulse might affect health. He became the first person to use the pulse as a diagnostic tool, a practice that continues today.

Galen's investigations into blood and air led to his discovery that the arteries carried blood, not air as had been propounded by earlier scientists and physicians. He hypothesized, too, that the liver removed waste products from the bloodstream and also transformed food into blood, which then traveled to various organs. These findings, which integrated blood, air, and food, were fundamental in paving the way for future studies of the circulatory system and metabolism, particularly those conducted by William Harvey of England (1578-1657).

Harvey demonstrated that although Galen was correct in reasoning that blood, air, and food were critical to life, his assumption that the liver converted food to blood was erroneous. Harvey also pursued Galen's emphasis on dissection, but took it a step further by dissecting live animals. This allowed Harvey to view the flow of blood and understand the fundamentals of the circulatory system. Harvey is recognized as the first scientist to determine that the heart's valves assist in blood flow into the heart rather than slowing it, as was commonly believed, and that blood flows from the venae cavae into the heart and then to the aorta. He also demonstrated that the pulse was caused by heart contractions, and that each contraction pumped blood.

Impact

Throughout his life, Galen asserted that experimentation was critical to the advancement of scientific study, and he demonstrated that view through his dissections, hypotheses, and writings. A teacher at heart, he conducted public dissections, wrote dictionaries to help other students grasp medical and philosophical concepts, and drafted hundreds of books and treatises explaining his findings and presenting his ideas.

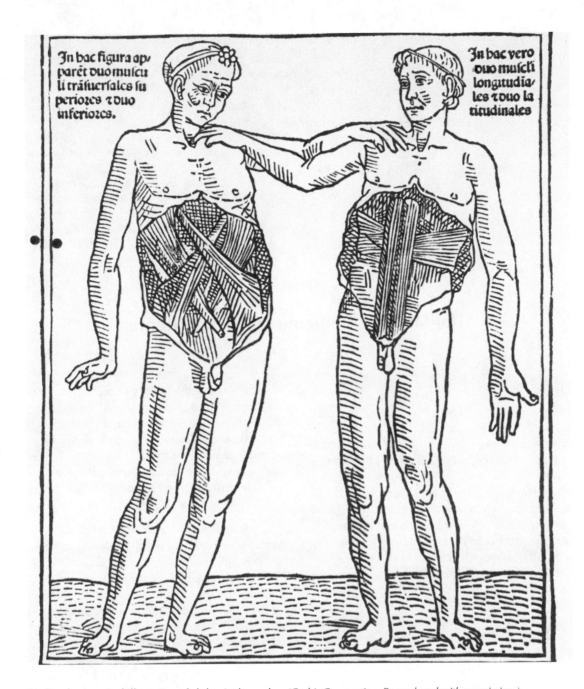

Medieval anatomical illustration of abdominal muscles. *(Corbis Corporation. Reproduced with permission.)*

This latter contribution firmly ingrained his studies and thoughts into the history of medicine. His writings were widely distributed during his lifetime, and his works were taught and collected at the library at Alexandria by A.D. 500. In the ninth century, learned Arabs began to collect Greek manuscripts, and in about 850, an Arab physician named Hunayn ibn Ishaq prepared a list of 129 of Galen's works. In this way, medicine in the Arab world became heavily based on the Galen's ideas. Then, beginning in the late eleventh century, Hunayn's translations, along with commentary on them written by other Arab physicians, were translated into Latin. These Latin versions became the basis of medical education in medieval universities in western Europe. Later Latin translations of Galen's works taken directly from the Greek influenced a new generation of medical students, who were eager to repeat his experiments. Unfortunately for Galen's reputation, this new period of interest in dissection revealed his shortcomings as an anatomist. When Brussels-born Andreas Vesalius (1514-1564) showed that many of the suppositions Galen had made based on dissecting animals were wrong, Galen's authority was undermined. However, Galen's ideas about physiology lasted until Harvey's work correctly explained the circulation of the blood.

Even though many of Galen's findings were eventually found to be incorrect, he was a dominant influence on medical theory and practice in Europe from the Middle Ages until the mid-seventeenth century, and his work was an important element in the rise of modern science.

LESLIE A. MERTZ

Further Reading
Books
Byers, Paula K. *Encyclopedia of World Biography*, 2nd ed. Detroit: Gale Research, 1998.

Magner, L. *A History of the Life Sciences*. New York: Marcel Dekker Inc., 1994.

May, M. T., trans. *Galen: On the Usefulness of the Parts of the Body*. 2 vols. Ithaca, NY: Cornell University Press, 1968.

Mayr, E. *The Growth of Biological Thought: Diversity, Evolution and Inheritance*. Cambridge, MA: The Belknap Press of Harvard University Press, 1982.

Simmons, J. *The Scientific 100: A Ranking of the Most Influential Scientists, Past and Present*. Secaucus, NJ: Citadel Press (Carol Publishing), 1996.

The Military Medicine of Ancient Rome

Overview

In his writing on surgery, famous ancient Greek physician Hippocrates (c. 460-377 B.C.) is credited with saying, "He who desires to practice surgery must go to war." What Hippocrates meant was that it was only during warfare that a physician could learn about closing wounds, infection, and human anatomy and become skilled at using the variety of surgical and other medical instruments in the doctor's kit. While caring for wounded soldiers was not always a priority for ancient military commanders, the medical care of wounded soldiers was a hallmark of Greek armies. Other armies learned from them. It was only after the heavy influences from the Greek tradition of military medicine that, by the first century A.D., the art of Roman military medicine rivaled the art of their conduct of warfare as it was carried out on three continents and over four hundred years.

Background

Before Greek influences, the Roman legions did not have any organized, professional medical services. Wounded soldiers were cared for by their fellow soldiers. Before the first century A.D., there were clear distinctions between sick Roman soldiers and wounded soldiers, and sick soldiers often did get some medical treatment. It was not until Emperor Trajan's time in the second century that wounded soldiers were cared for by *medici*, or doctors. The *medici* were not trained doctors but could dress wounds and perform simple surgeries.

Much about the history of military surgery can be understood through the artifacts discovered during archaeological excavation of military forts and camps. Many of the medical and surgical implements used by the Greek and Roman armies have been found at archeological sites and, interestingly, many are not greatly different from similar instruments in use today.

It is believed that Roman armies began marching with trained physicians and setting up field hospitals during the time of Galen (c. 129-216), a famous Greek physician who wrote extensively on the medical arts and the care of wounds. In addition to treating wounds, it appears that Roman army doctors also had great knowledge of pharmacology: they knew medicinal plants and regularly made medical preparations as well as recommending healing foods to return sick and wounded soldiers to the ranks.

Much of our knowledge about Roman army medicine comes from several sources: the writings of Galen, who attended gladiators; the writings of Celsus (25 B.C.-A.D. 50) and Paul of Aegina (c. 625-690); and the archeological excavations of Roman fortresses and battlefield sites.

Impact

The Roman Legions were highly structured and very efficient. By the time the Roman military medical service became a standard part of the legions, the Roman military medical service was also highly organized and efficient. Most sources

Ancient Greek tomb painting of a wounded soldier. *(Mimmo Jodice/Corbis. Reproduced with permission.)*

suggest that the Roman medical service expanded on the Greek military variety of purposes.

The Roman author Vegetius wrote that the Roman army should be "preserved" by providing a good, clean water supply, taking seasonal considerations into account, the use of medicine, and exercise for the troops. Specifically, Vegetius suggested that commanders not march the troops in the hot sun or in freezing weather and that they provide the troops with clean drinking water at all times. Once more, he said that sick soldiers should be "brought back to health by suitable food and cured by the skills of doctors." Likewise, it was recognized that soldiers could get sick from overeating after experiencing a famine. Most important, as recognized in today's best armies, the ranks of the Roman armies were filled with only the most physically fit soldiers,

who had to pass a medical examination before their service began.

In the early days of Roman military medicine, there was little distinction between medical and veterinary services, and human and animal hospital services were set up side by side. Later, when formal medical service became as highly organized as the rest of the Roman army, a *praefectus castrorum* was placed in overall charge of medical services. The *optio valentudinarii* were responsible for running the hospitals at legion fortresses. Wounded soliders were cared for by medics called *capsarii*, who carried bandage boxes.

The *medicus* was the Roman medical officer, a fully trained doctor who directed medical personnel. Many *medici* were either Greek or Greek-trained. Later Roman army surgeons were usually given the rank of *magister* or "master." Records show that medical supplies and carriages for bearing wounded were placed in the middle of marching columns.

One of the most common medical procedures was the removal of "missiles" from the body, generally arrows but also small lead beads or pebbles shot from slings. These often penetrated flesh and became embedded. The tools used by Roman military doctors, many used for extracting missiles, are themselves legion.

Archaeologists have found medical instruments at Roman forts and campsites that are classified as probes, spatulas, spoons, tweezers, scalpels, lances, curved and straight needles, medical glassware, small vessels, and ointment boxes. One of the most prolific Roman army medical service archaeology sites is at Baden, Germany, where the remains of a Roman military hospital were excavated. Artifacts described as earscoops, catheters, spoons, and other medical equipment were found. Coins found in association with the medical equipment showed the fort was active between A.D.100 and 200.

Another military hospital unearthed at Vetera (now Xanten, Germany) revealed hospital wards, rooms full of medical instruments, surgical suites, convalescence rooms, and possibly mortuaries. Among the discovered artifacts were levers and scoops thought to be used for extracting missiles from the body and notched probes that may have been used for extracting arrowheads after the arrow shaft was broken off. Although later than the heyday of the Roman legions, surgeon Paul of Aegina described how stones and other missiles from slings were to be removed by an ear probe, modified for the job at hand by adding a scoop.

Roman army records show that Celsus suggested using a "weapon extractor." Celsus noted that missiles that have entered the body and become fixed inside "are frequently troublesome to extract," because of their shape, size, or the way they have penetrated. "If the head of the weapon has fixed in the flesh," he wrote, "it is to be drawn out with the hands or by laying ahold of the appendage which is called the shaft, if it has not fallen off. When it has fallen out we make the extraction by means of a toothed forceps." Celsus went on to say that when he saw a curable wound he looked at two things: preventing hemorrhage and preventing inflammation. He suggested vinegar to staunch the flow of blood yet said that to prevent inflammation, blood must flow. He noted that inflammation was more likely when bone, sinew, cartilage, or muscle was injured. "If the wound is

in a soft part, it must be stitched," he noted. "But if the wound is gaping, stitching is not suitable." Celsius also wrote on bandages, preferring bandages of wide linen "sufficiently wide to cover in a single turn not only the wound but to a little extent the edges on both sides."

Roman industrial arts also aided in the cause of treating wounded soldiers. New metal alloys of bronze and silver provided sharper edges and were less expensive to manufacture. Because of rust, iron was not used for medical instruments. Artwork as well as poetry from just before and after the first century A.D. depict Roman army doctors removing arrows from soldiers.

Amputations were also performed by Roman military doctors. Celsus may have been one of the first military surgeons to discuss the merits of amputating above or through the damaged flesh. He advocated cutting only through good flesh and then sawing through the bone as close as possible to the good flesh, but leaving enough good flesh as a flap to cover the bone.

Most historians agree that there was not sufficient call for craftsmen who specialized in making medical instruments, but that physicians would likely have found craftsmen who could make what they described.

Roman army doctors also had a firm knowledge of pharmacology. Many medicines mentioned by Celsus are not unlike those of today made for the same purposes. Archaeologists have found the remains of five medicinal plants at Roman fort excavations. St. John's wort, used for blood ailments; fenugreek for poultices; figs for treating wounds; and plantain for dysentery have been found. Medicated wine was also thought to have been used. Some historians have suggested that the courtyard of each Roman army hospital was laid out as a garden for growing medicinal herbs.

The axiom "an army marches on its stomach" could well date to Roman times. Records show that Roman doctors knew that a balanced diet was necessary for healthy troops. Sources suggest that Roman soldiers were kept fed with corn, cheese, ordinary wine, fresh fruit, and vegetables. Bread was considered the most nutritious food, and each soldier received a ration of *panis militaris*—army bread made of wholemeal. Special diets were prepared for sick and wounded soldiers. Garden peas, lentils, and figs seemed to be popular for treating the sick.

Roman army doctors also understood that soldiers were prone to overeating after a battle or after going without food for a long time. The Roman historian Appian wrote after the siege of

Mutina in 43 B.C. that a number of soldiers fell sick after "excessive eating." The Roman remedy—drinking a concoction of wine and olive oil—probably cured many.

Not only did Roman military medical instruments influence military doctors for the next millennium, their experience and their records, particularly those kept by Celsus and Galen, directly influenced future military surgeons, who improved the art and science of military medicine in the eighteenth and nineteenth centuries.

RANDOLPH FILLMORE

Further Reading

Campbell, Brian. *The Roman Army, 31 B.C–A.D 337: A Sourcebook.* New York: Routledge, 1994.

Davies, Roy W. *Service in the Roman Army.* New York: Columbia University Press, 1989.

Milne, John Stewart. *Surgical Instruments in Greek and Roman Times.* New York: August M. Kelley, 1970.

Roth, Jonathan P. *The Logistics of the Roman Army at War.* Boston: Brill, 1999.

Zimmerman, Leo M., and Ilza Veith. *Great Ideas in the History of Surgery.* New York: Dover Publications, 1967.

Hospitals and Treatment Facilities in the Ancient World

Overview

Today's hospitals and facilities for medical treatment have a history that dates back to the Roman Empire, when military hospitals were organized to treat and repair the critically important Roman Army. Perhaps the most rudimentary form of the hospital prior to this were the healing temples of ancient Greece. Healing temples were sacred sites created for the sick to receive divine aid. They were often associated with public baths and spa-type facilities whose priest-physicians administered rituals of healing, massage, and herbal medicines. Overall, Greek medicine combined a philosophy of proper living, with regular exercise, proper diet, and massage, augmented by herbal drugs and regular visits to a healing temple for ritual devotions and tribute. Physicians, surgeons, and other specialists would typically provide specific treatments including surgeries and amputations at the home of the patient. Both Greek and Roman cultures were engaged in bloody wars and found it necessary to make advances in the medical treatment of wounds and illnesses. The Roman Empire was built by and dependent upon its powerful military, and although it was not a physician-friendly society, it did advance the concept of the military hospital, as well as the public water supply and sanitation systems. Later, as Christianity spread throughout the Roman world, charitable institutions for the poor and societal outcasts such as lepers and plague sufferers were established, and they gradually developed into charitable hospitals that offered various medical treatments for patients

and medical training for physicians in several Roman cities. When Roman civilization declined and fell, education, medicine, and medical facilities also declined, and they did not recover to even the levels reached in late Rome until well after the Renaissance.

Background

The historical time period from 2000 B.C. to the first few centuries A.D. is known to us from various forms of early written accounts, works of art and architecture, as well as many religious and mythical beliefs. The earliest civilizations of Mesopotamia, Egypt, Greece, and Rome grew from their military successes, and they were plagued by periodic wars, dependent upon slavery, and dominated by ritual pagan religions. Texts from the earlier Sumerian and Egyptian civilizations contain instructions for the treatment of physical wounds, and include herbal-based medicines for various internal illnesses. Their medicine and religion were closely interrelated and virtually inseparable. They sought divine aid from their gods of war and healing, and their drugs were herbal-based concoctions and likely of a limited effectiveness. They practiced haruspicy—divination by examination of animal entrails—chanted incantations, and displayed amulets, often in temples or at ritual altars. More elaborate medical treatments were performed by priest-physicians at the estate of the affluent patient, or at a temple visited by the lower-class patients. These temples were the closest approximation to any known medical facility in the Mediterranean and Middle Eastern civilizations prior to the first century B.C.

The Code of Hammurabi indicates that Sumerian medicine included prescribed medical treatments, use of medical instruments, and regulations such as the charging of fees and penalties for therapeutic failures. Egyptian medicine and religion were both state-controlled hierarchies, and their concept of health included the belief that diseases are caused by spirits, intestinal putrefaction, or worms. Medicine and religion were closely connected: priest healers treated the sick and injured, administered herbal medicaments with specific incantations and amulets, and mummified the bodies of the dead. Some of the Sumerian and Egyptian surgical instruments that have been found were used to lance abscesses, cauterize infections and wounds, and were combined with many drugs and antiseptics of vegetable and mineral origins. These civilizations did not establish any facilities for medical care or treatment.

By the fifth century B.C. Greek medicine was practiced by a range of folk healers and priest-physicians who also used a combination of ritual divination procedures and applications of herbal-based drugs. While clearly influenced by Egyptian medicine initially, the Greeks began to develop a more open and secular approach to health and disease. Greek culture emphasized the benefits of rigorous exercise and proper diet to develop a healthy body and mind, and the sacredness of healing is evidenced by their reliance on their gods of healing and disease, such as Apollo, and his son Asclepius. The cult of Asclepius gradually gained widespread following throughout the Greek world, and hundreds of Asclepian healing temples were built and visited by the sick and injured. The temples were located in salubrious locations near springs, and their priests were trained in the use of mineral baths, massage, diet, and herbal drugs. The patient paid tribute and made a request of the god, then slept within the confines of the temple, probably aided by an opiate, and the god would visit in a dream. The patient would wake in the morning and have his dreams interpreted by a priest-physician, who would then perform the proper animal sacrifice and administer an herbal medicine as a cure. The healing temples became associated with the bath or spa, where exercise, massage, mineral baths, and various herbal medicines were employed vigorously and combined with the Greek doctrine of proper diet and moral philosophy (perhaps even psychotherapy). The preventive and

therapeutic benefits of this approach to health were likely to do no harm at the least and probably offered substantial health improvements in many cases. While healers of various types had their shops and physicians and surgeons apparently associated with apothecaries on contract basis, there were no facilities that could be construed as public hospitals.

The Hippocratic school of medicine, which advocated the careful study of the patient and the illness, can be considered the first major advance in medical care and treatment, and it began the separation of medicine from religion. Hippocrates (460-377 B.C.) and his adherents developed a theory of disease based on natural causes, and created a medical system that sought to comfort and aid the patient. Hippocratic medicine was based on the bedside visit and careful physical examination, with cautious drug therapies and strict dietary regulations. Heroic intervention and risky procedures were rejected, although specialists trained in wound treatments, fractures, and amputations were allowed to perform their necessary functions in extreme cases. Hippocrates advocated a proper moral philosophy, bathing, sex, and sleep. However, for all the intellectual interest they had in medicine, the ancient Greeks had little interest in hospitals.

Roman culture gradually adopted the Greek system of medicine, but Rome's greatest contribution to medicine was the organization of medical schools, medical instruction, and public physicians, as well as the development of military and public hospitals. Roman cities and towns also constructed sanitary drainage systems and large aqueducts to provide clean water supplies. When Rome fell, it was plagued by a failure of its drainage systems and the subsequent increase in associated diseases, notably malaria. After Rome, it was not until the seventeenth century that Europe saw the existence of any public hospitals and medical facilities, and the standards of urban sanitation climbed back to the levels that Rome enjoyed at its pinnacle.

Impact

As the roots of Western medicine were being established in ancient Greece, a gradual demystification of illnesses and a separation of medicine and disease from religious doctrine and divine causes began. Greek culture was dominated by wars with its neighboring city-states, which necessitated improved medical treatment of arrow and sword wounds. Greek war heroes were also described as skilled healers, and a strong mythical and religious connection was attached to all healing and medical treatments.

Medicine practiced in the third century B.C. was Greek medicine with an Egyptian influence: human dissections performed in Alexandria by Herophilus (c. 335-c. 280 B.C.) and Erasistratus (fl. c. 250 B.C.) expanded the knowledge of human anatomy and advanced the understanding of physiology greatly. Medical treatment at this time was as advanced as any in seventeenth-century Europe, but in both cases the societies lacked any type of hospital facilities.

The medicine practiced in Rome was largely of Greek origin, with both male and female physicians. Initially, Roman leaders accused the newly arriving Greek physicians of being little more than paid killers. Romans believed disease, famine, and pestilence to be the wrath of vengeful and angry gods. As the need for improved health care continued to grow in Rome, Greek medicine became more prevalent as Greek medical sects and healing cults became more widespread. Celsus (25 B.C.-A.D. 40) advocated medical self help with his written volume *On Medicine,* a guide for the nonprofessional.

The Romans of course also valued the same preventive approaches endorsed by Greek culture and medicine: proper exercise, diet, and spa-type public baths, with some aspects of divine respect and tribute included. As towns enlarged into cities with large populations and trading centers, Roman public officials recognized the correlation between disease and hygiene. Public granaries were strictly regulated for cleanliness; public latrines and plumbed sanitation to control sewage were established in many Roman areas; and a good, clean, reliable water source was a prime directive of public works in Roman cities. Vitruvius (fl. first century B.C.) wrote of the importance of such a water supply, and the aqueducts built for this reason still function today in some areas. Though none of these constructs were hospitals or medical treatment facilities, they are the precursors of institutions built for the health benefits of the general public.

The Roman ethic of military organization and improvement extended from the battlefield to the development of the standard military hospital. The sick and wounded were treated in buildings set aside for this purpose, with large, well lit halls that had individual cells and larger rooms set off of the corridors, as well as baths, latrines, and food-preparation areas. Roman soldiers and officers were treated with respect and honor, and this was certainly extended to their proper medical treatment. Roman military hospitals have been found as far north as the Rhine river in Germany. Surgeries, splints, drainage tubes, and healing salves were applied to wounds. This was also true for Roman gladiators, who received care from the best Roman physicians, including Galen (129-c. 199), and were treated in designated clinics.

When Christianity began to replace the Roman pagan faith, one of its most powerful tools of persuasion was its appeal to the poor, sick, and lame, who were promised the miraculous healing powers of Christ. Though the Church doctrine subordinated medicine to theology and the physician to the priest, the obligation of charity became very strong. Alms for the needy led to houses for the poor and for lepers, such as the ones established by the Church of Rome in A.D. 250. Christianity employed the use of holy relics, oils, and baptisms, as well as the retelling of the Bible stories of healing miracles, but many Christian converts sought to do more. In Roman towns, churches and monasteries began to carry the burden of charity and medical care for the poor and needy. It is believed that at the Monastery of the Pantokrator in the Greek town of Caesarea (A.D. 369) in Roman-controlled Cappadocia, Saint Basil established the first true hospital (*nosocomia*). This facility could be considered a forerunner of today's charity hospital.

In Rome, a student of Saint Jerome became well known for her incredible acts of charity. Fabiola (d. A.D.399) was an affluent Christian convert who dedicated her time, energy, and resources to caring for the sick and indigent of Rome. She is said to have bodily carried even the most filthy and wretched off the streets, washing and caring for them herself, and she is credited with having established a public clinic for the poor and sick in Rome (A.D. 390). Important hospitals were also founded by the churches and monasteries of Edessa (375), Monte Cassino (529), Iona (563), Ephesus (610), and St. Albans (794). These hospitals became large complexes that included a hierarchy of physicians and specialists, with hundreds of beds, as well as teaching facilities, and homes for the poor, elderly, and lepers. The hospital in Jerusalem had over 200 beds by the mid-sixth century, and St. Sampson's Hospital in Constantinople was even larger, with surgical theaters and facilities for various medical specialists.

When the city of Rome was sacked by Alaric the Goth in A.D. 410, all of the Roman towns in the western part of the Empire suffered severe economic collapse. As the populations of its towns and cities dwindled, education and medicine suffered similar fates, and Europe slipped into many centuries of decline. The empire of the East, centered in Byzantium, remained strong and continued the traditions of medicine, as did the growing Islamic and Jewish communities. Medicine in these areas was reinvigorated by the addition of new ideas to the Greek and Roman traditions. But without the wealth of the Roman Empire, the hospitals of the Mediterranean area declined and disappeared, and did not return until the twelfth through the fourteenth centuries, when urban hospitals were created as charitable establishments by the ruling princes and patricians of Europe.

KENNETH E. BARBER

Further Reading

Clendening, Logan. *Source Book of Medical History*. New York: Dover Publications, Inc., 1942.

Porter, Roy. *The Greatest Benefit to Mankind: A Medical History of Humanity*. New York: W.W. Norton and Company, 1997.

Singer, Charles. *A Short History of Anatomy and Physiology from the Greeks to Harvey*. New York: Dover Publications, Inc., 1957.

Biographical Sketches

Alcmaeon

Sixth century B.C.

Greek Anatomist, Physician, and Philosopher

Alcmaeon is generally described as a pupil of Pythagoras (580?-500? B.C.), a member of the Pythagorean community, and a physician with a special interest in biological questions. However, very little is actually known about the life, work, and writings of Alcmaeon. Many of the discoveries attributed to him have been disputed by various scholars. Aristotle (384-322 B.C.) wrote a treatise entitled *Against Alcmaeon,* but only fragments of this text have survived. Alcmaeon of Croton (now Crotona) in southern Italy, the son of Perithous, was, according to Aristotle, a young man in the old age of Pythagoras. Most later writers assumed that Alcmaeon was a Pythagorean, but Aristotle did not specifically say that he was a member of that community. However, Aristotle suggested that either Alcmaeon derived his theory of opposites from the Pythagoreans or they acquired the theory from him. Alcmaeon apparently said that the gods have certainty about invisible and mortal things, but men could only speculate about such matters.

Although Alcmaeon mainly wrote about medical matters, some scholars argue that there is no direct evidence he was a physician. In addition to medical theory, physiology, and anatomy, Alcmaeon apparently wrote about meteorology, astronomy, philosophy, and the nature of the soul. There are many uncertainties about the scope of his anatomical and physiological investigations. Indeed, some commentaries assert that he was the first person to practice scientific dissections of human bodies and the first to attempt vivisection. Most authorities suggest that Alcmaeon was the first natural philosopher to carry out dissections and vivisections of animals for the sake of learning about their nature rather than for purposes of divination. Moreover, he may have introduced the practice of examining the developing chick egg as a way of studying embryology. For these pioneering studies of an important model system, Alcmaeon might be considered the founder of embryology.

In explaining the nature of health and disease, Alcmaeon contended that health required a harmonious balance of opposite qualities, such as moist and dry, hot and cold. He defined health as the establishment of isonomy, or equilibrium, of all the contrary qualities that make up the body. An excess of any one of the qualities, therefore, caused disease. His theory of health and disease had a great influence on the subsequent development of philosophy and apparently anticipated the teachings of Hippocrates (460?-377? B.C.).

There is little doubt that Alcmaeon conducted important studies of the anatomical structure of various animals, with a particular interest in using dissection as a means of understanding the nature of sense perception. He is said to have described the difference between arteries and veins, discovered the optic nerve and the Eustachian tube, and recognized the brain as the seat of the intellect. Some scholars dispute these assumptions and argue that it was Aristotle rather than Alcmaeon who first noted the duct passing from the ear to the throat that is now known as the eustachian tube (which was rediscovered by Eustachius in the sixteenth century).

Despite later criticism of Alcmaeon's theories, he was honored as one of the first to clearly define the difference between man and the animals. According to Alcmaeon, man is the only creature that has understanding and intelligence, while other animals perceive but do not understand. That is, Alcmaeon distinguished between sensation and understanding. By means of dissection and logical inference, Alcmaeon reached the conclusion that the brain was the center of intellectual activity. Only fragments of his theories of the special senses were preserved. Apparently, he attempted to develop a theory of vision that combined the ancient concept of vision as a radiation proceeding from the eye with the idea that vision involved an image reflected in the eye. He attempted to treat the study of the senses systematically and recognized the importance of air for the sense of hearing. Hippocrates and Plato (427?-347? B.C.) adopted Alcmaeon's theory of the brain as the common sensorium, but Empedocles (492?-432 B.C.), Aristotle, and the Stoics reverted to belief that the heart is the central organ of sense. The ancients did not actually recognize the nerves as specific anatomical and physiological entities, but Alcmaeon apparently knew that certain kinds of lesions could block some kind of "passages" involved in the transmission of sensations to the brain. Thus, Alcmaeon can be thought of as the founder of empirical psychology.

As would be expected of a Pythagorean, Alcmaeon apparently believed that the immortal soul was the source of life and was always in motion. His concept of the heavenly bodies was closely associated with his contention that the soul is immortal because it resembled immortal things and was always in motion like the heavenly bodies. Some scholars believe that Alcmaeon was the source of the concept Plato attributed to Timaeus (fl. c. 400 B.C.), a Pythagorean who said that the soul has circles in it revolving just as the heavens and the planets do. The orbits of the heavenly bodies always make perfect circles, but the circles in the human head may fail to complete themselves. Therefore, man dies because he cannot join the beginning to the end.

LOIS N. MAGNER

Aristotle. *(Corbis Corporation. Reproduced with permission.)*

Aristotle
384-322 B.C.
Greek Philosopher

By any measure, Aristotle ranks as one of the greatest geniuses who ever lived. He completely reworked Plato's philosophy and established it on a firm systematic basis. He formulated the disciplines of logic, psychology, and embryology, and made important contributions to the study of zoology, medicine, anatomy, physiology, and other life sciences.

Aristotle was born in the coastal town of Stagira in northern Greece. His father Nicomachus was court physician to King Amyntas III of Macedonia. His mother Phaestis was from a prominent family in Chalcis on the Greek island of Euboea. Both parents died when Aristotle was young. He was then raised by his scholarly uncle Proxenus, who gave the boy a wide-ranging education.

At the age of 17, Aristotle enrolled in the Academy of Plato in Athens. He was Plato's student and associate for 20 years, until Plato died in 347 B.C. Disappointed in Speusippus, who followed Plato as head of the Academy, Aristotle accepted the invitation of Hermeias to teach in Assos, Turkey. He married Hermeias's daughter Pythias and they had one daughter, also called Pythias. In 345 B.C. Aristotle moved to the Greek island of Lesbos where he began a collaboration with Theophrastus (372-287 B.C.), who became his most gifted disciple.

In 343 B.C. King Philip II of Macedonia hired Aristotle to tutor his 13-year-old son Alexander, who was later called Alexander the Great. Aristotle taught Alexander until 340 B.C., when the prince became king. Alexander remained Aristotle's friend and protector, and from 335 B.C. sent him biological specimens from all the lands he conquered.

Sometime between 340 B.C. and 336 B.C. Aristotle moved back to his hometown of Stagira, but he returned to Athens in 335 B.C. After his return, Aristotle founded his own school, the Lyceum, to rival the Academy. Aristotle's school of philosophy is known as Peripatetic, either because he had the habit of strolling around while he lectured (peripatetic is from the Greek verb *peripatein* meaning "to walk back and forth," or from the fact that his instruction was given in the *peripatos,* the covered walkway of the gymnasium. His wife Pythias having died, Aristotle had a liaison with a Stagirite woman, Herpyllis. They named their son Nicomachus after Aristotle's father.

When Alexander the Great died in 323 B.C., anti-Macedonian agitation broke out in Athens. Aristotle, who had long-standing Macedonian connections and was a friend of the Macedonian regent of Athens, felt himself in danger. He retired to his mother's family's home on the island of Euboea, reportedly stating that he was leaving Athens to save the Athenians from sinning twice against philosophy (referring to Socrates as the earlier victim). He died of a stomach disease a year later.

Although Aristotle wrote in Greek, we refer to the titles of his books in either Latin or English. No chronological ordering of his works is possible. Less than half of what he wrote survives, and much of it was probably written by students transcribing his lectures.

In the areas of natural science and its philosophy, Aristotle wrote *Physics, On Generation and Corruption, On the Sky, Meteorology,* and *On Breath.* His works on zoology include *History of Animals, On the Parts of Animals, On the Motion of Animals, On the Generation of Animals,* and *On the Gait of Animals.* Eight of his shorter works on life science (*On Sense and Sensible Objects, On Memory and Recollection, On Sleep and Waking, On Dreams, On Divination by Dreams, On Length and Shortness of Life, On Youth and Age,* and *On Respiration*) are collectively called *Parva Naturalia.*

Aristotle wrote four books about ethics, *Nicomachean Ethics, Eudemian Ethics, Magna Moralia,* and *Politics,* and two about the philosophy of art, *Rhetoric* and *Poetics.* His six books on logic, *Categories, On Interpretation, Prior Analytics, Posterior Analytics, Topics,* and *On Sophistical Refutations,* are collectively called the *Organon.* His *On the Soul* is regarded as the world's first book about psychology. His *Metaphysics,* a work of pure philosophical speculation grounded in empirical observations, has had a tremendous influence on Western philosophy and theology.

ERIC V.D. LUFT

Asclepiades of Bithynia
124?-40? B.C.
Greek Physician and Philosopher

Asclepiades was born in Prusa, Bithynia (in modern Turkey), in about 124 B.C. Bithynia was part of Greece at that time and Asclepiades went on to become one of the most influential Greek physicians in history. Like many of the Greek physicians, he also worked extensively in the areas of science and philosophy. As a predecessor of Galen (129-199?), Asclepiades exerted considerable influence over his peers until Galen began to dominate Greek medicine in A.D. 164.

Asclepiades was a follower of the philosopher Democritus (460?-370? B.C.), who had developed the atomic theory during the fifth century B.C. Atomic theory attempts to explain complex phenomena, including many aspects of nature, in terms of combined units of fixed particles. Asclepiades extended this basic concept to medical thought and his views on medicine were diametrically opposed to those of Hippocrates (460?-377? B.C.), who was perhaps the most important physician in history. Hippocrates firmly believed in the healing power of nature and supported a humoral doctrine of medicine. This principle states that there are four humors in the body: blood, phlegm, yellow bile, and black bile. A healthy person has these four humors mixed in exactly the right proportions, whereas someone who is ill has a deviation in the correct mixture. The goal of the physician was to restore the humors to the natural equilibrium and thereby restore health.

However, Asclepiades had his medical doctrine firmly rooted in atomism. This specific type of philosophy lends itself well to application within the natural sciences and is in fact a basic component of our known physical universe. This philosophy contends that the material universe is made up of minute, undetectable

Asclepius attending to a patient. *(Gianni Dagli Orti/Corbis. Reproduced with permission.)*

particles that have various forms, which account for observed differences. Because of this, it logically follows that any changes in the observed appearance must therefore be due to changes in them minute particles. Thus, Asclepiades firmly believed that the individual parts were more important than the sum of the whole organism.

Asclepiades specifically believed that disease resulted from either contracted or relaxed conditions of the minute solid particles that made up the individual. Because of this, he argued that the healing power of nature did not exist, but rather treatment should be started as quickly as possible. This treatment should be both agreeable to the patients and safe for them. He believed that proper treatment would restore the harmony within the impaired solid particles. The forms of treatment included the use of fresh air, corrective diet, hydrotherapy, massage, and exercise.

Asclepiades was an advocate for the mentally ill. He made extensive studies into mental conditions and clearly understood the differences between hallucinations and delusions. His theory of disease made it obvious to him that people who are mentally ill had those problems because of impaired particles. He attempted to treat these mental disorders in a humane manner. At that time, it was common for people who were considered insane to be locked up in dark, confining places. Asclepiades believed that this

only further misaligned the particles, so he advocated their release from these places. He further treated them with regimens that he believed would restore harmony, such as occupational therapy, music, wine, and exercise. He is considered to be a pioneer in the humane treatment those afflicted with mental disorders.

Another significant contribution that Asclepiades made was the integral role he played in gaining the acceptance of Greek medicine in Rome. He took up residence in Rome and practiced his healing art. He had such a positive impact that the Roman physician Aulus Cornelius Celsus (first century A.D.) gave a classic account of Greek medicine in his work *De medicina*. Asclepiades continued to make a significant contribution to his field until his death in Rome around 40 B.C.

JAMES J. HOFFMANN

Aulus Cornelius Celsus
25 B.C.-A.D. 50
Roman Medical Writer

Aulus Cornelius Celsus has been called the first important medical historian and one of the greatest Roman medical writers, as well as the creator of scientific Latin. He seems to have written an encyclopaedia dealing with agriculture, the art of warfare, rhetoric, philosophy, law, and medicine, but only the treatise known as *De medicina* (On medicine) survived. Today *De medicina* is universally regarded as an invaluable medical classic.

Yet almost nothing is known about the life of Celsus. Even his status as author of *De medicina* has been questioned. Because Celsus wrote in Latin, during an era in which Greek was considered the language of medicine and scholarship, Roman and medieval scholars ignored his work. In the fifteenth century two copies of *De medicina* were discovered. To Renaissance scholars, Celsus represented a pure source of first-century Latin grammar and medical philosophy, uncorrupted by medieval copyists. Therefore, *De medicina* was among the first medical works to be widely published after the introduction of the printing press.

The historical portion of *De medicina* is of great importance today because it is the primary or only source of much of what is now known about Hellenistic medicine and Alexandrian anatomy and surgery. *De medicina* is divided into three parts, according to the type of treatment appropriate to various diseases, that is, dietetic, pharmaceutical, and surgical. *De medicina* contains some of the earliest accounts of heart disease, insanity, and the use of ligatures to control bleeding. Celsus also offered excellent descriptions of hydrotherapy and operation for the removal of bladder stones (lateral lithotomy).

Historians are divided as to whether Celsus actually wrote or merely plagiarized the materials in *De medicina*. Some scholars say that Celsus was simply a compiler or translator, but others argue that the quality of the text and the display of critical judgment about medicine and surgery make this very unlikely. Indeed, many Renaissance scholars venerated Celsus as a master of organization, clarity, and style. It is known, however, that his contemporaries considered Celsus a man of quite modest talents. It is unlikely, given Roman customs, that Celsus would have been a professional physician. Traditionally, Roman landlords were expected to assume responsibility for the medical care of the sick on their estates. Generally this simply meant knowing enough about medicine to supervise the women or slaves who carried out the menial tasks actually associated with treating the sick. Like other Romans, Celsus asserted that physicians had been unnecessary in ancient Rome. The Greek art of medicine became a necessity only after indolence, luxury, and other Greek influences that led to illness had infiltrated Roman society.

After the death of Hippocrates (460?-377? B.C.), Greek medicine fragmented into various competing sects, such as the methodists, dogmatists, pneumatists, and empiricists. Without the analysis provided by Celsus, the origins and ideas peculiar to the sects that flourished in his time would be totally obscure. Celsus concluded that no sect was wholly right or wrong. Although medical practice primarily involved the selection of appropriate drugs, Celsus argued that medical practitioners must also master human anatomy and the art of surgery. While rejecting the Greek concept of the physician as necessary guide to proper regimen throughout life, Celsus offered a great deal of advice about healthful living. The best prescription for a healthy life, Celsus argued, was variety and balance, proper rest and exercise, and avoidance of a self-indulgent obsession with medical advice.

Surgery, according to Celsus, should be the most satisfying field for the practitioner because the surgeon knew that a cure was the result of his skill, not mysterious forces, accident, or good fortune. The surgeon must protect his patient from hemorrhage and infection by attention to cleanliness. Celsus defined the four cardinal signs of inflammation: *calor, rubor, dolor,* and *tumor* (heat, redness, pain, and swelling). He described surgical tools, techniques, and operations that were unknown to his Hippocratic predecessors, such as the use of the ligature for torn blood vessels, special spoons and dilators to remove barbed arrows from wounds, plastic surgery, and so forth.

LOIS N. MAGNER

Diocles of Carystus
fl. fourth century B.C.
Greek Physician

Diocles of Carystus was a philosopher and pioneer in Greek medicine, acclaimed by the historian Pliny to be second only to Hippocrates (c. 460-c. 377 B.C.) in reputation and ability.

Born in the late fourth century B.C. at Carystus, Euboea, he was the son of Archidamus, a physician. He moved to Athens and became a pupil of Aristotle (384-322 B.C.). Although Aristotle was known as a philosopher, he influenced many physicians of his time because of his inquiries into body physiology. His anatomy was adapted by Diocles and three prominent doctors of the Alexandrian school: Herophilus (c. 335-280 B.C.), Erasistratus (c. 304-250 B.C.), and

Praxagoras (fl. fourth cent. B.C.). The four raised Greek medicine to its highest point. Another of Aristotle's pupils, Alexander the Great, died in 323 B.C. exclaiming, "I die by the help of too many physicians"—showing there were a number of physicians in his court.

Physicians including Diocles became prominent in the late fourth century B.C., and when Alexander conquered Egypt and founded the city of Alexandria, he set the stage for the advent of the famous Alexandrian School of Medicine. His successor, Ptolemy I, collected a library of 70,000 manuscripts with information on medicine and drugs.

Diocles was a leading proponent of the dogmatic or logical school and sought to combine philosophy with the medical ideas of Hippocrates. While other physicians were engaged in speculation and superstition, Diocles formalized and organized medicine.

A prolific writer, Diocles was the first to use Attic Greek, the polished Greek of Athens, and showed the influence of Aristotle's literary style in writing. (Most physicians of the time wrote in Ionic Greek, a coarse vernacular style.) The subjects addressed were wide and varied. Only a few fragments of his writing are extant. Diocles carefully assembled the writings of Hippocrates. The Roman physician Galen (c. 129-216 A.D.) stated that Diocles was the first to use the term "anatomy." Like Aristotle, Diocles did not distinguish the nerves from veins and believed that the heart, not the brain, was the seat of intelligence.

Empedocles (c. 492-432 B.C.) greatly influenced the Greek physicians, including Diocles. For example, he was interested in reproduction and asserted that both man and woman furnished the seed that became the embryo, which was fully developed in 40 days. The male developed on the warmer, right side of the uterus and grew faster than the female. Menstruation was the same for all females—beginning at age 14 and ending at 60. Empedocles also influenced Diocles's physiology, in that the latter believed in the four basic elements: air, water, fire, and earth. Health was the proper balance of the system. The four humors—blood, phlegm, yellow bile, and black bile—that corresponded to the elements must also be in balance.

Diocles was close to political rulers. He dedicated a work on hygiene to the Macedonian prince Pleistarchus, son of the famous Greek general Antipater. He wrote a letter also on hygiene to King Antigone, a general of Alexander. The letter was preserved by the seventh-century Greek physician Paul of Aegina (c. 625-690 A.D.). A work called *Archidamus* was dedicated to his father. When the Greek classics were rediscovered in the sixteenth century, translations of the works of Diocles were made into Latin, French, and English. Also, large fragments of his work on diet were preserved by Oribasius (325-403 A.D.), physician to Emperor Julian.

Inspired by Aristotle and his studies of plants, Diocles was the first scientist to write on nutrition and the medical use of plants. He is considered to be the father of pharmacy. Two inventions were credited to Diocles: a bandage for the head and a spoon-like device for scooping arrows out of the flesh.

Diocles was second only to Hippocrates in the annals of Greek medicine. His work influenced many Greek physicians and scientists, such as Theophrastus (c. 372-287 B.C.) and Dioscorides (c. 40-90 A.D.).

EVELYN B. KELLY

Pedanius Dioscorides
40?-90?
Greek Physician and Pharmacologist

Dioscorides, the compiler of one of the first important Western herbals, is regarded as the founder of Western pharmacology. Dioscorides exemplifies the generalization that Roman medicine was usually carried out by Greek physicians. Little is known about Dioscorides, except that he probably studied at Alexandria before he became a *medicus* attached to the Roman army in Asia under the emperor Nero. "Military physician" is probably too specialized a translation for this term, because it is unlikely that the Roman legions of this period were accompanied by an organized medical or surgical staff. Military service gave Dioscorides the opportunity to travel widely and to study many exotic plant and animal species.

In his major work, a treatise now known as *De materia medica* (The materials of medicine), Dioscorides refers to nearly 600 plant species, including cannabis, colchicum, and peppermint. Only about 130 of these medicinal plants had been known to Hippocrates (460?-377? B.C.). The discussion of each plant includes its appearance, growth characteristics, place of origin and habitat, medical uses, and the proper method for preparing remedies. Drugs derived from animals and minerals were also described. In total, Dioscorides refers to about 1,000 simple drugs.

The text describes most of the drugs used in medical practice until relatively modern times.

Dioscorides, a keen observer and naturalist, was attempting to devise a more systematic treatment of botany and herbal medicine than his predecessors. Despite the presence of some bizarre ingredients, Dioscorides recorded recipes for many effective drugs, including purgatives, emetics, laxatives, analgesics, antiseptics, and so forth. Recent studies of *De materia medica* suggest that the arrangement adopted by Dioscorides reflects a subtle and sophisticated drug affinity system rather than traditional systems of organization, such as plant morphology or habitat. His classification scheme must have required close attention to the effects of drugs on a significant number of patients. According to Dioscorides, physicians needed information about all medically useful plants, their place of origin, habitat, growth characteristics, and proper uses. Physicians who failed to study the characteristics of medicinal materials were easily fooled by drug dealers who trafficked in worthless, cheap, and even dangerous substitutes. Unlike Theophrastus (372?-287? B.C.), his most important predecessor, who classified plants as trees, shrubs, and herbs, Dioscorides grouped plants under three headings: aromatic, culinary, and medicinal. According to Pliny the Elder (23-79), the Greek physician Crateuas (1st century B.C.) produced an herbal with colored illustrations, but this text has not survived. Thus the herbal of Dioscorides is generally considered the first systematic and illustrated treatise of medical botany.

Many of the herbs, spices, nuts, grains, and fermented beverages used as ingredients in remedies recommended by Dioscorides can be found in any modern grocery store. The medicinal properties assigned to them, however, would surprise modern cooks. Cinnamon and cassia, for example, were said to be useful in the treatment of internal inflammations, poisons, venomous bites, cough, diseases of the kidneys, menstrual disorders, and so forth. They were also said to induce abortion. Drinking a decoction of asparagus and wearing the stalk as an amulet supposedly induced sterility. Judging from the large number of remedies said to bring on menstruation and expel the fetus, menstrual disorders, contraception, and abortion must have been among the major concerns that patients brought to the physician.

Remedies were also made from minerals and animal parts and products, such as milk and honey. Chemical drugs included mercury, arsenic, lead acetate, and copper oxide. A remedy

Dioscorides. *(Library of Congress. Reproduced with permission.)*

for malaria included bed bugs mixed with meat and beans. A mixture of burnt river crabs, Gentian root, and wine was suggested as a treatment for the bite of a mad dog; eating the liver of the dog in question and wearing its tooth would prevent further complications. Stones found in a hippopotamus offered protection from the bites of serpents, but frogs could be used in an antidote if hippo stones were unavailable. For warming the joints and healing abrasions Dioscorides recommended grime from the gymnasium walls. He also described sleeping potions prepared from opium and mandragora and their use as surgical anesthetics.

Dioscorides's *De materia medica* constitutes the most authoritative source of information about the materia medica of antiquity and the most significant classical source of modern botanical terminology. In various translations and editions, *De materia medica* remained the leading pharmacological text for some sixteen centuries.

LOIS N. MAGNER

Erasistratus of Ceos
304?-250? B.C.
Greek Physician and Anatomist

A celebrated physician and anatomist, Erasistratus is considered in some circles the

Erasistratus of Ceos taking the pulse of Antiochus I. *(Corbis Corporation. Reproduced with permission.)*

founder of physiology for his advancement of anatomy as it related to medical knowledge during his time. He not only studied the structure of the body, but he also tried to explain it in terms of function. Among his many contributions were the ideas that sensory and motors nerves were functionally different and the identification of the proper operation of the epiglottis and the correct function of heart valves. Erasistratus was the first major proponent of a philosophy known as pneumatism. This idea was based on the premise that life depended on a subtle vapor, which was called pneuma. Along with Herophilus (335?-280? B.C.), Erasistratus is also considered to be the cofounder of autopsy, the science of studying human corpses.

Like many of his influential peers, few specifics are known regarding the life of Erasistratus, though it is believed that he was born in 304 B.C. in Ceos, which was part of Greece. Historical records from this time are scarce. Most of the knowledge that we have of him is in relationship to his work and practically nothing is known about his personal life. We can, however, get a rough sketch of his professional life through the historical documents that remain.

The center of Greek culture shifted to Alexandria, Egypt, soon after the time of Aristotle (384-322 B.C.). Erasistratus was one of Alexandria's most famous early residents. He was a teacher at a medical school, which broke with tradition and religious dogma by using dissection of human corpses for study. Up until that time, and also soon afterwards, none of the world's civilizations dissected the human body because it was held in such reverence and awe. Beliefs and fears regarding the soul, life after death, and even resurrection helped to inhibit the use of dissection for study. Knowledge of anatomy and physiology were slowly acquired through ordinary medical treatment, childbirth, and lower animals, but the field of anatomy remained largely incomplete in this regard. Erasistratus and Herophilus used dissection of human cadavers to give anatomy a scientific basis for the first time in history. This was the birth of autopsy as a medical science, effectively breaking an ancient barrier to progress in medicine. Unfortunately, the use of autopsy, and consequently the advancement of human anatomy and physiology, went into a great decline for the next 1,800 years.

Through his anatomical studies, Erasistratus described the major portions of the brain. From his study of nerves, he believed them to be hollow and filled with fluid. He understood the function of the heart valves and named the tricuspid valves based on their appearance. He was not, however, a strong proponent of the tripartite system of humors. This view held that there existed three distinct fluids: the nervous spirit (carried by nerves), animal spirit (carried by the arteries), and blood (carried by the veins). Based on this, Erasistratus expanded the philosophy called pneumatism.

Pneumatism was an ancient school of thought based on the idea that life is dependent upon a vapor or fluid called pneuma. It was an attempt to explain respiration in conjunction with what was believed to be the function of the blood, vessels, and nerves. Erasistratus held that life was intimately connected with pneuma, which was in the air we breathe. He further believed that health and disease resulted from the pneuma. There was a distinction between two types. The first was the "vital spirit" that was formed by the air the heart and transported through the arteries. The second, called the "animal spirit," was formed in the brain from the vital spirit and transported through the body by nerves. Erasistratus believed that any impediment to the action of pneuma would result in disease. While thousands of years of medical science have shown that many of his views were erroneous, Erasistratus remains a prominent figure in medicine because of his tremendous insight and influence.

JAMES J. HOFFMANN

Fabiola
d. c. 399
Roman Noblewoman and Hospital Founder

Fabiola was a Roman matron of noble birth who converted to Christianity and became a disciple of St. Jerome (c. 347-419/420), dedicating her considerable wealth and energies to the care of the indigent sick. She founded at Rome the first civilian public hospital in western Europe, and also cofounded at Porto the first hospice, which offered rest and refreshment to pilgrims and travelers.

Although her date of birth in Rome is not known, some facts about Fabiola's younger years are recorded. She belonged to the patrician Roman family of the Fabia and at some time in early adulthood became a Christian convert. Her first marriage was reputedly to a violent and abusive husband; she divorced him, according to her rights as a Roman citizen. Before he died she married again, a violation of Church law. After the death of her second husband, she performed public penance for this sin and was reinstated in the Church.

At this turning point in her life, Fabiola renounced all worldly pleasures and devoted herself to the care of the poor and the sick. Service to the sick, especially to the social classes held most in contempt by society, had assumed great importance in Christian teachings. It was nothing less than a religious and ethical duty for both the community and the individual, and for some a sacred vocation. It is no coincidence that the first hospitals were associated with Christian saints, healers, and philanthropists.

In 394 Fabiola financed the construction at Rome of the first civilian public hospital in western Europe. In addition, she used her country villa as a convalescent home for discharged hospital patients who needed shelter and rest, and donated large sums to churches and religious communities. Fabiola was no detached sponsor of charitable works, she tended personally to the wounded and diseased on a daily basis. St. Jerome, her teacher and spiritual advisor, remarked that there was no patient whose disease was so repulsive that Fabiola refused to nurse him herself. She is reported to have walked the streets of Rome in search of the sick, the dying, and the abandoned, sometimes carrying them to the hospital on her own shoulders. Her life became a model of Christian love and charity.

In 395 Fabiola, who was familiar with Hebrew, Greek, and Latin, traveled to Bethlehem to study the scriptures and participate in ascetic practices under the direction of St. Jerome. Following an attack by the Huns, which made life dangerous in the eastern provinces of the empire, Fabiola returned to Rome. Here she collaborated with the former Roman senator St. Pammachius (d. 409?), a boyhood schoolmate of St. Jerome and friend of Fabiola's, in another charitable project—the erection of a hospice (the first of its kind) for pilgrims coming to Rome. During the early Christian era, seafarers from Spain and Africa landed at Porto, a town close to Ostia at the mouth of the Tiber River. After the ruins of a large hospice were discovered at Porto in the nineteenth century, excavations revealed a building intelligently designed for housing travelers as well as caring for the sick. The hospice had five large wards, long connecting corridors, and large inner courts that could have held 400 beds.

For the remainder of her life, Fabiola continued her personal service to the poor and the sick, and also maintained her correspondence with St. Jerome. Following her death in 399 or 400, St. Jerome eulogized her, remembering that thousands had thronged to her funeral, and portraying her as a selfless and devoted advocate for the least fortunate members of her society.

DIANE K. HAWKINS

Galen

c. 129–c. 216

Greek Physician, Writer, and Philosopher

The work of Galen (Claudius Galenus) made him the primary authority on medical understanding and practice throughout the Middle Ages. Often described as the leading scientist of the time, he is particularly known for his contributions to physiology and often is noted as the father of experimental physiology. These contributions arose mainly from information he gathered during the many animal dissections he performed and from his insights and inferences regarding the functions of and relationships between various organs. Galen's reputation in the medical community was enhanced further by his role as physician to three Roman emperors, as well as by the more than 500 books and treatises he wrote about his findings and hypotheses.

Galen was born in Pergamum, off the east coast of the Aegean Sea in Asia Minor (present day Bergama, Turkey) in 129. In his early years, Galen was educated by his father, an architect, mathematician, and philosopher. When he turned 14, Galen began his studies at Pergamum, which continued for four years.

After Pergamum, Galen studied at Smyrna, at Corinth on the Greek Peninsula, and at Alexandria in Egypt. During this time, he also began writing and completed *On the Movements of the Heart and Lung* in Smyrna in 151. A strongly opinionated young man, Galen condemned those current-day teachers he felt were incompetent, calling them "ignoramuses" who presented "long, illogical lectures to crowds of 14-year-old boys who never got near the sick." In an effort improve the education of his fellow students, Galen began to write dictionaries on medicine and philosophy.

In 157 Galen moved back to Pergamum, where his interest in medicine continued. There he announced and demonstrated a self-described "cure for wounded tendons." The demonstration drew attention and helped him attain the position of chief physician to a troop of gladiators maintained by the high priest of Asia. Over the next three years, Galen treated their often massive injuries and gained new insights into the workings of the human body.

Galen left Pergamum for Rome in 162 when a war with the Galatians interrupted the gladiators' competitions, and thus his work. There he continued his studies, often publicly conducting demonstrations of anatomy and sharing his thoughts on medical treatments and on the workings of the human body, a topic that would eventually be called the field of physiology. He rose quickly in the medical profession because of his public demonstrations, his successes with rich and influential patients, and his great learning. His wealthy background, social contacts, and influential patrons also enhanced his reputation.

With his private and public studies and his understanding of physiology, Galen confirmed his belief in Hippocrates's hypothesis that health is governed by a balance of four bodily fluids, or humors. He became particularly interested in one humor, the blood. Not only did he establish that arteries carry blood instead of air, as previously thought, but he described how blood distributed food, or nutrition, to the organs in the body. Also during this period, Galen began a meticulous pharmacological investigation into the effects of various medicines on different illnesses, recording exactly how each medicine was made and the dosages given.

Galen returned to Pergamum briefly in 166, but returned to Rome in 169, where he served as physician to the emperor Marcus Aurelius and the later emperors Commodus and Septimius Severus. Galen was a prolific writer, producing about 500 tracts on medicine, philosophy, and ethics, many of which have survived in translated form. In his later years, he wrote two essays that listed all his works and provided a narrative of his career and evolution as a philosopher. Galen's life summaries represent one of the earliest recorded autobiographies of a scientist and possibly the first recorded bibliography.

The exact date of Galen's death is not known. Some scholars think that he died sometime after 210, but his Arab biographers asserted that he died in Rome in 216 or 217, at the age of 87. After his death, the Christian church accepted Galen as the standard authority on medicine throughout the Middle Ages. The ensuing lull in medical progress was broken more than 1,000 later, when scientists such as Andreas Vesalius (Belgian, 1514-1564) and the William Harvey (English, 1578-1657) began to follow Galen's premise that new knowledge and ideas are critical to the advancement of medical science.

LESLIE A. MERTZ

Herophilus of Chalcedon
335-280 B.C.
Greek Physician and Anatomist

Herophilus of Chalcedon was a Greek physician and anatomist who performed human dissections at the world-renowned Museum of Alexandria. Herophilus gained fame as a physician and medical instructor, and because of his careful human dissections, he has been called the "Father of Anatomy." Frequently quoted by the medical colossus Galen (129-216 B.C.), his careful and detailed works on the brain, eyes, nerves, liver, and arteries greatly advanced understanding of both human anatomy and physiology.

The Greek-controlled city of Alexandria, Egypt, was fast becoming a growing center of scholarly activities when Herophilus established himself there at the invitation of King Ptolemy (r. 323-285 B.C.), and he became the leading physician and anatomist at the Museum of Alexandria. He is believed to have produced at least nine written works that are known to have influenced his contemporaries and future generations of physicians, including the most influential of all physicians, Galen, whose writings dominated human medicine for centuries after his death. Unfortunately, none of the works of Herophilus survived directly, and we are left with only the frequent quotes of his writings made by others, especially Galen, Dioscorides (c. A.D. 40-90), Pliny (A.D. 23-79), and Plutarch (c. A.D. 46-after 119). Herophilus worked in Alexandria during a single brief period that saw the ruling Greeks relax their long-held prohibition on human dissections. This allowed Herophilus to study human internal anatomy in considerable detail, and he was able to advance overall knowledge of human anatomical design and function tremendously. This knowledge and learning came at the cost of being labeled a desecrating pagan and butcher of men in the minds of many subsequent Greek and Roman philosophers and writers in the generations that came after him.

Herophilus was a follower of the Hippocratic doctrine of medicine, which saw health and disease as a balance or imbalance of the four humors of the body. He believed that proper diet and exercise were the necessary ingredients to good health, and when a patient's health was out of balance, the most useful treatments available to physicians were a variety of herbal and mineral drugs, as well as bloodletting. This type of holistic medicine stressed proper nutrition, essential physical exercises, and a strong moral philosophy as keys to a healthy life. Physicians enforced these tenets in themselves and their patients, and when illness struck a patient, the physician sought to provide comfort, aid, and certainly some specific curative herbal concoctions, while doing no further harm to the patient. Herophilus was a student of Praxagoras (fl. fourth century B.C.), and he sought to advance understanding of the system of arteries and veins that had been described by Praxagoras. In his studies, Herophilus concentrated on the structures and functions of the three important organs that were subject to generations of great debate: the liver, the heart, and the brain.

Herophilus dissected human cadavers to find "the nature of the fatal malady," and probably performed these dissections in a public forum as instruction to other physicians. Herophilus expanded the known details of the human heart and its arteries and veins, and he focused on the pulse of the heart, measured it with a water clock and described various diagnostic values of the pulse. Herophilus also described the lacteals, the liver, the prostate gland, and named the duodenum. His greatest and most influential contribution lies in his details of the human brain and the many nerves of the body. He distinguished the cerebellum from the cerebrum, described the ventricles of the brain, and classified nerve trunks as containing sensory nerve connections into the brain and motor nerve connections from the brain to the rest of the body. He also described the structure of the eye and traced the arrangement of the brain's protective dura mater. Herophilus was one of the first to describe the brain not as a cooling structure, but as the organ of human intelligence and sensory control. Herophilus's work was well respected and influenced the Roman medical giant Galen, who became the most prominent physician of his time and dominated medicine for many subsequent generations.

KENNETH E. BARBER

Hippocrates of Cos
c. 460-c. 375 B.C.
Greek Physician and Philosopher

Hippocrates is rightly called the "Father of Medicine." His own writings and writings attributed to him, as well as legends about him and the oath of medical ethics that bears his name, continue to be important aspects of Western medicine today.

Hippocrates. *(Library of Congress. Reproduced with permission.)*

which is hot and dry; earth, which is dry and cold; water, which is cold and moist; and air, which is moist and hot; and from the four essential qualities: dryness, from a combination of fire and earth; cold, from a combination of earth and water; moistness, from a combination of water and air; and heat, from a combination of air and fire. The humors give rise to the four temperaments: choleric, when yellow bile predominates; melancholic, when black bile predominates; phlegmatic, when phlegm predominates; and sanguine, when blood predominates. The whole schema corresponds to the four seasons: summer, which is hot and dry; autumn, which is dry and cold; winter, which is cold and moist; and spring, which is moist and hot. When he treated a patient, Hippocrates sought to cure by restoring the balance between the humors.

The Hippocratic Oath, in which modern physicians still swear to uphold the ethics of their profession, was almost certainly not written by Hippocrates. It seems to be the product of a slightly later Pythagorean philosophy of medicine.

About 60 surviving works are attributed to Hippocrates, although it is impossible to decide which of the 60 were actually written by him. Ancient sources say that he wrote about 70. His topics cover every aspect of medicine—anatomy, physiology, and other medical sciences; clinical questions about internal medicine, diagnostic techniques, pathology, gynecology, pediatrics, pharmacology, and nutrition; medical ethics; and medical philosophy. Historians generally agree that the collection of works attributed to Aristotle at one time made up the library of a medical school, probably the library at Cos. The works were brought to the library in Alexandria, Egypt, during the second or third century B.C., where they were edited and made available to the medical community throughout the Hellenistic and Roman worlds.

The works of Hippocrates are marked by a deliberate and fatherly concern for his patients and his art, as well as an enlightened but serene assessment of the entire nature of things. In the *Aphorisms,* he wrote that life is short, art is long, opportunity is fleeting, experience is dangerous, and judgment is difficult. Authors as diverse as the Roman Stoic dramatist Seneca *(ars longa, vita brevis),* and the medieval English poet Geoffrey Chaucer ("The lyfe so short, the craft so long to lerne") have used this theme.

Hippocrates's ideas about medical treatment can be summed up by advice he gave in the *Epidemics.* He said that if physicians cannot cure a

According to Greek mythology, Asclepius, the god of medicine, settled his mortal son Podarios on the island of Cos in the Aegean Sea just after the Trojan War. Hippocrates is said to be a direct descendant of Podarios. All males in this line were physicians, and were known as Asclepiads. Hippocrates learned medicine from both Herodicus of Cnidus and Heraclitus. He probably lived and taught on Cos his entire life, but died in Larissa, Thessaly.

Hippocrates was esteemed during his lifetime as a medical practitioner, theorist, and scientist. His methods were gentle, simple, and often effective. His belief that many common diseases were caused by digestive disorders led him to prescribe simple dietary modification or mild herbal therapy for the majority of his cases.

Central to the Hippocratic teachings on physiology, or the functions of the body, is the humoral theory. Recorded by Aristotle and transmitted by Galen (A.D. 129-c. 216), it dominated Western medical thinking for 2,000 years. According to this theory, the body is healthy when the four fundamental liquids that govern it are balanced, but disease results when they get out of balance. These four liquids, or humors, are: yellow bile, which is hot and dry; black bile, which is dry and cold; phlegm, which is cold and moist; and blood, which is moist and hot. The humors derive from the four elements: fire,

disease, then they should at least be sure not to make it worse. Mistranslated into Latin as *Primum non nocere*, this is usually interpreted as "First, do no harm."

ERIC V.D. LUFT

Paul of Aegina
625-690
Greek Physician and Surgeon

Paul of Aegina was a Greek physician and surgeon whose writing contained almost everything that was known about Western medicine in his time. Because his work had such an influence on Arab medical practice, and Arab medical texts were the primary references for medieval medicine in Europe, it had a lasting influence on Western medieval medicine. Paul's greatest work was *Epitomae medicae libri septem* (Medical compendium in seven books). Abu al-Qasim (c. 936-c. 1013), one of the most important Islamic surgeons, borrowed extensively from the sixth book of *Epitome* when compiling his own work on surgery.

Paul was born on the Greek island of Aegina and lived during the reign of the Emperor Heraclius (575-641, r. 610-641). Very little is known about his life, but most sources agree that he was educated at the University of Alexandria. He is best known for his descriptions of surgical techniques in the *Epitomae*. Paul described amputation of a breast, lithotomy (surgical removal of a stone in the bladder), trepanation (removal of a disk of bone from the skull), tonsillotomy (removal of part of the tonsils), and paracentesis (removal of fluid from a body cavity). For a hernia in the groin, he recommended the removal of the testicle on the affected side. This treatment, though drastic, was acceptable during his time.

Paul knew how to use an emergency tracheotomy to open an airway and how to close it once the danger of suffocation was over. He described 62 types of pulses associated with disease and treatments for strokes and epilepsy, and gave the one of the first known descriptions of lead poisoning. The *Epitomae* also included a lengthy description of the removal of a fishbone from the throat. This possibly reflects the society in which he lived, for the diet at the time must have included a great deal of fish in order to warrant so much attention to this type of trauma.

The first book of *Epitome* contains information about the complaints of pregnant women, while the third book describes diseases of women. Perhaps Paul's best-known work is on difficult labor, a lengthy description which is probably compiled from Aëtius, an earlier Greek physician. The fact that he could describe pregnant women shows that Paul had experience in a field that was usually limited to midwives. He must have been a trusted and respected individual because few men at the time were allowed to examine a woman's body. Midwives reportedly asked Paul to help difficult cases, and he became known as a man-midwife. He was aware that obesity presented a risk for women in labor. He wrote on the different positions of the fetus in the uterus (known as presentations) and gave instructions for turning the baby. He also explained how to give psychological encouragement to a woman who was in "low spirits" or who was giving birth for the first time.

Paul was knowledgeable about menstruation, and advised that irregular or profuse bleeding during menstruation could be stopped by the use of ligatures around the limbs (another treatment that would seem rather drastic in modern times). He recommended the same treatment for hysteria, as hysteria was attributed to a "wandering" womb, which presumably would be controlled by the ligature. He also described the use of a speculum to gain visibility of the cervix.

LANA THOMPSON

Praxagoras of Cos
fl. fourth century B.C.
Greek Physician

Praxagoras was born on the island of Cos about 340 B.C. His father, Nicarchus, and his grandfather were physicians. Very little is known of his personal life, and none of his writings have survived. Between the death of Hippocrates in 375 B.C. and the founding of the school at Alexandria, Egypt, Greek medicine became entrenched in speculation with little advance in knowledge. During this period four men took up the study of anatomy: Diocles of Carystus (fl. fourth cent. B.C.), Herophilus (c. 335-280 B.C.), Erasistratus (c. 304-250 B.C.), and Praxagoras.

Galen (A.D. 129-216), the famous Greek physician, wrote of Praxagoras as an influential figure in the history of medicine and a member of the logical or dogmatic school. Galen also probably knew of the works of Praxagoras, which were extensive. He wrote on natural sciences, anatomy, causes and treatment of disease, and on acute diseases.

Praxagoras adopted a variation of the humoral theory, but instead of the four humors (blood, phlegm, yellow bile, and black bile) that most physicians held, he insisted on eleven. Like the other Greek physicians, he believed health and disease were controlled by the balance or imbalance or these humors. For example, if heat is properly present in the organism, the process of digestion is natural. Too little or too much heat will cause a rise in the other humors, which then produces certain disease conditions. He considered digestion to be a kind of putrefaction or decomposition, an idea that was held until the nineteenth century.

Praxagoras studied Aristotle's (384-322 B.C.) anatomy and improved it by distinguishing between arteries and veins. He saw arteries as air tubes, similar to the trachea and bronchi, which carried pneuma, the mystic force of life. Arteries took the breath of life from the lungs to the left side of the heart through the aorta to the arteries of the body. He believed the arteries stemmed from the heart, but the veins came from the liver. Veins carried blood, which was created by digested food, to the rest of the body. The combination of blood and pneuma generated heat. As one of the humors, thick, cold phlegm gathered in the arteries would cause paralysis. Also, he believed that arteries were the channels through which voluntary motion was given to the body, and that the cause of epilepsy was the blocking of the aorta by this same accumulation of phlegm.

Aristotle, Diocles, and Praxogoras insisted that the heart was the central organ of intelligence and the seat of thought. Praxagoras differed with the others in that he believed the purpose of respiration was to provide nourishment for the psychic pneuma, rather than to cool the inner heat.

His views of arteries were very influential on the development of physiology. Since the concept of nerves did not exist, Praxagoras explained movement to the fact that arteries get smaller and smaller, then disappear. This disappearance caused movement, a fact now attributed to nerves. However, he speculated about the role of movement and was satisfied that he had found the answer of the center of vitality and energy. His pupil Herophilus actually discovered both sensory and motor nerves.

Praxagoras was interested in pulse and was the first to direct attention to the importance of arterial pulse in diagnosis. He insisted that arter-ies pulsed by themselves and were independent of the heart. Herophilus refuted this doctrine in his treatise "On Pulses." In another area, Galen criticized Praxagoras for displaying too little care in anatomy. He suggested that Praxagoras did not arrive at his theories by dissection.

Praxagoras was very influential in the development of Greek medicine in general and the Alexandrian school in particular. After the death of Alexander the Great (356-323 B.C.), Egypt fell to the hands of General Ptolemy, who established a modern university with the first great medical school of antiquity. Human dissection was practiced, and although the university in Alexandria and its massive library were destroyed by bands of conquerors, later Arabic physicians made the efforts to preserve some of the writings. After the fall of the Byzantine Empire, Greek scholars brought back Greek medicine to the medical schools of the Western Renaissance.

The beliefs of Praxagoras held sway for centuries. For example, for nearly 500 years after his death, many still believed that arteries did not contain blood but pneuma. His most famous pupil, Herophilus, was instrumental in establishing the marvelous medical establishment at Alexandria.

EVELYN B. KELLY

Theophrastus of Eresus
372?-287? B.C.
Greek Scientist and Philosopher

Theophrastus was a scientist and philosopher who made significant contributions to nearly every area of thought and science, and in particular the study of botany and ecology. Originally named Tyrtamus, he received the nickname Theophrastus, meaning "divine speech," from Aristotle (384-322 B.C.). Theophrastus was an extremely prolific writer who tackled a variety of subjects. His treatises were in-depth and thorough. He is responsible for writing well over 200 documents on the subject of botany alone.

Born in Eresus on the island of Lesbos, Theophrastus was originally a pupil of Plato (427?-347 B.C.), though became attached to Aristotle after Plato's death. When Aristotle retired from the Lyceum (an academy founded in Athens by Aristotle), he appointed Theophrastus as his successor. While Theophrastus studied very diverse issues, he is best known for his work with plants. He has often been referred to as the first scientific botanist, and two of his practical, yet influential, books on the subject have survived into modern times.

Theophrastus wrote two seminal works in the field of botany called *On the Causes of Plants* and *On the History of Plants*. In these texts, he outlines he basic concepts of morphology, classification, and the natural history of plants. These works seemed to be aimed at the gardener, so they were quite practical in nature rather than being theoretical. His ideas on this subject were accepted without question for many centuries. Theophrastus was also a great advocate of science in his day and wrote many of his treatises with this in mind. Another area where he seemed to have a significant amount of influence was in the field of ecology.

It does not seem that the beginning of the field of ecology can be traced to any one historical event, but many authorities consider Theophrastus to be the father of ecology because he was the first to describe the interrelationships between organisms and the environment. While his descriptions are not ecology in the modern sense, he does deal with plants from within their ecological groups. Ecology was an extension of the field of natural history, and certainly he played a crucial role in its development. Theophrastus was not only interested in the living world, but he was also interested in studying the physical world as well.

Theophrastus wrote the oldest known document regarding the primary classification of over 70 different rocks and minerals. This work is still considered to have value, even though it is over 2,000 years old. It is considered by many to be the most influential text ever written in the field.

Theophrastus was one of the few students who fully embraced Aristotle's philosophy in all subject areas. He was the first person in the history of logic that was known to have examined the logic of propositions seriously. His contributions to this area and to other fields of philosophy were significant. As previously mentioned, Theophrastus wrote on a wide variety of subjects and even provided some of the most popular character sketches ever written.

His book, *Characters*, consists of 30 brief character sketches in which he defined an undesirable personal quality and then went on to describe the traits of a man who exemplified them. This was a widely read manuscript throughout the Renaissance and was even made into a popular play.

While much of his work has been lost to history, one thing that is certain is that Theophrastus was an industrious writer and thinker who made considerable contributions in nearly every area of philosophy and science. He died in around 287 B.C. after heading the Lyceum for 35 years. Upon his deathbed he is purported to have lamented that human life was too short, as he was just beginning to gain insights into its problems.

JAMES J. HOFFMANN

Biographical Mentions

Aemilius Macer
70-16 B.C.

Latin poet who wrote about birds and medicinal plants. He was a friend of Ovid's, with whom he traveled in Asia and Sicily, and his didactic poetry was influenced by the works of physician Nicander of Colophon. Aemilius Macer appeared as "Mopsus" in Virgil's Fifth Ecologue and, under the pseudonym Aemilius Macer, a medieval herbalist named Odo Magdunensis wrote *Macer Floridus de Viribus Herbarum,* a work that described in verse almost 90 different herbs.

Aetius of Amida
527-564

Physician to the Byzantine court who described a method for removing tattoos. Aetius advised applying saltpeter and terebinth resin (or turpentine) to the tattoo and then abrading the skin with salt a week later. The process could then be repeated if necessary. Variations of Aetius's method of tattoo removal were used for centuries until the 1990s, when laser became the method of choice to remove tattoos.

Agnodice
Third century A.D.

Greek female physician who disguised herself as a man to practice medicine, an occupation prohibited to women. According to legend, Agnodice once attempted to assist a woman crying out in the pain of childbirth. Thinking Agnodice was a man, the woman refused help, whereupon Agnodice lifted her tunic, revealing that she, too, was a woman. Agnodice was persecuted by male doctors, but her legend points to the realities of a male-dominated profession.

Aretaeus of Cappadocia
81?-138?

Greek physician who practiced Hippocratic medicine in Rome and Alexandria. He left many excellent accounts of disease, including *On the Causes and Indications of Acute and Chronic Diseases* and *On the Treatment of Acute and Chronic Diseases,* which contain description of asthma, epilepsy, diphtheria, pleurisy, pneumonia, and tetanus. Aretaeus was apparently the first to distinguish between spinal and cerebral paralyses and the first to provide a clear description of diabetes. Because the disorder was characterized by intense thirst and excessive urination, he named it diabetes, meaning "to pass through."

Artemidorus Daldianus
Second century A.D.

Greek mystic who lived in Roman Asia (now Turkey) and authored *Oneirocritica* (Interpretation of dreams). The text, one of the most often quoted works on dream interpretation, is a collection of earlier works. The first three books deal with dreams and divination; the fourth book addresses critics and contains an appendix.

Artemisia of Caria
fl. fifth century B.C.

Greek woman of Asia Minor who studied the medical sciences and botany. It was said of Artemisia, whose story is perhaps semilegendary, that she was an authority on medicinal herbs, and that the *Artemisia* genus of plant was named in her honor. It is possible that she was the same person as Artemisia, queen of Caria (d. c. 350 B.C.), who married her brother Mausolus (r. c. 377-c. 352 B.C.) and after his death erected the famous Mausoleum at Halicarnassus, one of the Seven Wonders of the Ancient World.

Flavius Magnus Aurelius Cassiodorus
490?-583?

Roman writer, statesman, and monk who established monasteries and encouraged the duplication of ancient texts by monastic scribes, thus ensuring the preservation of Roman culture as well as Greek, Christian, and pagan works that might have perished. He authored several of his own works, including a treatise on early Church music is still studied by scholars of ecclesiastical music and ancient musical instruments.

Charaka
fl. second century A.D.

Indian doctor who reportedly served as court physician for Kanishka (r. c. 78-103), or at least for the ruling Kushan dynasty. A student of the Hindu medical system known as Ayurveda, Charaka is said to have mentioned some 500 drugs in the *Charaka Samhita,* a well known treatise on medicine.

Chen Ch'uan
d. c. 643

Chinese physician who is recognized as the first person to identify the symptoms of diabetes mellitus. He noted that certain patients developed an intense thirst and had sweet urine.

Empedocles
492?-432 B.C.

Greek philosopher who gained fame as a physician, statesman, theologian, mystic, and democratic reformer. Only fragments of his long poem, *On Nature,* have survived. Galen called Empedocles the founder of the Italian school of medicine. Empedocles taught that the four ultimate elements—fire, air, water, and earth—were associated with the hot and the cold, the moist and the dry. According to Empedocles, we breathe through all the pores of the body, and respiration is closely connected with the motion of the blood. He is regarded as one of the first to suggest employing experimentation in physiology and medicine.

Epicurus
341-271 B.C.

Greek philosopher who introduced an ethical philosophy based on simple pleasure, friendship, and the avoidance of political activity and public life. He was the founder of schools of philosophy that competed with the Academy of Plato and the Lyceum of Aristotle. Unlike the more famous schools of his rivals, the Epicurean schools admitted women. Epicurus taught that the atomic theory of Democritus could serve as the basis of a philosophical system of ethics, as well as physics. The writings of Epicurus dealt with physics, meteorology, ethics, and theology.

Hsi Han

Chinese writer who in a work on agriculture mentioned what is now recognized as the first example of biological control of garden pests. His *Plants and Trees in the Southern Regions* (A.D. 304) notes that certain carnivorous ants could successfully protect mandarin oranges from insects that would normally infect the plants.

Hua T'o
145?-208

Chinese surgeon who is traditionally credited with the invention of anesthetic drugs, antiseptic ointments, medicinal baths, hydrotherapy, medical gymnastics, abdominal surgery, and other remarkable operations. Hua T'o was also allegedly a master of acupuncture and a brilliant diagnostician. Hua T'o burned his manuscripts while imprisoned by Emperor Ts'oa Ts'ao, and all of his surgical discoveries, except for his technique for castration, were forgotten after his death. He left no students or successors to carry on his art; it is possible that he existed only as a figure of legend.

Isidore of Seville
560-636

Spanish theologian and encyclopedist who is considered the last of the Western Latin Fathers. Isidore's *Etymologiae sive Originens*, an encyclopaedia of human and divine subjects, was considered an important reference book throughout the Middle Ages. The *Etymologiae* covers the liberal arts and medicine, the Bible and the Church, agriculture, warfare, shipping, and so forth. The etymological section is considered one of the first important landmarks in glossography, the compilation of glossaries. Isidore's other writings included linguistic studies, natural science, cosmology, history, biography, and theology. Isidore was canonized by Pope Clement VIII in 1598.

Lucretius (Titus Lucretius Carus)
99-44 B.C.

Roman writer and philosopher who is remembered for his poem *De rerum natura* (On the nature of things). *De rerum natura* provides an exposition of the theories of the Greek philosopher Epicurus, a philosopher revered by Lucretius. The poem encompasses the main principles of the atomic theory, refutes the rival theories of other philosophers, applies atomic theory to the structure and mortality of the soul, condemns of superstition, and describes the mechanics of sense perception, thought, and certain bodily functions, the creation of the Earth and the heavenly bodies, the evolution of life and human society, and the horrors of the plague of Athens.

Saint Luke
fl. first century A.D.

Syrian physician and Christian evangelist who is one of the only physicians mentioned in the Bible. Luke was converted to Christianity after meeting Saint Paul. Subsequently, in addition to traveling with Paul, he made a number of missionary trips on his own. A Greek, not a Jew, he was born in Antioch, capital of Syria. He is re-garded as the author of the Acts of the Apostles and the third Gospel.

Oribasius
325-403

Greek physician who dedicated his life to compiling the works of his predecessors and preserving existing medical knowledge. Oribasius was a prolific and influential medical writer, but he was not an original thinker. His compilations are of great importance because they include the works of ancient authors who would have been totally forgotten, if not for Oribasius. His compilations of works in medicine, hygiene, therapeutics, and surgery from Hippocrates to his own times contain selections from medical works that no longer exist and would otherwise be completely lost to history.

Phidias
493?-430 B.C.

Greek sculptor responsible for the Parthenon and Acropolis sculptures, including the statue of Zeus, one of the Seven Wonders of the Ancient World, and the statute of Athena, a massive ivory and gold sculpture. Hailed as the greatest sculptor of ancient Greece, Phidias was named by Pericles to oversee the erection of public works in Athens. After his enemies accused him of stealing gold from the Athena statue and carving pictures of himself and Pericles on Athena's shield, he fled into exile.

Philista
318-272 B.C.

Greek female medical lecturer. It was said that Philista was so beautiful she had to speak from behind a curtain—a story also told regarding a scholar about whom much more is known, Hypatia of Alexandria (370-415). It is possible that Philista, like her alleged contemporary Agnodice, was a legendary figure; even so, these legends illustrate the fact that a few women were attempting to find a place within the male-dominated world of Hellenic medicine.

Pliny the Elder
23-79

Roman scholar and writer who produced a comprehensive encyclopedia of zoology and botany, as well as astronomy, geography, and other topics. In the encyclopedia, *Historia Naturalis* (Natural History), Pliny compiled information gathered from some 2,000 written sources. In many cases, *Historia Naturalis* is the only remaining record of the works of ancient scholars. The 37-volume encyclopedia became a primary educational reference and remained in high regard until the fifteenth century, when critics determined that the books were rife with errors. It is now seen as an important literary work, but one that has little scientific merit.

Polyclitus
fl. fifth century B.C.

Greek sculptor who advanced knowledge and understanding of human anatomy and form. Although none of his artwork is known to have survived, his reputation as a creator of sculptural masterpieces is strongly supported by the large numbers of copies that were made of his famous works, such as *The Discus Thrower*. A written treatise on human form and proportion, called the *Kanon* and attributed to Polyclitus, enhances his reputation as a master of expressing proper human proportion and external anatomy in his sculpture. His work illustrates the earliest connection between forms of art that accurately detail the human body and their influence on the study of human anatomy and medicine.

Praxiteles
375-340 B.C.

Greek sculptor who is famous for his marble carvings of mythological gods and goddesses. What made him unique as a sculptor is the way he positioned the bodies of his figures. Rather than posing them in traditional rigid, face-forward, arms at the side, symmetrical positions, his statues are bent, inclined, or slightly twisted to appear more naturalistic. His best known works are *Hermes Carrying the Infant Dionysus, Apollo Sauroctonus,* and *Aphrodite of Cnidus.*

Pythagoras
c. 550-480 B.C.

Greek philosopher and mathematician who founded a philosophical and religious movement based on the belief that all natural systems are determined by mystical numerical relationships. He and his followers established the close relationship between science and mathematics and strongly influenced Plato. The Pythagoreans attempted to apply their principles to all areas of the natural world; it has been suggested that they may have written the Hippocratic Oath.

Rufus of Ephesus
fl. late first century B.C.-mid-first century A.D.

Greek physician whose writings on anatomy included the pulse and its relationship to the heartbeat. Rufus was skilled in physiology, clinical observation, surgery, and the treatment of disease. He described bubonic plague and gout; he studied the anatomy of the eye, the brain, and the placenta. He was first to describe the liver, although his description was based on a pig's liver. Rufus introduced medicines such as hiera, a strong cathartic containing bitter apple, and others. His surgical procedures to stop bleeding were advanced for his time. His treatise *On the Interrogation of the Patient* was important because it explained how medical and family histories contribute to an accurate diagnosis of a patient.

Soranus of Ephesus
fl. second century A.D.

Greek physician who was for centuries regarded as the leading authority on gynecology, obstetrics, and pediatrics. Supposed sixteenth-century innovations such as the obstetric chair and podalic version (delivery of the fetus feet first), were first described by Soranus, who also authored works on embryology, hygiene, pharmacology, chronic and acute diseases, surgery, and bandaging. Soranus belonged to the methodist school of medicine, viewing health and disease as a function of three possible states of the fine pores of the body: excessive constriction, excessive relaxation, or a mixture of the two.

Strato of Lampsacus
340?-270? B.C.

Greek philosopher and Aristotelian who succeeded Theophrastus as director of the Lyceum and head of the Peripatetic school of philosophy. Strato also studied physics and was the first person to note the acceleration of falling bodies. He was renowned for his work concerning the formation of a vacuum, or void, which later became the foundation of Greek designs for air and steam engines.

Susruta
380?-450?

Indian physician who has been called the greatest Indian surgeon in history. Susruta stressed the importance of sterilizing wounds and also introduced several types of surgery, including nasal plastic surgery and abdominal and cataract surgery. Susruta trained his students to practice surgical techniques on vegetables and dead animals before performing actual surgery on a patient. His medical texts were translated into Arabic during the late eighth century A.D.

Thucydides
c. 471-401 B.C.

Greek historian known for his eyewitness account of the plague that struck Athens in 429 B.C. at the outset of the Peloponnesian War

(431-404 B.C.) "The bodies of dying men lay one upon another," Thucydides wrote, "and half-dead creatures reeled about in the streets. The catastrophe became so overwhelming that men cared nothing for any rule of religion or law." Thucydides himself suffered from the illness but recovered, and noted that survivors were able to tend the sick without catching the illness again. The epidemic, which claimed the life of the city's great leader Pericles (c. 495-429 B.C.), was undoubtedly the result of poor sanitation and overcrowding exacerbated by the war. As for the nature of the illness itself, modern scientists have variously identified it as typhus or influenza.

Zeno of Citium
c. 335-c. 263 B.C.

Greek philosopher who founded the Stoic school. Among the areas addressed by the Stoics was health, and they were prescient in maintaining that a link exists between what modern people would call "mental health" and physical health. Thus the Stoics maintained that to be happy was to align oneself with nature. On the other hand, their teaching that individuals should patiently accept circumstances such as poverty, sickness, and death is certainly at variance with modern Western thinking.

Bibliography of Primary Sources

Aristotle. *De Anima* (On the soul, fourth century B.C.). Considered the world's first book about human psychology.

Aristotle. *Historia animalium.* Fourth century B.C. Identifies 495 species, including cephalopods, crustacea, testacea, and insects; discusses animal sounds, habitat, diet, migration, hibernation, and locomotion; analyzes animal diseases, health, and reproduction, distinguishing between oviparous and viviparous animals; and compares human and animal physiology. Aristotle was a systematist and his studies of morphology and structure may be described as preclassification. Based on his work, subsequent natural philosophers tried to organize nature according to classifications of comparative anatomy.

Aristotle. *Parva naturalia.* Fourth century B.C. Contains eight of Aristotle's shorter works on life science: *On Sense and Sensible Objects, On Memory and Recollection, On Sleep and Waking, On Dreams, On Divination by Dreams, On Length and Shortness of Life, On Youth and Age,* and *On Respiration.*

Artemidorus Daldianus. *Oneirocritica* (Interpretation of dreams, second century A.D.). An important and oft-quoted text on dream interpretation. Though probably a collection of earlier works, the first three books deal with dreams and divination, while the fourth book addresses critics and contains an appendix.

Charaka. *Charaka Samhita.* c. first century A.D. One of the principal medical texts of Ayurveda, a branch of traditional Hindu medicine, including mention of some 500 drugs. The date of the work is uncertain, as it was transmitted orally centuries before it was written down. Attributed to Charaka, a court physician, it is almost certain that other physicians added their own knowledge to the work. The *Charaka* probably obtained its present form in the first century A.D.

Ebers Papyrus. c. 1500 B.C. An ancient Egyptian papyrus consisting of a collection of medical texts. A large portion is devoted to the wide variety of drugs Egyptian physicians used to treat their patients. The sources of these drugs included plants, animals, and minerals. American Egyptologist Edwin Smith purchased the papyrus in 1869, then sold it to George Ebers in 1872.

Edwin Smith Papyrus. c. 1550 B.C. An ancient Egyptian papyrus, presumed to be a copy of an even earlier work, whose central focus is surgery, specifically involving wounds and fractures. Many of the cases discussed involve the closing of wounds and suggest the possible use of stitches. There is also evidence in the papyrus to suggest that Egyptian physicians used adhesive strips to close wounds. The papyrus was purchased in 1862 by American Egyptologist Edwin Smith. He made an attempt to translate it, but its importance was not realized until 1930, when it was fully translated by James Breasted.

Isidore of Seville. *Etymologiae* (Etymologies, seventh century A.D.). An encyclopedia of human and divine subjects that served as an important reference book throughout the Middle Ages. The *Etymologiae* cover the liberal arts and medicine, the Bible and the Church, agriculture, warfare, shipping, and other topics. The etymological section is considered one of the first important landmarks in glossography, the compilation of glossaries.

Paul of Aegina. *Epitomae medicae libri septem* (Medical compendium in seven books, seventh century A.D.). Contains descriptions of many surgical techniques, including the amputation of a breast, lithotomy, trepanation, tonsillotomy, paracentesis, and his recommended treatment for a hernia of the groin, involving the removal of the testicle on the affected side.

Pliny the Elder. *Historia naturalis* (Natural history, first century A.D.). A comprehensive encyclopedia of zoology and botany, as well as astronomy, geography, and other topics. Pliny compiled information gathered from some 2,000 written sources. In many cases *Historia naturalis* is the only remaining record of the works of ancient scholars. The 37-volume encyclopedia became a primary educational reference and remained in high regard until the fifteenth century, when critics determined that the books were rife with errors. It is now seen as an important literary work, but one that has little scientific merit.

Polyclitus. *Kanon.* Fifth century B.C. A treatise on human form, proportion, and the representation of external anatomy in sculpture. Though none of Polyclitus's artwork is known to have survived, his reputation as a creator of sculptural masterpieces is strongly supported by the large number of copies that were made of his famous works, such as *The Discus Thrower*. His work illustrates the connection between forms of art that accurately detail the human body and their influence on the study of human anatomy and medicine.

Rufus of Ephesus. *On the Interrogation of the Patient*. First century B.C.-first century A.D. Important medical treatise that explains how medical and family histories contribute to an accurate diagnosis of a patient.

Susruta. *Susruta Samhita*. c. seventh century A.D. One of the chief medical works of Ayurveda, a branch of traditional Hindu medicine, and an important Indian text on surgery. Includes description of more than 100 different surgical instruments, most made of iron, each with its own special purpose. The *Susruta Samhita* also contains some of the earliest practice exercises for medical students, which encouraged would-be surgeons to develop their techniques by working with nonhuman "patients," such as vegetables. The date of the work is uncertain, as it was transmitted orally centuries before it was written down. Attributed to Susruta, a surgeon, it is almost certain that other physicians added their own knowledge to the work. The *Susruta* probably obtained its present form in the seventh century A.D.

Theophrastus. *On the Causes of Plants* and *On the History of Plants*. Fourth-third century B.C. In these texts, Theophrastus outlines the basic concepts of morphology, classification, and the natural history of plants. Apparently intended for use by gardeners, these works were quite practical rather theoretical. His ideas on this subject were accepted without question for many centuries.

JOSH LAUER

Mathematics

Chronology

c. 2000 B.C. Mathematicians in Mesopotamia learn to solve quadratic equations, or equations in which the highest power is two.

c. 1350 B.C. The Chinese use decimal numerals—that is, a base-ten number system.

876 B.C. Date of the oldest document mentioning the concept and symbol of zero, which originated in India.

c. 500 B.C. Greek philosopher and mathematician Pythagoras derives his famous theorem; studies the relationship between musical pitch and the length of the strings on an instrument; and proposes the highly influential idea that the universe can be fully explained through mathematics.

c. 400 B.C. Greek mathematician and philosopher Archytas distinguishes between harmonic, arithmetic, and geometric progressions, and is the first to apply mathematics to mechanics.

c. 300 B.C. Euclid writes a geometry textbook entitled *Elements* which codifies all the known mathematical work of its time, and is destined to remain the authority on mathematics for some 2,200 years.

c. 250 B.C. Stone pillars erected by the Indian emperor Asoka contain the earliest preserved representation of the number system in use today.

c. 230 B.C. Greek astronomer Eratosthenes works out a system for determining prime numbers that becomes known as the "sieve of Eratosthenes."

c. 225 B.C. Archimedes, considered the greatest mathematical genius of antiquity, is the first to develop a reliable figure for π; develops a method for expressing large numbers similar to exponentials; and, in measuring curved surfaces, utilizes a type of mathematics akin to calculus.

c. 140 B.C. Hipparchus, a Greek astronomer, founds trigonometry.

c. 100 B.C. Negative numbers are used in China.

c. 250 Diophantus writes his *Arithmetica,* the earliest treatise on algebra still extant.

c. 390 Theon of Alexandria publishes Euclid's *Elements* with his own notes and commentary. He is the father of the noted female scholar Hypatia.

c. 400 Hypatia of Alexandria writes commentaries on Greek mathematicians Diophantus and Apollonius. The first woman mentioned in the history of mathematics, Hypatia is murdered by a fanatical Christian mob in 415. This event leads to the departure of many scholars from Alexandria and marks the beginning of its decline as a major center of learning.

499 Aryabhata, a Hindu mathematician and astronomer, writes his *Aryabhatiya,* which contains his significant description of the Indian numerical system.

Overview:
Mathematics 2000 B.C. to A.D. 699

The basic notions of number and magnitude can be seen in markings on cave walls and primitive tools. However, it took many thousands of years for the first use of quantities to evolve into the abstract concept of numbers as we use them today.

The cultures of the Mesopotamian region had developed the use of written numbers by about 1800 B.C. The Babylonians, in particular, developed a sophisticated number system, estimated π with some accuracy, used fractions, and solved complex quadratic equations. Mesopotamian mathematics was spread far and wide by the many conquerors of the region. Their ideas influenced other cultures from Europe to China.

Chinese mathematics also developed at an early date. The earliest surviving works, dating from about 300 B.C., contain detailed astronomical calculations, as well as surveying, agricultural, and other practical math problems. The Chinese were interested in number patterns, and the magic square was one of their innovations.

The oldest surviving Egyptian mathematical documents are the Moscow and Rhind Papyri. The Moscow Papyrus, dating from about 1890 B.C., contains examples of geometry that have great practical use for the calculation of areas and volumes. The Rhind (or Ahmes) Papyrus dates from around 1650 B.C., and contains copied information from earlier sources. It includes many examples of fractions, and is written in the form of arithmetic problems that relate to practical concerns. The text mainly discusses addition and subtraction, but also deals with a multiplication shortcut and proportions.

The Rhind Papyrus also contains the earliest example of an algebraic problem. Egyptian mathematics was absorbed by the Greeks, who revered the Egyptians as the fathers of mathematics. The first book on algebra was written by Diophantus of Alexandria (third century A.D.), a Roman scholar writing in the Greek tradition. Algebra was a minor field until the later work of Arab scholar Muhammad ibn-Musa (800?-847), also known as al-Khwarizmi. The term algorithm is derived from his name.

Numbers were often studied for their mystical properties, as well as their mathematical importance. Pythagoras (580?-500? B.C.) and his many followers were particularly interested in numbers that had special properties, such as the prime numbers. They also discovered the so-called perfect and amicable numbers. While Pythagoras's mystical teachings had a great impact on later mathematics, he is better remembered for the theorem regarding right-angled triangles that bears his name. Indeed geometry was the major preoccupation of ancient mathematics.

Greek mathematicians, such as Thales of Miletus (624?-548? B.C.) and Pythagoras, traveled to Egypt and Babylonia, where they learned basic geometry. The Greeks transformed the subject by insisting on deductive proofs. Thales is credited with the earliest geometrical theorems, and later thinkers used these to expand the field. Euclid (330?-260? B.C.) collected together all the theorems of previous mathematicians into his *Elements,* which was to remain the standard text in geometry for over 2,000 years. Later geometers, such as Archimedes (287?-212 B.C.) and Apollonius (262?-190? B.C.), continued to add new theorems.

There were three so-called Great Problems that dominated Greek mathematics and occupied the minds of many mathematicians for centuries to come. Doubling the cube was the most famous problem of its time, and one that many Greek thinkers attempted to solve. Hippocrates of Chios (fl. c. 460 B.C.) made some important early steps, and there was an elegant, if somewhat complex, solution proposed by Archytas (428?-350? B.C.). However, it was the work of Menaechmus (380?-320? B.C.) that is most remembered, for not only did he give two solutions to the problem, he also did foundational work on conic sections as a result.

However, the other two Great Problems, squaring the circle and trisecting an angle are both unsolvable problems. At first sight they seem tantalizingly possible, and squaring the circle, in particular, was to occupy the minds of many mathematicians for almost four millennia. It is impossible to construct a circle and a square of the same area because the value of π is an irrational number. Attempts to solve this, while futile, did lead to some very accurate approximations of π.

Trisecting an angle seems like an easy problem, and indeed for certain angles it is. However, there can be no method for trisecting an arbitrary angle with a ruler and compass. Even today many amateur mathematicians attempt to solve this unsolvable problem.

Practical problems in astronomy inspired the study of trigonometry. Indeed, many early mathematicians were most interested in astronomy. The first major work on trigonometric functions was by the Greek mathematician Hipparchus (180?-125? B.C.), using many Babylonian ideas. Eratosthenes (276?-194? B.C.) was able to use the rules of trigonometry to calculate the Earth's circumference with surprising accuracy. Later work by Menelaus (70?-130?), and particularly Ptolemy (100?-170?) developed the field of trigonometry and widened its practical applications.

Mathematics was also used to pose and solve logical puzzles. The Greek thinker Parmenides (515?-445? B.C.) and his pupil Zeno of Elea (490?-425? B.C.) raised a number of interesting logical puzzles with mathematical implications. The paradoxes of Zeno were concerned with ideas of infinity, something that worried many Greek philosophers. Zeno used mathematical logic to prove that motion was impossible. This was obviously not the case, but his mathematical proofs seemed unshakable, thereby causing many Greek thinkers to reevaluate their ideas of the world.

While the Roman world inherited much of Greek and Egyptian learning, they did not develop many mathematical fields. Roman numerals, which evolved from the early Greek and Egyptian number systems, represent each number by a unique collection of symbols. They remained the dominant number system of Europe for hundreds of years, and are still used in some situations today. However, the Roman system results in very long collections of symbols for some numbers, and made complex calculations very difficult.

One of the earliest mathematical tools was the counting board, a flat surface with marked lines on which counters were moved to represent numbers. Counting boards developed into the abacus, a frame with rods and sliding beads. Such devices made calculations quicker and easier, but also limited the development of mathematical theory, as the users of such devices were more interested in the results than the processes involved.

Counting boards and abacuses use a place-value system, where a small number of symbols (counters) are repeatedly used in different positions to represent all numbers. In our decimal system, based on 10, the first column represents the ones, the next the tens, the third is the hundreds, and so on. Some cultures developed written number systems using place values, such as the Babylonians, who used 60 as their base, and the Mayas of Central America, who used 20. One advantage of place-value notation is the ease with which multiplication and division can be carried out.

Place-value systems also generated a need for a symbol to represent nothing, the zero. When recording the results of a calculation from a counting board, there often arose confusion when the result had nothing in a certain place value, such as the number 203, which has no tens. In Roman notation there is no problem—it is CCIII—and so the need for the zero did not arise in the West. However, place-value notation required a placeholder in the empty spot, and so the zero was invented independently by the Babylonians, Maya, and Indians.

In India the zero became more than just a placeholder, it became a number in its own right. This meant that written calculation became just as easy to understand as the counting board, leading to a new appreciation of the rules of calculation. The Hindu zero also made it easier to deal with larger numbers. The idea of zero was to cause many philosophical headaches where ever it traveled, for the concept of nothing was frightening for many, and was related to the even more worrying notion of infinity.

The majority of ancient learning was lost to Europe after the collapse of Rome, but was studied, copied, and developed in the Arab world. Arab mathematicians absorbed the knowledge of many nearby cultures, including the Babylonians and the Hindus. This potent mix of mathematical learning was distilled, refined, and eventually transmitted to Europe from the time of the Crusades on. However, it took many centuries to break down the resistance to change encountered in Europe. When Eastern learning was at last accepted, the resulting revolution in mathematics was to redefine the field and lead to the mathematics of the modern era.

DAVID TULLOCH

Mesopotamian Mathematics

Overview

Mathematics and writing first appeared around 3000 B.C. in Mesopotamia. Originally used as a means of accounting, the abstract concept of numbers, independent of the things being measured, developed over the course of the next thousand years. In the last century of the third millennium, the powerful Ur III state (c. 2100-2000 B.C.) established a system of weights and measures and created a large class of bureaucrats trained in a standardized curriculum. By this time the sexagesimal (base-60) place-value number system was in use.

From the subsequent period, called Old Babylonian (c.2000-1600 B.C.), a large number of educational texts have survived. These reflect an interest in mathematics far beyond what would have been required in daily life. The problems emphasized procedure, leading to the computation of a number as the solution. Few sources have survived from after the Old Babylonian period until the Seleucid period (c.330 B.C. to A.D. 64) when astronomy became important. Due to a scarcity of sources, it is difficult to be sure of the influence of Mesopotamian mathematics on surrounding cultures.

Background

Mathematics in Mesopotamia began with accounting, evidenced by the earliest extant documents, which are receipts and stock records. During the second half of the fourth millennium (c. 3500-3000 B.C.) the Mesopotamians tried several different systems of record keeping before they settled on writing and numbers.

At first, numerical notation was tied to the types of objects being counted. They used one set of symbols for a herd of goats, another to represent measures of grain, and still others for the area of a field. Within each system the smaller units were bundled until they reached the next largest unit, just as we measure in inches, feet, yards and miles, or pints, quarts and gallons. The only difference was that the Mesopotamians had a special symbol for, say, "inch," and to write "4 inches" would write the inch symbol four times. In most modern metrological (weights and measures) systems, larger units are often big multiples of smaller units. The Mesopotamian systems were constructed so that there were never more than ten small units in a large one.

At first, there were about a dozen different systems using around 60 different "quantity" symbols. Gradually, this complex notation was simplified until by around 2500 B.C. most items were recorded in a single system, the one that had originally been used to count discrete objects. This system was very regular; with each larger unit alternately six or ten times the size of the previous one.

From the same time (c.2500 B.C.), from the ancient city of Shuruppak, comes the world's oldest math problem. Roughly translated, it says:

A granary. Each man receives 7 liters of grain. How many men?

Remarkably, two copies of this problem survive, one with the correct answer (164,571 men with 3 liters left over) and one quite incorrect. The tablet with the incorrect answer belonged to a student struggling to learn mathematics, and the mistakes he left on the tablet tell modern historians how mathematics was done 4,500 years ago. The excavations in which these tablets were found, at Shuruppak (modern Tall Fa'rah, Iraq), also uncovered the world's oldest multiplication table and the first geometrical exercise.

Over the next 400 years or so, writing gradually developed from a pictographic system into the cuneiform (wedge-shaped) writing that would be used for the next two thousand years. Early cuneiform writing required a separate stylus for numerical notation; by the late third millennium, however, the Mesopotamians began to draw number symbols with just one stylus. This eventually led to a reorganization of the notation using fewer symbols. In particular, the symbol for "60" was written as a larger version of the symbol for "1." At this point, the Mesopotamians were very close to having a place-value system.

The next step shows how closely the development of mathematics was tied to social and political developments. For most of the third millennium, Mesopotamia was divided into city-states of varying importance. During the Ur III dynasty (c. 2100-2000 B.C.), these city-states needed a large bureaucracy to run the empire's centralized state-run economy. To facilitate administration, there were wholesale reforms to the systems of weights and measures, designed to make calculation easier by linking the units in different systems. At some point during the Ur III period, it seems that most calculations were done in the new sexagesimal (base-60) system, although results were always converted to the proper metrological units. Unfortunately, we have very few actual mathematical tablets from this period. Our knowledge has been gleaned from the tens of thousands of economic documents the bureaucrats left behind.

There was no money in ancient Mesopotamia. At the beginning of the third millennium, the main currency was grain, measured in the capacity system. During the course of the millennium, however, silver became more widely used. As a result, a system of standard weights became necessary to replace the old capacity measures, and, since even small pieces of silver were valuable, it was necessary to be able to subdivide the weight units. Thus, the Mesopotamians had to be able to calculate fractions beyond the old halves and thirds. It seems that the full sexagesimal place-value system, extending both to the left and right of the "sexagesimal point" came as a generalization of the weight system.

The Ur III state established a centralized bureaucracy and standardized curriculum throughout Mesopotamia. The period after its collapse is known as Old Babylonian (c.2000-1600 B.C.). During this time power became more dispersed and mathematics grew beyond what would be needed for simple counting and accounting. In contrast to the scarcity of evidence for mathematics in Ur III, hundreds of Old Babylonian mathematical tablets have survived. These provide a good, though incomplete, picture of the period.

By the beginning of the Old Babylonian period scribes had a written heritage stretching back 1,000 years, and a school system that was probably 500 years old. The Mesopotamians used lists to organize knowledge, and they learned the lists by copying them. When studying mathematics, which was a central part of any scribe's training, they copied multiplication and reciprocal tables, often several times over on a single tablet. They learned to solve problems by copying out model solutions, the Mesopotamian equivalent of the worked examples in a textbook. In more advanced studies, they copied excerpts of Sumerian literature. Teachers had lists, too: lists of problems to set, sometimes with the answers given, but no workings; a few tablets appear to be lists of coefficients to put in certain types of problems.

The sexagesimal (base-60) place-value system of numeration used at this time was completely free of any metrological context. The numbers were abstract and could refer to anything. The sexagesimal system was used only for computation; final results were recorded in the standard metrological systems of the day. The sexagesimal system used only two symbols, a vertical wedge for 1 and a corner wedge for 10. Numbers up to 59 were made by bundling these signs; after that, 60 was written as a 1 in the 60s column. This allowed the place-value system to record numbers of any magnitude. In the Old Babylonian period, there was no symbol for an empty column (although it's needed less frequently in a base-60 system), and there was no equivalent of our decimal point to give an absolute size to the number. Absolute size was determined by the problem and needed to be kept in mind when converting back into the metrological units.

We don't know how the Mesopotamians did simple arithmetic. They may have had some sort of counting board or abacus, but there is no archaeological or textual evidence for this. They may have memorized their addition and subtraction facts; these computations are all taken for granted in the texts we have. For more complicated computations, it seems they wrote out intermediate calculations on "scratch tablets" which were then erased and reused. By their nature, very few such tablets have survived.

The key problem in arithmetic is how to do division. The Mesopotamians dealt with this by never doing division: they always multiplied by the reciprocal. (The reciprocal of a number n is the fraction $1/n$, so that a number times its reciprocal is always equal to 1.) The sexagesimal place-value system meant that fractions could be treated on the same footing as whole numbers, but it also meant that all fractions had to be expressed in sexagesimal notation. Just as some fractions, such as $1/3$ or $1/7$, cannot be written as finite decimals, so the same is true in the sexagesimal system. Since $1/3 = 20/60$, it can be written as a finite sexagesimal fraction, but $1/7$ cannot. Only numbers that are products of 2, 3, and 5 and no other primes have reciprocals with finite sexagesimal expressions. These are now called regular numbers.

The Mesopotamians made lists of all regular numbers between 2 and 81 together with their reciprocals. These were the reciprocal tables students had to copy. For each regular number (and also a few others, such as 7), they made tables of multiples of the number from 1 to 20, then 30, 40, and 50 times. These multiplication (or division) tables were learned by copying, although doubtless many scribes looked up specific products when they needed to. They also made tables of squares and square roots.

Mesopotamian mathematics was organized around problems. There were no abstract theories or theorems and proofs, and there were no equations to solve. Instead, there were procedures to follow. We don't know who developed these procedures, or how they did it. We don't know how they could be sure their procedures would always work.

The standard Old Babylonian mathematics exercise is a word problem requiring the computation of a number for its solution. The word problems drew on subjects that would be relevant to the scribes' lives and careers as bureaucratic overseers: building walls, digging canals, counting bricks, paying laborers, measuring grain. However, the teachers were not overly concerned with realism. In one case students had to find the area of a field that a given cistern of water would irrigate to an even depth of one finger-width. The area of the field is over 3,000 acres (1,214 hectares). In another example, a student has to calculate the volume of grain in a pile 60 meters (197 feet) long and 24 meters (79 feet) high. Instead of worrying about how realistic the problems were, the teachers chose parameter values designed to make the problems easy to work. For example, the cistern in the problem above is a cube 10 *nindan* on a side (a *nindan* was a standard measure of length, about 6 meters [20 feet] long). It's clear that the teachers wanted the students to focus on understanding and following the general procedures without being distracted by overly complicated coefficients.

Another indication of the emphasis on procedures is the tablets that have series of linked problems, where the student who had done the first problem would know the answers to all the remaining ones. For example, the first of a series of problems on excavation of a canal gives the length, the width, the depth, how much volume a worker could excavate in a day, and the daily wages in barley of a worker. The student is asked to find the area, the volume, the number of workers, and the total expenses. In the second problem, the expenses, the width, the depth, the work-norm and the wages are given with the same values, and the student has to find the length. The third problem asks for the depth. These tablets are always organized with a carefully graduated series of problems from easiest

to hardest (just like modern textbooks). There are 23 of them on this particular tablet.

Mesopotamian students understood addition, subtraction, reciprocals, and extraction of square roots. Hence, these were the only operations that could be used in solving problems. Many problems were what we would term linear, but there was a large number of quadratic ones, too. Linear problems could get complicated enough to test a student's mastery of the procedures, as in this example:

> *I found a stone. I did not weigh it. Eight times the weight and 3 gin I added. A third of a thirteenth of 21 times I added. I weighed it: 1 mana. What was the original weight of the stone?*

The highest expression of Old Babylonian mathematics, however, was the solution of quadratic, or rectangular, problems, which they solved without equations and without a quadratic formula. (They didn't use negative numbers, either.) Instead, they used standard forms for quadratic problems and standard procedures for solving them. Other types of problems were also usually reduced to a standard form, such as:

> *The length and width together of a field is 50. The area is 600. What are the length and width?*

The standard procedure is to take half the sum of the length and width. Call this the half-sum. Square it. Subtract the area and take the square root. Then the length is the half-sum plus the square root, and the width is the half-sum minus the square root. In the other standard type, the difference between the length and width are given instead of the sum. Interestingly, although hundreds of these rectangular problems are known, there are none that are these two basic types. Perhaps they were considered too easy to need writing down.

Rectangular problems, often involving fields, canals, excavations, walls, or gates were a staple of Mesopotamian mathematics. A natural problem in this setting involves finding the diagonal of a rectangle, or alternatively having the diagonal as one of the known parameters. Thus, Old Babylonian mathematicians had several procedures for working with Pythagorean (right-angled) triangles. One of the most famous Old Babylonian tablets, Plimpton 322, is a catalogue of parameters for triangle problems. The Mesopotamians did not use any angle measure, so all triangles were either right-angled, or composed of right-angled triangles. An equilateral triangle, for example, could be considered as two right-angled triangles back to back.

As in the rest of Old Babylonian mathematics, the purpose of a geometrical problem was always to calculate a number: the area or volume of a figure, or the lengths of its sides. Along with problems involving circles, triangles, rectangles and trapezoids, Old Babylonian scholars also constructed a complex series of area problems by inscribing certain figures inside each other. Most volume problems used cubes, rectangular prisms, or pyramids. Although Old Babylonian mathematics was grounded in the practical necessities, its scope and level went considerably beyond anything a scribe might meet in daily life. This is exemplified by its concern with quadratic problems, few, if any of which would have been needed to solve real problems. The problems' structures, too, betray a concern with elegance and pedagogical issues, not a desire for accurate modeling. Furthermore, the problems' parameters are chosen to make computations simple, not to reflect of the real world. The focus is clearly on having the student learn the procedures for solving the different classes of problems.

In contrast to the hundreds of mathematical tablets from the Old Babylonian period, we have only a very few from the succeeding thousand years. We know that the sexagesimal system endured, although there is little further record of the state of mathematical knowledge until after the end of the Assyrian and Babylonian civilizations. From the scant evidence we have, it appears that the broad outlines of mathematical knowledge remained about the same. There are differences in later tablets that are intriguing to experts, but no evidence of startling changes.

Impact

The next great collection of mathematical tablets that have survived dates from the Seleucid period (c. 330 B.C. to A.D. 64) when mathematical astronomy was on the rise. Mesopotamian astronomy was principally concerned with where and when certain astronomical events occurred. It was exclusively computational. Elaborate arithmetical schemes to determine a large number of phenomena were developed, but there was no accompanying geometrical or physical model. Computations had to be carried out with great precision, since rounding errors could accumulate quickly.

When doing long computations involving many-place numbers, there was a pressing need to keep the place-value columns aligned correct-

ly, and a sign for an empty column came into use. They also constructed many place tables of reciprocals of large numbers.These tables, the procedure texts, and the hundreds of years of observations they recorded were of enormous importance to the later development of astronomy. In particular, the astronomer Ptolemy (c. A.D. 100-175) adopted sexagesimal notation for fractions over the standard Greek and Egyptian use of unit fractions, as it was far superior for detailed calculations. This is how we came to use base-60 for fractional time measurement (the divisions of hours into minutes and seconds).

While much is known about Mesopotamian culture, it's difficult to assess the culture's influence on the development of mathematics. During its 3,000 years of written history, Mesopotamia was part of a large network of cultures with extensive diplomatic and trade ties. Many educated scribes in other parts of the world could read and write its language, cuneiform Akkadian. Merchants and traders were almost certainly aware of the Mesopotamian accounting systems. However, we lack clear evidence of any effect this had on other cultures. Egyptian mathematics as indicated in the Rhind and Moscow Papyri is roughly contemporaneous with Old Babylonian period, but developed in different directions. Early Indian and Chinese mathematics are shrouded in mystery. Greek mathematics first developed on the Ionian littoral, closest to the Persian Empire, but is renowned for its emphasis on abstract geometry rather than arithmetical computation. The great exception is the third century A.D. Greek mathematician Diophantus. Some of his procedures for solving quadratic equations are identical to Old Babylonian ones that were first written down two thousand years before. Without more archaeological evidence, it is impossible to assess the transmission of mathematical knowledge from Mesopotamia to other cultures.

DUNCAN J. MELVILLE

Further Reading

Aaboe, Asger. *Episodes from the Early History of Mathematics.* New York: Random House, 1964.

Melville, Duncan J. Mesopotamian Mathematics. http://it.stlawu.edu/~dmelvill/mesomath/index.html.

Neugebauer, Otto, and Abraham Sachs. *Mathematical Cuneiform Texts.* New Haven: American Oriental Society, 1945 (reprinted 1986).

Nissen, Hans; Peter Damerow, and Robert K. Englund. *Archaic Bookkeeping.* Chicago: University of Chicago Press, 1993.

Robson, Eleanor. *Mesopotamian Mathemaics, 2100-1600 B.C..* Oxford: Clarendon Press, 1999.

The Mathematics of Ancient India

Overview

It is widely believed that aside from the achievements of the ancient Greeks, Babylonians, and Egyptians, there was little progress in the area of early mathematics. In fact, many leading historians in this field have echoed that sentiment by either minimizing or ignoring the contributions of other cultures. However, India and its mathematicians were extremely important in the development of mathematical thought during antiquity, despite the reluctance of some authorities to question their contributions.

The contribution of early Indian mathematicians to contemporary mathematics cannot be overstated. Certainly their greatest contribution is our system of numbers. Since nearly all theories, principles, and constructs in this discipline rely on the number system, its development has had a tremendous impact on the field of contemporary mathematics. The novel Indian numerals were subsequently adopted by the Arabs, which eventually became known to Europe as Arabic numerals.

In addition, there were other notable contributions in the field of mathematics from this culture. While it is somewhat controversial, early Indian theorists are among the many groups who are credited with the invention and application of zero. The Indians were the first to use zero as a placeholder. Among the other important achievements were the estimation of π and the length of the solar year, each to four decimal places.

Many important documents from the early period of Indian mathematics have not survived or have survived only in the form of translations. Because of this, there are questions regarding the relationship between Greek and Indian mathematics. Original references are quite scarce, so it is difficult to credit one group with specific ideas. There are still many unanswered questions about the relationship between early mathematics in India and Greece. This becomes problematic in later centuries as well because of the proliferation of original materials, so that it is difficult to distinguish between ideas of European or Indian origin.

Background

The earliest known mathematics of the region arose in the Indus valley in present-day Pakistan. It was associated with the Harappan civilization, which consisted of a few cities and many smaller villages scattered throughout the valley. The civilization was established around 2,500 B.C. and survived at least 800 years. The inhabitants were literate and had adopted a uniform system of weights and measures. Analysis of their system revealed that it was surprisingly similar to the one currently used in the United States. The "Indus inch" was 1.32 in (3.35 cm), which could be strung together in units of 10 to make a 13.2-in (33.5-cm) "Indus foot." An artifact made of bronze has also been discovered with discreet marks every 0.367 in (0.932 cm). One hundred of these would make the distance 36.7, which is a fair approximation of a yard, meter, and a grown man's stride length. Archeological evidence suggests that these units of length were used accurately by the Harappan civilization in their construction.

Perhaps the most famous evidence of the mathematical prowess of the early Indian civilizations is contained in the Sulbasutras. The Sulbasutras are ancient religious texts that contained, among other things, detailed geometrical knowledge in the form of instructions for building altars. While these were practical guides that were intended as mathematical texts, they nevertheless show how adept these early civilizations were at math.

Around the third century B.C., when the Sulbasutras are believed to have been written, the Brahmi numerals were developed. After several modifications, these numerals evolved into the standard 1, 2, 3, 4, 5, 6, 7, 8, 9 still used in modern times. While the Brahmi numerals themselves did not have place values, the Indian numbering system eventually developed a place value system, which turned out to be extremely simple, yet very elegant. It should also be noted that the Indian number system is almost exclusively base-10, as opposed to the other systems that developed at that time that used other base systems, such as base 20 or base 60. While the origins of the Brahmi numerals are questioned, there is ample evidence that subsequent cultures modified this system for their own use.

The next major civilization to have a significant impact on the development of mathematics in India was the Gupta. The Gupta period marked the time when the Gupta dynasty ruled over much of India from the early fourth century

A.D. to the late sixth century A.D. This period is thought of as the Classical Age of India, characterized by new ideas and prosperity. The Gupta numerals developed from the Brahmi numerals, and these became known over large areas as the Gupta expanded their empire by conquering new territory. However, the greatest influence of the Gupta civilization was not the numerals themselves, but the utilization of a place-value system with them.

A place-value system is one that uses numerals standing for different values depending on their position relative to the other numerals. Although the Babylonians are credited with developing the first place-value system in about the nineteenth century B.C., the Indian number system was unique in that it used base-10, making our own numbering system a direct descendent of it. A historical document indicates that this system has been used in India since before 594 B.C. It is not known whether this system was developed independently of other cultures, or if its inception was due to the influence of the Greeks, Babylonians, or possibly even the Chinese. However, we do know that this system was conveyed to other cultures, where it had a profound impact on the development of mathematics.

The religious and philosophical group called the Jains also shaped the mathematics of India. The Jains were founded in the sixth century B.C. They worked on problems such as number theory, cubic equations, quadratic equations, and statistics. They also had an understanding of advanced ideas such as that of infinity. Their work was later summarized and expanded by Aryabhata (476-550?), the most important ancient mathematician from that region.

Aryabhata headed the classical era of Indian mathematics. He helped to ignite a new era in mathematics, which in turn spurred on other sciences, such as astronomy. He recognized the importance of scientific investigation and established research centers to fulfill that goal. Among his many accomplishments were the introduction of the concept of trigonometry, the most precise estimation of π up to that date, and an accurate estimation of a solar year.

Impact

India's contributions to the development of mathematics played a vital role in establishing our system of numbers, and provided many other fundamental concepts as well. Throughout antiquity, no other cultures surpassed those of the Indian subcontinent when it came to the development and implementation of the science of mathematics.

The tremendous impact that Indian mathematicians had on the development of mathematical concepts is often obscured because of unsubstantiated evidence, inappropriate credit to other cultures, prejudice, and, finally, just plain ignorance. Evidence is mounting, however, that India was at the forefront of mathematical thought. It has been well established that our system of numbers has its roots firmly planted in India. Not only did we inherit the written appearance of our numerical characters from ancient India, we also borrowed their base-10 place-value system of numbers, which is still used today. Without this, mathematics as we know it would simply not exist. The ancient Indians provided a useful, flexible, and intuitive model for us to use.

It is hard to imagine what mathematics would be today without our current number system. Things we take for granted in every day life, such as money, are based on this system. The implementation of a place-value system is so elegant and practical that we often take it for granted. It is one of the most important mathematical developments in history.

Another important element that India contributed to the world of mathematics is the work of Aryabhata, who extolled the virtues of scientific research. This spurred on further scientific study, which served as a model for future generations of scientists. By establishing research centers, Aryabhata provided the impetus and desire to further knowledge in the field. As the knowledge of mathematics grew, so did the knowledge of physical sciences by its application. As an example, great advances were made in geography and astronomy as a direct result of having the mathematics necessary to solve problems.

Among the significant contributions made by individuals, Aryabhata himself provided the answers to many important questions and helped developed the branch of mathematics called trigonometry. He is often credited with the development of zero as a placeholder, although some historians credit one of his pupils with that achievement. This had tremendous implications for mathematics and made it much easier to signify certain numbers. Modern society certainly owes a huge debt to those ancient scientists and mathematicians.

JAMES J. HOFFMANN

Further Reading

Murthy, T. S. Bhanu. *A Modern Introduction to Ancient Indian Mathematics.* New Delhi: Wiley Eastern, 1992.

Neugebauer, Otto. *The Exact Sciences in Antiquity.* New York: Dover Publications, 1969.

Thibaut, G. *Mathematics in the Making in Ancient India.* Calcutta: K.P. Bagchi, 1985.

Mayan Mathematics

Overview

The Maya of Central America designed one of the most complex and accurate calendar systems known. This calendar system was based heavily on Mayan mathematics, including the Mayan base-20 numbering system. Arising in isolation from the world's other great intellectual traditions, it is no surprise that the Mayan calendar and numbering systems differed substantially from those that developed in China and the Middle East. But different does not necessarily mean inferior, and the Maya had a rich tradition in mathematics and astronomy.

Background

It is possible that only three civilizations independently developed mathematics: the Chinese, the peoples of the Middle East, and the Maya Indians of Central America. Of these, there is the possibility that Chinese and Middle Eastern mathematicians may have had some influence on each other via the Hindu, but this is not certain. What is certain is that the mathematical tradition of the New World arose in complete isolation from any other because of the geographic isolation of the Central American civilizations.

The Maya themselves arose in Central America about 800 B.C. as a group of linguistically related farmers. Like other ancient empires, they became stronger and began to expand geographically, largely through wars and diplomacy, until they were the dominant political force in a part of Central America centered around the Yucatan Peninsula and the surrounding mountains and jungles of what are now Mexico, Belize, and Guatemala. Scholars recognize three major periods in Mayan history: the Preclassic (about 800 B.C. until about A.D. 300), Classic (about A.D. 300 to 900) and the Postclassic (A.D. 900 to 1520). In all, the Maya ruled their part of the New World for nearly 2,000 years.

During their political domination of Central America, the Maya developed an intense interest in astronomical phenomena (as did most ancient civilizations and, for that matter, our current civilization). This necessarily led to attempts to predict the occurrence of such natural phenomena as lunar and solar eclipses, the rising and setting of planets, the movement of the Sun and Moon across the sky, and more. As was the case elsewhere in the world, this astronomical interest led, in turn, to the construction of a calendar and the mathematical infrastructure that permitted a calendar to be developed.

A calendar is, to some extent, an attempt to understand nature by seeking some sort of pattern that can be described mathematically. For example, our current calendar describes the length of time it takes the Earth to circle the Sun—a little less than $365\frac{1}{4}$ days, with corrections such as leap years and other calendar rules. We use our calendar to predict the ending of seasons and the completion of an orbit around the Sun, and we find that relatively simple mathematics is sufficient to accomplish this. For the Maya, this alone was not enough, and they developed a much more elaborate calendar that required far more sophisticated mathematics to make it work. Sometime during their Preclassic stage, the Maya invented a system of mathematics, a numbering system, and a calendar sufficient for their purposes that were unique in the ancient world.

Impact

The Maya arose, flourished, and began their decline in isolation from the intellectual and cultural traditions of Europe and Asia. Their first contact with Europeans was during Spain's conquest, a hostile interaction that hardly facilitated cultural and intellectual exchange. Because of this, Mayan mathematics has had very little practical impact on the rest of the world and is of primarily intellectual interest. However, Mayan mathematics and the Mayan calendar had a profound impact on Mayan life, and it is from this perspective that it will be discussed. In addition, the Maya independently developed some important mathematical concepts, and an examination of what we can learn from this is also warranted.

To a large extent, the Mayan calendar and numbering system are difficult to discuss separately, because it was their astronomical observations that led to the development of both. Therefore, the following discussion will address both.

Mayan mathematics invented two very important concepts: place (or positional) notation and the idea of zero. These concepts allowed Hindu-Arabic numbering systems to supplant Roman numerals; they were vital to the development of advanced mathematics. Some, in fact, have termed these concepts among the most brilliant intellectual achievements in human history because neither is obvious and both are necessary for complex calculations as we know them.

Positional notation is what we use for writing numbers such as 195. The rightmost number, 5, occupies the "ones place," the 9 is in the "tens place," and the 1 is in the "hundreds place." This is our way of saying that we use a numbering base of ten, so each "place" can have any of ten values (0, 1, 2, 3, 4, 5, 6, 7, 8, 9). As a number's position moves to the left, its value increases by a factor of ten, so a 9 in the tens place is worth ten times as much as a 9 in the one's place, and one tenth as much as a 9 in the hundreds place. In short, a number's true value depends on its position.

The other concept, that of using a zero, is perhaps more radical. To some extent, using a base of ten makes sense because we have ten fingers. Given enough thought, this suggests a sort of positional notation. However, nothing intuitively leads us to use a symbol to represent a value of zero. In our notation, zero is called a placeholder—it does not, in and of itself, have a value, but it indicates that a place still exists. Consider this example: if we did not use a zero, the number 19 could mean 19, 190, 109, 1009, or much more. Using a zero tells us exactly what 19 means, removing all ambiguity, while making mathematical calculations possible. Incidentally, it is arguable that positional notation requires the use of zero for it to have any mathematical utility, so these two concepts likely evolved closely together in time.

That any civilization discovered both of these concepts is remarkable. That two should have discovered them independently is even more so, especially when we note that the great

mathematical cultures of the ancient world, the Egyptians, Romans, Greeks, and Babylonians, all failed to do so. This speaks very highly of the Mayan mathematicians and the mathematical sophistication of which they were capable.

The Mayan calendar seems to have been based on the combined cycles of the Sun, the Moon, and Venus. The Mayan lunar calendar was accurate to about five minutes per year (as compared to the Julian calendar, which had an annual error of about 11 minutes). It is perhaps best to view the Mayan calendar as a series of three intermeshed wheels, like gears, each of a different size. The innermost wheel contained the 13 days of each Mayan month. With 20 months in the Mayan religious year, the religious cycle was 260 days long. This was the "Sacred Round." A separate calendar was the Mayan "Vague Round," which consisted of 18 months, each with 20 days. Added to this were five "nameless days" of bad luck, giving a calendar year of 365 days. The Sacred Round and the Vague Round intermeshed, too, and the Maya defined a "Long Count" of about 52 years, defined as the amount of time needed for both cycles to return to the original configuration.

When all was said and done, the Maya developed a calendar with an initial date (that is, the date from which all else is reckoned, much as Western nations reckon dates from A.D. 1) of about 3114 B.C. It is not likely that this dates the beginning of the Mayan civilization, but probably a date from Mayan mythology. Their calendar went forward from this point, and seems most accurate at predicting celestial events with the Sun, Moon, and Venus.

Finally, it must be noted that the Maya likely discovered many of the properties of the number pi, although this is a supposition based on their other mathematical achievements. In particular, Petr Beckmann, in his book *A History of* π, notes that the remarkable accuracy of the Mayan calendar almost demands that they both understood and used π in many of their calculations. Beckmann, a mathematician, bases this supposition on two facts; the accuracy of Mayan astronomical calculations and the fact that many other less mathematically sophisticated civilizations developed an appreciation of the existence and utility of π. While this has not been confirmed archeologically, Beckmann also comments that, from a mathematical standpoint, it is easier to believe the Maya reached an understanding of π rather than that they learned more circuitous and more difficult methods of making their calculations. However, he also admits that there is no direct evidence of such work, so this question, like so many others may never be answered definitively. What can be said is that, during their time, the Maya developed a remarkably accurate and sophisticated system of numbers that gave them one of the most complex, and accurate calendars of the ancient world.

P. ANDREW KARAM

Further Reading

Beckman, Petr. *A History of* π. New York: St. Martin's Press, 1976.

Sabloff, Jeremy. *The New Archaeology and the Ancient Maya.* New York: Scientific America Library, 1990.

Ancient Chinese Mathematics

Overview

China has one of the world's oldest traditions in mathematical discovery, comparable to those of Egypt and the Middle East. The first Chinese mathematics text is of uncertain age, some dating it as early as 1200 B.C. and others over a thousand years later, but there is little doubt that relatively advanced mathematical concepts were discovered and practiced in China well before the birth of Christ. For much of its history, China has been in contact with the West, albeit intermittently, and Chinese and Western mathematicians influenced each other for centuries. Although it is sometimes difficult to determine who influenced whom, some Chinese contributions clearly predate those of the West, or are so obviously different that it is apparent they arose independently. In any event, the history of Chinese mathematics is both long and distinguished.

Background

There is no way to know when mathematics first appeared as a separate discipline in China. In fact, even dating the first Chinese civilization is difficult; various accounts place the first Chinese empire between about 2700 B.C. and 1000 B.C., and the earlier date is not considered unreasonable. These dates make it clear that Chinese civilization is probably neither much older nor much younger than those of the Middle East, although many aspects of the latter are better documented and, thus, easier to verify.

It is unlikely that a mathematical tradition could develop in the absence of some sort of civilization and, indeed, nothing more complex than counting seems to have developed anywhere in the world except where civilization already existed. This may be because there is little need, much

less time, to develop sophisticated mathematical structures in the absence of cities and some sort of government. For example, many early mathematical documents deal with issues such as the proper measuring and allocation of properties, constructing buildings, calculating taxes, business computations, and the like. None of these sorts of transactions are likely to occur in the simpler (and smaller) tribal structures that predated the first cities and civil governments. In addition, in the absence of larger structured civil organizations, virtually everyone was responsible for sustaining themselves and their family. It was only with the rise of more centralized and larger governments that agriculture could progress to the point where a relatively large number of people could devote themselves to administration, business, crafts, or research. Thus, mathematics as a discipline would not have developed in the absence of civilization for another reason: there would be no need for it.

Given the above, it is also likely that mathematics began to advance rapidly once civilizations were established. One of the apparent hallmarks of civilization is the appearance of cities, a central government, and coordination of the activities of much larger populations. These, in turn, prompt the need for businesses, taxes to support the government, formal rules for land use, and so forth. It was these that seem to have driven mathematics to appear in short order in Egypt, China, Mesopotamia, India, and Latin America. And, once established as a formal discipline, in each of these places mathematics seems to have migrated to issues such as keeping calendars, making sense of astronomical phenomena, and so forth, becoming more abstract over time.

This certainly seems to have been the pattern in ancient China. Many of the first works in Chinese mathematics, including the first book,

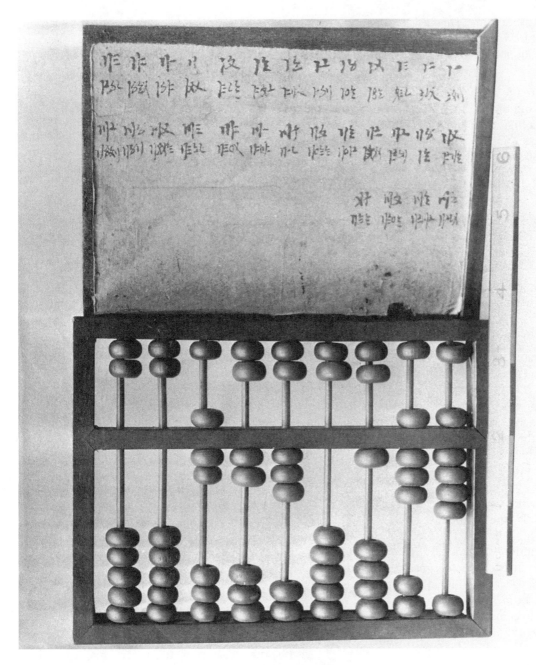

A Chinese abacus. *(Corbis Corporation. Reproduced with permission.)*

Nine Chapters on the Mathematical Art, and followed by *The Ten Classics of Mathematics,* concentrate on methods for solving practical problems in mathematics. However, these books also make it clear that many of these problems had existed (and been solved) for many years, or even centuries. What is also clear in these and other contemporaneous books is that Chinese mathematicians had already reached a high level of abstraction and sophistication by a few centuries B.C. In fact, by this time the Chinese had far surpassed Western mathematics in many areas. Tragically, much of this knowledge was lost on at least a few occasions when Chinese emperors ordered books

burned and libraries destroyed. These actions not only caused Chinese mathematics to regress, but even kept many later Chinese scientists and mathematicians from knowing what their predecessors had done. Only recently have these early Chinese advances come to light; unfortunately, too late for them to have had the impact that would otherwise have been their due.

Impact

As noted above, for several centuries Chinese and Western mathematicians exchanged information, in spite of the slow rate of communications at the time. This makes assigning proper credit for various mathematical discoveries difficult, but not necessarily impossible.

The list of accomplishments that can be credited to Chinese mathematicians is impressive, and most of these appear to have arisen in China first, or at least independently. These include an impressive estimate of the value of π, use of zero, use of decimals and decimal fractions for calculations, use of negative numbers, the algebraic treatment of geometric problems, methods for solving problems with many factors, and methods for extracting square and cube roots. There are more, but even this short list is impressive. In the third century A.D. Chinese mathematicians calculated the value of π to 10 decimal places, a feat that would not be matched elsewhere for another 1,400 years.

In its simplest form, simply writing a number as 123 is decimal notation. Decimal refers to the fact that each place represents a larger factor of ten, so the 3 represents 3×1, the 2 represents 2×10, and the 1 means 1×100. This is similar to what is meant by positional notation, although having one does not necessarily dictate the other. Other civilizations, such as the Maya and the Hindu, developed these concepts as well, although the Maya used a base other than 10. It's interesting to note that some advanced civilizations did *not* develop these concepts. For example, the Greeks used letters to represent their numbers and the Romans never advanced beyond Roman numerals. To appreciate the difference between the decimal system and Roman numerals, consider the difficulty of multiplying XXVIII times XIII as compared to 28 times 13. This, of course is a simple problem to which the answer is 364, but calculating even this relatively simple problem in Roman numerals is not a trivial task.

Decimal fractions are simply extending this use to numbers less than 1. For example, the number 0.5 is a decimal fraction (for ½) as is the number 2.25 or 10.4. Developed in the first century A.D. in China, they were not widely used in the West for over 1,500 years.

Chinese mathematicians were very much in advance of their Western counterparts in the use of negative numbers. As late as the fifteenth century, Western mathematicians felt that negative numbers simply could not exist and many refused to even consider them. Today, of course, we recognize that negative numbers do exist, as anyone who has overdrawn their bank account knows. This has not always been the case. To

their credit, the first mention of negative numbers in Chinese mathematics dates to at least the second century B.C., while they do not appear in Western mathematics for another 1,700 years. However, some texts suggest that Chinese mathematicians, while not worried about using negative numbers in calculations, did not view them as having an actual physical meaning. Therefore, they should be given due credit for the mathematical advance, while continuing to recognize their seeming unwillingness to fully accept the implications of this advance.

The other Chinese advances can, to some extent, be grouped together into developments in problem solving through algebra. There is a great deal of evidence that many algebraic techniques were developed in China, spreading to India and, from there to the Islamic scholars of the seventh and eighth centuries. This is not to take credit from Islamic mathematicians, who did a superb job in seeing the utility of these techniques, gathering them together, and adding to them their own unique contributions. The result is that, what we call algebra today is effectively a collaboration of Chinese, Hindu, and Islamic mathematical insights and advances, compiled and given utility by Islamic scholars.

Among the techniques developed by Chinese mathematicians are methods for extracting the roots of equations that were to be rediscovered in Europe up to a millennium later. They also learned to solve systems of linear equations, developed some very basic matrix algebra, and described geometric problems using equations instead of pictures. This last advance stands in interesting counterpoint to some of the Western mathematical traditions. In Egypt and Mesopotamia, for example, mathematics seems to have sprung from geometry, as early mathematicians attempted to find new ways of solving some problems. In China, by comparison, it appears as though geometry may have arisen from algebra as Chinese mathematicians tried to find ways to illustrate their mathematical discoveries. What makes this contrast particularly interesting is that both mathematical traditions arrived at fundamentally the same conclusions, giving a great deal of credence to the final results.

It is likely that many Chinese discoveries made their way to the West in during ancient and early medieval times. In particular, China is known to have had extensive contact with the West during the Han and T'ang dynasties (about 200 B.C. to A.D. 200 and A.D. 618 to 906, respectively). During these times, it is likely that China exported some mathematics to the West, although it is not certain what. Some claim that the ideas of decimal notation and zero originated in China, spreading to the West during these periods, but this may never be known for certain. What is certain is that China seemingly forgot her mathematical achievements towards the end of the first millennium A.D. and, by the Renaissance, had been eclipsed by Europe. In one of history's ironies, subsequent Chinese mathematicians learned mathematics from Europe that had been originally discovered in China and either exported to Europe or lost. There is no way to know what world mathematics would be like today if this episode of mathematical amnesia had not occurred, but it is certain to have been very different.

P. ANDREW KARAM

Further Reading

Boyer, Carl, and Uta Merzbach. *A History of Mathematics.* New York: John Wiley & Sons, 1991.

Temple, Robert. *The Genius of China: 3000 Years of Science, Discovery, and Invention.* New York: Simon & Schuster, 1986.

The Moscow and Rhind Papyri

Overview

The contents of the Moscow and Rhind Papyri shed considerable light on the nature and extent of ancient Egyptian mathematics. Both papyri provide vivid documentary evidence of geometrical reasoning in the Egyptian Twelfth Dynasty and insight into the practical applications of arithmetic prior to the more formal development of mathematical theory in ancient Greece. A careful analysis of the mathematical presentation and content of the two documents, however, limits the claims of Egyptian influence upon the later rise of theory in Greek mathematics.

Background

The archaeological record leaves little doubt as to the use and influence of mathematics on ancient Egyptian culture. Temples and other cultural artifacts provide extensive evidence of mathematical reasoning that predates the existing documentary record. The arrangement of pillars and stones in temple monuments, such as those found at Karnak, are lasting tribute to the careful calculations of ancient priests and astronomers in their attempt to provide accurate calendars based upon the movements of the Sun.

Whatever the initial need for a written record—whether its first use was as a more portable means of recording and deciphering astronomical data, or whether the general rise of civilization provided a swelling and multifaceted need to record the methods of mathematical reasoning—the earliest existing documentary records embodied in the Moscow Papyrus and the Rhind Papyrus disclose that the ancient Egyptians utilized considerable practical skill in the use and application of mathematics.

The Moscow Papyrus refers to a document, originally purchased by V.S. Golenishchev that is now maintained at the Moscow Museum of Fine Art. The papyrus, discovered in 1890 and translated by 1930, dates to approximately about 1850 B.C. The author is unknown. The papyrus is approximately 15 feet (4.6 m) long and 3 in (7.6 cm) wide and relates 24 practical problems, such as that required to calculate the volume of a frustum (base of a pyramid). Not all of the contents of the Moscow Papyrus, however, remain readable; portions of some of its problems are damaged, lost, or illegible.

The Rhind Papyrus, named after Scottish Egyptologist A. Henry Rhind, physically dates to approximately 1650 B.C. The papyrus is also known as the Ahmes Papyrus, a name recognizing the scribe who is credited with authoring the document. The content of the Rhind Papyrus, however, comes from a more ancient document that dates back 200 years to around the time of the composition of the Moscow Papyrus. The Rhind Papyrus was originally purchased by Rhind in Luxor in 1858, then translated and published by 1927. The Rhind Papyrus measures approximately 18 feet (5.5 m) long and 13 in (33 cm) wide. The Rhind, maintained at the British Museum, consists of 85 problems, all of which remain legible.

Within the papyri, the mathematical applications extend to everyday problems such as those dealing with feed mixtures to be used with cattle and questions relating to the allocation and storage of food. For example, problem 3 on the Rhind Papyrus concerns the division of six loaves of bread among 10 men. Other topics include exchange rates, inverse proportions, and harmonic means.

The Rhind also contains problems that utilize exponents. Problem 79, for example, states:

> *An estate consisted of seven houses; each house with seven cats; each cat eating seven mice; each mouse eating seven heads of wheat; each head of wheat capable of yielding seven hekats of grain. Houses, cats, mice, heads of wheat and hekat measures of grain, how many of these in all were in the estate.*

This requires calculating the powers of seven up to 7^5 (7 houses, 49 cats, 343 mice, 2401 heads of wheat, 16807 hekats of grain) then adding them all to reach a total of 19607. A version of this problem survives today in the nursery rhyme, "As I was going to St. Ives."

Both papyri present mathematical problems centered around everyday life with examples related to common situations, such as agriculture and building, computations of land areas and granary volumes.

Impact

To a large extent, our knowledge regarding the development of ancient Egyptian mathematics comes both from the architectural record and from the problems and solutions contained in several ancient documents, especially the Moscow Papyrus and the Rhind Papyrus. Moreover, it is evident that although the Moscow and Rhind Papyri date to the Middle Kingdom, the origin of the mathematics utilized in solving the problems presented is of a more ancient origin, perhaps representing a mathematics common in heritage to mathematics developed in other cultures.

The Rhind Papyrus provides abundant evidence of the extreme practicality of Egyptian mathematics. The Rhind was inscribed in the hieratic (cursive) writing of the Egyptian Middle Kingdom: A translation reveals that the document was intended to be a studied collection of exemplary mathematical problems and what are termed "obscure secrets" for students. The mathematical problems contained on the papyrus are presented in a series of exercises, in form similar to a teaching text, and are formulated in a nonalgebraic rhetorical manner with little abstract notation.

Of considerable interest to scholars is the mathematical range evidenced in the papyri. The problems utilize basic arithmetic, measurement, fractions, algebraic methodology, and geometry. It is clear from their content that the ancient Egyptians could solve problems equivalent to a modern linear equation with one unknown by using a method that is now known as the method of false position. Moreover, the geometrical methods displayed show a strong emphasis on measuring and include solutions (not all of them correct) for problems regarding the determination of the areas for circles, isosceles triangles, isosceles trapezoids, and quadrilaterals. Although there are varying interpretations of the hieroglyphs contained on the papyri, it is clear that the Egyptians formed numbers by grouping, and that the operations of multiplication and division were basically binary. The geometric problems also show a primitive understanding of the concept of similarity.

It is important to note that the papyri indicate an attempt at a standardization of units in Egyptian mathematics. The common area unit was the *setat,* or the square *khet.* The basic unit of length was the royal cubit, a distance some scholars estimate was derived from the distance from the elbow to the tip of the middle finger. Volume was measured in units of *hin.*

Most importantly, the papyri evidence a cleverness in the solution of problems. For example, the Egyptians recognized that the volume of a cylinder was quantitatively similar to the volume of a rectangular container (the volumes of both are equal to the area of their respective bases times their height). They used this observation to develop a method to determine the approximate area of a circle (e.g., given the diameter of the circle, subtract its ninth part, and square the result). By using only doubling and adding, the Egyptians could also construct columns of numbers from which the product of two numbers could be determined with great accuracy. The Moscow Papyrus contains the only known numerical example in the ancient world of the correct formula for the determination of the volume of a frustum of a square pyramid.

Although later Greek scholars lauded the use of the unit fractions in solving geometric problems, Egyptian mathematics was hampered from further development by their handling of fractions. The record indicates that the ancient Egyptians utilized unit fractions with the numerator equal to one. The fraction $^6/_{10}$, for example, would be broken into the sum of the unit fractions of $^1/_2$ and $^1/_{10}$. Because all fractions contained the number 1 as a numerator, in script this was written simply as 2 + 10 with a symbol above the numbers to symbolize a unit fraction.

The similarities of the problems and solutions found in the Moscow and Rhind Papyri establish that Egyptian mathematics remained remarkably uniform throughout its early history. Only very modest mathematical abstractions are present in the texts and the methodology evident is solidly arithmetic (i.e., built around addition) with elementary congruencies used only in relation to measuring.

The emphasis on practical application, indicating that the texts were apparently intended for students of mathematics, provides little support or evidence of the theoretical contributions once assumed by scholars of Greek mathematics. In fact, the extant record reveals a lack of Egyptian emphasis on discerning differences between exact and approximate calculations that often drive the development of theory.

Without question, there is abundant evidence in the later documentary record that the problem-solving techniques of the Egyptians influenced several prominent Greek scholars who studied in Egypt, including Pythagoras (580?-500? B.C.), Eudoxus (400?-350? B.C.), and Thales (624?-547? B.C.). Though these Greek mathematicians credited ancient Egyptian mathematics with great contributions toward the development of mathematical theory, the content of the Moscow and Rhind Papyri provide evidence of only the fundamental practicality and utilitarian application of Egyptian mathematics. There is little evidence that there was a theoretical foundation awaiting extension by Greek scholars. Instead, the preoccupation with every-

day and practical mathematics found in the papyri stands in stark contrast to the development of abstract mathematical thought in subsequent Greek culture.

It would be a mistake, however, to completely ignore or dismiss the development of theory in Egypt. For example, the calculation of the surface of a circle, evident in the Rhind Papyrus, and the calculation of the surface of a hemisphere, in the Moscow Papyrus, are earliest documented formulations related to the quadrature of the circle and the leveling of a curved surface. Implicit in the practical application of Egyptian mathematics was a conscientious effort to build algorithmic and numerically efficient techniques that enabled the solution of concrete problems that inspired broader theoretical development.

K. LEE LERNER

Further Reading

Books

Chace, A.B., et al. *The Rhind Mathematical Papyrus.* Oberlin, Ohio: Mathematical Association of America, 1927-29.

Eves, Howard. *An Introduction to the History of Mathematics.* New York: Holt, Rinehart, and Winston, 1964.

Gillings, R.J. *Mathematics in the Time of the Pharaohs.* Cambridge, MA: MIT Press, 1982.

Katz, V.J. *A History of Mathematics: An Introduction.* New York: Addison-Wesley Educational Publishers, 1998.

Robins, G., and C. Shute. *The Rhind Mathematical Papyrus: An Ancient Egyptian Text.* London: British Museum Publications, 1987.

Toomer, G.J. "Mathematics and Astronomy." In *The Legacy of Egypt.* Edited by J.R. Harris, pp. 27-54. Oxford: Clarendon Press, 1971.

Periodical Articles

Gerdes, P. "Three Alternate Methods of Obtaining the Ancient Egyptian Formula for the Area of a Circle." *Historia Mathematica* 12, no. 3 (1985): 261-8.

Early Counting and Computing Tools

Overview

Early civilizations developed a variety of ways of representing and manipulating numerical quantities in addition to the use of written symbols. Systems based on counting fingers and sometimes toes were replaced by tokens with a variety of numerical values, and devices for doing elementary arithmetic by the manipulation of

physical objects followed. Counting boards and the abacus remained in widespread use through the Middle Ages and into the Renaissance.

Background

Long before the invention of writing, humans appear to have developed some sense of number. Carvings made on bones found in Africa dating from the Upper Paleolithic period have been interpreted as a rudimentary means of counting the days for each phase of the moon. Also ancient and nearly universal is the tally stick, a piece of wood on which notches could be carved as animals or other objects were counted. For business transactions the stick could be split lengthwise, so that both parties could carry away a record of the process.

Small clay tokens begin appearing at archaeological sites during the eighth millenium B.C., sometimes appearing in hollow vessels where they may well have been placed during a counting process. The Babylonian civilization made extensive use of tokens and then developed a system of cuneiform numerical notation, pressing a stylus into wet clay. Many of the surviving cuneiform mathematical tablets appear to be arithmetic texts.

Among the first effective counting tools must certainly be listed the fingers and toes of the human body. Virtually every number system known is based on the numbers ten, five or twenty. The modern languages often include verbal fossils that reflect these origins. The word for 80 in modern French, for example, is literally translated as "four twenties."

The earliest counting and calculating tools include a variety of counting boards and several variations of the abacus. The earliest counting boards were probably wooden trays that could be covered with a thin layer of dust or sand in which numerical notations could be traced with a finger or stylus. Early references to the Hindu-Arabic numerals as "dust numbers" seem to support this conclusion, as does the term "abacus," which appears to be derived from a Semitic expression "to wipe the dust."

Few examples of pre-Roman abaci or counting boards have survived, so our understanding of their use depends on written records. The basic idea seems to have been discovered by the Babylonians. The Greek historian Herodotus (484?-425? B.C.) wrote that "The Egyptians move their hand from right to left in calculation, while the Greeks move it from left to right." Presumably he was referring to the use of some form of counting board. A fourth-century B.C. marble counting board has been preserved in Greece, while a Greco-Roman abacus of a somewhat later date features markers that were slid among slots cut through the board.

The abacus may have begun as a natural development of the counting board, with the counters held in place by being strung along wires. Eventually, at least two different varieties emerged. In one, later adopted by the Arabs, 10 beads were strung along each string. In the other, there was a divider or center bar, with (usually) five beads on one side and either one or two beads on the other.

Ancient civilizations differed greatly in the written number systems they developed. The Babylonians used a positional notation based on the number 60. The Egyptians developed two separate systems, a hieroglyphic system based on the repetition of symbols and the later hieratic (cursive) system, which used nine digits somewhat like ours. The Greeks invented two systems, both using letters of the Greek alphabet and neither based on position. The Romans likewise developed their system based on the repetition of letters representing different quantities. The majority of these systems were quite unsuited to practical arithmetic, especially multiplication and division, and most historians of mathematics believe that practical mathematics was generally done with the abacus or counting board in these societies.

In about the year 300 B.C. a system of "rod numerals" appeared in China. This system was a decimal one, or more precisely a centesimal one, with 9 separate symbols for the integers 1 through 9 and nine different symbols for multiples of ten between 10 and 90. The symbols from one to five were written as a series of vertical strokes, while those for six to nine employed a single horizontal stroke to represent five units, along with one to four vertical strokes. For the multiples of 10 the role of the vertical and horizontal rods was reversed. Larger numbers were written by alternating the vertical and horizontal stroke symbols from right to left. Chinese civil servants carried bags of bamboo, ivory, or iron rods, with which they did calculations, often with great facility, by manipulating the individual rods of the rod numerals. Numbers expressed in rod numeral can be very easily translated into the position of beads on an abacus with a dividing bar, but there is considerable uncertainty about the date of the adoption of the abacus in China, because the rod manipulation system also worked very well. Historians can only say that use of the abacus in China is at least 1,000 years old. The abacus was not introduced into Japan until the sixteenth century.

Impact

The counting board and abacus spread rapidly throughout Europe and Asia, undoubtedly through the contacts between merchants and traders. The use of the abacus persisted in Europe, only gradually supplanted by calculations on paper. Indeed, the use of the abacus may have helped foster the spread of the Hindu-Arabic system, since it was natural to transcribe the results of a calculation onto paper in this system, with its positional notation and its place-holding zero. Calculations on paper did not fully replace the abacus in Europe until the seventeenth century. The abacus remains popular in Asia and there are still individuals considered experts in its use.

The use of mechanical devices to perform arithmetic and more general mathematical operations has been a recurrent theme since the Renaissance. With the invention of logarithms by the Scottish mathematician John Napier (1550-1617), multiplication could be accomplished by adding logarithms and the taking of exponents by multiplying logarithms by numbers. While this development greatly simplified many of the calculations that would be required in astronomy and navigation, it brought new importance to the avoidance of errors in the calculation and typesetting of mathematical tables. For this reason, Napier sought to automate the process. He developed a calculating machine using a set of round white rods, which came to be called Napier's bones, to do accurate calculations. While the operation of the "bones" was too complex to be practical, the slide rule, based on the addition of logarithms, became a standard tool for engineering calculations until it was replaced by the handheld calculator in the 1970s.

In 1642 the great French mathematician Blaise Pascal (1623-1662) invented an adding machine with numerous rotating cogged wheels. Though too complex for its times, its basic mechanism would reappear in the earliest cash registers.

Another early mathematical machine was the "difference engine" conceived in 1833 by English mathematician Charles Babbage (1792-1871). Babbage managed to get initial financial support for his project from the English government, but had to turn to private sources when his government funding was cut off. Among his supporters was Ada Lovelace (1815-1852), daughter of Lord Byron, whose efforts to popularize the project are our primary source of information about it. Had the machine been completed, it would have been much like a modern programmable computer that could store numbers and instructions in its own memory. Babbage is thus considered the "grandfather" of the modern computer.

The modern age of computing began with the rather abstract analysis of the general process of symbol manipulation conducted by British mathematician Alan Turing (1912-1954). Turing determined that a machine or "automaton", which could read and write one symbol at a time on an endless tape, could implement any mathematical process, even the proving of a theorem. Which calculation would be done was determined by a set of rules that was realized by the machine. Turing found that there was at least one such machine

THE ABACUS VS. THE ELECTRIC CALCULATOR

In Tokyo, Japan, on November 12, 1946, a competition took place between the Japanese abacus (*soroban*) and an American electronic adding machine. Mr. Kiyoshi Matsuzaki, a champion operator of the abacus in the Savings Bureau of the Japanese Ministry of Postal Administration, was pitted against Private Thomas Nathan Wood of the 20th Finance Disbursing Section of General MacArthur's headquarters, who was considered the best operator of the electric calculator in Japan. The U. S. Army Newspaper, *Stars and Stripes*, reported the results, stating: "The machine age tool took a step backward yesterday ... as the abacus, centuries old, dealt defeat to the most up-to-date electric machine." A Japanese newspaper commented that "Civilization, on the threshold of the atomic age, tottered Monday afternoon as the 2,000-year-old abacus beat the electric calculating machine in adding, subtracting, dividing.... Only in multiplication alone did the machine triumph." Yet despite this victory for tradition, the Japanese became, and remain, the largest manufacturers of electronic calculators in the world. While calculating machines have become more advanced, and much faster, even today the speed and accuracy of any calculating device is still determined a great deal by its user.

DAVID TULLOCH

that could accept instructions given as a string of numbers, which would then make it function as another machine. The instructions would then serve as the machine or computer's "program."

The first computers were built near the end of the Second World War. They were enormous, expensive, and generally could only be used for at most a few hours at a time before one of the vacuum tubes burned out. With the invention of the transistor in 1947, and a process of miniaturization that continues to the present, they have grown smaller, less expensive, more powerful, and more reliable. By the 1970s enough components could be incorporated into a small microprocessor chip that handheld electronic calculators became commonplace. In their processing of numerical data, one bit at a time, some family resemblance to the ancient abacus may still be recognized.

DONALD R. FRANCESCHETTI

Further Reading

Bell, Eric Temple. *Development of Mathematics*. New York: McGraw-Hill, 1945.

Boyer, Carl B. *A History of Mathematics*. New York: Wiley, 1968.

Cajori, Florian. *A History of Elementary Mathematics*. New York: Macmillan, 1930.

Grattan-Guiness, Ivor. *The Rainbow of Mathematics: A History of the Mathematical Sciences*. New York: Norton, 1997.

Kline, Morris. *Mathematical Thought from Ancient to Modern Times*. New York: Oxford University Press, 1972.

The Philosophy of the Pythagoreans

Overview

The way of thinking about the world that came to be known as philosophy emerged in the sixth century B.C. among groups of Greek thinkers scattered around the Mediterranean region. The Pythagoreans were one of the most influential of these groups. During their approximately two hundred years as an organized community, the Pythagoreans spread ideas about numbers, nature, and man that were profoundly important to the subsequent study of mathematics, music, and astronomy.

Background

As with most figures of the ancient world, factual evidence about Pythagoras and the community he founded is quite minimal. Historians agree, however, that Pythagoras of Samos (c. 560-c. 480 B.C.) founded a community of like-minded men among a Greek colony on the southern coast of Italy around 530 B.C. While it is tempting to describe this group in a modern way as a "school," in fact it began as a kind of religious community or cult. Their primary motivation for intellectual pursuits was as a means of achieving spiritual purification. By emphasizing contemplation and study, especially of numbers and numerical relationships, and practicing physical asceticism, the Pythagoreans hoped to bring their souls into harmony with the greater cosmos and therefore to escape the cycle of the transmigration of souls. This doctrine of reincarnation, or the "wheel of birth," held that souls were immortal but would be forced to pass through life again and again in various animal forms until sufficient purity was achieved. The Pythagoreans sought to achieve this purity directly through their philosophy and contemplative life.

Their beliefs in the transmigration and immortality of souls and in the purifying potential of philosophy were central to the religion of the Pythagoreans. They additionally believed in abstinence from various physical and dietary practices, in strict measures of loyalty and secrecy, and in the mystical significance of certain symbols. But the most influential of the Pythagorean beliefs was certainly their contention that all of reality was mathematical. Their elevation of numbers and numerical relationships to a position of such philosophical importance brought them to unique insights about subjects from acoustics to astronomy.

The precise history and activities of the Pythagoreans are not clearly established. It seems, however, that their society flourished into the fifth century B.C. throughout a portion of southern Italy near Crotona. Their power and influence had negative consequences, as in the middle of the fifth century they became entangled with political disputes in the region and ultimately were violently suppressed. The surviving members of the Pythagoreans then scattered throughout Greek-speaking regions outside of Italy. While their period as an influential religious brotherhood had come to an end, individuals and small groups continued to perpetuate the Pythagorean philosophy well into the following century.

Impact

While Pythagorean ideas about ethics and metaphysics appear in many important Greek treatises, including the works of Plato (c. 427-347

B.C.) and Aristotle (384-322 B.C.), the mathematical ideas promulgated by the Pythagoreans had the greatest influence on the subsequent history of philosophy and science. Because the Pythagoreans maintained an oral rather than a written tradition, we have no direct evidence of their beliefs or discoveries and must rely on the accounts written by later philosophers who made use of or commented upon Pythagorean ideas. Aristotle, for example, contended that the Pythagoreans believed the entire universe to be essentially musical, and therefore mathematical. Whether the Pythagoreans believed this literally or merely as a model of some sort, we cannot know. But we can be certain that numbers and musical harmony were the most significant concepts in the Pythagorean understanding of the world. The most common characterization of the worldview of the Pythagoreans is the complete identification of reality with numbers.

The Pythagoreans studied the properties of numbers, the reflection of numerical properties in geometrical figures, and the existence of numerical relationships in the natural world. They were the first to study sums of series. By representing these series geometrically, they were able to classify different patterns resulting from summing different sets of numbers. Among students of elementary mathematics, the Pythagoreans are best known for one such geometrical relationship. They observed that in right-angled triangles, the square of the hypotenuse is equal to the sum of the squares of the other sides. This "Pythagorean theorem" remains one of the touchstones of elementary geometry, and is a clear example of the style of investigation and discovery typically attributed to the Pythagoreans. Another Pythagorean contribution important to elementary geometry was their recognition that the diagonal and the side of a square are incommensurable (that is, their ratio will be an irrational number).

In contrast to our modern understanding of these mathematical ideas, the Pythagoreans came to many of their discoveries by manipulating pebbles to represent numbers. These pebbles have their own legacy in modern mathematics; the Greek word for pebble was *calculus* and gave us the English term "calculate." The Pythagoreans represented numbers as triangular, square, or rectangular, depending on whether that number of pebbles could be arranged symmetrically in one shape or the other. For example, three is a triangular number, while four is a square number.

No set of ideas was more important to the worldview and philosophy of the Pythagoreans

Pythagoras of Samos discovered the importance of numbers in describing the real world. *(Corbis Corporation. Reproduced with permission.)*

than those relating numbers to musical harmony. The Pythagoreans identified the fundamental musical intervals of the octave, the fifth, and the fourth by using ratios, another conceptual discovery, and by demonstrating the production of harmonics on their stringed instruments. That is, the harmonic of the octave is produced by touching the string at one-half of its length, while the fifth is produced at two-thirds of the length. Building on this foundation, the Pythagoreans produced a system of musical scales and chords.

The Pythagoreans extended their understanding of musical intervals to their study of the heavenly bodies. An earlier philosopher, Anaximander (c. 610-c. 546 B.C.), suggested that the heavenly bodies consisted of three movable wheels. The Pythagoreans built upon this idea by identifying the intervals between the three wheels with the musical intervals of the octave, fifth, and fourth. They were the first to distinguish the diurnal revolution of the heavens from east to west from those of the Sun, Moon, and planets from west to east. Pythagoras is also credited with discovering the sphericity of Earth. Centuries later, Nicolaus Copernicus (1473-1543) credited the Pythagoreans for these astronomical concepts as important precursors to his own hypothesis that Earth and other planets rotate about the Sun, rather than the Sun rotating around Earth.

Tracing the impact of the Pythagoreans is somewhat frustrating. Later Greek philosophers including Plato, Aristotle, and Euclid (c. 330-c. 260 B.C.) explicitly and tacitly credited the Pythagoreans with many ideas, beliefs, and discoveries. But since these very sources are in many cases our only documentation of the work of the Pythagoreans, it is impossible to disentangle original ideas from their subsequent interpretations. If we are to give the Pythagoreans their broadest possible due, then they would get credit for the birth of the study of mathematics for its own sake rather than in support of commercial or other activities. The formal study of music and harmony, and the practice of music theory, are also achievements that can be confidently attributed to the Pythagoreans.

The boldest ideas of the Pythagoreans—their characteristic views that nature is made up of numerical relationships and the musical harmonies that correspond to those relationships—have reappeared in different forms throughout the history of science. The concept of the "harmony of the spheres" of the universe, for example, was a guiding principle for astronomer Johannes Kepler (1571-1630). Efforts to explain nature by mathematics have often yielded descriptions of nature as inherently mathematical, and these have at times shaded into purely

mathematical accounts of natural phenomena. Most intriguing of all, however, may be the science of digitization, which electronically represents sound, images, and information of all kinds as numbers. The digital world can be seen as the ultimate realization of the Pythagorean ideal of nature as number. Although no causal connection could ever be drawn from Pythagoras in the fifth century B.C. to twenty-first century digital music, the latter no doubt owes some kind of conceptual debt to the imagination of the Pythagoreans.

LOREN BUTLER FEFFER

Further Reading

Dreyer, J.L.E. *A History of Astronomy from Thales to Kepler.* New York: Dover Press, 1953.

Furley, David. *The Greek Cosmologists.* Cambridge: Cambridge University Press, 1987.

Lindberg, David. *The Beginnings of Western Science.* Chicago: University of Chicago Press, 1992.

Lloyd, G.E.R. *Early Greek Science: Thales to Aristotle.* Cambridge: Cambridge University Press, 1970.

Neugebauer, Otto. *The Exact Sciences in Antiquity.* Princeton: Princeton University Press, 1952.

O'Meara, Dominic. *Pythagoras Revived: Mathematics and Philosophy in Late Antiquity.* Oxford: Clarendon Press, 1989.

The Birth of Number Theory

Overview

Number theory is the mathematical study of the properties of numbers and the relationships between numbers. The study of prime numbers, for example, is a focus of much of number theory, and Fermat's famous Last Theorem was, until recently, one of the major unsolved problems in the theory of numbers. A Greek mathematician, Diophantus of Alexandria (third century A.D.), did some of the very first work in number theory, and, in so doing, set the stage for nearly two millennia of future work.

Background

A prime number is a number that can be evenly divided by only itself and the number one. There is no way to know with certainty when prime numbers were first discovered, but it was certain-

ly near the dawn of mathematics. They have been known of for at least 2,500 years, and undoubtedly much longer. Prime numbers were studied by the Pythagoreans in the first few centuries B.C., but not as an end in and of themselves.

It is also certain that ancient mathematicians studied the properties of nonprime numbers. Here again the Pythagoreans played an important role, describing the properties of even and odd numbers, theorizing about prime numbers, and examining some different types of equations. At the same time, and in subsequent years, other mathematicians began reporting the manner in which certain equations could be solved, while still other mathematicians attempted to apply these equations to geometrical problems.

In spite of these advances, the study of numbers was not really a distinct subdiscipline of mathematics. It was of abstract interest, producing some useful results, but no mathematicians devoted their careers to such studies.

After reaching a zenith in the third century B.C., Greek mathematics went into decline for several centuries. Greek mathematicians continued to produce original contributions to the field, but not at the rate or with the same originality as earlier. However, by the second century A.D., Greek mathematics began to revive, entering a "silver age." During this period, they never really returned to previous glories, but briefly became, again, some of the world's preeminent mathematicians. During these centuries, the city of Alexandria in particular was a dominant force in mathematics, producing such great mathematicians as Euclid (330?-260? B.C.), Diophantus, and Hypatia (370?-415?).

It was during this time that the great mathematician, Diophantus of Alexandria reached the peak of his powers. Diophantus was a renowned algebraist, geometer, and the founder of the field of number theory. His most significant contribution was his book *Arithmetica,* written around A.D. 250. This work apparently consisted of 13 volumes, though only six have survived, and it appears as though the others were lost to history shortly after his death. In the surviving books, Diophantus described several propositions regarding number theory. One of these suggested that the difference of the cubes of two rational numbers is equal to the sum of the cubes of two separate rational numbers (mathematically, $a^3 - b^3 = c^3 + d^3$). More important by far was another statement that, when explored in later centuries, led to Fermat's Last Theorem. In all, Diophantus's *Arithmetica* must be considered not only one of the greatest mathematics works of its era, but also the first coherent work in number theory in the history of mathematics.

Impact

Although earlier mathematicians, including the Egyptians and Babylonians, had explored many of the number theory problems posed by Diophantus, they had never before been assembled together in a single work. It was this compilation of problems in number theory that helped launch it as an independent segment of mathematics. In particular, Diophantus's work had these impacts on mathematics:

1. It inspired the work leading to Fermat's famous Last Theorem which, in turn, led to major advances in mathematics.

2. It led to the formal study of the field of number theory.

3. Resulting work on the theory of equations led to methods of solutions that set the stage for what was to become the field of algebra.

Perhaps the most far-reaching problem given in Diophantus's *Arithmetica* was Problem Number 8 in Volume II. In this problem, Diophantus asked for a method of representing a given square number as the sum of two other squares. Reading a translation of this problem, Pierre de Fermat (1601-1665) was inspired to wonder if such a problem with higher exponents could be answered. His theorem, that a solution could not be found with any exponents higher than two, served as a puzzle and an inspiration to mathematicians for over three centuries.

This problem was hardly unique. The Babylonians had solved it satisfactorily over a millennium earlier, as had the Egyptians and the older Greek Pythagoreans several centuries before Diophantus's time. However, a problem need not be new to be interesting. This same type of problem continues to be taught in schools and universities today because it is important and useful, and Diophantus likely included it in his book for exactly this reason. What Diophantus had no way of knowing is that over 1,000 years later Pierre Fermat would read a recent translation of this work and it would inspire him to jot his now-famous note for the Last Theorem in the margin of Diophantus's book.

Attempts to prove or disprove Fermat's Last Theorem occupied the attention of number theorists for well over 300 years, spurring many advances in mathematics. With any historical event, it is tempting to ask "What if?" In this case, we can ask ourselves, "What if this volume of Diophantus's work was among those lost to history?" "What if it were not found and translated until after Fermat's death?" "What if Fermat just never had a chance to read it?" Or even, "What if Diophantus had not thought to include this particular problem in his book, thinking it too elementary or not sufficiently interesting for inclusion?" It is quite possible that, had any of these events come to pass, Fermat would not have made his famous observation, leaving mathematics, and the world, the poorer for it.

In addition to Fermat's work, number theory has enriched mathematics immeasurably. As noted above, number theory is the study of numbers and their interrelationships. The search for patterns in the occurrence of prime numbers is one example of number theory in action. For many centuries, mathematicians have struggled to construct a formula for finding ever-larger prime numbers. Even today an occasional story appears in the news noting that a new supercomputer has "discovered" the largest prime number to date. Though of little but academic interest to most, it must be noted that the encryption algorithms on most computers and most encryption software depends on very large prime numbers to securely encrypt everything from Internet purchases to military communications to illicit communications between criminals. This is but one example of the impact that number theory has had on our everyday world.

The largest part of Diophantus's work, indeed the largest part of early number theory, dealt with solving different types of equations. A particular type of equation continues to be known as "Diophantine." These equations are defined as equations in which the only operations are multiplication, division, addition, and subtraction; all of the constants are natural numbers; and the only interesting solutions are positive or negative natural numbers. An example of a Diophantine equation is the famous $a^2 + b^2 = c^2$. This is the basis of the Pythagorean Theorem as well as Fermat's Last Theorem.

The study of equations has proven fruitful over the centuries, and has led to a deeper understanding of nature as these equations are put to work describing the phenomena surrounding us. For example, consider the equation mentioned above. This equation was known to the ancient Egyptians, the Babylonians, and other ancient mathematicians, and they used it to solve basic and applied problems in geometry. They also realized early that there was no easy solution for some problems, and that other problems seemed to have no solution at all. As an example, consider a triangle with two sides that are three and four meters long. Using the Pythagorean theorem, finding a solution is simple, and a quick calculation shows the remaining side must be five meters long. This is written as $3^2 + 4^2 = x^2$, so x must be equal to the square root of 25, which is 5. Now consider a triangle where both sides are one meter long. In this case, x is equal to the square root of two, a number that neither terminates nor repeats. Thus, one need not look far to find a simple equation with an answer that cannot be stated with perfect precision.

Carrying this example a bit further, consider the equation $x^2 + 1 = 0$, another simple equation. In this case, x is equal to the square root of -1, and the solution for this problem was not really appreciated or solved until more than 1,000 years after Diophantus's death. Once solved, however, with the "invention" of imaginary numbers, our ability to understand and describe our world took a huge step forward. Imaginary numbers (and complex numbers, formulated by Caspar Wessel in 1799), have since become a vital tool for engineers, scientists, and electronics designers worldwide.

Of course, most of these developments were far from Diophantus's thoughts. Indeed, it is certain that he and his contemporaries could not conceive of the uses to which their findings in number theory would someday be put. Luckily, this degree of foresight is not a necessary condition for great work, because we have all benefited from their discoveries.

P. ANDREW KARAM

Further Reading

Aczel, Amic. *Fermat's Last Theorem: Unlocking the Secret of an Ancient Mathematical Problem*. New York: Four Walls Eight Windows, 1996.

Boyer, Carl, and Uta Merzbach. *A History of Mathematics*. New York: John Wiley & Sons, 1991.

Dunham, William. *Journey through Genius: The Great Theorems of Mathematics*. New York: Penguin Books, 1990.

Number Systems

Overview

The earliest number systems for which we have written records include the Babylonian, Egyptian, Greek, Roman, Chinese, Hindu, and Mayan. Some systems employed the repetition of symbols to express larger quantities, while others involved either explicit multiplication of quantities by powers of a basic unit or the specification of powers by position in a string of symbols. Full development of the positional system, the basis of modern arithmetic calculation, could not occur before a symbol for zero and the computational rules for its use were established, a situation that did not fully materialize in Europe until after the Renaissance.

Background

What we know about the number systems of earlier civilizations is largely the result of the archaeological research of the past 200 years and is necessarily incomplete. Far more effort has gone into understanding the societies of the ancient Near East and the Mediterranean world than those of the Americas, Africa, and the greater part of Asia. Quite clever and useful systems may yet remain to be discovered in these less studied areas.

The earliest detailed records of arithmetic notation appear in the clay tablets of Babylon, the area between and around the Tigris and Euphrates Rivers, site of the Sumerian (4000-2500 B.C.) and Akkadian (2500-1000 B.C.) kingdoms. From the Akkadian period we have a wealth of records recorded in cuneiform—that is, as imprints of a wedge-shaped tool on clay tablets that were subsequently baked.

In Akkadian arithmetic the digits from one to nine were recorded using the corresponding numbers of vertical impressions of the stylus, while multiples of 10 up to 50 involved repetitions of a horizontal wedge shape. A different vertical stroke was used to denote the number 60, and higher numbers were expressed as combinations of 60, 10, and units. Thus the number 144 would be written with the symbol for 60 repeated twice, the symbol for 10, repeated twice and a cluster of four vertical strokes.

By clustering symbols, the Babylonians went on to invent what may have been the first positional notation. While the symbol for 60 repeated three times close together stood for three times 60, or 180, the three identical symbols written with a space after the first of them would be interpreted as the square of 60 added to twice 60, or 3,720. To reduce the possibility of misreading, a special symbol for the required space, consisting of two short oblique wedge impressions, was introduced in about 300 B.C. The symbol, which filled one of the roles of the zero in modern mathematics, was not understood as a quantity, however.

It is not certain why the number 60 was so significant in Babylonian mathematics. Astronomical interests may have played a role, as there are roughly half of 60 days in the month and just over six times 60 days in a year. The Babylonians also had special symbols for the common fractions $\frac{1}{2}$, $\frac{1}{3}$, and so on, and extended their positional notation to express fractions as so many sixtieths plus so many thirty-six-hundredths and so on.

The ancient Egyptians developed two separate systems of writing and, correspondingly, two separate systems of numerical notation. In the earlier, hieroglyphic, system separate symbols were used for units, tens, hundreds, and other powers of 10. Numbers were written from right to left and multiples of each power of 10 were indicated by repetition. Thus the number 144 would be written with the symbol for unit repeated four times followed by the symbol for 10 repeated four times followed by the symbol for 100. Unit fractions, that is fractions with 1 as the numerator were written by placing an oval above the symbol for the number in the denominator.

Beginning in about 2500 B.C. the Egyptians adopted a simplified script now called "hieratic" for writing on papyrus. The hieratic number system had separate symbols for the numbers one through 10, and thus marks the first appearance of ciphers, a set of digits like our 1 through 9. It also used a positional notation reading from right to left for numbers larger than 10. Unit fractions were indicated by a dot placed over the denominator. The Egyptians, unlike the Babylonians, did not develop a formalism for fractions in general, but expressed fractions like $\frac{2}{5}$ as a sum of unit fractions.

The Greeks also developed two main systems of notation for numbers. The Attic system, the earlier of the two, resembled Egyptian hieroglyphics in that it was based on powers of 10 and used repetition to indicate multiples. The system used groupings of vertical strokes for the numerals 1 to 4, and the letters π (pi) for five, δ (delta) for 10, η (eta) for 100, and χ (chi) for 1,000, so that the number 144 would appear as ηδδδδ||||. The Attic system adopted special symbols for 50 and 500 that combined the letters for five and 10 or five and 100, respectively.

The Ionian or alphabetic system of the Greeks used letters in alphabetical order, including the archaic (unused then and now) letters digamma for 6, koppa for 90, and sampi for 900. Thus this system represented 1 by α (alpha), 2 by β (beta), 4 by δ (delta), 9 by θ (theta) and 10 by ι (iota), then increased by 10s so that κ (kappa) represented 20, λ (lambda) 30, μ 40, up to ρ (rho), which represented 100, and then increased by hundreds ending with 800, represented by ω (omega), and 900 by sampi. In this system thus 144 would appear as δμρ. Units of 1,000 were indicated by a comma-like symbol preceding one of the first nine letters, while multiples of 10,000 were indicated by M (capital Mu).

The Roman system, still used for the occasional recording of dates or to indicate volume numbers of books, used the letters I, V, X, L, C, D, and M to represent the numbers 1, 5, 10, 50, 100, 500 and 1,000, respectively. Roman fractions were based on denominators that were multiples of 12. Thus there were special symbols for $1/_{12}$, $2/_{12}$, $1/_{24}$, and so on.

By about 300 B.C. a system of "rod numerals" had been developed in China, with sets of one to five vertical strokes indicating the numbers from 1 to 5 and sets of one to four strokes under a single horizontal stroke representing the numbers 6 through 9. The multiples of ten from 10 through 50 were written with from one to five horizontal strokes, and from 60 to 90 with from one to four horizontal strokes under a single vertical stroke. Larger numbers were written by alternating the vertical and horizontal stroke symbols from right to left. Thus the number 144 would be written as a group of four vertical strokes to the left of a group of four horizontal strokes to the left of a single vertical stroke. The system of rod numerals lent itself to the performance of arithmetic through the manipulation of actual rods, an art at which some civil servants excelled. At the same time, the Chinese also employed a written system based on ciphers for the digits one through nine and for the various powers of 10. In this decimal, but not positional system, each number appeared as a sum of multiples of powers of 10, much like our present system, but without the need for place-holding symbols since the powers of 10 were shown explicitly.

Number symbols also appear in India by the third century B.C. There is considerable variation in the symbols used over time. The Brahmi symbols use separate ciphers for the numbers 1 through 9, several of which bear some resemblance to the modern numerals. The Brahmi set also included distinct symbols for multiples of 10 up to 60. After 600 B.C. and many variations, including abandoning symbols for words at times, Hindu mathematicians returned to the Brahmi set, and augmented it with a zero character that had both the place-holding and arithmetic properties of the modern number. The concept of the zero so greatly facilitated computation that it has become indispensable. Archaeological research has now shown that the Maya of Central America may have made a similar discovery independently at an even earlier date.

Impact

Echoes of the Babylonian number system can be recognized in our traditional measurements of angle, with 360° in a full circle and each degree divided into 60 minutes of arc and each minute into 60 seconds. The same can be said of our measurements of time by minutes and seconds.

Our modern number system is derived from that of the Hindus with the addition of the symbol for zero. This system was introduced into Renaissance Europe through Latin translations of the work of Arab scholars, notably, Muhammad ibn-Musa al-Khwarizmi (780?-850?), whose book, now known only as *Algoritmi de numero Indorum*, or "Al-Khwarizmi on the Hindu method of calculation," explains the Hindu number system and how it can be used in arithmetic calculations. It is from the title of this book that we obtain the word "algorithm" for a systematic method of calculation. The method was further popularized by Leonardo of Pisa (1170?-1140?), also known as Leonardo Fibonacci, the son of a government official who traveled throughout northern Africa and learned from merchants about the Hindu-Arab system of numbers and methods of calculation, which he described in the *Liber abaci,* or book of calculations.

That the modern system is based on the number 10 is no doubt a reflection upon human anatomy with its 10 fingers and 10 toes. For electronic computation the binary system, in which numbers are represented as a sum of powers of two, is more natural as it requires only two distinct symbols, a one or a zero, which can be represented by the "off" and "on" states of an electronic switch.

DONALD R. FRANCESCHETTI

Further Reading

Boyer, Carl B. *A History of Mathematics.* New York: Wiley, 1968.

Grattan-Guiness, Ivor. *The Rainbow of Mathematics: A History of the Mathematical Sciences.* New York: Norton, 1997.

Kline, Morris. *Mathematical Thought from Ancient to Modern Times.* New York: Oxford University Press, 1972.

The Historical Relationship
of Logic and Mathematics

Overview

Above the gateway to Plato's *Academy* appeared the following inscription: "Let no one who is ignorant of geometry enter here." Plato (427?-347 B.C.)—an ancient Greek philosopher thought by many to be the most influential of all philosophers—regarded an understanding of the principles of geometry essential to the training of philosophers. Philosophers had begun to recognize a relationship between mathematical proofs, such as those exemplified through the study of geometry, and logic, the science of identifying the formal principles of reasoning. It was not until Aristotle (384-322 B.C.), a student of Plato's, however, that logic was systematized into a form that remains, for the most part, intact today.

Background

The need for definitive proof in philosophy became apparent early in history, especially in those instances where philosophical positions contradicted each other. Perhaps the most notable of the era's debates illustrating the need for resolution by way of proof concerned whether reality was essentially unchanging in nature (and therefore change was an illusion) or whether reality was in constant flux (and therefore permanence was an illusion). Parmenides (b. 515 B.C.) argued in favor of the former position, and his student, Zeno of Elea (495?-430? B.C.), tried to demonstrate that the opposite position led to absurdities. The opposite position, however, had earlier been adopted by Heraclitus (540?-480? B.C.). Aristotle believed that by studying the inferences employed in these arguments (i.e., by analyzing the arguments), flaws in the arguments (if there were any) could be determined. If none were found, then the argument could be deemed acceptable, or valid.

Thus, for Aristotle, logic was a tool that philosophers could use either to strengthen their arguments or weaken their opponents'. In short, logic is the instrument by which we can acquire knowledge and, in the process, be sure of our conclusions. Aristotle used this tool for developing his various theories about the nature of the world and of man's place in it.

Impact

Aristotle is commonly credited with the invention of the syllogistic argument form. In his *Prior Analytics* Aristotle proposed the following syllogism:

(i) *Every Greek is a person.*
(ii) *Every person is mortal.*
(iii) *Every Greek is mortal.*

According to Aristotle, each of these three sentences exhibit proportional unity, that is, they are capable of having a truth value—of being either true or false. Such sentences are known as *statements*. The first two statements, in turn, are known as *premises*; they provide evidence for the third statement, which is known as the *conclusion*. What Aristotle demonstrated with this argument is that the conclusion follows with *necessity* from the premises. In other words, assuming the premises to be true, the conclusion must follow—it cannot be otherwise. Such arguments are said to be *deductively valid*.

It is easy to see, however, that the conclusion to the following syllogism does not follow with necessity.

(i) *Every Greek is a person.*
(ii) *Every person is mortal.*
(iii) *Every mortal is Greek.*

In this example, the conclusion does not follow because there are many mortals who are not Greek. Such arguments are said to be *deductively invalid*; although they are syllogistic in form, their conclusions are not supported by the premises.

In in a series of works known as the *Organon,* Aristotle identified the rules by which arguments of the first type—valid deductive arguments—can be distinguished from arguments of the second type—invalid deductive arguments. Indeed, as noted by one scholar, Aristotle believed that it was essential to science itself that such rules be identified:

The sciences—at any rate the theoretical sciences—are to be axiomatised. What, then, are their axioms to be? What conditions must a proposition satisfy to count as an axiom? Again, what form will the derivations within each science take? By what rules will theorems be deduced from axioms?

Those are among the questions which Aristotle poses in his logical writings, and in particular in the works known as *Prior* and *Posterior Analytics* [which comprise part of the *Organon*].

Not coincidentally, many of the illustrations of the scientific method that Aristotle proposed were taken from mathematics. The conclusions of mathematical inferences, after all, like the conclusions of deductively valid arguments, follow with necessity. For example, it follows with necessity that two things added to two other things must total four things; it is impossible for two and two to equal three, five, or any other number; and what is more, four equals two and two not accidentally or coincidentally, but because it must. So, just as science is concerned with the rules by which theorems (or conclusions) can be deduced from axioms (or premises), so too is the field of mathematics.

One of Aristotle's most revolutionary contributions to logic was the introduction of symbolic notation, which is also firmly rooted in mathematics. Variables enabled Aristotle to express logical principles directly, instead of describing them through examples, like those indicated above. So, rather than trying to express the logical principle of *conversion* by a linguistic example—"If no pleasure is good, then no good will be pleasure"—by using variables, Aristotle could directly illustrate the principle of conversion thusly: "If *A* belongs to none of the *B*s, then *B* belongs to none of the *A*s"; or even more simply as, "If no *A*s are *B*, then no *B*s are *A*."

By incorporating variables into logic, Aristotle set the stage for both mathematics and science. In this way, the axioms of deductive reasoning could be identified more easily and more readily applied to science and mathematics, thereby furthering both enterprises. Centuries later, Aristotle's project still was alive, as British philosophers Bertrand Russell (1872-1970) and Alfred North Whitehead (1861-1947), in their three volume classic *Principia Mathematica* (1910-13), set out to establish that

> *all pure mathematics deals exclusively with concepts definable in terms of a very small number of fundamental logical concepts, and that all its propositions are deducible from a very small number of logical principles.*

Today, the project begun by Aristotle to identify and systematize the principles of reasoning—the science of logic—remains as important as ever to man's acquisition of knowledge and understanding of the natural world.

MARK H. ALLENBAUGH

Further Reading

Hurley, Patrick J. *A Concise Introduction to Logic.* 7th ed. Belmont, CA: Wadsworth, 2000.

Lejewski, Czeslaw. "History of Logic." In *The Encyclopedia of Philosophy.* Edited by Paul Edwards. Vol. 4. New York: Macmillan, 1967.

The Three Unsolved Problems of Ancient Greece

Overview

The geometry of ancient Greece, as characterized by Euclid's famous book, the *Elements*, has formed the basis of much of modern mathematical thought. For example, the Greek insistence on strict methods of proof has survived to this day. The methods and theorems found in the *Elements* were taught to schoolchildren almost unchanged until the twentieth century. Even today, school geometry is essentially the same geometry as that composed by Euclid (c. 325-c. 265 B.C.) well over two millennia ago.

It became the practice in traditional Greek mathematics to accept geometrical constructions only if they could be performed with an unmarked straightedge and a compass. This custom is derived from the first three postulates of Euclid's *Elements*. A *postulate* is a statement that is accepted as true without proof. In the *Elements*, Euclid gives five postulates that are the starting points for the propositions or theorems given in the body of the book. The first three of these postulates address the construction of a straight line and a circle:

1. A straight line can be drawn between any two points.

2. A finite straight line can be extended indefinitely.

3. A circle can be drawn with any center point and any line segment as a radius.

Although none of these postulates (or any others) refer directly to a straightedge or a compass, this tradition, usually attributed to Plato (427-347 B.C.), became an integral part of Greek geometry. The Greeks referred to constructing geometric figures using only a straightedge and a compass as the *plane method.*

Although the bulk of Greek geometry was constructed using plane methods, three problems defied solution by these methods for centuries. The ancient problems of squaring the circle (or quadrature of a circle), trisecting an angle, and doubling a cube (or duplicating a cube) have produced countless attempted solutions. These attempts have come from the great mathematicians in history, as well as from numerous amateurs and cranks. Each attempted solution had one thing in common; they all failed.

Although the Greeks were unable to prove the impossibility of solving the three problems by plane methods alone, they were certainly aware of the difficulties in solving each problem. Instead of despairing over their futile attempts to solve the problems using only straightedge and compass, the Greek mathematicians attacked the problems using other, less traditional means. In fact, we now know that none of the three problems can be solved using only a straightedge and a compass. In these failed attempts, however, we find important advances made in many areas of mathematics.

Squaring the Circle

Background

To "square" a given geometric figure (such as a triangle or a circle) means to construct a square whose area is equal to the area of the given figure. Greek geometers succeeded in squaring figures bounded by straight lines, such as rectangles and triangles. Finding the exact areas of such figures was a relatively easy task. The next logical step, according to the Greek author Proclus (411-485), was to square regions bounded by nonlinear curves, the simplest such figure being the circle. The problem called "squaring the circle," often referred to as "quadrature of a circle," is to construct a square equal in area to a given circle; a simple-sounding explanation for a problem that has intrigued mathematicians for several millennia. Although the ancient Babylonians, Indians, and Chinese attempted the prob-

lem of squaring the circle, it was the work of the classical Greek mathematicians that made the problem famous for centuries.

The problem of squaring the circle takes on different meanings depending on how one approaches the solution. Beginning with the Greeks, many geometric methods were devised that allowed the construction of a square whose area was equal to a given circle. However, none of these methods accomplished the task at hand using the plane methods requiring only a straightedge and a compass. Of all the methods through the centuries that have been found to square the circle, none has involved the exclusive use of the straightedge and the compass. Rather, all required more sophisticated geometric methods such as the use of conic sections or complicated mechanical devices.

Therefore, when we say that the problem of squaring the circle is unsolved, what we mean is that it has never been solved using plane methods. In fact, it was not until late in the nineteenth century that it was proven that the problem of squaring the circle could not be solved by plane methods. In spite of the history of futility behind the problem, the work done in squaring the circle by means other than plane methods has proven to be fertile soil for the growth of mathematics. Additionally, as so often happens in science and mathematics, even the unsuccessful attempts at squaring the circle using only a straightedge and a compass have proven valuable to the development of mathematics.

We owe the credit for giving the problem of squaring the circle an important place in mathematics to the Greeks. The Greek mathematician Anaxagoras (499-428 B.C.) was among the first to attempt to solve the problem (while in prison, no less), but his work on squaring the circle has not survived to modern times. The first recorded progress made comes from two Greek mathematicians named Antiphon and Bryson. Antiphon (480-411 B.C.) approximated the area of a circle by inscribing first a square inside of the circle, then an octagon, then a 16-sided polygon, and so forth. Inscribing a figure inside a circle means to draw the figure, such as the square, so that the vertices just touch the inside of the circle. As the number of sides of the inscribed polygon doubled, the area of that polygon became closer to the area of the circle. Obviously, no matter how many sides the inscribed polygon was composed of, the area would always be smaller than that of the circle. Bryson (fl. 450 B.C.) improved Antiphon's approximation by circumscribing (drawing the figures around the outside of) the circle with polygons, thus guaranteeing the correct answer to be between the area of the inscribed polygon and the circumscribed polygon.

Hippocrates of Chios (c. 470-c. 410 B.C.) made what seemed to be important progress on the problem when he was able to construct a square equal in area to a region called a lune. Since a lune is a region bounded by the arcs of two circles, Hippocrates (not the same man as the physician for whom the Hippocratic Oath is named) seemed headed in the right direction. In fact, Hippocrates was able to solve the problem of squaring three kinds of lunes in his lifetime. Alas, his work did not lead to a successful solution to the problem of squaring a circle. In fact, the eighteenth-century Swiss mathematician Leonhard Euler (1707-1783) was the next to successfully square a new kind of lune, actually squaring two such figures that had eluded mathematicians since the time of Hippocrates. This, it turns out, marked the end of "lune-squaring," as it was eventually proven that only five such figures (the three found by Hippocrates plus the two by Euler) were squarable.

It is important to remember that the "unsolved" label attached to the problem of squaring the circle comes from the countless attempts throughout history to solve the problem using only a straightedge and a compass. However, many methods involving other geometric techniques have been used throughout history to successfully solve the problem of squaring the circle. For example, Greek mathematicians such as Dinostratus (c. 390-c. 320 B.C.) and Nicomedes (c. 280-c. 210 B.C.) used a curve called the quadratix to square the circle. The quadratix could not be constructed, however, using only a straightedge and compass.

Archimedes (287-212 B.C.), considered the greatest mathematician of ancient times, made several advances on the problem of squaring the circle. In his treatise *On the Measurement of a Circle*, Archimedes stated and proved a theorem that equated the area of a circle to that of a right triangle. It would seem that this had solved our problem, since it is a relatively simple matter to construct a square whose area is equal to any triangle. This is not the case, however. Archimedes had not solved the problem of squaring the circle, because his method did not actually allow the *construction* of a triangle equal in area to a circle using only straightedge and compass. Although to us this might seem like an insignificant technicality, it was very important to the Greeks.

The figures must actually be constructable for the problem to be considered solved. Archimedes also used a curve that he invented, now called the spiral of Archimedes, to square the circle. Unfortunately, the spiral of Archimedes, like the quadratix, could not be constructed using only a straightedge and compass. Therefore, the problem of squaring the circle remained unsolved in the tradition of Greek geometry.

Impact

Squaring the circle became a popular problem wherever it was introduced in the world. There is evidence that the problem was attempted in India, China, and in the Arabic empires during the medieval period. European mathematicians of the Renaissance, including Leonardo da Vinci (1452-1519), were also interested in the problem of squaring the circle. Famous mathematicians such as Carl Friedrich Gauss (1777-1855), Gottfried Leibniz (1646-1716), and Isaac Newton (1642-1727), all of whom searched for better and more accurate methods for approximating the value of π, were in essence working on the problem of squaring the circle. The reason that the calculation of π is so closely tied to squaring the circle is that the problem essentially reduces to constructing a square whose side is the square root of π times the radius of the circle. Therefore, to square the circle requires that a line of length π be constructed using only straightedge and compass.

The number π is defined as the ratio of the circumference of a circle to its diameter. It was also understood by ancient mathematicians to be intimately related to the area of a circle, thus its importance to the problem of squaring a circle. Ancient cultures, long before classical Greek times, were concerned with the value of π. The famous Rhind Papyrus, an Egyptian relic (c.1650 B.C.), essentially gives an estimate of π at 3.16. Babylonian clay tablets, from roughly the same era, give an estimate equivalent to $3\frac{1}{8}$. Archimedes found an excellent approximation for π (about $3\frac{1}{7}$) by inscribing and circumscribing polygons with 96 sides. Over two millennia later, the Indian mathematician Srinavasa Ramanujan (1887-1920) found several remarkably accurate approximations to π and his techniques continue to interest research mathematicians to the present day. Using modern computers, mathematicians have calculated π to millions and even billions of decimal places. (Remember, π is an irrational number, which means the decimal never ends or repeats.)

In modern times, the huge number of supposed solutions to the problem of squaring the circle, submitted to various scientific societies throughout Europe from amateurs and cranks, prompted the Paris Academy of Sciences and the Royal Society of London to cease considering the "solutions" sent to them by fame-seeking amateur mathematicians. But this was not before many respected would-be mathematicians tried their hand at solving the problem. For instance, in the seventeenth century the British political philosopher Thomas Hobbes (1588-1679) claimed to have solved the problem and to have revolutionized geometry in the process. Only after a protracted battle of words with the mathematician John Wallis (1616-1703) were all the fallacies in Hobbes's argument exposed and Hobbes himself became a castaway from the mainstream mathematical community. At one point, the quest to square the circle became such an obsession that the British mathematician Augustus De Morgan (1806-1871), coined the term "morbus cyclometricus," or circle-squaring disease.

Eventually, in the nineteenth century, the German mathematician Ferdinand von Lindemann (1852-1939) proved that π was a transcendental number (a transcendental number is one that cannot be the root of an algebraic equation with integer coefficients). This in turn confirmed that no straightedge and compass construction of a solution to the circle-squaring problem was possible. However, unconvinced amateurs continued to seek a solution.

The numerous attempts to square the circle, both with plane methods and using other geometric constructions, have led to many advances in mathematics. One of the most important modern advances that have grown at least partially from circle-squaring work is the calculus technique called integration. In fact, the technique of approximating the area of a circle by inscribing and circumscribing polygons around a circle, first used by the ancient Greeks, is a forerunner of integration.

Trisecting an Angle

Background

The construction of regular polygons (polygons that have equal sides and equal angles) and the construction of regular solids (solids whose faces are equal regular polygons) was a traditional problem in Greek geometry. Some regular polygons, such as equilateral triangles and squares, and some regular solids, such as a cube, were rel-

atively easy to construct. In fact, the Greeks were able to construct any regular polygon with an even number of sides, as well as some with an odd number of sides (such as a triangle and a pentagon), using only plane methods, in other words a straightedge and a compass. They could even construct complicated solids like regular hexagons (6 sides), octagons (8 sides), decagons (10 sides), dodecagons (12 sides), and pentadecagons (15 sides). In 1796 Carl Friedrich Gauss was able to construct a regular polygon with 17 sides using only a straightedge and a compass. But in order to construct a regular polygon with an arbitrary number of sides, it was required that an arbitrary angle be divided an arbitrary number of times. For instance, to construct a regular polygon with 9 sides requires that a 60° angle be trisected. The Greeks knew that any angle could be bisected using only a straightedge and a compass. Cutting an angle into equal thirds, or trisection, was another matter altogether. This was required to construct other regular polygons. Hence, trisection of an angle became an important problem in Greek geometry.

The Greeks found that certain angles could be trisected rather easily. The problem of trisecting a right angle is a relatively simple process. There are other angles that can be trisected with relative ease. In fact, Hippocrates of Chios, who we have already seen was instrumental in finding solutions to the problem of squaring the circle, found a relatively simple method to trisect any given angle. Unfortunately (at least for traditional Euclidean geometry), Hippocrates's method did not use only a straightedge and compass in its construction. Others succeeded in solving the problem, but never by using the plane methods that required only straightedge and compass.

The methods that the Greek mathematicians did find to trisect an angle involved curves such as conic sections or more complicated curves requiring mechanical devices to construct. The curve called the quadratix, which we have seen was used to square the circle, was also used to trisect an angle. Curves such as Nicomedes's (c. 280-c. 210 B.C.) conchoid and the spiral of Archimedes were also used to trisect an angle. The Greeks found several methods for trisecting an angle using curves called conic sections. A conic section is a curve obtained by the intersection of a cone and a plane. Examples of conic sections are circles, ellipses, parabolas, and hyperbolas. The method of solving the trisection problem by the use of conic sections lived on for many centuries. Even the great French mathematician and philosopher René Descartes (1596-

1650) found a way to trisect an angle using a circle and a parabola. None of these curves, however, could be constructed using the restrictions required by traditional Greek geometry.

Impact

The problem of trisecting an arbitrary angle using only a straightedge and a compass generated interest for centuries just as the problem of squaring the circle had. If *bisecting* an arbitrary angle was so easy, and *trisecting* certain angles was also relatively simple, surely, thought some, the problem of trisecting an arbitrary angle could be solved. Mathematicians found that very close approximations to trisecting an angle could be made by continued bisections. In fact, if this process were repeated an infinite number of times, an exact trisection could be made. In addition to the straightedge and compass restrictions, however, the Greeks required that a process be accomplished in a finite number of steps to be valid. Continued bisection, then, did not represent an acceptable solution to the trisection problem.

Like the problem of squaring the circle, it had become evident to most trained mathematicians by the eighteenth century that a solution to the problem of trisection probably did not exist. François Viète (1540-1603), the mathematician credited with introducing systematic notation into algebra, warned in his lectures about the many flawed proofs from enthusiastic amateurs. In fact, in 1775 the Paris Academy of Sciences discontinued examination of angle trisection methods submitted by the public, much like they had done with solutions to the circle-squaring problem. It was not until 1837 that Pierre Wantzel (1814-1848) completed a proof that the problem was impossible using only a straightedge and a compass. Wantzel essentially showed that trisecting an angle could be reduced to solving a cubic equation. Since most cubic equations could not be solved with straightedge and compass, neither could the trisection problem. This put a stop to attempts by serious mathematicians to solve the problem, but unconvinced amateurs continue to seek fame by looking for methods to trisect an angle.

Doubling a Cube

Background

For a cube with a given volume, can one construct another cube whose volume is double the original using only a straightedge and compass? This is the third problem of ancient Greek geom-

etry. Like the problems of squaring the circle and trisecting an angle, the origin of the problem of doubling a cube (also referred to as duplicating a cube) is not certain. Two stories have come down from the Greeks concerning the roots of this problem. The first is that the oracle at Delos commanded that the altar in the temple (which was a cube) be doubled in order to save the Delians from a plague. After failing to solve the problem, the men from Delos questioned Plato as to how this might be done. Plato's response was that the command was actually a reproach from the gods for neglecting the study of geometry. Needless to say, the plague at Delos continued. The problem of doubling a cube is often referred to as the Delian problem after the citizens of Delos who suffered for their ignorance.

A second version of the origin of the cube-doubling problem relates that King Minos commanded that a tomb be erected for his son, Glaucus. After its completion, however, Minos was dissatisfied with its size, as its sides were only 100 feet (30.5 m) in length. He commanded that the cubic tomb be made twice as large by doubling each side of the tomb. Since the volume of a cube is the length multiplied by the width multiplied by the height, the volume of the original cube was

$$V_1 = l \times w \times h$$

By doubling the length of each side, the new cube would have a volume given by

$$V_2 = 2l \times 2w \times 2h = 8(l \times w \times h)$$

So if each side were doubled, the resulting volume would not be double the original, but eight times the original volume. Like the men of Delos, King Minos's subjects could not solve the problem of doubling the cube. Although both of these stories may contain as much myth as fact, the problem of doubling a cube using only a straightedge and a compass became important in Greek geometry.

Eratosthenes (276-194 B.C.) is credited with one of the first solutions to the problem of doubling a cube, using a mechanical instrument of his invention to construct the required cube. Interestingly, Plato is also credited with a mechanical solution to the problem, although he is said to have abhorred the use of mechanical devices in geometry. Archytas of Tarentum (c. 428-c. 350 B.C.) produced a remarkable construction of the problem, relying on the intersection of several three dimensional objects, including a cylinder, a cone, and a surface known as a tore. Eudoxus (408-355 B.C.), the inventor of an important mathematical technique called the method of exhaustion, is also said to have constructed a solution to the problem. In addition, Nicomedes (c. 280-c. 210 B.C.) used the same curve (conchoid) to solve the problem of doubling the cube as he had used to solve the problem of trisecting an angle.

Just as with the other two problems of Greek geometry, the problem of doubling a cube was solved using conic sections. Menaechmus (c. 380-320 B.C.) was able to find two solutions using the intersection of conic sections. In fact, it is said that Menaechmus discovered conic sections while attempting to solve the problem. Many other famous Greek mathematicians, including Apollonius, Heron, Philon, Diocles, Sporus, and Pappus, constructed their own solutions to the problem of doubling the cube. None, however, succeeded in solving the problem using the straightedge and compass alone.

Hippocrates of Chios, who was an important figure in the history of the problems of squaring the circle and trisecting an angle, also worked on the problem of doubling a cube. Hippocrates found that the problem of doubling a cube could be solved if the related problem of finding two mean proportionals between one line and its double was solved. This means, in modern notation, to find two values, x and y, such that $a/x = x/y = y/2a$. This leads to the equation $x^3 = 2a^3$, which means that a cube with side x has twice the volume of a cube with side a. Although Greek mathematicians were able to find ways to perform this construction, none met the requirements of the exclusive use of straightedge and compass.

Impact

Attempts to solve the problem of doubling the cube have led to many other important discoveries in mathematics, just as the same has happened with the other two problems of Greek geometry. Conic sections, discovered by Menaechmus while he was trying to solve the problem of doubling the cube, have been an extremely important part of mathematics throughout history. The Persian mathematician (and poet) Omar Khayyam (1048-1131) used the intersection of conic sections to solve cubic equations, a problem closely related to duplicating the cube.

There is a very good reason that neither the Greeks nor anyone else were ever able to find a solution to the problem of doubling the cube using plane methods; such a solution does not exist. In the sixteenth century, François Viète was able to show a relation between the solutions of

cubic equations and the problems of duplication of a cube and trisection of an angle. René Descartes later showed that any cubic equation could be solved using a parabola and a circle, but not with a line and a circle (the beginning point of straightedge and compass constructions). The final nail in the cube-doublers coffin came from Pierre Wantzel in 1837. Wantzel proved that a geometric construction to double a cube using only straightedge and compass could not exist, just as he had proved that a similar construction for trisecting an angle was impossible.

Summary

The history of the three unsolved problems of Greek geometry is interesting in itself, but it is made even more interesting by the impact that these three problems have had on mathematics through the centuries. These problems, and related problems, were addressed by Arabic mathematicians of the medieval period such as Omar Khayyam, by great European mathematicians of the Renaissance and early modern period such as Leonardo da Vinci and René Descartes, and by modern mathematicians like Ramanujan. Usually when these problems were addressed, new advances in mathematics resulted. The great German mathematician Carl Friedrich Gauss was at least partially motivated by these problems in his work on the solutions of algebraic equations. Gauss asserted a connection between a certain type of equation, called a cyclotomic equation, and the construction of regular polygons. And this sort of construction was tied closely to our three Greek problems. In fact, Wantzel's proof of the impossibility of trisecting an angle or doubling a cube using plane methods was the culmination of the work started by Gauss.

The problems of squaring the circle, trisecting an angle, and doubling a cube are three of the most famous mathematical problems in history. They challenged the minds of the greatest mathematicians of ancient Greece and have interested mathematicians well into modern times. Their influence on mathematics through the ages makes these problems an important part of the history of man's search for answers to the questions of science.

TODD TIMMONS

Further Reading

Books
Beckmann, Petr. *A History of π*. New York: Barnes & Noble Books, 1971.

Dunham, William. *Journey Through Genius: The Great Theorems of Mathematics*. New York: Wiley and Sons, 1990.

Dunham, William. *The Mathematical Universe*. New York: Wiley and Sons, 1994.

Heath, Thomas. *A History of Greek Mathematics*. New York: Dover Publications, Inc., 1981. (Republication of original 1921 edition.)

Jesseph, Douglas M. *Squaring the Circle*. Chicago: University of Chicago Press, 1999.

Katz, Victor J. *A History of Mathematics*. Reading, MA: Addison-Wesley, 1998.

Internet Sites
O'Connor, J. J. and Robertson, E. F. "Squaring the Circle." www-groups.dcs.st-and.ac.uk/~history/HistTopics/ Squaring_the_circle.html.

O'Connor, J. J. and Robertson, E. F. "Trisecting an Angle," www-groups.dcs.st-and.ac.uk/~history/HistTopics/ Trisecting_an_angle.html.

O'Connor, J. J. and Robertson, E. F. "Doubling the Cube," www-groups.dcs.st-and.ac.uk/~history/HistTopics/ Doubling_the_cube.html.

The Foundations of Geometry: From Thales to Euclid

Overview

Ancient Greek mathematicians systematized their knowledge about geometry, with the compendium by Euclid providing definitions, axioms, theorems, and proofs in final form. His *Elements* dictated the standard for communicating mathematical results and served as the basis for elementary mathematics education for centuries.

Background

Traditionally, Thales of Miletus (c. 624-547 B.C.) is credited with bringing geometry to Greece from Egypt. Stories about him relate that he proved such theorems as the equality of base angles in an isosceles triangle, the equality of vertical angles, and the angle-side-angle criterion for triangle congruence. These stories are important not because they are incontrovertibly true, which is impossible to ascertain, but because they demonstrate that Greeks of several hundred years later saw themselves as approaching geometry in a new way after Thales's contributions.

For example, the Pythagoreans of the fifth century B.C. found that the side and diagonal of a square do not have a common measure. Their discovery of incommensurability forced a rethinking of the Pythagorean philosophy that the fundamental units of the universe were numbers, but it also reinforced the Greek tendency to develop logical arguments for evaluating the truth of their mathematical assertions. By the end of the century, Hippocrates of Chios (c. 460-380 B.C.) had begun to prove elementary theorems about plane figures and to organize those theorems into a systematic presentation.

Mathematicians associated with Plato's Academy continued to define the content of the discipline as well as the place of geometry relative to other fields of knowledge. Plato (429-347 B.C.) wrote in *The Republic* that plane and solid geometry were two of the five subjects necessary to the education of the philosopher-king. Mastering geometry trained the student in logical argument and taught him to search for and love truth, which Plato represented with eternal and ideal geometric forms. In other words, knowledge of geometry and logic became prerequisite to the study of philosophy. Plato also recorded the research efforts of Theodorus (fl. 425 B.C.) and Theaetetus (417-369 B.C.) to geometrically demonstrate the existence of the irrational numbers between $\sqrt{3}$ and $\sqrt{17}$.

Eudoxus (408-355 B.C.) studied at the Academy and visited Egypt before opening his own school in Cnidus. He further developed the theory of proportion for magnitudes, which were now distinguished from numbers. His method of exhaustion helped him measure the area of a circle by placing the circle between two polygons and then increasing the number of sides of the polygons. Eudoxus also gave a variation of the methodology for mathematics set down by Aristotle (384-322 B.C.) in *Posterior Analytics*. Eudoxus laid out definitions explaining geometrical terms, postulates enunciating presuppositions peculiar to geometry, and axioms stating self-evident truths before he tried to prove the validity of new results.

By the time Euclid (c. 330-c. 270 B.C.) lived, then, much of the material of plane geometry had been discovered. Mathematicians had also made significant progress in solid geometry. Euclid taught at the Museum and Library in Alexandria, the flourishing Egyptian city under Greek control, where he could compile results from earlier writings on geometry. He outlined his *Elements of Geometry* in thirteen sections, or books: six books on plane geometry and proportions of magnitudes, three on the theory of numbers and proportions of numbers, one examining incommensurable magnitudes, three on solid geometry, and one constructing the five regular polyhedra. Although Euclid did not claim to include any new topics, the systematic structure of his work was so popular with other mathematicians that all other previous geometrical compendiums were lost to history.

Euclid's format also began with definitions, postulates, and axioms (which he called "common notions"). He drew upon previous works for some of this preliminary material, but he apparently wrote the fifth, or parallel, postulate himself. Euclid believed that he could not prove propositions such as that the sum of the angles of a triangle was equal to two right angles or the Pythagorean theorem without taking for granted that lines that eventually converge are not parallel. Then, he attempted to organize the fundamental theorems and problems, or constructions, of geometry into deductive order. The proof of each proposition was to depend only on the statements that had already been proven.

Impact

The ancient Greek approach to geometry, as codified in Euclid's *Elements*, has been such a pervasive facet of Western civilization that its role cannot be overstated. Its cultural function can be classified into at least four overlapping areas. The work influenced the development of mathematics, obviously, but it also served as a geometry textbook for more than 2,000 years. To use *Elements* as a reference or with students, it was necessary to keep the work in circulation through new copies or editions. Intellectuals both in and outside of mathematics additionally have held up *Elements* as a model of the ideal process for reasoning.

First, then, *Elements* inspired the next generation of mathematical researchers in Greece. Archimedes (c. 287-212 B.C.) and Apollonius (250-175 B.C.) were probably trained by Euclid's students. Among his considerable achievements,

Archimedes used geometrical demonstrations and the method of exhaustion to prove many theorems about conic sections. Apollonius carried this work even further. In addition, Eratosthenes (276-194 B.C.) used Euclidean geometry to calculate the size of the Earth. While some ancient mathematicians used *Elements* as a building block toward new results, others established a tradition of copying and commenting upon the *Elements* themselves. Remarks by Proclus (A.D. 410-485) and Pappus (fl. A.D. 300) and an edition by Theon (fl. A.D. 350) became the manuscripts most often referred to by later generations. Generally, *Elements* had established the form for any methodical study of nature—complete proofs needed a general enunciation of the proposition under consideration, a complete demonstration following deductive principles, and a conclusion restating what was sought and accomplished and giving the formulaic expressions for "what was to be proved" (for theorems) or "what was to be constructed" (for problems).

Medieval Arabic scholars continued the tradition of closely examining *Elements* and translated the work into Arabic. Several writers were inspired to discuss different types of proofs and to decide which styles were most convincing. European authors, on the other hand, had to reconstruct Euclidean geometry on their own, since only lists of Euclid's propositions had been preserved in the West. There, the practical geometry of surveyors was appreciated first before Gerard of Cremona (c. 1114-1187 A.D.) and others finally translated Arabic and then Greek full versions of *Elements* into Latin.

In part because they had mastered *Elements*, scholars such as Roger Bacon (1214-1294), Jordanus Nemorarius (fl. 1225), and Nicole Oresme (c. 1323-1382) discovered original results in optics and geometry. Euclid's work again provided impetus to new mathematics in the Scientific Revolution. Although the content of *Elements* was no longer unfamiliar to experienced mathematicians, researchers followed the form of the text when presenting their own results. The most famous example of this may be Isaac Newton's 1687 *Principia Mathematica*, where he set out the mathematical laws governing the universe in the language of classical geometry rather than with his new fluxional calculus.

Throughout its long history, *Elements* also functioned as a textbook. To many readers, its deductive framework was an effective pedagogical device for helping students proceed from basic to more complex concepts. Thus, young scholars in the ancient world often gathered around a master to discuss *Elements*. As early medieval monastery schools were replaced by universities, professors turned back to Plato's list of the liberal arts for the curriculum. *Elements* was the obvious source to teach plane and solid geometry through dictation and eventually printed books when they became available and affordable. As important as the mathematical content was the work's system of logical deduction. Memorizing *Elements* made students' minds orderly and disciplined.

This mental discipline was believed to benefit all educated people, not just mathematicians. In fact, the form of *Elements* had already long been an ideal for nonmathematical disciplines. For example, Thomas Aquinas (1225-1274) wrote *Summa theologiae*, a rational system of all secular and divine knowledge, in the style of a geometrical treatise. René Descartes (1596-1650) separated himself from medieval scholasticism but still appealed to Greek mathematical methods in his philosophical and scientific writings. Even the celebration of rationalism of the Enlightenment philosophers was informed by the classical heritage in general and *Elements* in particular.

Ever since Euclid wrote the parallel postulate, though, mathematicians had worried about the logical status of that statement. They believed that one should not have to accept the theory of parallels without proof, but repeated attempts to treat the concept of parallel lines as an axiom or to prove the parallel postulate as a theorem were unsuccessful. Girolamo Saccheri (1667-1733), Johann Lambert (1728-1777), John Playfair (1748-1819), and Adrien-Marie Legendre (1752-1833) were some of the last mathematicians to attack the problem before Carl Friedrich Gauss (1777-1855), Nicolai Lobachevsky (1793-1856), and Janos Bolyai (1802-1860) independently realized that a system of geometry based only upon the first four postulates was structurally as valid as *Elements*. The identification of non-Euclidean geometries had consequences, though—scientists and teachers had always assumed that Euclidean geometry was a perfect representation of the physical world. One of the areas that was most unsettled was British mathematics education. Teachers there had ignored the alternative or simplified geometry textbooks written in the eighteenth and nineteenth centuries to insist on *Elements* as the sole reference for university students and schoolchildren. There was a protracted struggle between traditionalists and the reformist Association for the Improvement of Geo-

metrical Teaching before teachers finally decided in the early twentieth century to emphasize geometry's practical applications rather than to focus on mental discipline.

Meanwhile, professional mathematicians developed modern axiomatics during the nineteenth century. After putting algebraic disciplines such as the differential and integral calculus on foundations acceptable under new standards of rigor, some people turned their attention back to Euclidean geometry. While several people worked on the foundations of geometry, David Hilbert (1862-1943) is the most notable name in this regard. He tried to avoid the implicit assumptions made by Euclid by beginning with three undefined terms—the point, straight line, and plane—and defining the relationships between them with axioms rather than with more definitions. He wanted geometry to be a wholly abstract science, and he set consistency, independence, and completeness as the criteria for the axiomatic system that appeared in the 1899 *Grundlagen der Geometrie*.

Finally, *Elements* has been one of the world's most influential historical documents. The claim is often repeated that only the Bible has appeared in more editions than *Elements*. Indeed, one of the earliest printed books (1482) was also the first printed edition of Euclid's *Elements*, a Latin translation by Johannes Campanus (fl. 1260). Robert Simson (1687-1768), J. L. Heiberg (1854-1928), and Thomas Heath (1861-1940) were some of the mathematicians and philologists who searched for the oldest and most accurate manuscript copy of *Elements*. Today, some of the oldest fragments and most significant manuscripts include Egyptian potsherds dating from 225 B.C., an A.D. 888 copy in the Bodleian Library at Oxford University, and a tenth-century A.D. manuscript in the Vatican Library. *Elements* has also helped spark some of the most heated controversies in the writing of the history of mathematics, including whether the discovery of incommensurables was truly a "revolutionary" event and whether Greek mathematicians actually did algebra in documents such as Book II of *Elements*.

Although contemporary historians disagree about when and to what extent Greek geometry was formalized, the perception of Greek mathematics as a model for logical deduction continues to shape Western thinking. The foundations of geometry established in ancient Greece have inspired readers of *Elements* to become mathematicians and discover new mathematical results, set the standard for scientific demonstration, and modeled proper reasoning for scholars in all fields of knowledge. Despite the discovery of non-Euclidean geometries, school geometry textbooks are still based largely on *Elements*. The historical prominence accorded geometrical argument additionally has given rise to a tradition of editing and commenting on *Elements*. In sum, *Elements* is essential reading for all those who consider themselves heirs of ancient Greek civilization.

AMY ACKERBERG-HASTINGS

Further Reading

Books

Artmann, Benno. *Euclid: The Creation of Mathematics.* New York: Springer-Verlag, 1999.

Fowler, David. *The Mathematics of Plato's Academy: A New Reconstruction.* 2d ed. Oxford: Clarendon Press, 1999.

Gray, Jeremy. *Ideas of Space: Euclidean, Non-Euclidean, and Relativistic.* 2d ed. Oxford: Clarendon Press, 1989.

Heath, Thomas L., trans. and intro. *The Thirteen Books of the Elements*, by Euclid. 3 vols. 2d ed. New York: Dover Publications, Inc., 1956.

Heilbron, John L. *Geometry Civilized: History, Culture, and Technique.* Oxford: Clarendon Press, 1998.

Knorr, Wilbur Richard. *The Ancient Tradition of Geometric Problems.* New York: Dover Publications Inc., 1986.

Lindberg, David C., ed. *Science in the Middle Ages.* Chicago and London: The University of Chicago Press, 1978.

Richards, Joan L. *Mathematical Visions: The Pursuit of Geometry in Victorian England.* Boston: Academic Press, Inc., 1988.

Periodical Articles

Grattan-Guinness, Ivor. "Numbers, Magnitudes, Ratios, and Proportions in Euclid's *Elements*: How Did He Handle Them?" *Historia Mathematica* 23 (1996): 355-75.

Knorr, Wilbur Richard. "The Wrong Text of Euclid: On Heiberg's Text and Its Alternatives." *Centaurus* 38 (1996): 208-76.

Advances in Algebra

Overview

A number of important algebraic results had been calculated by Babylonian mathematicians around 2000 B.C. The Egyptians also addressed the solution of algebraic problems, but did not advance as far, possibly due to their more cumbersome number system. The Greeks used algebraic methods in support of their interests in geometry and in the theory of numbers. The Arabs preserved Greek mathematical manuscripts and integrated Greek and Hindu ideas on algebra in books that would reach Italy during the Renaissance and stimulate a development of algebraic ideas and important new results.

Background

In its modern and elementary sense, algebra is the branch of mathematics concerned with finding the values of unknown quantities defined by mathematical relationships. The Babylonians were the first group to concern themselves, or at least to have left a record of their concern, with algebraic problems. The term "Babylonian" denotes peoples living in and around the Tigris and Euphrates river valley and actually includes several successive cultures with different languages and scripts. The earlier Sumerian (4000-2500 B.C.) kingdom was overrun by the Akkadi-

ans, and it is from the later Akkadian (2500-1000 B.C.) kingdom that we have numerous mathematical texts preserved in cuneiform writing on clay tablets.

Babylonian mathematical texts posed the problem of finding a number that, when added to its reciprocal, yields a set number. This problem leads directly to a quadratic equation and the texts provide a verbal prescription for finding the two solutions of the equation as they are given by the quadratic formula now taught in secondary school mathematics courses.

The Babylonians often posed their problems in geometric terms, describing the product of two quantities as an area and the product of three as a volume. They were able to solve a limited number of problems that involved taking a cube root. They were also able to solve simple sets of simultaneous equations in more than one unknown. They were able to solve special cases of five simultaneous equations in five unknowns, and even one case, arising in astronomy, of 10 equations in 10 unknowns.

The later Babylonian texts, written in Akkadian, introduced special symbols for unknown quantities by using two of the old Sumerian picture-symbols to represent them. While the Babylonian accomplishments are noteworthy, Babylonian algebra had its limitations as well. The Babylonians made no use of negative numbers and simply discarded solutions that were less than zero. Although their number system came to include a special place-holder symbol, much as our number "0" is used to establish that "17" means 17 and not 107, they did not understand the mathematical properties of zero and thus were constrained to use rather inefficient methods of calculation.

Our knowledge of Egyptian mathematics is drawn from two manuscripts on papyrus, the Moscow Papyrus and the better-known and longer Rhind Papyrus, discovered by Englishman Henry Rhind in 1858. While both papyri were apparently written about 1700 B.C., it is believed that their mathematical knowledge dates back to about 3500 B.C.

The problems discussed in the Egyptian documents are mainly single equations in one unknown, of the type now solved in the first few weeks of a first course in algebra. The solution of such equations was, however, much more challenging in an Egyptian number system, which did not allow easy division or a general representation of fractions. Only the simplest quadratic equations were described. The Egyptians appear, from the documents at least, to be somewhat behind the Babylonians in their mathematics, possibly as a result of their relative independence from foreign influence.

The mathematical high point of the classical world occurred in the Greek civilization, especially during the Hellenistic or Alexandrian period that began around 300 B.C. The Greeks brought a new attitude and a new emphasis to mathematics. To Greek thinkers, practical matters were the business of slaves, servants, and merchants, while the contemplation of abstract truths was considered the proper occupation of the upper class. The philosopher Plato (427?-347 B.C.) considered mathematical objects as prototypes of ideal objects, of which objects in the real world were a pale reflection. Aristotle (384-322 B.C.), Plato's student, codified the methods of deductive logic. Euclid (330?-260? B.C.), a Greek mathematician living in Alexandria, Egypt, codified all existing geometrical knowledge in the *Elements*, the first textbook of geometry to deduce all its conclusions in a systematic way from a small group of axioms and postulates. So impressive was the Euclidean synthesis that versions of the *Elements* were still used for teaching even in the early part of the twentieth century.

Greek mathematics therefore emphasized geometry over algebra, which would not have its axiomatic formulation until the late nineteenth century. Euclid's *Elements* did make some use of algebraic methods, particularly in the parts dealing with proportions and similar figures. With their enthusiasm for theoretical arguments, the Greeks also developed the area of number theory, which emphasized the mathematical properties of the integers. Many of the issues in number theory had been introduced by Pythagoras (580?-500? B.C.) who, with his disciples, developed a theory of music based on ratios, and investigated the relationships between numbers that could be represented as square or triangular arrays of dots. Pythagoras is best known as the author of the theorem that states that the square of the hypotenuse of a right triangle equals the sum of the squares of the other two sides. The Babylonians and Egyptians also had developed lists of "Pythagorean triples"—that is, triples of whole numbers that satisfied the Pythagorean relationship.

The most important Greek mathematician to deal with algebraic problems for their own sake was Diophantus, who was active in Alexandria around A.D. 250. Very little is known about the life of Diophantus. He may have been born in Babylonia or studied there. Only about half of Diophantus's great work, the *Arithmetica*, has survived to the present time. The focus of this work was not algebra as we know it, nor arithmetic, but rather number theory. It was nonetheless important for the development of algebra in its use of letters to represent unknown quantities and its introduction of a simple notation for powers of the unknown, so that reference to a geometric "square" or "cube" was no longer implied. Diophantus investigated the conditions under which sets of equations in more than one unknown quantity would have whole number solutions. Such equations are known as Diophantine equations to this day. In this, he generalized the Pythagorean triples to other sets of numbers satisfying other relationships.

Outside the classical world and the ancient Near East, there is evidence of sophisticated mathematical knowledge in China, India, and among the Mayans of Central America. These cultures made important achievements in number systems, arithmetic, and geometry, often in the service of astronomy and religion. Specific writings on algebra are not apparent prior to the work of the Hindu Brahmagupta (598-660), whose work would influence the algebraic interests of medieval Arab scholars.

Impact

Our word "algebra" is derived from the Arabic *Kitab al-jabr w'almuqabala,* a book that described the art of manipulating statements about an unknown quantity into one of a number of standard forms that could then be solved by a stated procedure. The book is the work of the Arab mathematician and astronomer Muhammad ibn-Musa al-Khwarizmi (780?-850?), from whose name we also obtain the word "algorithm," which describes any systematic procedure for obtaining a mathematical result. Al-Khwarizmi was one of the scholars working at the "House of Wisdom," established at Baghdad by the Caliph al-Ma'mun (809-833), where he had access to both Greek and Hindu mathematical texts. His work was translated into Latin by Robert of Chester, an English scholar living in Muslim Spain, in 1145.

The revival of interest in algebra in Europe begins with the fall of the Byzantine Empire in 1543, which brought an influx of scholars and mathematical manuscripts into Renaissance Italy. In 1545 the Italian physician and mathematician Girolamo Cardano (1501-1576) published a book entitled *Ars Magna,* or "The Great Art." This book incorporated significant new results, including the solution of the cubic and quartic equations. The next major step forward in algebra occurred with the work of the French lawyer and writer Francois Viète (1540-1620), who reintroduced the Diophantine practice of letting a letter represent the unknown. Viète's new notation set the stage for the study of functions of variables and the study of transformations of functions caused by introducing new variables, ideas important throughout modern mathematics.

The existence of whole number solutions to Diophantine equations has inspired much mathematical research over the centuries. In 1900 the great German mathematician David Hilbert (1862-1943) included on his list of the 23 most important unsolved problems in mathematics determining whether whole number solutions of a given Diophantine equation exist. Undoubtedly the most important Diophantine equation in the history of mathematics is a generalization of the Pythagorean relationship to higher powers:

$$a^n + b^n = c^n,$$

where n is 3 or 4 or a larger whole number. French mathematician Pierre de Fermat (1601-1665) indicated in a marginal note in his copy of Diophantus's *Arimetica* that he had found a proof that these equation had no whole number solutions, but did not include a proof. Numerous mathematicians would seek a proof or disproof of Fermat's Last Theorem before British mathematician Andrew Wiles (1953-) produced a satisfactory proof in 1994.

DONALD R. FRANCESCHETTI

Further Reading

Bell, Eric Temple. *Development of Mathematics.* New York: McGraw-Hill, 1945.

Boyer, Carl B. *A History of Mathematics.* New York: Wiley, 1968.

Grattan-Guiness, Ivor. *The Rainbow of Mathematics: A History of the Mathematical Sciences.* New York: Norton, 1997.

Kline, Morris. *Mathematical Thought from Ancient to Modern Times.* New York: Oxford University Press, 1972.

The Development of Trigonometry

Overview

Even in the time of the ancient Babylonians and Egyptians, theorems involving ratios of sides of similar triangles were used extensively for measurement, construction, and for an attempt to understand the movement in the heavens. The Greeks began the systematic study of angles and lengths associated with these angles, again in the service of astronomy. The history of trigonometry is intimately associated with that of astronomy, being its primary mathematical tool. Trigonometry will eventually become its own branch of mathematics, as the study of the modern trigonometric functions.

Background

The people in the ancient civilizations in Egypt and Mesopotamia looked to the heavens. The heavens told the people when to plant and when to harvest. In order to commemorate important events, people needed a yearly calendar. Observing the position of the Sun is necessary for calendar making. In order to tell the time of day, one must look at the lengths of shadows. These shadow lengths are measured using an upright stick in the ground and measuring the length of its shadow. Trigonometry has its origins in the calculation of these measurements. Indeed, the modern degree measurement for arcs and angles has its origins in Babylonian measurement.

The famous Babylonian clay tablet now known as Plimpton 322 dates from approximately 1700 B.C. This tablet is best known for its listing of Pythagorean triples, a listing of sides and the corresponding hypotenuse of right triangles. Less well known is the fact that one of the columns on the tablet contains the square of the ratio of the diagonal to one of the sides, and as one moves down the column this ratio decreases almost uniformly. If one looks at the squares of these ratios as the square of the cosecant of the adjacent angle, this angle increases almost uniformly from around 45° to 58°. Does the Plimpton 322 serve in part as a trigonometric table?

Most of our knowledge of early Egyptian mathematics comes from the Rhind (or Ahmes) Papyrus. The papyrus was bought in 1858 in a Nile resort town by a Scot, Alexander Henry Rhind, hence its name. It is sometimes called the Ahmes papyrus in honor of the scribe who copied it about 1650 B.C. Problems 56 through 60 of this papyrus illustrate the origins of trigonometry in ancient Egypt. These problems concern square pyramids, naturally a subject of great interest to the Egyptians. When constructing a pyramid, the faces must maintain a constant slope. The word *seqt* or *seked* appears frequently in these problems, and to the modern reader it means the ratio of the horizontal distance of a slanted line from the vertical to the height, much like *run* over *rise* in the modern notion of slope. One can also think of the *seqt* as the cotangent of the angle between the base of a pyramid to its face. However, the Egyptians did not think of the *seqt* in this way. To them, it was simply expressed as a length. Trigonometric functions as ratios (as we think of them today) had not appeared at this time.

The use of the geometry of similar right triangles rather than trigonometry is also a feature of the mathematics of early Greece. In fact these ideas may have come directly from the Babylonian civilization. It is said that Thales of Miletus (c. 624-547 B.C.) computed the height of a pyramid by comparing the length of its shadow to that of a vertical rod. Book II of Euclid's *Elements* (c. 300 B.C.) contains propositions equivalent to the law of cosines, although in geometric language. Archimedes's (c. 287-212 B.C.) theorem on the broken chord can be restated as a formula for the sine of the difference of two angles.

Greek astronomers were also working with these geometric tools. Aristarchus of Samos (310-230 B.C.) observed that when the Moon is half full, the angle between the lines of sight to the Sun and the Moon is almost a right angle, at 87°. Using this observation, Aristarchus estimated that the Sun was between 18 and 20 times as far from Earth as the Moon. While this would be a trigonometry problem for us today (using a 3°-87°-90° right triangle made up of Earth, the Moon, and the Sun), to Aristarchus it was a geometry problem. Using these distance measurements, Aristarchus could also find the ratios of the sizes of the Sun and Moon to Earth. To calculate the sizes of the Sun and the Moon, the size of Earth was needed. This measurement was provided by Eratosthenes of Cyrene (276-195 B.C.), who found it using the relationships between angles and arcs of a circle and angles and chords of a circle.

These relationships between angles, arcs, and chords form the basis of observational astronomy. They help give a method for measuring the positions of the stars and planets in the sky, or alternately, on the "celestial sphere," the boundary of the spherical universe, with Earth at its center, which rotates around Earth and where the objects in the sky are located. The Babylonians were responsible for this "ecliptic system" of locating celestial bodies, and the Greek astronomers made use of it as well. In order to work with this system, one needs spherical trigonometry. To understand this, however, it was first necessary to understand plane trigonometry. For example, Menelaus's theorem gives the relationship among arcs of great circles on a sphere and is proven by first showing how the relationship works with segments in the plane.

The subject of trigonometry proper originated with the astronomer Hipparchus of Bithynia (190-120 B.C.) To begin with, Hipparchus introduced the Babylonian degree measurement of angles and arcs of circles to the Greeks. He was also the first to give a tabulation of lengths associated with angles that would allow for the solution of plane triangles. Hipparchus created this table in order to perform his astronomical calculations. Hipparchus's table is unfortunately lost, but from other works that refer to him and his works, we can reconstruct what his table must have looked like.

Hipparchus considered every triangle to be inscribed in a circle with a fixed radius. This meant that that each side of the triangle is a chord of the circle. The only trigonometric function in Hipparchus's trigonometry was the chord function, now abbreviated crd(α), where α is the central angle opposite the chord. In his table, Hipparchus calculated the chord length of every angle from 7.5° to 180° in steps of 7.5°. These chord lengths also vary with the radius of the circle, so a radius had to be fixed. Hipparchus used a fixed radius of $R = 57 + {}^{18}/_{60}$. This radius comes from the calculation $2\pi R = 360 \times 60$, which corresponds to the number of minutes in the circumference of a circle. The approximation of $3 + {}^{8}/_{60} + {}^{30}/_{60}{}^{2} = 3.1416...$ for π is used in the calculation. This means that the chord function is related to the modern sine function by the relation crd(α) $= 2R \sin(\alpha/2)$. Note that the chord is a length, not a ratio of lengths.

Hipparchus began his chord table with the chord of a 60° angle. This angle creates an equilateral triangle with a vertex at the center of the circle, so this chord has the same length as the radius of the circle. From there Hipparchus used two results from geometry: crd$\alpha^2(180 - \alpha) = (2R)^2 -$ crd$\alpha^2(\alpha)$, which is merely a restatement of the Pythagorean theorem; and crd$\alpha^2(\alpha/2) = R(2R -$ crd$(180 - \alpha))$, which comes from the similarity of a right triangle inscribed in a semicircle with the triangles formed by dropping a perpendicular from a vertex to the hypotenuse.

After Hipparchus, the next major figure in the development of trigonometry was the astronomer Menelaus of Alexandria (c. A.D. 100), who wrote *Chords of a Circle,* which is lost to us, and *Sphaerica,* a work on spherical trigonometry. In *Sphaerica,* Menelaus establishes many propositions for spherical triangles (formed from arcs of great circles on a sphere) that are analogous to propositions for plane triangles. Menelaus's theorem appears in this work, and this theorem is fundamental to the field of spherical trigonometry.

The culmination of Greek trigonometry occurs in a work that is also the culmination of Greek astronomy, the *Syntaxis Mathematikos* (Mathematical compilation) of Claudius Ptolemaeus, better known at Ptolemy (c. A.D. 100-178). This work was referred to by Islamic scientists many centuries later as *al-magisti,* "the greatest," and this name translated into Latin became *Almagest,* the name by which it has been known ever since. Ptolemy realized that the chord in a circle is related to the sides of both spherical and plane triangles, so his work required him to construct a table of chords.

Ptolemy's table was more complete than that of Hipparchus. Ptolemy's table calculated chords of arc from 0.5° to 180° in steps of 0.5°. Ptolemy used the value of 60 for his fixed radius. Ptolemy started his table by calculating the chords using regular polygons inscribed in a circle of radius 60. The side of a regular triangle gives crd(120°), that of a regular square gives crd(90°), a regular pentagon, crd(72°), and a regular decagon, crd(36°). Ptolemy then developed formulas for crd(180 − α), crd($\alpha/2$), and crd($\alpha \pm \beta$) (the last using Ptolemy's theorem) and used these formulas to complete his table. This table allowed Ptolemy to solve plane triangles, much as trigonometric tables allow us to solve plane triangles now. He could also calculate the values necessary in the eccentric model of the heavens, where the Sun revolves around Earth not in a circle, but around a circle with center away from Earth. With this calculation, Ptolemy could predict the position of the Sun and other celestial objects in the sky at various times. Although Ptolemy's trigonometry would be modified by mathematicians in the East, his concepts of astronomy held for another 1,400 years, until the time of Copernicus (1473-1543).

During Ptolemy's time, northern India was conquered and ruled by the Kushan Empire which established trade routes with Rome. Historians believe that Greek astronomy was transmitted to India over these routes. The Gupta rulers of India (fourth to the sixth century A.D.) also had regular communication with the inheritors of Greco-Roman culture. The earliest known Hindu work with trigonometry is the *Paitāmmaha Siddhānta,* written early in the fifth century. This work contains a table of half-chords, or in Sanskrit *jyā-ardha.* Interestingly, this word eventually would become our modern term *sine.* The word *jyā-ardha* was shortened to *jya* or *jiva.* When this was translated later in Arabic by Islamic scholars, they translated it as *jiba,* and wrote it *jb,* since vowels were not written. This was later interpreted at *jaib,* meaning "bosom" or "fold," and was translated into Latin as *sinus.* This table of half-chords seems to be derived from the table of chords of Hipparchus, since it uses the same value of the radius as he did.

The first work to mention the sine function (although still as a length in a circle of fixed radius, not as a ratio of lengths) was the *Aryabhatiya* of the Hindu astronomer Aryabhata I, written around 510. His sine table was actually a table of values of Rsin(α). Aryabhata was aided in the completion of his table by developing the equivalent of the modern identity $\sin[(n + 1)\alpha] - \sin(n\alpha) = \sin(n\alpha) - \sin[(n - 1)\alpha]-^1/_{225} \sin(n\alpha)$. Hindu tables are constructed using such recursion relations for sine. Aryabhata also introduced the cosine and versine functions, again as lengths and not as ratios; using the modern $R\cos(\alpha)$ for their cosine function. The *versine* or "versed sine" function is defined as vers(α) = $R(1 - \cos(\alpha))$. Once again, the emphasis in this work was using trigonometry in service of astronomy and the making of calendars.

These Greek and Hindu works were translated into Arabic in the eighth century and afterward. Islamic astronomers furthered the study of trigonometry for astronomical use as well as for religious use, since Muslims must know the direction of Mecca from any place they find themselves. These works in turn made their way back into the West, and trigonometry developed into the branch of mathematics that is studied today.

Impact

With the development of trigonometry, astronomy was transformed from an observational qualitative science to a predictive quantitative science. This transformation promoted the idea that a mathematical description of natural phenomena is possible. For the first time, observational data could be converted into mathematical models. Ptolemy's *Almagest* was a milestone event in this regard, serving for applied mathematics the role Euclid's *Elements* played for theoretical mathematics. Mathematics had been applied to solve problems before, but trigonometry

was developed to do applied mathematics, and has continued to do so in more and more sophisticated ways.

Although the trend towards applications coincided with the decline of Greek mathematics, these applications attracted Hindu and Islamic scholars, mainly because they were relevant to religion and calendar making. Mathematics in these two cultures was on the rise, and thanks to them much of the heritage of the classical Greeks was returned to the West.

As trigonometry and astronomy developed as a predictive science in both East and West, it was important that calculating tables became more and more accurate, and that calculations became less cumbersome. As a direct result, later centuries witnessed advances in Hindu-Arabic numeration, decimal fractions, and the invention of logarithms, all designed to ease the computational burden. The French astronomer and mathematician Pierre-Simon de Laplace (1749-1827) said that the invention of logarithms, "by shortening the labors, doubled the life of the astronomer."

GARY S. STOUDT

Further Reading

Boyer, Carl and Uta Merzbach. *A History of Mathematics.* New York: John Wiley & Sons, 1991.

Evans, James. *The History and Practice of Ancient Astronomy.* New York: Oxford University Press, 1998.

Maor, Eli. *Trigonometric Delights.* Princeton, NJ: Princeton University Press, 1998.

Neugebauer, Otto. *A History of Ancient Mathematical Astronomy.* New York: Springer-Verlag, 1975.

Eratosthenes Calculates the Circumference of the Earth

Overview

The measurement of distance became increasingly important to ancient and classical civilizations as their territorial and cultural horizons expanded. Using elegant mathematical reasoning and limited empirical measurement, in approximately 240 B.C., Eratosthenes of Cyrene (276-194 B.C.) accurately measured the Earth's circumference. This feat was more than simply a scientific achievement. Eratosthenes's calcula-

tion, and others like it, contributed to the field of geodesy (the study of the shape and size of the Earth) and helped spur subsequent exploration and expansion. Ironically, centuries later the Greek mathematician and astronomer Claudius Ptolemy would reject Eratosthenes's mathematical calculations, which, when combined with other mathematical errors that he made, produced a mathematical estimation of a smaller Earth that, however erroneous, made extended seagoing journeys and exploration seem more feasible.

Background

Eratosthenes, who served under Ptolemy III and tutored Ptolemy IV, was the third librarian at the Great Library in Alexandria. This post was of considerable importance because the library was the seat of learning and study in the ancient world. Ships coming into the port of Alexandria, for example, had their written documents copied for inclusion in the library. Over the years, the library's collection grew to encompass hundreds of thousands of papyri and scrolls that contained much of the intellectual wealth of the ancient world.

In addition to managing the collection, reading, and transcribing documents, Eratosthenes studied and wrote on many topics. Although all of his writings and calculations have been lost, we know from the work of other Greek scholars that Eratosthenes studied the fundamental concepts and definitions of arithmetic and geometry. One of his discoveries was the "sieve of Eratosthenes" a method for determining prime numbers that's still used. Eratosthenes also compiled a star catalogue that included hundreds of stars, devised a surprisingly modern calendar, and attempted to establish the date of historical events, beginning with the siege of Troy. So diverse were his abilities that his contemporaries apparently referred to him as "Beta"—the second letter of the Greek alphabet—implying that he was well versed in too many scholarly disciplines to be the best at any one of them.

Eratosthenes is best known for his astonishingly accurate and ingenious calculation of the Earth's circumference. Although his own notes on the method of calculation have been lost, there are tantalizing references to them in the works of Strabo and other scholars, including in Pappus's *Synagoge* or "Collection," a compilation and summary of work in mathematics, physics, astronomy, and geography published in the third century A.D. Beyond accurately estimating the Earth's circumference, based on observed differences in the Sun's zenith position, Eratosthenes also made a amazingly precise measurement of the tilt of the Earth's axis.

Impact

Apparently inspired by observations in the scrolls he reviewed as librarian, Eratosthenes noticed subtle differences in the accounts of shadows cast by the midday summer Sun. In particular, he read of an observation made near Syene (near modern Aswan, Egypt) that at noon on the summer solstice the Sun shone directly into a deep well and that upright pillars were observed to cast no shadow. In contrast, Eratosthenes noticed that in Alexandria on the same day the noon Sun cast a shadow upon both pillars and a stick thrust into the ground.

Based on his studies of astronomy and geometry, Eratosthenes assumed that the Sun was at such a great distance its rays were essentially parallel to the Earth by the time they reached it. Although the calculated distances of the Sun and Moon, supported by measurements and estimates made during lunar eclipses, were far too low, the Eratosthenes's assumptions proved essentially correct. Assuming the parallel incidence of light rays, he needed to determine the difference between the angles of the shadows cast at Syene and Alexandria at the same time on the same day. In addition, he had to calculate the distance between the two cities.

Viewed from the perspective of modern science, it seems intuitive that Eratosthenes would try to determine precise values for the angles and distances needed to complete his calculations. In the ancient world, however, this type of objective science was far different from the prevailing scholarly tradition, which took a more philosophical or mathematical approach to problems. Moreover, Eratosthenes's belief that the Earth was round was itself subject to debate.

To perform his calculation, Eratosthenes determined the angular difference between the shadows at Syene and Alexandria to be about 7°. He determined the distance to Syene at about 500 miles (805 km), possibly, as some legends hold by paying a runner to pace it off. Eratosthenes reasoned that the ratio of the angular difference in the shadows to the number of degrees in a circle (360°) must equal the ratio of the distance to the circumference of the Earth. The resulting estimate, about 25,000 miles (40,234 km), is astonishingly accurate.

In making his calculations Eratosthenes measured distance in *stadia,* a unit of measure based on the Greek footrace, or *stade.* These units varied from place to place in the ancient world. Eratosthenes almost certainly used the Attic stade, which was based on one circuit of the track in the Athens stadium, 606 feet, 10 inches (185 m), or a little over a tenth of a mile. Using this measure, Eratosthenes was able to calculate a circumference that varies only a few percent from the modern value of 24,902 miles (40,076 km) at the equator. It is necessary to specify that this is the circumference *at the equator* because the Earth is actually an oblate (slightly compressed) sphere with a bulge in the middle, making the circumference at the equator greater than it would be if measured around the poles.

Eratosthenes's theories and calculations were published in his *Geography,* a title that reflects the first-known use of the term, which means "writing about the Earth." Although his calculations were disputed in his own time, they allowed the development of maps and globes that remained among the most accurate produced for over a thousand years. This, in turn, sparked interest in geography and geodesy, and emboldened regional seafaring exploration using only the most primitive navigational instruments. Eratosthenes's work, moreover, helped solidify belief in a round Earth, and promoted an early theory that the relative warmth or coolness of a locations climate was determined by its distance from the equator. *Geography* also supported the concept of antipodes—undiscovered lands and peoples on the "other side" of the world.

Eratosthenes's work may have inspired the Greek astronomer and geographer Claudius Ptolemy to make his own determination of the circumference of the Earth in the second century A.D. Unfortunately, he rejected Eratosthenes's calculations and substituted errant values asserted by the geographer Posidonius (130-50 B.C.). In this system, a degree covered what would now equal approximately 50 miles (80 km), in-

stead of Eratosthenes's more accurate estimation of about 70 miles (113 km) per degree at the equator. Although Ptolemy went further than Eratosthenes in his calculations, measuring the movement of shadows over varying time intervals, his inaccurate assumptions and measurements skewed his final values to a less accurate and much smaller circumference of approximately 16,000 miles (25,750 km) .

Ptolemy published these inaccurate numbers in his *Almagest,* which was written about A.D. 150 and remained the world's most influential work on astronomy and geography throughout the Middle Ages. Ptolemy's well publicized error of a smaller Earth eventually the possibility of surviving a westward passage to India seem more possible. Although the point is disputed by many scholars, Ptolemy's mistake may have played a part in Columbus's decision to seek a westward route to India.

K. LEE LERNER

Further Reading

Clagett, Marshall. *Greek Science in Antiquity.* Abelard-Schuman, New York, 1955.

Dutka, J. "Eratosthenes's Measurement of the Earth Reconsidered." *Archive for History of Exact Sciences* 46 (1), 1993:1. 55-66.

Fowler, D. H. *The Mathematics of Plato's Academy: A New Reconstruction.*Oxford: Clarendon Press; New York : Oxford University Press, 1987.

Goldstein, B.R. "Eratosthenes on the Measurement of the Earth." *Historia Mathematica* 11 (4), 1984: 411-416.

Heath, T. L. *A History of Greek Mathematics.* Oxford: The Clarendon Press, 1921.

Roman Numerals:
Their Origins, Impact, and Limitations

Overview

The numeral system developed by the Romans was used by most Europeans for nearly 1800 years, far longer than the current Hindu-Arabic system has been in existence. Although the Roman numeral system provided for easy addition and subtraction, other arithmetic operations proved more difficult. Combined with the lack of an effective system for utilizing fractions and the absence of the concept of zero, the cumbersome nature of the Roman numeral system,

while it served most of the needs of the Romans, hindered future mathematical advances.

Background

The Roman numeral system for representing numbers was developed around 500 B.C. As the Romans conquered much of the world that was known to them, their numeral system spread throughout Europe, where Roman numerals remained the primary manner for representing numbers for centuries. Around A.D. 1300, Roman numerals were replaced throughout most of Europe with the more effective Hindu-Arabic system still used today.

Before examining the limitations posed by the use of Roman numerals, it is necessary to understand how Roman numerals are utilized. A numeral is any symbol used to represent a number. In the Hindu-Arabic numeral system, the numeral 3 represents the number three. When the numeral 3 is held in place by one or more zeros, the value increases by an order of magnitude, e.g., 30, 300, 3000, and so on. In the Roman numeral system, numerals are represented by various letters. The basic numerals used by the Romans are: I = 1, V = 5, X = 10, L = 50, C = 100, D = 500, M = 1000. These numerals can be strung together, in which case they would be added together in order to represent larger numbers. For example, the number 72 would be represented as LXXII (L + X + X + I + I, or 50 + 10 + 10 + 1 + 1 in Arabic numbers).

In order to prevent numbers from becoming too long and cumbersome, the Romans also allowed for subtraction when a smaller numeral precedes a larger numeral. Therefore, the number 14 would be represented as XIV instead of XIIII. Under this system, a numeral can only precede another numeral that is equal to ten times the value of the smaller numeral or less. For example, I can only precede and, thus, be subtracted from V and X, which are equal to five and ten times the value of I, respectively. Under this rule, the number 1999 cannot be represented as MIM, because M is equal to one thousand times the value of I. The Roman representation of 1999 is MCMXCIX, or M (1000) + CM (1000-100) + XC (100-10) + IX (10-1). Most of these rules, while often used by the Romans, were not standardized until the Middle Ages. Thus, one might find 9 represented on some older documents as VIIII instead of IX.

Because the largest numeral used by the Romans was M, or 1000, it proved impractical to write extremely large numbers, such as 1,000,000, as a string of 1000 Ms. To avoid this problem the Romans wrote a bar, called a *vinculum*, over numerals to express that numeral as a number 1000 times its original value. Instead of writing 6000 as MMMMMM, 6000 could simply be written as $\overline{VI}$ and 1,000,000 as $\overline{M}$. Using this form of notation, the Romans were capable of writing large numbers.

Impact

The Romans adopted the symbols that they used for their numerals from a variety of sources, including their Greek counterparts. The origin of I to represent one is straightforward, derived from counting on one's hand, where one finger, which resembles I, equals one of whatever was being counted. The V came to represent five because when five items are counted on the hand, a V is formed by the space between the thumb and first finger.

Originally the Romans adopted the Greek letter X, or chi, to represent 50. Through the study of monument transcriptions historians have been able to determine that L replaced X as 50, and X came to represent 10. How X came to represent 10 is not entirely clear. One theory suggests that X was derived from one V, or five, placed on top of another, upside-down V. Thus, the two Vs formed an X. Another theory suggests that when counting to 10, Romans did so by making ten vertical marks and then crossing them out with an X in order to easily count groups of ten. This is similar to the manner in which Americans keep tallies by groups of five in which four vertical marks are crossed through with a fifth diagonal mark. Eventually the Romans adopted just X to be the numeral for 10. The symbol C came to represent 100, because it is the first letter of the Latin word for one hundred, *centum*. Likewise, M was adopted for 1000, because the Latin word for one thousand is *mille*.

Unlike the Greeks, the Romans were not concerned with pure mathematics, such as number theory, geometric proofs, and other abstract ideas. Instead, the Romans preferred utilitarian mathematics. The Romans primarily used mathematics to figure personal and government accounts, keep military records, and aid in the construction of aqueducts and buildings. The Roman numeral system allowed for simple addition and subtraction. For addition, Romans simply lined up all of the numerals from the numbers being added, and simplified. For example, in order to solve the problem 7 + 22, or VII + XXII, the numerals were first arranged in descending order, or XXVIIII. Because VIIII, or 9, is not in acceptable form, this was changed to IX, the generally recognized manner of writing 9. The correct answer remains, XXIX, or 29. Subtraction can be done in a similar manner by crossing out similar numerals from the two different numbers.

The fact that multiplication and division were fairly difficult operations for the Romans spurred development of counting boards to aid with these operations. The counting boards, which resembled the familiar abacus, could also be used for addition and subtraction. Counting boards based on the Roman design were used throughout Europe until the Middle Ages. Even with these counting boards, multiplication and division of large numbers remained a difficult task. Therefore, Romans developed and often consulted multiplication and division tables to solve problems involving large numbers.

In addition to difficulty with the multiplication and division of numbers, several other problems severely limited the use and effectiveness of Roman numerals. One flaw of the Roman numeral system was the absence of a way to numerically express fractions. Romans were aware of fractions, but putting them to use was difficult, as they were expressed in written form. The Romans would have written three-eighths as *tres octavae*. The Romans usually expressed fractions in terms of the *uncia*. An uncia originally meant $1/12$ of the Roman measure of weight (English derived the word "ounce" from uncia). Soon, however, uncia evolved to mean $1/12$ of anything. Although basing the use of fractions on $1/12$s, the Romans were able to express one-sixth, one-fourth, one-third, and half. While the modern numerical expression of one-fourth is $1/4$, the Romans would have expressed one-fourth as three unciae ($3/12 = 1/4$). This system allowed the Romans to approximate measures, but they could not easily express exact measures.

Another flaw that limited Roman mathematics was the absence of the concept of zero. As with the previous number systems of the Sumerians, Babylonians, and Egyptians, the Romans did not have a place-value system that included the concept of zero as a placeholder for numerals. This forced the Romans to adopt the cumbersome system with numerals that represented 1, 5, 10, 50, 100, 500, and 1000, as described above. Unlike the ancient Greeks, the Romans also did not understand or explore the concept of irrational numbers. This severely limited the Romans in geometry, because much of geometry rests on an understanding of π, the ratio of the circumference of a circle to its diameter.

Although not limiting from a practical engineering standpoint, these flaws in the Roman mathematical systems limited the advancement of mathematical theory in Rome. In the wake of Roman conquests, most of Europe adopted the Roman numeral system and used it throughout the Middle Ages. Accordingly, theoretical mathematical advances were likewise also stunted throughout most of Western civilization for nearly 1,000 years. The absence of zero and irrational numbers, impractical and inaccurate fractions, and difficulties with multiplication and division prevented the Romans and the Europeans who later used the system from making advances in number theory and geometry as the Greeks had done in the Pythagorean and Euclidean schools.

During these mathematical Dark Ages, advancements in these fields were made by Middle Eastern and Indian subcontinent civilizations. With the innovation of zero place use within the Hindu-Arabic place-value system, great advances were made in these regions in the fields of geometry, number theory, and the invention and advancement of algebra.

Regardless of the Roman numeral system's limitations, the existing archaeological record establishes that the Romans were able to overcome many of those limitations with regard to the practicalities of construction. Roman roads and aqueducts remain as a testament to the engineering feats that the Romans were able to accomplish with their flawed system. Although Roman numerals are no longer a necessary component of mathematics, they are an important part of the history of the development of Western civilization. Modern numerals remain aesthetically important because of their widespread artistic use in art, architecture, and printing.

JOSEPH P. HYDER

Further Reading

Helfman, Elizabeth. *Signs and Symbols around the World.* New York: Lothrop, Lee & Shepard Co., 1967.

Lindberg, David C. *The Beginnings of Western Science.* Chicago: University of Chicago Press, 1992.

The Origins of the Zero

Overview

The zero was invented three times in the history of the mathematics. The Babylonians, the Maya, and the Hindus all invented a symbol to represent nothing. However, only the Hindus came to understand the importance of what the zero represented. Today we use a descendant of the Hindu zero, which had a long journey and encountered much resistance until finally accepted in the West.

Background

Before any invention can be made and accepted in a society, there has to be a need. One of the reasons the zero was not developed along with other numbers is that many early number systems had no real need for a symbol to represent nothing. This may sound strange, but even today we rarely use the term zero in everyday speech. We say, "There are no apples," not "There are zero apples."

The Egyptian, Greek, and Roman number systems represented each number by a unique collection of symbols. For example, in the Roman system the number 23 is XXIII—two tens and three ones—while 203 is CCIII. There are two major faults with such a system. First, the number of symbols needed to represent some numbers can be very large. For example, in Roman numerals 338 is CCCXXXVIII. More importantly, however, such a number system makes complex calculations very difficult. It is hard enough to add and subtract Roman numbers, but just try multiplication and division.

There were other types of number systems that did, eventually, generate a need for a symbol to represent nothing. Place value number systems use a small set of numerals in different positions to represent numbers. The modern decimal system we use today is such a system, with a place for the ones, a place for the tens, a place for the hundreds, and so on. So the number six hundred and twenty-seven is written as 627, or 6 hundreds, 2 tens, and 7 ones.

Day-to-day calculation in the ancient world was done on a counting board or abacus, which allowed for quick and simple computation of addition and subtraction. Each column on an abacus represented a different place in the number system. In a decimal abacus the first column would be the ones, the next the tens, the third the hundreds, and so on. The amount of counters in each column represents the number in a visual manner.

However, place value systems have a problem when representing numbers that do not have any value in a particular place. For example, you may calculate a result on an abacus as three hundred and two, or three counters in the hundreds column, none in the tens, and two in the one column. In the Roman system the number is easily written as CCCII, but in a place value system, without a symbol for nothing, you run into trouble.

Impact

The zero was invented separately three times. In each case it was needed as a placeholder in a place-value number system. The Babylonians used numbers based on 60, a sexigesimal system. We still use their system for measuring the minutes in an hour, and the degrees in a circle ($6 \times 60 = 360°$).

Without a zero symbol, Babylonian scribes had problems recording numbers that had no value in a certain place. To begin with they left a gap between numerals, which would be like writing two hundred and four as 2 4. However, not everyone followed this convention, and when copies were made the gaps were often left out. Even when this system was followed, it was difficult to tell a number like 204 from 2004, as they would both be written as 2 4.

Then, sometime in the third century B.C., an unknown scribe started to use a symbol to represent a place without a value, and so the first zero was invented. With a symbol for the zero there was no longer confusion over numbers like our 204 and 2004.

Yet, while the Babylonians can claim to be the first users of the zero, they did not understand it in the modern sense. Their zero did not represent a number in itself, it was just a placeholder. The idea of zero was still a little vague.

The Babylonians also refused to end numbers with zeroes. In our system that would be like writing 3,000 as 3. However, you would also write 30 as 3, and 3 as 3, which makes them hard to tell apart. Babylonian readers relied on context to determine the value of such numbers.

We do this as well to some extent. If someone tells us an orange costs 15, we assume 15 cents, yet if we are told a new book costs 15 we assume they mean 1,500 cents, $15. Babylonian astronomers could not rely on context, and so they used the zero at the end of numbers just as we do, as it enabled then to note degrees and minutes of arc more accurately. Their innovation was not, however, accepted by wider society.

The Greek world encountered the Babylonian zero as part of the spoils of the conquests of Alexander the Great (356-323 B.C.). However, most Greeks had no use for it, as their number system was not a place value system. The concept of zero also raised some unsettling philosophical questions, and contradicted the teachings of Aristotle (384-322 B.C.). Again, like the Babylonians, only Greek astronomers used the zero, for the benefits it gave them outweighed the problems it caused.

Seven centuries after the Babylonian discovery, and half a world away, another culture using a place value number system also invented a zero of their own. The Maya of South America developed a sophisticated and complex time-keeping system. They used a number of different calendars for various purposes, one of which was their religious fear that time might one day come to an end if the calendars ran out.

The Maya used a place-value system using 20 as a base, although the second place only went up to 18. Sometime in about the fourth century the need to hold a place that had no value in their number system led them to develop the zero. They had a number of different symbols for the zero, from a bowl-like object, to a complex face. However, like the Babylonians, the Maya did not use the zero in any wider sense. It remained just a placeholder, and their complex number system limited calculations.

The third invention of the zero occurred in India, although some scholars still debate whether the Babylonian zero could have traveled to India. Again, it was created by the need for a placeholder.

In Hindu mathematics numbers were also written as symbolic words, which made mathematics a little like poetry, and had the added advantage of making copying very accurate. The first use of a Hindu mathematical word for zero dates from a 458 cosmology text, and the first surviving use of a numeral for zero in India dates from 628. In the intervening period, the idea of the zero appears to have become widely accepted in Hindu mathematics.

However, unlike the Maya and the Babylonians before them, the Hindus understood the zero as more than just a placeholder. Perhaps because of the practice of representing numbers with symbolic words, they realized that the zero represented the absence of a quantity. This was a big step, for it enabled mathematics to begin to use the zero in written calculations.

Previously, all calculations had been done on counting frames and abaci. With the zero used as a placeholder in any place value, written numbers became just as easy to understand as those on a counting frame. With the ability to write down all numbers accurately on a surface, it became possible to record the steps of a calculation, not just the result. This led to an interest in the rules of calculation, which marked the birth of algebra.

From India the use of the zero spread to Cambodia by the seventh century, and made it to China shortly after. By the eighth century the Hindu zero had traveled to Arab lands, and was adopted along with the entire Hindu numeral set. It is these numbers that we use today, and they are often called Arabic numbers, despite their origin in India. The Arab acceptance of the Hindu zero caused a revolution in the Arab sciences, and not just for the ease of calculation it offered. The Arabs had absorbed much of Greek learning, and had to deal with the same philosophical problems that had led the Greeks to reject the zero. In particular, the zero contradicted Aristotle's rejection of a vacuum, so to accept the zero meant denying a cornerstone of Greek philosophy.

The zero finally reached Europe in the twelfth century, though, again, it was not accepted with open arms. The Christian Church had united the Bible with the teachings of Aristotle, reinterpreting the pagan Greek's ideas so they became a proof of the existence of God. However, European merchants found the zero to be an invaluable tool in business, and eventually the zero became an accepted part of European mathematics.

The zero still causes arguments today. Asking someone what any number divided by zero is can often provoke debate. How many zeroes go into five, for example? The simple answer is that you cannot, under any circumstances, divide by zero, but why that should be the case is not always obvious. The late development of the zero, and its slow and difficult journey across the world is related to the surprisingly complex philosophical and intellectual questions the number raises.

DAVID TULLOCH

Further Reading

Kaplan, Robert. *The Nothing That Is: A Natural History of the Zero.* Oxford: Oxford University Press, 1999.

Reid, Constance. *From Zero to Infinity: What Makes Numbers Interesting* 4th ed. Mathematical Association of America, 1992.

Seife, Charles. *Zero: The Biography of a Dangerous Idea.* New York: Viking, 2000.

Biographical Sketches

Apollonius of Perga
c. 262-c. 190 B.C.
Greek Mathematician

Though he is known as "The Great Geometer," even that title fails to do justice to Apollonius of Perga and his career. His *Conics* laid the foundations for Newtonian astronomy, ballistics, rocketry, and space science—all 2,000 or more years in the future when he wrote—with its discussion of conic sections, which describe the shape formed by the path of projectiles. Along the way, Apollonius developed his own counting system for large numbers, and put forth a new mathematical worldview that opened the way for the infinitesimal calculus many centuries later.

Born in the town of Perga in southern Asia Minor (now Turkey), Apollonius later studied Euclidean geometry in Alexandria. He also visited Pergamum and Ephesus, both important cities in Asia Minor. In addition to the *Conics,* he wrote a number of other works, all of which have been lost, but whose English titles include *Quick Delivery, Vergings, Plane Loci, Cutting-Off of a Ratio,* and *Cutting-Off of an Area.* Pappus (fl. c. A.D. 320), the principal source regarding these lost works, also summed up the material contained in them. Other ancient writers referred to lost writings of Pappus, such as his discussion of "burning mirrors" for military purposes, in which he disproved the claim that parallel rays of light could be focused on a spherical mirror.

By far the most influential of Apollonius's works, however, was the *Conics,* which consisted of eight books with some 400 theorems. In this great treatise, he set forth a new method for subdividing a cone to produce circles, and discussed ellipses, parabolas, and hyperbolas—shapes he was the first to identify and name. In place of the concentric spheres used by Eudoxus (c. 400-c. 350 B.C.), Apollonius presented epicircles, epicycles, and eccentrics, concepts that later influenced Ptolemy's (c. 100-170) cosmology. Even more significant was his departure from the Pythagorean tendency to avoid infinites and infinitesimals: by opening up mathematicians' minds to these extremes, Apollonius helped make possible the development of the infinitesimal calculus two millennia later.

In the first four volumes of his *Conics,* Apollonius examined notions of geometry passed down by Euclid (c. 325-c. 250 B.C.) and others, and maintained that he had made it possible for the first time to solve Euclidean problems such as finding the locus relative to three or four lines. The second half of the *Conics* discussed conic sections, and confronted problems such as that of finding a "normal" on a point along a curve.

The *Conics* also presented what became known as the "problem of Apollonius," which calls for the construction of a circle tangent to three given circles, and discussed a means of finding the point at which a planetary orbit took on an apparently retrograde motion. The most important factor in this monumental work, however, was not any one problem, but Apollonius's overall approach, which opened mathematicians' minds to the idea of deriving conic sections by approaching the cone from a variety of angles. By applying the *latus transversum* and *latus erectum,* lines perpendicular and intersecting, Apollonius prefigured the coordinate system later applied in analytic geometry.

Apollonius, whose work influenced mathematicians beginning with Hipparchus (fl. 146-127 B.C.) and Hypatia of Alexandria (c. 370-415), has continued to inspire thinkers throughout the ages. The last book of his *Conics* was lost, and among those who have attempted to recreate it were al-Haytham (Alhazen; 965-1039), Edmond Halley (1656-1742), and Pierre de Fermat (1601-1665). Even today, mathematicians are still examining the work of Apollonius, and finding in it applications to problems and situations they could scarcely have imagined.

JUDSON KNIGHT

Archytas of Tarentum
c. 428-350? B.C.
Greek Mathematician

His achievements as a mathematician by themselves give Archytas of Tarentum distinction: not only was he first to integrate mathematics and mechanics, but he formulated the harmonic mean as a method for solving the problem of doubling the cube. But Archytas's achievements went beyond the realm of mathematics: he was not only a philosopher, but a great statesman and military leader. As a man, too, Archytas gained admiration for his acts of kindness, one of which was destined quite literally to change history.

Probably born at Tarentum (now Taranto, Italy), Archytas grew up during a time when Greek states still dominated the Mediterranean. Among the most powerful of those was Syracuse in Sicily, whose leader, the tyrant Dionysius the Elder (c. 430-367 B.C.), had begun to conquer parts of mainland Italy. He had thus driven most Pythagoreans from the southern part of the Italian Peninsula, and the school at Tarentum was the last local institution dedicated to the ideas of Pythagoras (c. 580-c. 500 B.C.).

It is possible that while teaching at the school in Tarentum, Archytas may have had for a pupil a man destined to become one of the most influential thinkers of all time: Plato (427-347 B.C.), who was almost his exact contemporary. In any case, Archytas played a key role in the history of Western thought by saving the great philosopher from execution by Dionysius the Younger, who succeeded his father in 367 B.C.

Plato had gone to Sicily as tutor for the young tyrant, but the latter had proven more interested in drinking and revelry than in learning. Looking for an excuse to dismiss his teacher, Dionysius had readily believed a story by a jealous official that Plato's teaching was part of an Athenian plot to gain control of Sicily. Only a letter from Archytas saved Plato's life at a time when most of his greatest works remained unwritten.

But this was far from Archytas's only contribution to thought. By applying mathematical principles to the use of pulley and screw, he became the first to wed mathematics to mechanics, and he later wrote a treatise on the marriage of the two disciplines. He also discussed the theory of means, and differentiated between an arithmetic, geometric, and harmonic means. Though he did not develop the last concept, Archytas

Archytas of Tarentum. *(Bettmann/Corbis. Reproduced with permission.)*

was the one who gave the harmonic mean—previously called a "subcontrary"—its name.

The investigations of the harmonic mean arose in the course of Archytas's work on the problem of doubling a cube. This was a challenge much harder than it seemed at first, because if one simply doubled all sides of the cube, the resulting shape would be not twice, but eight times as large as the original. Archytas's innovation lay not in his use of two mean proportionals to deal with this question, but in his application of a semicircle rotating in three-dimensional space as a way of finding the harmonic mean. Thus he was among the first to introduce movement into geometry.

He was also credited with adding greatly to the number of known geometrical theorems, and has been credited with most of the ideas contained in Book VII of Euclid's (c. 325-c. 250 B.C.) highly influential *Elements*. In line with Pythagorean principles, Archytas was particularly interested in the application of mathematics to music, and discussed numerical ratios between notes. Scholars have often criticized him, however, for not always presenting clear and consistent explanations of his findings.

Part of the explanation for this may well be that Archytas was distracted. A highly influential figure in Magna Grecia, as Greek-controlled Italy was called, he was elected commander-in-chief by a confederation of city-states, and reputedly went on to an unbroken series of military victories. He also wrote philosophical works, but few of those writings have survived—nor have three works by Aristotle (384-322 B.C.) on the philosophy of Archytas.

Archytas was also widely admired for his nobility and kindness, which he displayed in particular by his gentle treatment of his slaves and his fondness for babies and small children. The latter led him to apply his interest in mathematics and mechanics to the creation of two toys: a flying mechanical pigeon, and a rattle for amusing babies. According to the poet Horace (65-8 B.C.), Archytas died in a shipwreck on the Adriatic Sea.

JUDSON KNIGHT

Aristaeus the Elder
c. 370-c. 399 B.C.
Greek Mathematician

The details of Aristaeus the Elder's life are scanty, and the list of his writings—all of them lost—is in question. Some of this may be accounted for by a confusion with a figure of whom even less is known, if indeed he actually existed: Aristaeus the Younger. Of the elder Aristaeus, however, historians do know that he was among the originators of conics and conic section theory, a man judged a "worthy mathematician" by no less a figure than his contemporary Euclid (c. 325-c. 250 B.C.).

The few known facts about Aristaeus the Elder come from the writings of Pappus (fl. c. A.D. 320), who lived six centuries later. In his *Treasury of Analysis,* Pappus referred to Aristaeus as "the Elder," leading to the inference that there must have been another Aristaeus born later; but this is the only indication that the other Aristaeus ever lived.

In Pappus's time, copies of *Five Books Concerning Solid Loci* still existed, and the later author used this work as a resource when discussing Aristaeus's ideas. The curves, lines, and points of cones were the subject matter of the book, which Euclid later credited as the source for much of his own writing on conics in Book XIII of his *Elements.* (Indeed, the latter may represent a version of at least part of Aristaeus's *Five Books,* edited and greatly amended by Euclid.)

According to Hypsicles (c. 190-c. 120 B.C.), Aristaeus also wrote another book called *Comparison of the Five Regular Solids,* in which he supposedly developed a theorem later applied by Apollonius (c. 262-c. 190 B.C.) in introducing his own comprehensive theory of conic sections. Some historians, however, believe that *Comparison* was written by the hypothesized younger Aristaeus.

Adding to the confusion surrounding Aristaeus the Elder is the fact that the "five" in the title of that second work calls to mind the name of the *Five Books.* Even more reminiscent of the title of Aristaeus's one confirmed work, however, is the name *Five Books of the Elements of Conic Sections.* The latter is allegedly yet another book by Aristaeus, and if it existed, it may have been written to further elucidate concepts explored in *Five Books Concerning Solid Loci.*

Whatever the details of Aristaeus's life, career, and writings, it is certain that he held a high place in the history of Greek geometry prior to Euclid. With the latter and Apollonius, he is regarded as one of the leading figures in the development of methods for analyzing conic sections.

JUDSON KNIGHT

Aryabhata the Elder
476-550
Indian Mathematician and Astronomer

His *Aryabhatiya* assumed a level of significance among Indian mathematicians comparable to that of Euclid's *Elements* in the West, but as was typical of many Hindu thinkers, Aryabhata considered mathematics of secondary importance to astronomy. Indeed, most of his achievements in math were in service to his study of the planets, yet it was as a mathematician that he had his greatest impact on the thinking of scholars in India, and later in Arabia. Thanks largely to Aryabhata, Indian mathematics passed out of the "S'ulvastra period," when math fell primarily under the control of priests, and into the more scientifically oriented "astronomical period" that lasted until about 1200.

During the sixth century, at a time when Europe was descending into darkness and Arabia had not yet awakened, India had the beginnings of a thriving scientific community at the city of Ujjain in the central part of the subcontinent. Yet Aryabhata, one of India's greatest mathematicians, came from Patna or Pataliputra in eastern India. Already a millennium old, the city, capital of the Mauryan Empire centuries before, had long since fallen into ruins. Symbolic of its state of disrepair was the fact that Pataliputra was a center of superstition where priests taught that Earth was flat and that space was filled with invisible and demonic planet-like forms. The persistence of these ideas made the achievements of Aryabhata all the more impressive.

The form of the *Aryabhatiya* (499), written in the same sort of verse used for social amusements, further reflected the climate of Indian learning in his time: in one famous passage from his great work, Aryabhata used the poetic conceit of commanding a "beautiful maiden" to answer an inversion problem. In fact inversion—which involves starting with a solution and working backward, developing the steps whereby one reached that solution—would be one of many new concepts introduced in the *Aryabhatiya*.

Bringing together teachings from ancient Greek and Indian astronomers, as well as new ideas from Aryabhata himself, the *Aryabhatiya* developed various rules for arithmetic and trigonometric calculations. It also contained a number of important "firsts" or near-firsts, including one of the first recorded uses of algebra. Furthermore, it was one of the first texts to include the idea of number position or place value (i.e., tens, hundreds, thousands, etc.). These concepts would have enormous impact as they moved westward, as would another idea implemented by Aryabhata in his text: the Hindu numeral system.

In addition, Aryabhata calculated the most accurate number for π up to that point in history, and in his *Ganita*—a poem in 33 couplets—he correctly stated the formula for finding the areas of a triangle and circle. He developed a solution to the indeterminate quadratic $xy = ax + by + c$, a solution that would be rediscovered by Leonhard Euler (1707-1783) some 1,200 years later.

As an astronomer, Aryabhata proved highly prescient in his suggestion that the reason the stars and planets seem to move around Earth is that Earth is in fact rotating on its axis as it moves around the Sun. It would be nearly a thousand years before a Western astronomer, Nicolaus Copernicus (1473-1543), recognized the same fact.

JUDSON KNIGHT

Anicius Manlius Severinus Boethius

480?-524

Roman Philosopher and Scholar

Though he is remembered primarily as a philosopher, Boethius had a powerful im-

pact on the history of mathematics. This was in large part because progress in mathematical learning among Western Europeans had come to a halt several centuries before his time, and would not resume for another half- millennium or more. Thus Boethius's writings remained among the principal guides to mathematics during the early medieval period.

Four years before his birth, the Western Roman Empire came to an end when the Germanic chieftain Odoacer (433?-493) removed the last Roman emperor from power and declared himself "king of Italy." At the time, the Romans did not perceive the fall of the Western Empire as an earth-shattering event, and many believed that business would continue as usual. Certainly that was the impression among Boethius's family, a distinguished line that could trace their roots back more than six centuries.

Before he reached his teens, Boethius lost his father, and therefore another prominent Roman named Symmachus became his guardian. Boethius later married Symmachus's daughter Rusticana, with whom he had two sons. In the course of his education, Boethius fell under the influence of Plato (427?-347 B.C.), whose *Republic* offered a model for participation in government by philosophers, and this—combined with family traditions—influenced his choice of a career in public life. He soon rose through the ranks, reaching the position of consul in 510.

By that time, Italy had come under the control of the Ostrogoth chieftain Theodoric (454?-526), who was to have a tragic effect on Boethius's career. As a Roman official, Boethius busied himself with his political duties—including service in the senate—and his studies and writings. The latter included an examination of the quadrivium, a group of four subjects (arithmetic, music, geometry, and astronomy) studied by Romans for ages.

In writing on the quadrivium, Boethius discussed the relationship between music and science, specifically that between a note's pitch and the frequency of the sound. This reflected the influence of Pythagorean ideas, and indeed Boethius's *Arithmetic* constituted medieval scholars' principal source regarding Pythagorean number theory. Boethius also translated *Categories* and *De interpretatione* by Aristotle (384-322 B.C.), as well as Porphyry's (234?-305?) *Isagoge*, and planned to translate and write commentaries on all the works of Plato and Aristotle. Events, however, prevented him from the fulfillment of this extremely ambitious goal.

Ancius Manlius Severinus Boethius and Philosophy
(crowned and carrying a scepter). *(Corbis Corporation.
Reproduced with permission.)*

His misfortunes began in 522, when he came to the defense of a senator accused of treason. It seems the senator, Albinus, had written to Justin I, ruler of the Eastern Roman (Byzantine) Empire, asking for the aid of the mainstream Christian emperor against Theodoric, who adhered to the Arian heresy. Boethius apparently thought Albinus guilty, but wanted to protect the reputation of the senate, and this exposed him to charges of suppressing evidence. He was then accused of treason himself, and imprisoned in the town of Pavia.

Boethius would spend the remaining two years of his life in jail, where he wrote his most enduring work, *The Consolation of Philosophy,* which set the tone for a thousand years of European history. Though he was a devout Christian, the *Consolation* makes little direct reference to Christian principles. Indeed, the text seems more closely tied to pagan Stoicism than to Christianity, with its principles of free will and redemption. Built on the conceit of a message delivered to Boethius in his cell by the allegorical figure of "Lady Philosophy," the *Consolation* teaches that divine justice can be seen even in the most random and seemingly arbitrary misfortune.

Boethius died in prison in 524, either by execution or as the result of torture. He quickly came to be regarded as a martyr, was later canon-

ized, and gained a further measure of immortality through his influence on medieval thought.

JUDSON KNIGHT

Conon of Samos
280?-220? B.C.
Greek Astronomer and Mathematician

Conon of Samos is known primarily for his work as an astronomer, in particular his discovery of the constellation Coma Berenices. He also made significant contributions to mathematics in his discussion of conics, which influenced Apollonius of Perga (262?-190? B.C.). A friend of Archimedes (287?-212 B.C.), he may have had an impact on the work of that great mathematician and scientist as well.

As court astronomer to Egyptian ruler Ptolemy III Euergetes, Conon named the famous constellation after Ptolemy's queen Berenice II. (These names, along with Cleopatra, were a fixture of the Ptolemaic dynasty, and continued through the era of the last Ptolemaic pharaoh, Cleopatra VII—the famous Cleopatra—two centuries later.) It was said that when Ptolemy returned from a campaign in Syria in 246 B.C., Berenice cut off a lock of her hair and offered it at the temple of Arsinoë Zephyritis. This was significant because the Greek Ptolemies, taking their cue from an Egyptian tradition that went back some 2,500 years, were regarded as gods themselves. Thus it posed something of a crisis when the lock of Berenice's hair was lost.

Conon redeemed the situation somewhat, from a public relations standpoint at least, by naming his recently discovered constellation "the lock Berenice." Coma Berenices consists of seven faint stars near the tail of Leo, and between that constellation and those of Virgo and Boötes. In time, the constellation would be a topic of poets such as the Greek Callimachus, as presented in his *Berenikes plokamos* (The lock of Berenice). Nearly 2,000 years after Conon, Alexander Pope satirized this tradition in *Rape of the Lock*, a mock epic in which Berenices's stolen lock appears in the sky as a new star.

In the meantime, Conon himself had become a figure celebrated by poets of Rome's golden age, who depicted him as the model of an astronomer. According to Catullus (84-54 B.C.), he "discerned all the lights of the vast universe, and disclosed the risings and settings of the stars, how the fiery brightness of the sun is darkened, and how the stars retreat at fixed times."

Seneca (3? B.C.-A.D. 65) wrote that Conon "recorded solar eclipses observed by the Egyptians," and Ptolemy (100?-170) claimed he had revealed 17 "signs of the seasons" in his *De astrologia*. The latter, like all of Conon's other writings, has long since disappeared. As for his friendship with Archimedes, it is likely that they met in Alexandria. Archimedes became a great admirer of Conon, and, according to Pappus (fl. c. A.D. 320), based his famous spiral of Archimedes on a shape discovered by Conon.

Apollonius maintained that Conon's *Pros Thrasydaion* (In reply to Thrasydaeus), a work that addressed the points of intersection of conics, including circles, contained numerous mistakes. Yet Apollonius himself based much of the discussion of conic sections in his *Conics*, Book IV, on the work of Conon.

JUDSON KNIGHT

Dinostratus
390?-320? B.C.
Greek Mathematician

For many years, historians of mathematics cited Dinostratus as the first to achieve something approaching the squaring the circle—that is, finding a square equal in area to a given circle, using only a compass and straight edge. In fact it is impossible to do so, but Dinostratus came close by using a curve called a quadratix.

His life is a mystery, though Proclus (410?-485) maintained that Dinostratus was a close friend of Plato's (427?-347 B.C.) in Athens. Other than this, the only thing known about him was his use of the quadratix, a curve discovered by Hippias (fifth century B.C.).

In order to describe the quadratix, one must first imagine a square *ABCD*. Point *A* marks the center of a circle, the radius of which is labeled as *AE*. *E* is an arbitrary point of the curve *BED*, which forms the quarter-arc of the circle with *A* at its center. If *AE* moves uniformly from *AB* to *AD* (a path equivalent to a clock going from 12 to 3), and *BC* descends straight down toward *AD,* the radius *AE* and the line *BC* will intersect at point *F.* As these two keep moving downward, the point of their intersection forms an arc, *BG.*

Having established point *G,* where the arc meets the base of the square, it can be shown that the length of arc *BED* is in the same proportion to the length of *AB* as the latter is to the

length of *AG*. That ratio in turn makes it possible to map the curve BED onto a straight line.

Pappus (fl. c. A.D. 320), the principal source regarding Dinostratus's use of the quadratix to square the circle, later proved these ratios by a *reductio ad absurdum,* showing that the other possible proposition (*BED : AG = AB : AG*) led to false conclusions. He also reported that Sporus (240?-300) criticized Dinostratus's method for several reasons, among them the fact—correct but perhaps not completely useful—that the descending segment *BC* would never be in the same spot as *AD*, and thus *G* would constitute a limit rather than a point.

Aside from Pappus, there is no clear evidence that Dinostratus applied the quadratix to the squaring of the circle, and even if he did so, his method requires more than a compass and straight edge; thus it does not really address the classical problem of squaring the circle. In 1882, nearly 22 centuries after Dinostratus's time, Ferdinand von Lindemann (1852-1939) showed conclusively that π is a transcendental number, meaning that it would be literally impossible to square the circle using only a compass and straight edge.

JUDSON KNIGHT

Diocles
c. 240-c. 180 B.C.
Greek Mathematician

The details of his life are virtually unknown, and Diocles is remembered almost entirely for a fragmentary manuscript entitled *On Burning Mirrors*. In it he discussed not only the physical problem referred to in the title, but such subjects as cutting sphere with a plane, as well as the famous Delian problem of doubling the cube.

On Burning Mirrors may actually have been a collection of three separate short works, combined under a single title that does not reflect the whole. In any case, the book consisted of 16 geometric propositions, most of which involved conics.

In the first proposition, Diocles put forth the focal property of the parabola, and in propositions 2 and 3 presented the properties of spherical mirrors. The next two propositions show the focus directrix parabola construction, which as with the statement on the focal property of that shape are innovations credited to Diocles. In propositions 7 and 8, Diocles examined a problem first put forth by Archimedes (c. 287-212 B.C.), on how to cut a sphere with a plane.

Propositions 10 through 12 offered Diocles's solution to the Delian problem. The latter had long perplexed Greek mathematicians who, using only a straightedge and compass, attempted to double the volume of a cube. Today this problem could be expressed in simple terms: find length x for a cube such that $x^3 = 2a^3$, where a is the length of the known cube. The Greeks, however, possessed no such sophisticated algebraic notation, and attempted to work the problem geometrically. Diocles's solution to the problem involved a special curve called a cissoid, which he used to find two mean proportionals. The use of the cissoid, however, involved more than a straightedge and compass, and in fact by the nineteenth century A.D. mathematicians recognized that it was impossible to solve the problem using only those tools.

In a 1976 article for a journal on the history of mathematics and the sciences, G. J. Toomer made a case for Diocles, and not his contemporary Apollonius of Perga (c. 262-c. 190 B.C.), as the originator of such familiar terms as *hyperbola, parabola,* and *ellipse.* In any case, *On Burning Mirrors* discussed a number of interesting subjects, not least of which was the one contained in the title. In this regard, Diocles put forth the problem of finding a mirror such that the reflected light forms certain curves, questions that would have implications for the creation of sundials.

JUDSON KNIGHT

Diophantus of Alexandria
210-290
Greek Mathematician

The work of Diophantus is particularly remarkable in light of the time in which it took place. By his era, in the third century A.D., the Roman Empire—of which his hometown, the Greek city of Alexandria in Egypt, was a part—had long since entered its slow decline. Little in the way of original mathematical work took place during this period, yet in his *Arithmetica,* Diophantus explored the frontiers of mathematics. This new, abstract arithmetic involved the use of special symbols, and centuries later it would come to be known as algebra.

From a sixth century Greek collection of brainteasers comes this one, concerning Diophantus: "...his boyhood lasted $^1/_6$th of his life; he married after $^1/_7$th more; his beard grew after $^1/_{12}$th more; and his son was born five years later; the son lived to half his father's age, and

the father died 4 years after the son." Unraveling the puzzle yields the information that Diophantus married at age 26 and became a father at 31, and that his son died at age 42, four years before Diophantus died at the age of 84.

As for his *Arithmetica,* this consisted of 130 problems in 13 books, of which only six survive. The book offers numerical solutions to determinate equations, or ones with a single solution, and seems to have been largely concerned with equations that have rational integer solutions. Apparently Diophantus, perhaps drawing on the influence of the ancient Pythagorean school, believed that irrational solutions (i.e., solutions that resulted in numbers that cannot be expressed as a fraction) were impossible.

More important than his beliefs about irrationals was Diophantus's use of the first symbolic notation, including abbreviations for an unknown and its powers. Prior to that time, mathematicians had simply written out all elements of the problem. Most significant of all, however, was the nature of the problems themselves: for instance, "...a sixfold number increased by twelve, which is divided by the difference by which the square of the number exceeds three." Such a problem would have been difficult indeed to solve with the limited notation Diophantus introduced; by contrast, mathematicians today would write $(12 + 6n)/(n^2 - 3)$. Obviously Diophantus had gone past the realm of arithmetic, but not until al-Khwarizmi (c. 780-c. 750)—from whose *Kitab al-jabr* the word "algebra" is drawn—did the implications of this become clear.

It is not correct to refer to Diophantus as the "father of algebra," because some of the concepts he used go back to the Babylonians; but it is obvious that he greatly advanced the mathematics of unknowns, and was first to give form to their study. In addition to his other achievements, Diophantus seems to have been the first mathematician to treat fractions as numbers. Centuries later, his work would attracted the admiration of such luminaries as Regiomontanus (1436-1476) and Pierre de Fermat (1601-1665).

JUDSON KNIGHT

Eratosthenes of Cyrene
c. 285-c. 205 B.C.
Greek North African Mathematician, Astronomer, and Geographer

Most famous of the librarians at Alexandria, Eratosthenes provided a measurement for

Earth's circumference within 1% of the actual number. He also developed a method for finding prime numbers, and made contributions as an astronomer, geographer, philosopher, and poet.

Born in Cyrene, now part of Libya, Eratosthenes's father was named Aglaus, but these are the only known facts of his origins. He studied with the grammarian Lysanias, the philosopher Ariston of Chios, and the poet Callimachus (c. 305-c. 240 B.C.), second librarian of Alexandria. In his teens he traveled to Athens, where he may have studied both at the Academy established by Plato (427-347 B.C.) and the Lyceum of Aristotle (384-322 B.C.).

Eratosthenes, who at a later time might have been called a Renaissance man, displayed his talents as a poet in *Hermes* and *Ergion*. These in turn attracted the attention of Ptolemy III Euergetes (r. 246-221 B.C.), Greek ruler of Egypt, who invited Eratosthenes to Alexandria as tutor for the crown prince. Soon afterward, Ptolemy appointed Eratosthenes director of the city's famed library.

During his fruitful career, Eratosthenes contributed to mathematical knowledge with a technique for finding prime numbers that came to be known as the "sieve of Eratosthenes." Using this formula, a type of algorithm, he could pull out primes from an ordered list of the natural numbers. Not only did this method spark later investigation in number theory, but it could perhaps be cited as an early manifestation of computing techniques.

In working on the Delian problem, which involved finding mean proportionals as a way of doubling the cube, Eratosthenes developed what was called the mesolabe or "mean-finder," which made it possible to determine an infinite number of mean proportionals between two lines. He also wrote a number of mathematical texts, but the only piece of his writing that survives is a letter to Ptolemy III explaining the mesolabe.

Most famous among Eratosthenes's many achievements was his measurement of Earth's circumference. This he did by measuring the distance from Alexandria to the city of Syene in the south, which he determined to be 5,000 *stadia* (489 miles; 783 kilometers). By using calculations of the Sun's position over both cities during the summer solstice, he was then able to estimate that the difference between them represented 2% of Earth's circumference, the total of which he calculated at 252,000 stadia (24,662 miles; 39,459 kilometers).

In fact the true figure at the equator is 24,901.55 miles (39,842.48 kilometers), meaning that Eratosthenes was astonishingly close. At his time, of course, no expedition had ventured below the equivalent of 10° north; nor did anyone in the Old World know about the existence of the New. Certainly his figure implied the existence of a very large unexplored area on the other side of the world, and had Christopher Columbus (1451-1506) been using the numbers derived by Eratosthenes when he made his historic landing in the Caribbean some 1,700 years later, he would have realized that he had not reached Asia. Unfortunately, however, Hipparchus (190-126 B.C.) had rejected Eratosthenes's measurement in favor of a much smaller number, and Hipparchus's follower Ptolemy (c. A.D. 100-170) had propagated this misconception, which remained the received wisdom throughout the Middle Ages.

As for Eratosthenes, his now-lost manuscript *On the Measurement of the Earth* marked the foundation of geodesy, the branch of mathematics that deals with determination of Earth's size and shape, and the location of points on its surface. Among the topics of study within geodesy is the system of latitude and longitude, which Eratosthenes seems to have pioneered in his maps, the most accurate in the world at the time.

Believed to have developed the calendar that remained unchanged in the Greco-Roman world until the time of Julius Caesar (102-44 B.C.), Eratosthenes also established the first reliable method of dating events, by reference to the years of the Olympic festivals. In his later years, he went blind, and perhaps as a result of despair over his inability to any longer read the manuscripts of his beloved library, he starved himself to death in the ninth decade of his life.

JUDSON KNIGHT

Euclid
325?-250? B.C.
Greek Mathematician

Euclid's *Elements*, the bible of geometry for 2,000 years, remains the most influential textbook in history, and one of the essential works of human civilization. Therefore it is all the more ironic that the man who wrote it is such a figure of mystery that at various times historians have suggested that he never lived—or that "Euclid" was actually the name for a group of mathematicians.

In fact the historical existence of a man named Euclid of Alexandria, author of the *Elements,* seems safely established. Most of what scholars know about his life, however, comes from a short summary in Proclus's (410?-485) commentary on the *Elements.* As for his place of origin, this has been variously identified as Tyre (now in Lebanon), Greece, or Egypt, with the last two being the most likely candidates. Whatever the case, he was almost without a doubt ethnically Greek, and certainly brought up in the Greek language, culture, and civilization.

More difficult is the placement of Euclid in time. Some sources indicate that he traveled as an adult to Alexandria in 322 B.C., but other information on the dates of his life suggests that he would have been a small child in that year. It seems likely that he first studied in Athens at the Academy established by Plato (427?-347 B.C.), though long after Plato's death, and that soon afterward he moved to Alexandria, where he became involved with that city's great library as its first teacher of mathematics. This, too, suggests a later date for his birth: in 322 B.C., the establishment of the Alexandrian library was still several decades in the future.

The one clearly established aspect of Euclid's life—passing over speculation that he was either a myth or the name for a committee—was his authorship of the *Elements.* Though the book contained concepts of his own, it is primarily a summation of mathematical knowledge passed down from Pythagoras (580?-500? B.C.) onward, and its genius lies in its cogent explanation of basic principles, as well as its clear and thorough explication of geometric proofs.

Consisting of 13 books in which Euclid elaborated on some 450 propositions, the *Elements* begins with a definition of points, lines, planes, angles, circles, triangles, quadrilaterals, and parallel lines. In Book II, Euclid addressed rectangles and squares; in Book III, circles; and in Book IV polygons. He continued with a discussion of proportion and area (Book V), followed by an application of this theory to plane geometry (Book VI). Book VII covers arithmetic, Book VIII irrational numbers, and Book IX rational numbers, while the remainder of the volume is devoted to three-dimensional, or solid, geometry.

Among Euclid's original contributions was a new proof of the Pythagorean theorem, which included a proof of the existence of irrational numbers. He also developed a means of showing that the number of primes is infinite, and created an exhaustion method for measuring area

Euclid. *(Bettmann/Corbis. Reproduced with permission.)*

and volume, later adopted by Archimedes (287?-212 B.C.).

Euclid's five postulates are among the most important aspects of his work. The first three of these focus on construction with the straight edge and circle or compass, the only tools of Euclidean geometry, while the fourth states that all right angles are equal. This seems like an easy conclusion, but in reaching it Euclid was forced to adopt a view that was far from obvious, treating space as a homogeneous entity in which a figure is independent of its position.

Most controversial, however, was the fifth postulate, which discussed the relationship between two straight lines placed side by side. If a line placed at a 90° angle to the first line does not also intersect the second line at a 90° angle, Euclid indicated, the first two lines *must* eventually meet in the direction of the angle that is less than 90°. Later Proclus would develop a well known formulation of the Fifth Postulate: "Through a given point in a plane, one and only one line can be drawn parallel to a given line." Euclid and other mathematicians recognized that the preceding four postulates did not constitute a proof for the fifth, and in later centuries the Fifth Postulate came under increasing challenge. This would culminate with the development of non-Euclidean geometry in the nineteenth century.

Given the significance of his *Elements,* it may be surprising to learn that Euclid is credited as the author of numerous other texts, including works on plane geometry, spherical geometry, and perspective. Furthermore, he wrote a number of books that have been lost, including a volume called *Conics* that apparently influenced the more famous work of that name by Apollonius of Perga (262?-190? B.C.).

JUDSON KNIGHT

Eudoxus of Cnidus

c. 408-c. 355 B.C.

Greek Astronomer and Mathematician

First to apply mathematics properly in the study of astronomy, Eudoxus also contributed directly to mathematical study with his theory of proportion and his method of exhaustion. In addition, he worked as a medical doctor, and gained renown as a philosopher and political writer. He also wrote a seven-volume description of his travels entitled *Circuit of the Earth.*

Born in the Greek colony of Cnidus in Asia Minor, Eudoxus came from a long line of physicians, and trained as doctor. By the age of 23, he had moved to Athens to work as a physicians' assistant, and while in the great city attended lectures by Plato (427-347 B.C.) at the latter's Academy. He then returned to Cnidus, where he completed his studies before going to Egypt with another physician.

It was there, at the observatory in Heliopolis along the Nile, that Cnidus discovered his second calling. As a physician, Eudoxus was accustomed to making detailed observations, one of the few areas in which ancient medicine excelled, and the record of data he compiled at Heliopolis was quite thorough. This he presented in the *Phaenomena,* a rather straightforward astronomical study containing lists of stars that rise or fall below the horizon at the beginning of each month, as well as locations for all constellations relative to one another.

Returning to Asia Minor with the new course of his career set, Eudoxus founded a school in the town of Cyzicus. There he wrote *On Speeds,* a much more important work than *Phaenomena* in which he presented a new theory of the motion made by the Sun, Moon, and planets. Given the spherical shape of Earth, Eudoxus imagined a series of concentric spheres around it, and eventually developed a description of 27 spheres necessary for picturing the movement of all known bodies.

The idea of the concentric spheres seems obvious today: hence people think of a planet's orbit around the Sun—though it is actually elliptical rather than circular—as taking place along the equator of an imaginary sphere. But in Eudoxus's time, this idea, which would eventually be represented physically in a variety of astronomical instruments, was far from obvious.

His theory of proportions and his method of exhaustion would later find their way into the *Elements* of Euclid (c. 325-c. 250 B.C.). According to Eudoxus, if a given object is larger than a second one, then its ratio to a third object will also be larger than the ratio of the second to the third—which again is a seemingly self-evident point from the perspective of the twenty-first century A.D., but was not so in the fourth century B.C. As for Eudoxus's method of exhaustion, Archimedes (c. 287-212 B.C.) wrote that he gave the first proofs of two propositions already known at the time: that the volume of a pyramid is one-third that of a prism with the same base and height, and that the same relationship is the case for a cone and cylinder.

Eudoxus moved to Athens, but when the people of Cnidus overthrew the oligarchy there and established a democracy, they asked him to come back and write a constitution for the new state. He then made his way home, where after completing his work for the government he founded an observatory and school. According to Aristotle (384-322 B.C.), Eudoxus also made a reputation for himself as a philosopher.

JUDSON KNIGHT

Geminus

c. 130-c. 70 B.C.

Greek Philosopher, Astronomer, and Mathematician

The most important contribution of Geminus to mathematics was his classification of mathematical disciplines, and his efforts to define both mathematics and science. He also gave his own version of Euclid's parallel postulate, and wrote a survey of Greek astronomy. Much of his work as mathematician and astronomer, however, was in service to his primary career as a Stoic philosopher intent on justifying the Stoic worldview against attacks on the sciences by Epicureans and Skeptics.

Most likely Geminus came from the Greek island of Rhodes, a center of astronomical study. Though his name was Latin, he was not of Italian background; rather, the world in which he lived was one completely dominated by Rome. Geminus studied under Posidonius (c. 135-c. 51 B.C.), and though the latter was only slightly older than he, Posidonius had a reputation as, if not the wisest man of his day, certainly the most learned of the Stoics. (It is possible Geminus's dates were c. 10 B.C.-c. A.D. 60, in which case he would have been studying in Posidonius's school, but not under Posidonius. The dates of his life are based on the calendar in his *Isagoge* or *Introduction to Astronomy,* but the details of that calendar are subject to at least two possible interpretations.)

Geminus's *Theory of Mathematics,* now lost, presented an overview of geometry. The latter was by then a long-established discipline among the Greeks, and this gave him a certain perspective that would have been beyond the reach of his predecessors. Thus he undertook to define mathematics as a whole, and to classifying it within the context of the sciences.

At least from the time of Aristotle (384-322 B.C.), Greek schools of thought tended to view the world hierarchically; thus the Stoics viewed all of existence as a continuum from the lowest inanimate object to God, or Absolute Reason. This hierarchical conception naturally informed Geminus's schema of classification, as did the Aristotelian notion that pure science is above applied—or as Geminus called it, "mixed" or "impure"—science.

Pure science dealt with what Geminus referred to as the intelligible, whereas applied science concerned that which he described as the tangible. He wrote that "geometry and arithmetic, sciences which deal with... the eternal and unchangeable... [were] extended by later writers to cover what we call 'mixed' or applied mathematics...which has to do with sensible objects, e.g. astronomy or optics." Because they were primarily concerned with what he regarded as the intelligible, arithmetic and geometry were therefore at the apex of Geminus's system. As for the mathematics of the tangible, he divided it into six classes: mechanics, astronomy, optics, geodesy, canonics, and logistics.

This system says as much about the Greek mind—for instance, the emphasis on the ideal over that which can be experienced, a legacy of Plato (427-347 B.C.)—as it does about mathematics itself. Nonetheless, it constituted a valuable early attempt to give shape to the mathematical discipline, and to place it within the context of scientific study.

Elsewhere in Geminus's mathematics text, pieces of which survive only in quotes from others, is his discussion of Euclid's (c. 325-c. 250 B.C.) parallel postulate. Proclus (410?-485) quotes him thus: "...when the right angles are lessened, [that] the straight lines converge [as Euclid had indicated] is true and necessary; but the statement that, since they converge more and more as they are produced, they will sometime meet is plausible but not necessary...."

In his *Introduction to Astronomy,* a work that does survive, Geminus dealt with a "tangible" discipline. His approach to astronomy was not original, however, but rather based on the work of Hipparchus (190-126 B.C.).

JUDSON KNIGHT

Hippias of Elis
c. 460-c. 400 B.C.
Greek Mathematician

Rarely does the personality of a mathematician or scientist play a significant role in his biography, but Hippias was so widely known as a braggart that he can hardly be mentioned without making note of this fact. Yet he also seems to have had cause for boasting, being a man knowledgeable in a wide variety of areas, including several with a mathematical application. In this regard it is notable that he may have developed the quadratix as a means of doubling the cube, trisecting angles, and squaring the circle nearly a century before Dinostratus (c. 390-c. 320 B.C.) used it for the latter purpose.

From Elis on the Greek mainland, Hippias made his living as a traveling philosopher, perhaps a member of the Sophists. Among the areas on which he lectured were poetry, grammar, history, politics, and archaeology, as well as mathematics and related fields. The latter included calculation, geometry, astronomy, and the application of math to music as pioneered by the Pythagorean school.

He was also, according to Plato (427-347 B.C.), a boastful man. T. L. Heath, in *A History of Greek Mathematics* (1921), related that Hippias "claimed...to have gone to the Olympian festival [i.e., the Olympic Games] with everything that he wore made by himself, ring and sandal (engraved), oil-bottle, scraper, shoes, clothes, and a

Persian girdle of expensive type; he also took poems, epics, tragedies, dithyrambs, and all sorts of prose works." Clearly Hippias was something of a showoff, but he was also quite talented, having developed a mnemonic system that made him able to repeat a catalogue of 50 names after hearing it only once.

His single contribution to mathematics, which certainly would have been a great one, seems to have been the quadratix. This curve, which Dinostratus later used for the squaring of the circle, could also be used to trisect angles, which was Hippias's purpose. Also, as Heath wrote, Archytas (c. 428- 350? B.C.) used it for the duplication of the cube.

The quadratix may be described in terms of a square *ABCD*. Point *A* marks the center of a circle, the radius of which is labeled as *AE*, and the curve *BED* forms a one-quarter arc of the circle with *A* at its center. If *AE* moves uniformly from *AB* toward *AD* (rather like a windshield wiper moving clockwise from top to bottom), and *BC* descends straight downward toward *AD*, the radius *AE* and the line *BC* will intersect at point *F*. A segment may then be drawn from *F* downward to intersect *AD* at point *H*.

Using these materials, Hippias developed a set of ratios that established an equivalence between angle *EAD*, arc *ED*, and ($FX \times \pi/2$). He was then able to graphically illustrate means for dividing angle *EAD* by a given ratio. However, Sporus (c. 240-300) criticized Hippias's method as a circular one, and in any case, to use it involved more than the two basic tools of compass and straightedge.

JUDSON KNIGHT

Hippocrates of Chios
c. 470-c. 410 B.C.
Greek Mathematician

Designated as Hippocrates of Chios to distinguish him from the better-known physician of the same name, Hippocrates has been cited as the greatest mathematician of the fifth century B.C. He wrote the first textbook on geometry, in which he addressed problems such as squaring the circle and doubling the cube.

Hippocrates's achievements were particularly notable in light of the fact that he started his career as a mathematician late in life. Certainly he does not seem to have set out in his youth to pursue such a career; on the contrary, his involvement in mathematics was the indirect outcome of

misfortunes. Apparently he had a successful business as a merchant until he was attacked near Byzantium by Athenian pirates, though another version of the story depicts his assailants as corrupt customs officials who seized his goods and threatened imprisonment if he complained.

It seems that Hippocrates then went to Athens itself to seek legal redress, and while waiting for his case to come to court, he attended lectures on mathematics and philosophy. During this time, he came under the influence of the mathematical school based on the principles of Pythagoras (c. 580-c. 500 B.C.), and in time he reached such a degree of proficiency as a mathematician that he opened a Pythagorean school of his own. Pythagoras had forbidden his students to earn money through their mathematical knowledge, but in view of Hippocrates's recent financial hardships, the Pythagoreans of Athens made an exception.

Among the legacies of Hippocrates was a mathematical textbook, long since lost, called the *Elements of Geometry*. The first work of its kind, it would have an enormous impact on another book of a similar title, the highly influential *Elements* of Euclid (c. 325-c. 250 B.C.). In his work, known through the writings of Aristotle (384-322 B.C.), Proclus (410?-485), and others, Hippocrates became the first mathematician to adopt scientifically precise and logical methodology for developing geometrical theorems from axioms and postulates. It is also likely that *Elements of Geometry* contained the first written explanation of Pythagorean principles, since the Pythagoreans who preceded him did not believe in committing their ideas to writing.

Other issues addressed in the book included the Delian problem of doubling the cube. In attempting a solution, Hippocrates became the first mathematician to use a method of reduction: that is, altering a difficult problem to a simpler form, which, once it was solved, made it possible to apply the solution to the original problem. Dinostratus (c. 390-c. 320 B.C.) would later apply this idea to the construction of the quadratix for squaring the circle, and Tartaglia (Niccolò Fontana; 1499-1557) would use much the same principle to the solution of cubic equations two millennia after Hippocrates. In Hippocrates's case, by reduction he discovered that finding mean proportionals was the key to doubling the cube—a principle that revolutionized Greek mathematicians' approach to the problem.

Hippocrates's quadrature of the lune, a crescent shape, represented an attempt to solve another problem of long standing among Greek mathematicians, the squaring of the circle. Starting from the principle that the ratio of area between two circles is the same as that of the squares of their diameter, Hippocrates set out to square the lune—that is, to find a square equal in area to a given lune. The process he followed was a complicated one, but in essence it put to work his idea of reduction by finding a correlation between a given straight segment and a curved one.

Also significant was Hippocrates's use of *reductio ad absurdum,* a method he may have introduced. The *reductio,* which is reflected today in the scientific method, sets out to prove the opposite of what is desired; and by so doing, it shows that the original proposition is true. In similar fashion, a scientist makes every effort to disprove his or her propositions, and only if an idea stands this test is it formalized as a theory. (By contrast, a pseudo- or anti- scientist starts with a conclusion and then looks for evidence that corroborates it, simultaneously ignoring all facts that do not.)

In addition to his mathematical work, Hippocrates also conducted astronomical studies. In this he was hampered, however, by his allegiance to Pythagorean ideas such as the claim that there was only one comet, which simply reappeared from time to time.

JUDSON KNIGHT

Hypatia of Alexandria
c. 370-415
Egyptian Mathematician, Astronomer, and Philosopher

Hypatia of Alexandria was, in her prime, the leading mathematician and philosopher in Western civilization. Although she is not credited with new theorems in mathematics, Hypatia's work was crucial in preserving and explaining the work of earlier mathematicians and astronomers. For a woman of her time, Hypatia's fame and stature were unprecedented, and in recent years she has been adopted as a symbol for feminists, religious partisans, and those interested in her Egyptian heritage.

Hypatia was the daughter of Theon, the last known head of the museum at Alexandria, Egypt, an ancient center of classical learning. By the late fourth century, Alexandria was the intellectual center of the Western world, a center of Hellenic scholarship and science.

Hypatia was taught by her father and worked with him. Historians do not know whether she

Hypatia of Alexandria. *(Bettmann/Corbis. Reproduced with permission.)*

traveled or whether she stayed in Alexandria all her life, but she may well have studied in Athens. Hypatia taught mathematics and philosophy in Alexandria, perhaps through the museum and perhaps on her own. She had many influential students and associates, including the Roman prefect of Alexandria and a future Christian bishop. It is quite certain that she never married.

Hypatia and Theon are not credited with new theorems in mathematics, but their work was crucial in preserving and explaining the work of earlier mathematicians and astronomers. Theon's editions and commentaries are the primary source for our knowledge of many of the works of Euclid (c. 300 B.C.) and Ptolemy (second century A.D.). Scholars know that Hypatia prepared at least one of Theon's commentaries on Ptolemy's *Almagest,* as Theon credits her in his introduction to the work. The *Almagest* was the basis of astronomical work in both the Western and the Islamic world until astronomy was rewritten in the sixteenth century by Copernicus (1473-1543). Theon's commentaries were a way of offering a teacher's explanation of the text, and Hypatia was renowned for her teaching. So it is logical that she would expand the influence of her teaching through clarifying and expounding on texts.

None of Hypatia's writings exist today, but ancient reports tell us that she wrote commentaries on Diophantus (third century A.D.), a Greek mathematician whose major innovation was the introduction of symbolism into algebra. Some previously lost chapters of Diophantus's text have been found in an Arabic translation. Because these chapters are more carefully written than those in Greek and have clear explanations, it may be that Hypatia's commentary survives as part of this translation. The work of Diophantus is unique in the Greek tradition because of its emphasis on algebra rather than geometry, so Hypatia's contribution to preserving it was a true service that helps us understand the historical basis of today's algebra.

Hypatia also wrote a commentary on the *Conics* of Apollonius of Perga (c. 262-c. 190 B.C.), another Greek mathematician. Apollonius's work contains some of the most advanced and difficult work in Greek geometry. His *Conics* was the inspiration for Ptolemy's description of planetary orbits in the second century A.D., and in the seventeenth century for Descartes (1596-1650) and Fermat (1601-1665) as those two mathematicians developed analytical geometry. Existing texts have been analyzed for signs of Hypatia's writing and may provide some clues of her editing.

Much legend has been built around Hypatia, but we do have reliable reports of her death in 415 at the hands of a Christian mob, which seized her on the street, beat her, then dragged her body to a church, where they scraped the flesh from her bones with sea shells. At the time, Alexandria was experiencing violent disputes between Christians and non-Christians as well as between religious and civil leaders. Hypatia's horrible death was probably brought about by her high standing in the community, her friendship with Orestes, the leading Roman official in Alexandria, and the fact that she was a non-Christian in an increasingly Christian atmosphere. She also represented learning and science, which the early Christians associated with paganism. After his appointment in 412, the Christian bishop Cyril attempted to assert his control over the city and over Orestes, his rival. If Cyril did not order Hypatia's death, he certainly created the climate that would vilify a pagan woman allied with Orestes. Fortunately, Hypatia's writings were not singled out for destruction. Instead, her work helped maintain the mathematical and astronomical tradition of earlier scholars, and mathematicians continue to examine the ways in which her commentaries enhanced this tradition.

EDITH PRENTICE MENDEZ

Liu Hui
fl. c. 263
Chinese Mathematician

The first non-Greek mathematician of distinction, Liu Hui developed an early approximation of π. He is also known for his commentary on an ancient Chinese mathematical work and for his use of calculating rods, a form of computing developed by mathematicians in China centuries earlier.

Liu Hui lived in the era of the Three Kingdoms (221-265), a period of anarchy immediately following the collapse of the Han Dynasty. Despite the confusion of the times, the age produced a great number of cultural and scientific advancements; in the fourteenth century a popular book entitled *Romance of the Three Kingdoms* would celebrate this period in Chinese history.

The three kingdoms were Wu, Shu, and Wei, in whose government Liu Hui served as an official. The only certain date of his life is 263, when he wrote a commentary on *Chiu-chang Suan-shu* or *Jiuzhang Suanshu* (Nine chapters of mathematical art). Two years later, in 265, Wei came under the control of the Western Ch'in Dynasty, which eventually absorbed Wu and Shu. The Western Ch'in would maintain power until 316, when China experienced one of the many invasions by nomadic tribes from the north that characterized its history in the premodern period. Only in 589, nearly four centuries after the collapse of the Han Dynasty, would China reunify.

As for the "Nine Chapters," this was a text of unknown authorship dating back to the first century B.C. The oldest known Chinese mathematical text, it contained 246 problems, which as the title suggested were presented in nine chapters. The first of these concerned arithmetic and the fundamentals of geometry, and included a discussion of the counting rods.

The latter, which probably originated in the fifth century B.C., were small bamboo sticks positioned in such a way as to provide something like a decimal place-value system. A space signified zero, a concept that had yet to be formalized even by the mathematicians of India, who are typically credited with this idea. The counting-rod system arranged numbers from left to right—a notable fact in a country where people read from right to left.

Liu Hui is often credited with using red counting rods for positive numbers and black ones for negatives—a concept that, like zero, had yet to gain formal definition. (The articulation of these and other fundamental ideas is typically credited to mathematicians of India's Gupta Empire, c. 320- 540.). In fact the use of red and black counting rods can be traced to Chapter 4 of the "Nine Chapters."

The second, third, and sixth chapters involved applications of mathematics for the purposes of governing—for instance, calculation of fair taxes in chapter 6—and chapter 5 explored measurements of various figures. The seventh chapter examined mathematical logic; the eighth simultaneous linear equations and negative numbers; and the ninth the applications of what mathematicians in the West would have called the Pythagorean Theorem. (The latter may have been discovered by the Babylonians as many as a thousand years before Pythagoras [c. 580-c. 500 B.C.], and it is likely that the Chinese discovery of the principle took place independently.)

Chapter 5 also included a figure of 3 for π, but in his commentary on the "Nine Chapters," Liu Hui presented a much more accurate figure. This he did by using polygons to approximate circles: by working from a 96- to a 192-sided polygon, he reached a figure of 3.141014. Today, of course, mathematicians have calculated the value of this irrational number to well over 1 million places, but 3.141592 is the abbreviated form. A century before Liu Hui, Ptolemy (c. 100-170) had calculated it at 3.1416, and two centuries later, the Chinese mathematician Tsu Ch'ung-chih (Zu Chongzhi; 429-500) offered the figure of 355/113, very close to the number accepted today.

In addition to his commentary on the "Nine Chapters," Liu Hui wrote *Haidao suanjing,* or "Sea Island Mathematical Manual." This began as an appendix to the earlier work, but eventually grew to include, among other things, nine surveying problems.

JUDSON KNIGHT

Menaechmus

c. 380-c. 320 B.C.

Greek Mathematician

In the course of trying to solve the problem of doubling the cube, Menaechmus discovered the conic sections, which would have an enormous impact on mathematics in modern times. He was also responsible for distinguishing between the two meanings of the word *element,*

which up to that point had caused confusion among Greek geometers.

Whereas historians know almost nothing about the life of Dinostratus (c. 390-c. 320 B.C.), who applied the quadratix of Hippias (c. 460-c. 400 B.C.) in an attempt to square the circle, information about Menaechmus is much more readily available—a salient point, since he was the younger brother of Dinostratus. Nonetheless, his family life is otherwise a mystery, as are the facts of his birth.

Various writers have provided details on Menaechmus's career, allowing a historian to piece together a picture of sorts. There was a Menaechmus, from the region of either Alopeconnesus or Proconnesus, who wrote three commentaries on the *Republic*. The first of these is in Thrace, the second along the Sea of Marmara, and both are close to Cyzicus in Asia Minor, where Menaechmus studied under Eudoxus of Cnidus (c. 408-c. 355 B.C.) After completing his schooling, Menaechmus apparently worked as tutor for Alexander the Great (356-323 B.C.).

Like many mathematicians of his time and place, Menaechmus became involved in attempts to solve the Delian problem, or the challenge of doubling the cube. This would be quite simple today, using the formula $x^3 = 2a^3$, where a is the length of the known cube and x that of the doubled one. The Greeks, however, lacked not only this notation, which makes it possible to conceive of the problem algebraically; they also lacked the algebraic formula itself, and approached the challenge with the simple geometric tools of compass and straightedge. Hippocrates of Chios (c. 470-c. 410 B.C.) had discovered that the solution lay in finding the mean proportionals between two given lines; Archytas of Tarentum (c. 428-350? B.C.) improved on this method; and Menaechmus took the solution one step further with his work in what came to be known as conic sections.

Eventually Menaechmus offered two methods for finding a solution, both of which involved slicing cones with planes as a way of finding the mean proportionals between two numbers. The greatest importance of his work, however, lay not in its application to the Delian problem, but in his introduction of conic sections. Later, Apollonius of Perga (c. 262-c. 190 B.C.) would develop terminology for the shapes made by slicing cones—hyperbola, parabola, and so on—and in modern times these forms would find application in everything from calculus to rocket science.

Also useful was Menaechmus's distinction of the two meanings inherent in the term *element*. According to Proclus (410?-485), "he discussed for instance the difference between the broader meaning of the word *element* (in which any proposition leading to another may be said to be an element of it) and the stricter meaning...." This "stricter meaning" is encompassed in the modern term *elemental* (in its everyday, non-mathematical sense), which fits Proclus's definition as "something simple and fundamental standing to consequences drawn from it in the relation of a principle, which is capable of being universally applied and enters into the proof of all manner of propositions." As for Menaechmus, he placed an emphasis on the first meaning, which more clearly has an application to the structure of mathematics.

In his recognition that at least some of the challenges facing thinkers could be explained in terms of language, Menaechmus calls to mind modern thinkers such as Ludwig Wittgenstein (1889-1951). He also rejected the prevailing distinction between "problems" and "theorems," stating that both terms describe problems, the only difference being their purpose and the nature of the challenge they posed.

JUDSON KNIGHT

Nicomachus of Gerasa
c. 60-c. 100
Roman Syrian Mathematician and Philosopher

In a play by the satirist Lucian, one character says to another, "You calculate like Nicomachus." The latter, a philosopher of the neo-Pythagorean school, is best remembered for his *Arithmetike eisagoge,* or *Introduction to Arithmetic,* a highly influential if rather unusual mathematical text.

Gerasa is now the town of Jerash, Jordan, then part of Roman Syria, and it is likely that Nicomachus was ethnically related to the peoples of this area while being thoroughly Romanized in terms of culture and speech. It can be inferred that he studied in a school that espoused the ideas of Pythagoras (c. 580-c. 500 B.C.), because those ideas pervade his *Introduction to Arithmetic.*

In the latter text, Nicomachus examined odd, even, prime, composite, and perfect numbers. He also presented an interesting theorem in which he showed that by adding consecutive odd numbers, successively including one additional number, it was possible to produce a series of the sums of all cubed numbers. Thus $1^3 = 1$; $2^3 = 3 + 5$; $3^3 = 7 + 9 + 11$; and so on.

On the other hand, Nicomachus sometimes made false statements that seemed to be true because the information he furnished supported his original assertion. Hence his statement that all perfect numbers—ones in which the divisors add up to the number itself (e.g., $1 + 2 + 3 = 6$)—end in 6 or 8 alternately. This is not the case, but it seemed to be so, due to the fact that the only perfect numbers he knew were 6, 28, 496, and 8,128.

The book also shows the strange, highly unscientific, side of Pythagorean mathematics, which for instance assigned personalities to numbers. In discussing an abundant number (one in which the sum of its divisors is greater than the number itself), Nicomachus wrote that these called to mind an animal "with ten mouths, or nine lips, and provided with three lines of teeth; or with a hundred arms, or having too many fingers on one of its hands...." The opposite of an abundant number was a deficient number, whose divisors add up to less than the number, and these resembled a creature "with a single eye...one-armed, or one of his hands has less than five fingers...."

Pappus (fl. c. A.D. 320) and other mathematicians of the late ancient world despised Nicomachus's book, and Boethius (c. 480-524) perhaps revealed himself as a true medieval with his admiration of it. He turned it into a school book, and despite—or perhaps because of—its peculiarities, the *Introduction to Arithmetic* became the standard arithmetic text of the Middle Ages. Not until the Crusades (1095-1291), when western Europeans increasingly became exposed to Arab versions of more significant ancient works, was it replaced.

In addition to the *Introduction to Arithmetic,* Nicomachus wrote a book of musical theory, the *Manual of Harmonics.* Here, too, he applied Pythagorean notions, though in this case he was on much firmer ground with the relation of music to mathematics.

JUDSON KNIGHT

Nicomedes
c. 280-c. 210 B.C.
Greek Mathematician

Nicomedes is remembered for developing the conchoid, a special curve he used for solving two of the famous problems that perplexed ancient Greek mathematicians, trisection of the angle and duplication of the cube. Also notable was his lemma, a minor theorem he discovered in the course of working on the latter problem.

Based on the fact that he knew about (and criticized the methodology of) the measurement of Earth's circumference as performed by Eratosthenes (c. 285- c. 205 B.C.), it is possible to date Nicomedes with some accuracy, but nothing else is known of his life, except that he may have come from the Greek city-state of Pergamum in Asia Minor. *On Conchoid Lines,* his most notable—and perhaps only—written work, has been lost.

Portions of it that have survived in the writings of others, however, provide historians with knowledge of the conchoid and lemma. The first of these looks rather like what a modern person would describe as a very flat Bell curve, but to the Greeks it appeared like a sea creature; hence the derivation of the name from *konche,* or mussel shell. Below this curve was a line, and beneath that a point parallel to the apogee of the curve. By determining the length of a segment from the apogee to the lower point, it was possible to find segments of equal length—both of which also intersected the curve—on either side of that one. This in turn yielded a trisected angle, providing a solution of sorts to one of the great problems of antiquity.

The conchoid could also, Nicomedes maintained, be used for finding mean proportionals and thus for solving the Delian problem of doubling the cube. In working out the latter problem, Nicomedes developed his lemma. Using the traditional instruments of straightedge and compass, he began with two lines that bisect and form part of numerous triangles. Eventually this process yielded the lemma, which identified two particular sides of the triangles as equal, and this in turn gave him a way of finding mean proportionals. Centuries later, François Viète (1540-1603) would use the lemma for solving equations of the third and fourth degrees.

JUDSON KNIGHT

Proclus Diadochus
410-485
Byzantine Philosopher

The life and career of Proclus represent both a summing-up of classical Greek mathematics,

and a throwback to a time—already ancient in Proclus's era—when Greek mathematical study was at its zenith. It is thanks to his writing, particularly a commentary on Euclid's *Elements,* that modern scholars know of many long-lost works of antiquity.

Son of a prominent couple named Patricius and Marcella, Proclus grew up in the town of Xanthus on the southern coast of Lycia in Asia Minor (modern Turkey). His father intended him to study law in Alexandria, but on a visit to the capital of the Eastern Roman Empire at Byzantium or Constantinople (now Istanbul), he decided instead to pursue a career as a philosopher.

By that point Proclus had already been pursuing his legal education in Egypt, but he returned to Alexandria with an entirely different aim. Over the years, he studied with Leonas of Isauria, a Sophist; the Egyptian grammarian Orion; the philosopher Olympiodorus the Elder; various instructors in Latin and rhetoric; and a mathematician named Heron, though not the better-known figure who went by that name.

Eventually he decided to go to Athens to pursue the study of neo-Platonic philosophy in the Academy established by Plato (427-347 B.C.) himself. (In the sixth century the Byzantine emperor Justinian closed down the Academy, by then some 900 years old, as a pagan institution.) At the Academy he studied under Plutarch of Athens, founder of the Athenian school of Neoplatonism and director of the Academy, as well as Plutarch's successor Syrianus. When the latter stepped down, he named Proclus head of the Academy, and it was around this time that the scholar acquired the title *Diadochus,* meaning "successor."

In the years that followed, Proclus wrote a variety of works, none more important than his commentary on the *Elements* of Euclid (c. 325-c. 250 B.C.). Euclid's work is valuable in a number of regards, but perhaps most of all for the fact that it constitutes the principal source of information regarding mathematicians and works that would otherwise have long been forgotten. Also of interest to the history of science was *Hypotyposis,* an overview of astronomical theories put forth by Hipparchus (fl. 146-127 B.C.) and Ptolemy (c. 100-170). Works on physics include *Liber de causis* or "Book of causes" and *Elements of Physics,* which discusses ideas put forward by Aristotle (384-322 B.C.).

Proclus also wrote books of poetry and theology. His poetry was quite good, and the seven

of his poems that survive are acclaimed as examples of an accomplished late-classical style. These poems are mostly hymns to the gods, an interesting fact because by his time, Christianity had triumphed throughout both the Eastern and Western empires. Yet Proclus remained unabashedly pagan, and his writings on religion concerned not Jesus Christ, but the gods of Greece and the Orient.

In many regards Proclus seemed to recall the ways of the mathematicians who followed the school of Pythagoras (c. 580-c. 500 B.C.). Like them he was a vegetarian, and he mixed with his scientific study heavy doses of superstition and bizarre beliefs that had nothing to do with science. (For instance, he maintained that Earth is the center of the universe because a Chaldean priest of old had said so.)

He also practiced magic, and though these tendencies might seem a harbinger of the medieval period that followed his time, in Proclus's case they can more accurately be seen as harkening to the Pythagorean past. Greece, after all, never experienced a medieval dark age; instead it simply preserved the advanced learning of the past, without adding a great deal to it, as a dusty relic of antiquity. Much the same was true of Proclus, who is remembered not for his original contributions, but for the manner in which he summed up the mathematical and scientific knowledge acquired during the energetic millennium that preceded his own time.

JUDSON KNIGHT

Pythagoras
569?-475? B.C.
Greek Philosopher and Mathematician

Despite the fact that he is famous for the discovery of the theorem that bears his name, Pythagoras did not view himself primarily as a mathematician; nor did the members of the society he founded, whose principles addressed nonscientific subjects such as reincarnation, or metempsychosis. Yet Pythagoras and his followers, the Pythagoreans, viewed the concept of the number as fundamental to the universe, and with their focus on numerical properties virtually inaugurated the serious study of mathematics in the West.

The son of Mnesarchus, a merchant from Tyre (now in Lebanon), and his wife Pythias, Pythagoras grew up in Samos, Ionia, on what is now the western part of Turkey. His father's profession gave him reason to travel widely, and apparently the boy accompanied him on trips as far away as Italy. During his youth, Pythagoras fell under the influence of several great teachers, most notably Thales (625?-547? B.C.).

Around the year 535 B.C., Pythagoras visited Egypt, where he became interested in various mystical rites while studying at the temple of Diospolis. Following the Persian invasion of Egypt in 525 B.C. he was taken back to Babylon as a prisoner, which placed him in that city alongside the Israelites during the Captivity. Like the Israelites, who adopted the idea of Satan from Zoroastrian scriptures, Pythagoras came under the influence of Zoroastrianism and the much older religion of the Magi.

By 520 B.C., he had made his way back to Samos, where he founded a school based on his emerging mystical worldview. It appears that Samian students were not interested in his rather unusual, Egyptian-influenced teaching style, and in an effort to avoid being forced into a life of public service in his hometown, Pythagoras used this lack of interest as an excuse to move to Italy. In 518 B.C. he settled in Croton, at the eastern tip of the peninsula's boot heel, where he established the Pythagorean society.

His followers, a group that grew steadily after his arrival in Croton, called themselves *mathematikoi.* They believed that mathematics was at the heart of reality, and that symbols possessed a mystical significance that drew the human closer to the divine. One of the world's first secret societies, the Pythagoreans lived communally, practiced vegetarianism, and practiced vows of secrecy and loyalty. Aside from all their other unusual qualities, they stood apart from much of the ancient world in that women were allowed to enjoy full participation in their society, and did so as full intellectual equals of men.

From a mathematical standpoint, one of the most significant contributions made by Pythagoras was his treatment of number as an abstract entity separable from all specifics. Perhaps for the first time, 2 was just two—not two pebbles or two horses or two ships. As for his famous theorem—that the square of a right triangle's hypotenuse is equal to the sum of the squares of its other two sides—it appears that the Babylonians a millennium before him recognized this principle, and that Pythagoras was simply the first to prove the theorem. He and his followers also showed that the sum of the angles in a triangle is equal to two right angles.

Theaetetus of Athens
c. 417-c. 369 B.C.
Greek Mathematician

Believed to have influenced Euclid's work in books X and XI of the *Elements,* Theaetetus studied what Pappus described as "The commensurable and the incommensurable, the rational and irrational continuous quantities"—that is, rational and irrational numbers. Student of Theodorus of Cyrene, he was a friend and associate of both Socrates and the latter's pupil Plato.

In fact Plato (427-347 B.C.), who clearly admired him, is the principal source regarding Theaetetus, who became a central figure in two Platonic dialogues, *Theaetetus* and the *Sophist.* In the first of these, Plato recorded a discussion between Socrates (c. 470-390 B.C.), Theodorus (465-398 B.C.), and Theaetetus that apparently occurred in 399 B.C.

Plato noted that Theaetetus's father was a wealthy man named Euphronius of Sunium who had left his son a large fortune. Trustees of the will had cheated Theaetetus out of most of his wealth, however, but he remained generous, and Plato described him as the essence of a gentleman. The great philosopher also described Theaetetus's mind as a beautiful one, though apparently his outward appearance—he had a flat nose and bug eyes—did not match his inward one.

As is the case with all too many ancient thinkers, all traces of Theaetetus's work have disappeared, but an echo of them is preserved in the most influential geometry text ever written: the *Elements* of Euclid (c. 325-c. 250 B.C.). In his introduction to the latter book, Pappus (fl. c. A.D. 320) noted that Euclid's discussion of irrational numbers—those that continue indefinitely without any repeating pattern, and cannot be expressed as fractions—had its origins in Theaetetus's interpretations of Pythagorean ideas.

Theodorus had first shown that the square roots of nonsquare numbers from 3 to 17 were irrational, but as Plato wrote in *Theaetetus,* it was the student who managed an early generalization of these results. In this version, Theaetetus and Socrates used the term "square" as it is known today, and described numbers with irrational roots as "oblong numbers." (Of course the latter term is meaningless in a geometric sense, because a number with an irrational root can still be represented in theory by a square—with the caveat that the sides are of indefinite or approximate length.)

Pythagoras of Samos. *(Corbis Corporation. Reproduced with permission.)*

The Pythagoreans—and specifically Hippasus of Metapontum (fl. c. 500 B.C.)—are also credited with the discovery of irrational numbers, or infinite decimals with no indefinitely repeating digits. The idea of an irrational, however, went against Pythagorean precepts, which maintained at all things can be expressed in terms of whole numbers, or the ratios of whole numbers. Similar reasoning led to the Pythagorean rejection of concepts such as the infinite and the infinitesimal. These positions highlight the fact that though he and his followers made many mathematical discoveries, Pythagoras—who believed that each number had a "personality"—was at heart a mystic and not a mathematician.

In 508 B.C., a noble named Cylon tried to force his way into the society, and Pythagoras rejected him because he did not regard Cylon as having a pure interest in mathematics for its own sake. Cylon then set out to destroy the society, and Pythagoras fled to the Italian city of Metapontum, where according to some accounts he committed suicide rather than allow Cylon to take over the society he had founded. In later years, the Pythagoreans became a powerful force in southern Italy, so much so that in the mid-fifth century B.C. they came under severe attack from enemies, and had to flee to Thebes and other cities in Greece.

JUDSON KNIGHT

Sometimes credited for the entirety of Euclid's Book X (though it is almost certain the latter is Euclid's own work, a development of ideas put forth by Theaetetus and others), Theaetetus investigated a number of other mathematical questions. These included a theory of proportion; studies of the octahedron and isocahedron; and work on the medial, binomial, and apotome.

In 369 B.C. Theaetetus fought for Athens in a battle with Corinth. He gained honor on the battlefield, but was wounded and contracted dysentery. He died in Athens, having been brought back to his home.

JUDSON KNIGHT

Theodorus of Cyrene
465-398 B.C.
Greek North African Mathematician

Known as much for his pupils as for his other work, Theodorus of Cyrene worked on questions involving the square roots of 3 and 5. This led him to discoveries concerning irrational numbers, or numbers that continue indefinitely without any repeating pattern.

Theodorus was born in Cyrene, now part of Libya, which at that time was a Greek colony. Though he was in Cyrene when he died 67 years later, he must have spent part of his life in Athens, where he studied under Protagoras (c. 485-c. 410 B.C.) and interacted with Socrates (c. 470-390 B.C.). He later taught both Theaetetus of Athens (c. 417-c. 369 B.C.) and Plato (427-347 B.C.), the principal source of information regarding his work.

Plato later wrote that Theodorus showed his students that "the side of a square of three square units and of five square units" was "not commensurable in length with the unit length"—in other words, the square roots of 3 and 5 are irrational numbers. Today it is possible to view the first 1 million digits of both (1.732... and 2.236... respectively) at http://antwrp.gsfc.nasa.gov/htmltest/rjn_dig.html, a website containing calculations by NASA mathematicians Robert Nemiroff and Jerry Bonnell for the square roots of these numbers, as well as 2, 3, 6, 7, 8, and 10.

As for Theodorus, little else is known about him aside from Plato's brief reference to his teachings regarding the irrational roots of 3 and 5. It is possible to infer from the Plato passage, located in his *Theaetetus*, that the irrational nature of $\sqrt{2}$ had already been established. Some claim that Pythagoras (c. 580-c. 500 B.C.) showed that $\sqrt{2}$ was irrational, and certainly Theodorus must have used the Pythagorean theorem to construct lines of length $\sqrt{3}$ or $\sqrt{5}$.

Most interesting, perhaps, is the role Theodorus played in the generalization of the idea of irrational numbers. This is indicated by this sentence in Plato: "The idea occurred to the two of us [Theaetetus and Socrates], seeing that these square roots appeared to be unlimited in multitude, to try to arrive at one collective term by which we could designate all these roots...." Using a method of reduction known at the time, he apparently proved that the roots of nonsquare numbers from 3 to 17 were irrational, and from this probably arrived at a general theorem that the square roots of all nonsquare numbers are irrational.

JUDSON KNIGHT

Theon of Alexandria
c. 335-c. 400
Greek Mathematician and Astronomer

The career of Theon, whose daughter was Hypatia of Alexandria, signifies the twilight of scholarship in antiquity. His commentaries on the work of Euclid and Ptolemy, written for the less adept among his students, indicate the declining quality of the ancient academies. Likewise his writings are largely bereft of original thinking, containing rather a summation of the progress that had taken place in much more intellectually vital eras.

Theon's life can be dated by a number of occurrences, including the solar eclipse he witnessed on June 16, 364, as well as a lunar eclipse on November 25 of that year. He also compiled a list of Roman consuls, which he continued to update until 372. A resident of Alexandria for his entire life, he apparently was a member of the Museum, a scholarly society dedicated to scientific research. The latter ceased to exist during Theon's time or shortly thereafter, but it appears he was spared the pain of seeing his daughter Hypatia (370-415) killed by a fanatical mob reacting to her "pagan" teachings.

Apparently Hypatia assisted him with writing a version of Euclid's (c. 325-c. 250 B.C.) *Elements*. In this regard Theon's work—though his interpretation of Euclid revealed him to be far from the equal of the earlier mathematician—proved crucial to the development of mathematics: because of the massive destruction that attended the fall of the Western Roman Empire, his remained the earliest known version of the *Elements* during the Middle Ages. (Only in the nineteenth century, in fact, was an earlier version discovered.) Theon performed the service of correcting some mistakes on Euclid's part, and of elaborating in areas where he thought Euclid had been too brief; but in some cases he simply misunderstood the original, and in "correcting" it passed on misinformation.

Almost without exception, the contribution of Theon to mathematical knowledge consists not in the brilliance of his own work, but in his service as a chronicler of that written by others. Aside from helping preserve the *Elements*—which during the Middle Ages became the most influential math text ever written—he also provided commentaries on Euclid's *Optics* and *Data*. A useful aspect of the writings of Theon (and of many other ancient writers) are their references to works by others that have long since been lost. In many cases the information they provide constitutes all that is now known regarding those books. However, it is likely that the *Catoptrica,* purportedly a book by Euclid, was in fact a commentary on Euclid by Theon.

He also wrote on the works of Ptolemy (c. 100-170), most notably the *Almagest*—destined to exert as much impact on astronomy in later years as the *Elements* did on geometry—but also the *Handy Tables* and the *Syntaxis*. The latter contains a valuable account of sexagesimal fractions, or fractions in which the denominator is 60, which the Greeks used for finding square roots.

JUDSON KNIGHT

Zeno of Elea
c. 490-c. 430 B.C.
Greek Philosopher

The accomplishments of the Eleatic philosopher Zeno illustrate the law of unintended effect. Setting out to justify the propositions put forth by his teacher Parmenides, namely that change is impossible, Zeno achieved something quite different. Intentionally or not, his famous paradoxes showed both the power and the limitations of logic, and philosophers' interest in the challenges posed by these problems led to the formalization of logic, or dialectic, as a discipline.

Like Parmenides (b. c. 515 B.C.), Zeno came from Elea, a Greek colony in southern Italy. At about the age of 40, in 449 B.C., he accompanied his teacher to Athens, where he met and impressed Socrates (c. 470-390 B.C.) In fact, much of what is known about Zeno's career comes from the writings of Socrates's own pupil, Plato (427-347 B.C.).

Parmenides taught that nonbeing is an impossibility; only being—timeless, changeless, and all of one substance—exists. Of course this account of reality, tempting as it might seem in some regards, is hard to maintain in the face of sensory data that forcefully suggests not only the variety of substance within the world, but also the changing nature of that substance.

Parmenides had attempted to deal with this by maintaining that the only real knowledge resides in the intellect itself, not in the world of experience; but in Zeno's generation it became apparent that the Eleatic school would have to furnish some sort of proof. This Zeno attempted to do with his paradoxes, which, rather than prove the Eleatic position directly, sought to make ridiculous its opponents' propositions regarding the reality of change and motion.

Zeno reportedly wrote the paradoxes—of which Plato claimed there were 40 or more—in a work called the *Epicheiremata*. Both it and the majority of the paradoxes have long since disappeared, though it is likely that the other problems in principle resembled the four that survive.

In one paradox, Zeno referred to an arrow being shot from a bow. At every moment of its flight, it could be said that the arrow was at rest within a space equal to its length. Though it would be some 2,500 years before slow-motion photography, in effect he was asking his listeners to imagine a snapshot of the arrow in flight. If it was at rest in that snapshot, then when did it actually move?

Another paradox involved Achilles, hero of the *Iliad* and "swiftest of mortals," in a footrace against a tortoise. Because he was so much faster than the tortoise, Achilles allowed the creature to start near the finish line—a big mistake, because as Zeno set out to prove, Achilles could then never pass the tortoise. By the time he got to the point where the tortoise started, it would have moved on to another point, and when he got to that second point, it would have moved on to yet another point, and so on. There would be no point at which Achilles could pass the tortoise.

These and Zeno's other two paradoxes, which were similar in concept, failed to prove that motion was impossible, but they did impress philosophers with the importance of logic itself.

Through use of logic, Zeno seemingly created a series of statements that could not possibly be true. Thus was born the scientific study of the dialectic, which has captured the imagination of philosophers and mathematicians ever since.

Advances in the study of calculus by Karl Weierstrasse (1815-1897), and in logic and language by Ludwig Wittgenstein (1889-1951) and his friend Bertrand Russell (1872-1970), helped thinkers unravel the mystery of Zeno's paradoxes and their hidden assumptions. In each problem, Zeno treated either space or time as though they were made up of an infinite number of points—for instance, infinite arrow "snapshots." This is true in the ideal world of geometric theory, where a line does indeed have an infinite number of points; but those points are without extension, whereas a "point" in the real world has some length.

Of course Zeno did not deliberately build in this paradox-within-a-paradox, any more than he intentionally created the revolution in thinking that his work spawned. Undoubtedly, however, he possessed a certain "pugnacity" (Plato's term) in defending his point of view. Later this fighting spirit would put him at odds with the Eleatic tyrant Nearchus, who had him put to death.

JUDSON KNIGHT

Biographical Mentions

Ahmes
c. 1680-c. 1620 B.C.

Egyptian scribe credited with writing down the material on the Rhind Papyrus, named for Alexander Henry Rhind, the Scottish Egyptologist who discovered it in 1858. A secretary rather than author, as he himself stated in the manuscript, Ahmes indicated that he was recording mathematical information from an earlier work, one dated to perhaps 2000 B.C. The Rhind Papyrus contains 87 problems in basic arithmetic, and illustrates simple fractions.

Ammonius Saccus
175-242

Alexandrian Greek philosopher known as the founder of Neoplatonism, which was the culmination of seminal Greek philosophy and a real synthesis of previous Greek metaphysical philosophy. Either his writings are not extant, or he never left any from his reputed extensive teaching. Ammonius's most important disciple in the furthering of Neoplatonic thought was Plotinus, the dominate theorist of the philosophy and the teacher of Porphyry, who simplified Plotinus's Neoplatonic thought, which in turn influenced Macrobius and Augustine.

Ammonius Hermiae
fl. c. 500

Byzantine North African philosopher who served as director of the school in Alexandria. A student of Proclus (410?-485), Ammonius taught Eutocius of Ascalon (c. 480-c. 540), who dedicated his commentary on Archimedes's (c. 287-212 B.C.) *On the Sphere and Cylinder* to him. His own writings primarily concerned logic and the physical sciences, and a number of these works—including several commentaries on Aristotle (384-322 B.C.)—remained in widespread use among European scholars throughout the Middle Ages.

Anaxagoras of Clazomenae
499-428 B.C.

Greek philosopher who made the first known efforts toward squaring the circle, and who provided an early scientific cosmology. According to the Roman architect Vitruvius (first century A.D.), Anaxagoras also wrote a treatise on how to paint objects such that those in the background appeared smaller than those in the foreground—in other words, how to use perspective. Anaxagoras anticipated the scientists of the Enlightenment by some 2,000 years with an account of the Solar System as a material, rather than an ethical, entity; and was the first thinker known to have stated that the Moon's light is a reflection of the Sun's.

Antiphon the Sophist
480-411 B.C.

Greek orator and statesman who first proposed the method of exhaustion for squaring the circle. Antiphon suggested that a regular polygon be inscribed in a circle, and the number of its sides successively doubled until the difference in areas between the polygon and the circle would have been exhausted. His method received criticism from Aristotle (384-322 B.C.) and others, and it appears likely that Antiphon believed a circle to be a polygon with a nearly infinite number of sides; nonetheless, he greatly advanced efforts to equate the area of a circle with that of a square.

Archimedes
c. 287-212 B.C.

Greek mathematician and scientist who, by improving on previous methods of exhaustion in squaring the circle, developed the first reliable figure for π. Archimedes is best known for his numerous discoveries, such as the principle of buoyancy; and for his inventions or improvements, including pulley systems and the catapult. He also worked as an astronomer and physicist, conducting studies of the Solar System and of the principles of gravity and equilibrium. A student of Euclid (c. 325-c. 250 B.C.) and Conon (fl. c. 245 B.C.), Archimedes developed the first mathematical exposition of the principle of composite movements, and was able to calculate square roots by approximation. Thus though he is known primarily for his contributions to physics and technology, Archimedes was also considered one of the great mathematical geniuses of antiquity. His work with curved surfaces anticipated calculus by some 2,000 years.

Aristarchus of Samos
c. 310-c. 230 B.C.

Greek astronomer and mathematician who applied geometric theory to calculate the relative sizes, and distance between, the Moon and Sun. The first scientist to propose a heliocentric, or Sun-centered, model of the universe, Aristarchus has often been regarded solely as an astronomer, but in fact much of his work was in pure mathematics. With regard to his Sun–Moon measurements, he used the angle between the half-illuminated Moon and the Sun to estimate that the Sun is about 20 times as large as the Moon, and about 20 times as distant from Earth. That both estimates are too small is the fault of his measuring instruments rather than of his methodology.

Autolycus of Pitane
c. 360-c. 290 B.C.

Greek astronomer and mathematician who wrote the two earliest surviving mathematical works. *On the Moving Sphere* is a study of spherical geometry with a clear astronomical application, and *On Risings and Settings* a work on astronomical observations. It is possible that the first of these was based on a now-lost geometry textbook by Eudoxus of Cnidus (c. 408-c. 355 B.C.), whose theory of homocentric spheres Autolycus supported.

Brahmagupta
598-670

Indian astronomer and mathematician whose discussions of number theory prefigured ideas now accepted in mathematics. His *Brahmasphutasiddhanta* (628), translated as *The Opening of the Universe,* defined zero as the result obtained when a number is subtracted from itself—by far the best definition of zero up to that time. Brahmagupta also provided rules for "fortunes" and "debts" (positive and negative numbers), and used a place-value system much like that which exists today. In addition, *The Opening of the Universe* offered an algorithm for computing square roots, a method for solving quadratic equations, and rudimentary forms of algebraic notation.

Bryson of Heraclea
b. c. 450 B.C.

Greek mathematician who improved on the method of exhaustion, first introduced by Antiphon the Sophist (480-411 B.C.), for squaring the circle. Bryson's method involved inscribing a square in a circle; circumscribing a larger square; and placing a third square between the inscribed and circumscribed ones. It is not clear exactly how he used these forms, but he seems to have maintained that the circle was greater than all inscribed polygons and less than all circumscribed ones. By increasing the number of sides to the polygons, he appears to have suggested, it would be possible to diminish the difference between them and the circle.

Callippus of Cyzicus
c. 370-c. 310 B.C.

Greek astronomer and mathematician who used mathematical astronomy for a number of purposes, including the creation of a 76-year cycle that aligned the solar and lunar years. Later astronomers used the 940-month calendar of Callippus, who accounted for the inequality of the seasons by developing a model in which the Sun had a variable velocity at different times of the year. His 34-sphere system improved on the system of homocentric spheres proposed by his teacher Eudoxus of Cnidus (c. 408-c. 355 B.C.)

Chang Ts'ang
d. 152 B.C.

Chinese mathematician sometimes credited as author of the *Chiu-chang Suan-shu* or *Jiuzhang Suanshu* (Nine chapters of mathematical art). The latter, the oldest known Chinese mathematical text, contained 246 problems, which as the title suggested were presented in nine chapters. Some four centuries after Chang Ts'ang, Liu Hui (fl. c. 263) wrote a famous commentary on the "Nine Chapters."

Cleomedes
fl. first century A.D.

Greek astronomer whose *On the Circular Motions of the Celestial Bodies* provides valuable information regarding the work of his more distinguished predecessors. *On the Circular Motions* is made up almost entirely of ideas taken from others, most notably Posidonius (135-51 B.C.), and contains a mixture of accurate and wildly inaccurate data. Of particular interest is the fact that the text serves as the principal source regarding the methodology used by Eratosthenes (c. 285-c. 205 B.C.) in making his famous measurement of Earth's circumference.

Democritus
c. 460-c. 370 B.C.

Greek philosopher, best known for his atomic theory, who also contributed to the study of geometry. It was Democritus who first stated that the volume of a cone is one-third that of a cylinder with the same base and height, and that the same relationship exists for a pyramid and prism. Half a century later, Eudoxus of Cnidus (c. 408-c. 355 B.C.) proved this proposition. Among Democritus's mathematical writings were *On Numbers, On Geometry, On Tangencies, On Mappings,* and *On Irrationals.* All these works have been lost.

Dharmakirti
fl. seventh century A.D.

Indian Buddhist philosopher and logician who developed a system of syllogistic reasoning that became highly influential in the East. In his "Seven Treatises," Dharmakirti outlined a precise form of syllogism that, like its Western counterpart, consisted of three parts; however, the purpose of those three parts—and indeed the methodology governing their use—was radically different. Dharmakirti's logic had an enormous impact on thought in Buddhist countries, and though the influence of Buddhism waned in his native India, eventually his principles became an established if seldom acknowledged part of Hindu logic as well.

Dionysodorus
c. 250-c. 190 B.C.

Greek mathematician who, according to Eutocius (c. 480-c. 540), developed a method for solving cubic equations. Using a parabola and a rectangular hyperbola, Dionysodorus constructed an elegant model for cutting a sphere in a given ratio. He also wrote *On the Tore,* a lost work, and appears to have invented a conical sundial.

Domninus of Larissa
c. 420-c. 480

Syrian Jewish philosopher who wrote on number theory and geometry. He studied at the Academy in Athens, where he developed a rivalry with Proclus (410-485), then returned to his home in Syria and wrote several works, of which two are known. The *Manual of Introductory Arithmetic* and *How to Take a Ratio Out of a Ratio* were both translated and published some 1,400 years after Domninus's death. *Introductory Arithmetic* seems a reaction to the quasi-mystical principles put forth by another Syrian writer, Nicomachus of Gerasa (c. 60-c. 100), in a work by the same name three centuries earlier.

Dositheus of Pelusium
fl. c. 230 B.C.

Greek mathematician who succeeded Conon of Samos (c. 280-c. 220 B.C.) as director of the mathematical school at Alexandria. It appears that Dositheus studied under Conon, and after his teacher's death replaced him as head of the school. Archimedes (c. 287-212 B.C.) had earlier carried on a correspondence with Conon, and continued to do so with Dositheus, to whom he dedicated at least four treatises.

Eudemus of Rhodes
c. 350-c. 290 B.C.

Greek philosopher and the first major historian of mathematics. Eudemus studied under, and became close friends with, Aristotle (384-322 B.C.). Later he returned to his native Rhodes, where he wrote a number of works, including *History of Geometry* and *History of Astronomy.* Though these books have been lost, much of what they contained was passed on to other ancient writers, and collectively they constitute a principal source of information on numerous ancient thinkers and their achievements. Also highly significant is the fact that Eudemus kept notes on his great mentor's classes, thus preserving much of Aristotle's teaching.

Eutocius of Ascalon
c. 480-c. 540

Byzantine Middle Eastern scholar whose writings on the history of mathematics constitute valuable sources of information. Eutocius wrote commentaries on Archimedes' (c. 287-212 B.C.) *Measurement of the Circle, On the Sphere and Cylinder,* and *On Plane Equilibria,* as well as the first four books of the *Conics* by Apollonius of Perga (c. 262- c. 190 B.C.). It is thanks to Eutocius that modern historians know about solutions to the problem of duplicating the cube by a number of thinkers, as well as a wide variety of other material that would otherwise have been lost.

Heraclides of Pontus
387-312 B.C.

Greek philosopher mistakenly credited with putting forth a heliocentric, or Sun-centered, model of the universe. Heraclides, who studied under Plato (427-347 B.C.) and Aristotle (384-322 B.C.), maintained that Venus was at times "above" or "below" the Sun—terminology a number of later scholars interpreted as meaning that the planets revolved around it. In fact he seems to have meant "ahead of" or "behind" the Sun, and it appears clear that he believed Earth to be at the center of the universe. He was, however, the first thinker to state that Earth rotates on its axis once daily.

Hero of Alexandria
c. 10-c. 75

Greek physicist, also known as Heron, who in addition to his famous work in mechanics wrote a number of mathematical texts. Among these are *On the Dioptra*, which examines surveying; *Mechanica*, a work for architects; *Metrica*, on measurement; and *Stereometrica*, on the measurement of three-dimensional objects. An example of the mathematical information included in Hero's writing is a formula for finding square roots first developed by the Babylonians almost two millennia before his time.

Hipparchus of Rhodes
190-120 B.C.

Greek astronomer sometimes credited as the father of trigonometry. Hipparchus discovered the precession of the equinoxes, or the shift in direction of Earth's axis of rotation, and produced a calculation of the year's length to within 6.5 minutes of the actual figure. Though his role as father of trigonometry is subject to some dispute, it appears that he laid the foundations for the discipline with his table of chords for solving triangles. He was also the first Greek mathematician to divide the circle into 360 degrees.

Hippasus of Metapontum
fl. c. 500 B.C.

Greek Pythagorean philosopher credited with the discovery of irrational numbers—that is, infinite decimals with no indefinitely repeating digits. Hippasus apparently discovered that the length of an isosceles triangle's shorter side must be expressed as an irrational number if the length of the two equal legs is a whole one. Ac-

cording to one legend, he made this discovery while on board ship with a group of other Pythagoreans, and the idea of an irrational proved so antithetical to Pythagorean views of wholeness and harmony that they threw him overboard. A conflicting tale maintains that other Pythagoreans discovered the secret of irrationals. By revealing this information to outsiders for pay, according to this version, Hippasus violated two rules of the Pythagorean society: the vow of secrecy and the prohibition against profiting from mathematical wisdom.

Hypsicles of Alexandria
c. 190-c. 120 B.C.

Greek mathematician and astronomer who wrote a number of works, including the so-called "Book XIV" of the *Elements*. In this book, often mistakenly included with the original writings of Euclid (c. 325-c. 250 B.C.), Hypsicles improved on Apollonius's (c. 262-c. 190 B.C.) approach to problems involving a dodecahedron and an icosahedron inscribed in the same sphere. Hypsicles is also credited with works on polygonal numbers, regular polyhedra, and arithmetic progressions. The latter appears in his *On the Ascension of Stars*, the first astronomical text to divide the zodiac into 360 degrees.

Iamblichus
fl. c. 320

Syrian philosopher whose work emphasized the mystical aspects of Pythagorean number theory. According to Iamblichus, Pythagoras (c. 580-c. 500 B.C.) himself had discovered "amicable" numbers, pairs in which each is the sum of the other's proper divisors. By expanding on the nonscientific aspects of Pythagorean mathematics—aspects that indeed did go back, in many respects, to ideas of Pythagoras himself—Iamblichus helped perpetuate interest in numerology, magic, and astrology.

Leucippus
480?-420? B.C.

Greek philosopher best known as the founder of atomism, the notion that all matter is composed of indestructible atoms; this concept was further elaborated in the writings of his pupil Democritus of Abderra. It is possible that Leucippus studied under the philosopher Zeno of Elea, whose paradoxes about motion stimulated much Greek physical speculation. The one surviving fragment of his written work expresses a deterministic view of the world with no room for chance events.

Li Ch'un-feng
fl. c. 650

Chinese mathematician noted for his commentaries on ancient mathematical works. Li Ch'unfeng led a group of scholars who in about 650 created commentaries on the "Nine Chapters of Mathematical Art," the "Mathematical Classic of the Gnomon of Chou," and other works. These came to be known collectively as the "Ten Classics of Mathematics," which remained the definitive mathematical text in China for at least four centuries.

Liu Hsin
fl. 6 B.C.-A.D. 23

Chinese scholar and astronomer credited as the first to use decimal fractions. Liu Hsin is also known for a treatise on the calendar and for his calculation of π at 3.1547. About five centuries later another Chinese mathematician, Tsu Ch'ung-chih (Zu Chongzhi; 429-500), established the formula of 355/113, which is very close to the number—3.141592 followed by an infinite number of digits—accepted today.

Manava
fl. c. 750 B.C.

Indian priest and author of one of the *Sulbasutras,* early Hindu mathematical texts. Mathematics in ancient India primarily served the purposes of priestly rites, and Manava's Sulbasutra concerns accurate construction of altars for making sacrifices. Long before Greek mathematicians began trying to square the circle, the Manava Sulbasutra offered information on converting squares or rectangles to circles. Among the various values for π given in the text was the figure of 25/8 or 3.125.

Marinus of Neapolis
c. 450-c. 500

Byzantine philosopher and student of Proclus (410?-485) who wrote commentaries on mathematics and astronomy. Marinus apparently came from Samaria in Judea, and studied in Athens before taking Proclus's place as director of the Academy. His works include an introduction to Euclid's (c. 325-c. 250 B.C.) *Data; Life of Proclus;* and two astronomical texts, one on the Milky Way and the other a correction of Theon's (c. 335-c. 400) rules for direction of parallax in longitude.

Menelaus of Alexandria
70?-130?

Greek mathematician and astronomer who is considered the founder of spherical trigonometry, which was later used by Ptolemy in his *Al-gamest.* From Ptolemy we know that Menelaus made astronomical observations in Rome in A.D. 98, and used these to calculate the rate of movement of the equinox. He may have written on spherical propositions, the weight and distribution of objects, geometry, the triangle, and possibly on mechanics, but only his work on spherical trigonometry, the *Sphaerica,* survives.

Oenopides of Chios
490?-420? B.C.

Greek mathematician and astronomer credited by some sources with discovering that the Sun moves in an oblique circle from east to west (the obliquity of the zodiac). He also calculated the length of the Great Year of the Sun and Moon, the time it takes for them to return to the same relative position in the sky. He speculated, incorrectly, that the cause of the Nile river floods was underground temperature changes, and may have written on geometrical constructions.

Panini
fl. fifth century B.C.

Indian grammarian whose system of rules for the Sanskrit language is regarded as a precursor of modern formal language theory. In his *Astadhyayi,* Panini gave some 4,000 aphoristic rules for the language, which—in part thanks to his systematization—remained largely unchanged for the next two millennia. Panini's linguistic formulae have often been likened to mathematical functions, and it has been suggested that the Hindu number system and mathematical reasoning are linked to the structure of the Sanskrit language.

Pappus of Alexandria
290?-350?

Greek mathematician, astronomer, and geographer whose major work was the *Synagoge* (or Collection). While serving as a handbook of geometry, the *Synagoge* also incorporated the work of earlier mathematicians, and in many cases is the only existing source for these ideas. It also contains influential work on astronomical and projective geometry, and both René Descartes and Isaac Newton used Pappus's work. He also wrote a geography of the world, and may have written on Euclid's *Elements,* music, and water devices.

Parmenides of Elea
c. 515-480 B.C.

Most important of ancient Greek philosophers prior to Socrates. He was the first to address the nature of being and the problem of change. He argued that all existing things do not change inasmuch as all things that exist always have existed, and nothing can become anything other than it is. Although our senses indicate otherwise—in other words, that things change—Parmenides argued that such perceived change was an illusion.

Perseus
fl. second century B.C.

Greek mathematician known chiefly through Proclus's (410?-485) comments on his development of spiric surfaces and sections. A spiric surface, as defined by Proclus, is one in which a circle revolves around a straight line (the axis of revolution) but always remains in the same plane as the axis. Depending on whether the axis cuts the circle, is tangent to it, or is outside the circle, three distinct varieties of spiric surface are possible. (Visually these resemble an oval, a figure 8 with a broad waist, and a figure 8 with a narrow waist.) Proclus compared Perseus's work on spiric sections—formed when a plane parallel to the axis of revolution cuts the spiric surface—to Apollonius's (c. 262-c. 190 B.C.) studies of conics.

Philon of Byzantium
c. 280-c. 220 B.C.

Greek physicist and inventor known for his treatise on mechanics, and for his work on the duplication of the cube. In his *Mechanics* Philon discussed the lever, catapults, and pneumatics. His approach to the Delian problem (cube doubling) was stated in terms of creating a catapult capable of hurling a missile twice as large as the payload of an existing catapult.

Plato
c. 428-c. 348 B.C.

One of the most influential of ancient Greek philosophers and a student of Socrates. He Founded the Academy, a school of higher education that lasted nearly 1,000 years. Aristotle studied there for nearly 20 years. Plato wrote many texts, most notably the *Republic*, wherein the ideal society is described as one led by philosopher-kings, and democracy is rejected as a valid form of government. He argued that reality is divided into a world of senses, of which we only can have incomplete knowledge, and a world of ideas, of which we can have true and complete knowledge. For example, every circle we see contains some imperfection—in other words, it is only an approximation of an ideal circle. We know what a circle is by comprehending the idea, or "form," of the circle.

Porphyry Malchus
233-309

Syrian philosopher noted for his mathematical commentaries. A student and close associate of Plotinus (205-270), the founder of Neoplatonism, Porphyry tried to bring together the ideas of Plato (427-347 B.C.) and Aristotle (384-322 B.C.). His commentary on the latter's *Categories* influenced the development of logic, while his writing on the *Elements* of Euclid (c. 325-c. 250 B.C.) apparently provided Pappus (fl. c. 320) with a principal source. Porphyry also wrote a biography of Pythagoras (c. 580-c. 500 B.C.) Among his students was Iamblichus (fl. c. 320), who took a very different, and less scientific, approach to mathematics than his teacher.

Ptolemy (Claudius Ptolemaeus)
100?-170?

Greek mathematician, astronomer, and geographer whose major work, the *Almagest,* is a large mathematical text that describes the movement of the heavens. A Roman citizen of Greek descent living in Alexandria, Egypt, he was the most influential ancient astronomer. The mathematics used in the *Algamest* is the geometry of Euclid, applied to explain astronomical observations. Ptolemy's ideas dominated astronomy for 14 centuries. He also wrote on other mathematical topics, astrology, and geography.

Severus Sebokht
fl. c. 650

Syrian cleric and scholar who made the first known mention of Hindu numerals outside of India. Sebokht also wrote the earliest known treatise on the astrolabe, as well as works on other aspects of mathematics, astronomy, geography, philosophy, and theology. Of Hindu computations he wrote that these "excel the written word ... and are done with nine symbols."

Serenus
c. 300-c. 360

Greek mathematician known both for his work in geometry, and for his commentaries on the texts of others. Of the former, both *On the Section of a Cylinder* and *On the Section of a Cone* survive. Most notable among his commentaries was one on the *Conics* of Apollonius (c. 262-c. 190 B.C.), which unfortunately has been lost.

Shotoku Taishi
573-621

Japanese prince who greatly influenced the development of his country, and who is traditionally regarded as the father of Japanese arithmetic.

Shotoku authored the "Seventeen-Article Constitution" (604), which provided the governing principles of Japan, and played an important role in encouraging Japanese acceptance of the Buddhist religion. He was present when in c. 600 the Korean Buddhist priest Kanroku presented a set of works on astrology and the calendar to the Japanese empress. Legend has it that as a result of this experience, Shotoku became fascinated with mathematics, and exhibited a talent for performing calculations.

Simplicius
c. 490-c. 560

Byzantine philosopher who wrote commentaries on several early mathematics texts. A student of Ammonius Hermiae (fl. c. 500), Simplicius later studied at the Academy in Athens under Damascius (480-c. 550), who was still serving as director in 529 when the emperor Justinian I (r. 527-565) closed down all pagan schools. Thereafter Simplicius, Damascius, and other scholars spent a brief period as scholars in the court of Persia's Khosrow I (r. 531-579) before returning to Athens. Simplicius's writings include commentaries on Aristotle's (384-322 B.C.) *De caelo* and *Physics,* as well as Euclid's (c. 325-c. 250 B.C.) *Elements.* All these works involve extensive references to the work of various mathematicians, and Simplicius serves as a source regarding long-lost works as well as the men who wrote them.

Sporus of Nicaea
c. 240-c. 300

Greek mathematician and teacher of Pappus (fl. c. A.D. 320), noted for his work on classic problems and his critiques of other mathematicians' approaches. Sporus developed his own solutions to the problems of doubling the cube and squaring the circle, and it was in the latter context that he produced the most notable among his critical comments regarding the work of others. It was his view—and numerous historians since have agreed—that the quadratix of Hippias (c. 460-c. 400 B.C.) required the user to know the radius of the circle in question, thus assuming the knowledge it was supposedly designed to provide. In addition to his work as a mathematician, Sporus wrote on the Sun and the comets.

Thales of Miletus
625?-547? B.C.

Greek natural philosopher, engineer, and mathematician whom later Greek mathematicians credit with bringing Egyptian mathematics to Greece, thereby founding the study of Greek geometry. He is considered the first philosopher to suggest natural, rather than supernatural, causes for events such as earthquakes. He thought water was the basic element that all others were created from. A wealthy engineer and politician, he is said to have successfully predicted an eclipse in 585 B.C., and was possibly the teacher of Anaximander. None of his writings survive.

Theodosius of Bithynia
c. 160-c. 90 B.C.

Greek mathematician and astronomer whose surviving works include the *Sphaerics.* Intended to provide a mathematical grounding for astronomy, the *Sphaerics* expands on the *Elements* of Euclid (c. 325-c. 250 B.C.) with regard to spheres. In *On Habitations,* Theodosius discussed the view of the universe due to Earth's rotation, and in *On Days and Nights* he examined rules governing the length of day and night as a function of location and season.

Theon of Alexandria
335?-405?

Greek mathematician and astronomer who wrote a commentary on Ptolemy's *Algamest* that also provides information on older lost books. A member of the Museum of Alexandria, Theon produced rewritings of texts by earlier authors. His version of Euclid's *Elements* was more popular than the original, but he added errors that were to remain for centuries. His lost works include texts on an astronomical instrument (the astrolabe), omens, and the rising of the Nile. He was the father of Hypatia.

Thymaridas of Paros
c. 400-c. 350 B.C.

Greek mathematician known for his work on Pythagorean number theory. Thymaridas described prime numbers as "rectilinear" because they could only be represented as one-dimensional segments, whereas a figure such as 8 equated to a rectangle of 2 units by 4 units. He also developed a method of solving simultaneous linear equations called the "flower of Thymaridas."

Tsu Ch'ung-chih (Zu Chongzhi)
429-500

Chinese mathematician and astronomer who developed a remarkably accurate calculation of π. Tsu's figure of 355/113 or 3.1415926 is correct to six decimal places, and remained the most reliable approximation for many centuries. He wrote a mathematical text, now lost, with his son, and in 463 created a calendar that was never used.

Wang Hs'iao-t'ung
fl. c. 625

Chinese mathematician whose *Ch'i-ku Suan-ching* includes the first known use of cubic equations in a Chinese text. The work offers 20 problems involving mensuration, but does not provide any rule for solving cubic equations. Foremost mathematician in China during the seventh century, Wang Hs'iao-t'ung was regarded as an expert on the calendar.

Won-wang
1182-1135 B.C.

Chinese mystic and author of the *I-king* or *I-ching*, which includes some mathematical information. The book discusses the magic square, a matrix in which the sum of all figures added along any straight line is the same. The magic square, which the text attributes to the semi-legendary Emperor Yu (c. 2200 B.C.), is in turn related to the faces of a die.

Xenocrates of Chalcedon
396-314 B.C.

Greek mathematician and philosopher who wrote 70 works, of which only the titles remain. A student of Plato, Xenocrates was head of the Academy from 339 until his death. He tried to organize and popularize Plato's ideas, but understood them differently than others at the Academy. He wrote two mathematical works, *On Numbers* and *The Theory of Numbers,* and supposedly calculated the total number of syllables that can be made from the letters of the Greek alphabet at 1,002,000,000,000.

Zeno of Sidon
c. 150-c. 70 B.C.

Greek Near Eastern philosopher noted for his penetrating critique of Euclid (c. 325-c. 250 B.C.). A thinker of the Epicurean school, Zeno was predisposed against mathematics and the sciences, but unlike Epicurus (341-270 B.C.) himself, he made his attack from a position of knowledge. It was Zeno's position that geometry contained hidden assumptions, and though his work gained little attention at the time, modern scholars have tended to regard him as a brilliant critic of mathematical reasoning.

Zenodorous
200?-140? B.C.

Greek mathematician who wrote about plane and solid figures of equal perimeter or surface, but with different areas and volumes (isometric figures). While his texts are lost, it is known from other writers that he proved the sphere to be the solid figure with the greatest surface area for a given volume. He is also said to have developed the burning mirror, a device that focuses the sun's rays.

Zhang Heng
78-139

Chinese mathematician and astronomer who developed the world's first seismoscope, as well as one of the earliest rotating globes. His seismoscope, an ingenious invention that he unveiled in 132, consisted of a cylinder surrounded by dragons with frogs beneath. When the earth shook, a ball would drop from a dragon's mouth into that of a frog, making a noise. Zhang Heng also produced a corrected calendar in 123, and discussed cosmology in several texts. He provided a figure of $\sqrt{10}$ for π, which at 3.162 was only correct to one decimal place.

Bibliography of Primary Sources

Apollonius of Perga. *Conics* (c. 200 B.C.). This work consisted of 8 books with some 400 theorems. In this great treatise, he set forth a new method for subdividing a cone to produce circles, and discussed ellipses, parabolas, and hyperbolas—shapes he was the first to identify and name. In place of the concentric spheres used by Eudoxus, Apollonius presented epicircles, epicycles, and eccentrics, concepts that later influenced Ptolemy's cosmology. Even more significant was his departure from the Pythagorean tendency to avoid infinites and infinitesimals: by opening up mathematicians' minds to these extremes, Apollonius helped make possible the development of the infinitesimal calculus two millennia later. The most important factor in this monumental work, however, was not any one problem, but Apollonius's overall approach, which opened mathematicians' minds to the idea of deriving conic sections by approaching the cone from a variety of angles. By applying the *latus transversum* and *latus erectum,* lines perpendicular and intersecting, Apollonius prefigured the coordinate system later applied in analytic geometry.

Archimedes. *On the Equilibrium of Planes* (c. 240 B.C.). Here Archimedes considered the mechanics of levers and the importance of the center of gravity in balancing equal weights.

Archimedes. *On the Sphere and Cylinder* (c. 240 B.C.). In this work Archimedes built on the previous work of Euclid to reach conclusions about spheres, cones, and cylinders. As described in *The Scientific 100* (Simmons 1996): "He showed that if these figures have the same base and height—imagine a cone inscribed in a hemisphere which itself is inscribed within a cylinder—the ratio of their volumes will be 1:2:3. In addition, the surface of the sphere is equivalent to two-thirds of the surface of the cylinder which encloses it." Archimedes was immensely pleased with this discovery, even requesting that his family have a sphere and cylinder engraved on his tombstone.

Archimedes. *On Floating Bodies* (c. 240 B.C.). Archimedes used *On Floating Bodies* to recount his theory regarding water displacement and help found the science of hydrostatics. In this book, he demonstrated that when an object of any shape and weight is floated in water, its vertical, buoyant force is equal to the weight of the water it displaces. One legend of Archimedes holds that he first understood this connection between the weight of a floating object and the resulting increase in water level while watching bath water rise as he sunk his body into a tub. He was said to have been so excited by this insight that he jumped from the tub and ran stark naked through the streets proclaiming his discovery.

Aristaeus the Elder. *Five Books Concerning Solid Loci* (c. 390 B.C.). The curves, lines, and points of cones were the subject matter of this book, which Euclid later credited as the source for much of his own writing on conics in Book XIII of his *Elements*.

Aristotle. *Organon* (c. 340 B.C.). A series of works, including the *Prior* and *Posterior Analytics,* in which Aristotle established many of the fundamental rules of logic, which were applied to philosophical inquiry, mathematics, and many other branches of science. In particular, he delineated the rules of deductive argument and developed symbolic notation to express such arguments.

Aryabhata. *Aryabhatiya* (A.D. 499). Bringing together teachings from ancient Greek and Indian astronomers, as well as new ideas from Aryabhata himself, this work developed various rules for arithmetic and trigonometric calculations. It also contained a number of important "firsts" or near-firsts, including one of the first recorded uses of algebra. Furthermore, it was one of the first texts to include the idea of number position or place value (i.e., tens, hundreds, thousands, etc.). These concepts would have enormous impact as they moved westward, as would another idea implemented by Aryabhata in his text: the Hindu numeral system.

Autolycus of Pitane. *On the Moving Sphere* and *On Risings and Settings* (c. 310 B.C.). The two earliest surviving mathematical works. *On the Moving Sphere* is a study of spherical geometry with a clear astronomical application, and *On Risings and Settings* a work on astronomical observations. It is possible that the first of these was based on a now-lost geometry textbook by Eudoxus of Cnidus, whose theory of homocentric spheres Autolycus supported.

Boethius. *Arithmetic* (c. A.D. 520). This work would later become medieval scholars' principal source regarding Pythagorean number theory.

Brahmagupta. *Brahmasphutasiddhanta* (The opening of the universe, A.D. 628). This work defined zero as the result obtained when a number is subtracted from itself—by far the best definition of zero up to that time. Brahmagupta also provided rules for "fortunes" and "debts" (positive and negative numbers), and used a place-value system much like that which exists today. In addition, the work offered an algorithm for computing square roots, a method for solv-

ing quadratic equations, and rudimentary forms of algebraic notation.

Chang Ts'ang. *Chiu-chang Suan-shu* or *Jiuzhang Suanshu* (Nine chapters of mathematical art, c. 160 B.C.). Chang's authorship of this work is in not certain. In any case, the work, the oldest known Chinese mathematical text, contained 246 problems, which as the title suggested were presented in nine chapters. Some four centuries after Chang Ts'ang, Liu Hui wrote a famous commentary on the "Nine Chapters."

Cleomedes. *On the Circular Motions of the Celestial Bodies* (first century A.D.). This work provides valuable information regarding the work of Cleomedes's more distinguished predecessors. *On the Circular Motions* is made up almost entirely of ideas taken from others, most notably Posidonius, and contains a mixture of accurate and wildly inaccurate data. Of particular interest is the fact that the text serves as the principal source regarding the methodology used by Eratosthenes in making his famous measurement of Earth's circumference.

Dharmakirti. *Seven Treatises* (seventh century A.D.). Presented a system of syllogistic reasoning that became highly influential in the East. In this work, Dharmakirti outlined a precise form of syllogism that, like its Western counterpart, consisted of three parts; however, the purpose of those three parts—and indeed the methodology governing their use—was radically different.

Diocles. *On Burning Mirrors* (c. first century B.C.). Diocles is remembered almost entirely for this fragmentary manuscript. In it, he discussed not only the physical problem referred to in the title, but such subjects as cutting a sphere with a plane, as well as the famous Delian problem of doubling the cube. *On Burning Mirrors* may actually have been a collection of three separate short works, combined under a single title that does not reflect the whole. In any case, the book consisted of 16 geometric propositions, most of which involved conics.

Diophantus of Alexandria. *Arithmetica.* (c. A.D. 250). Considered one of the greatest mathematics works of its era and the first coherent work in number theory in the history of mathematics. It apparently consisted of 13 volumes, though only 6 have survived, in which Diophantus describes several propositions regarding number theory, including one that inspired Pierre de Fermat to pose his famous "Last Theorem." Though earlier mathematicians, including the Egyptians and Babylonians, had explored many of the number theory problems posed by Diophantus, they had never before been assembled together in a single work. It was this compilation of problems in number theory that helped launch it as an independent segment of mathematics.

Eratosthenes. *Geography*. A lost work in which Eratosthenes published his theories and calculations. The title reflects the first-known use of the term *geography,* which means "writing about the Earth." Although his calculations were disputed in his own time, they allowed the development of maps and globes that remained among the most accurate produced for over a thousand years. This, in turn, sparked interest in geography and geodesy, and emboldened regional seafaring exploration using only the most primitive navigational instruments. Eratosthenes's work, moreover, helped solidify belief in a round Earth, and promoted an early theory that the relative warmth or coolness of a locations climate was determined by its distance from the equator. *Geography* also supported the concept of antipodes—undiscovered lands and peoples on the "other side" of the world.

Eratosthenes. *On the Measurement of the Earth* (c. 225 B.C.). Now lost, this work marked the foundation of geodesy, the branch of mathematics that deals with determination of Earth's size and shape, and the location of points on its surface. Among the topics of study within geodesy is the system of latitude and longitude, which Eratosthenes seems to have pioneered in his maps, the most accurate in the world at the time.

Euclid. *Elements* (c. 300 B.C.). Considered the bible of geometry for 2,000 years, this work remains the most influential textbook in history, and one of the essential works of human civilization. The book is primarily a summation of mathematical knowledge passed down from Pythagoras onward, and its genius lies in its cogent explanation of basic principles, as well as its clear and thorough explication of geometric proofs. Consisting of 13 books in which Euclid elaborated on some 450 propositions, the *Elements* begins with a definition of points, lines, planes, angles, circles, triangles, quadrilaterals, and parallel lines. In Book II, Euclid addressed rectangles and squares; in Book III, circles; and in Book IV polygons. He continued with a discussion of proportion and area (Book V), followed by an application of this theory to plane geometry (Book VI). Book VII covers arithmetic, Book VIII irrational numbers, and Book IX rational numbers, while the remainder of the volume is devoted to three-dimensional, or solid, geometry. Euclid's five postulates are among the most important aspects of his work; the first three of these focus on construction with the straight edge and circle or compass, the only tools of Euclidean geometry, while the fourth states that all right angles are equal. Most controversial, however, was the fifth postulate, which discussed the relationship between two straight lines placed side by side.

Eudemus of Rhodes. *History of Geometry* and *History of Astronomy* (fourth century B.C.). Though these books have been lost, much of what they contained was passed on to other ancient writers, and collectively they constitute a principal source of information on numerous ancient thinkers and their achievements.

Eudoxus of Cnidus. *On Speeds* (c. 375 B.C.). Here the author presented a new theory of the motion made by the Sun, Moon, and planets. Given the spherical shape of Earth, Eudoxus imagined a series of concentric spheres around it, and eventually developed a description of 27 spheres necessary for picturing the movement of all known bodies.

Geminus. *Theory of Mathematics* (c. 100 B.C.). Now lost, the work presented an overview of geometry. The latter was by then a long-established discipline among the Greeks, and this gave Geminus a certain perspective that would have been beyond the reach of his predecessors. Thus he undertook to define mathematics as a whole, and to classifying it within the context of the sciences. It constituted a valuable early attempt to give shape to the mathematical discipline, and to place it within the context of scientific study.

Hero of Alexandria. *Metrica* (c. first century A.D.). An important work on geometry that was lost and not rediscovered until 1896. It contains formulas to compute the areas of things like triangles, cones, and pyramids. The area of the triangle is often attributed to Hero, but it is likely he borrowed it from Archimedes or the Babylonians.

Hippocrates of Chios. *Elements of Geometry* (c. 460 B.C.). Now lost, this was a mathematical textbook. The first work of its kind, it would have an enormous impact on another book of a similar title, the highly influential *Elements* of Euclid. In his work, known through the writings of Aristotle, Proclus, and others, Hippocrates became the first mathematician to adopt scientifically precise and logical methodology for developing geometrical theorems from axioms and postulates. It is also likely that *Elements of Geometry* contained the first written explanation of Pythagorean principles, since the Pythagoreans who preceded him did not believe in committing their ideas to writing.

Hypsicles of Alexandria. *Elements, Book XIV* (c. 150 B.C.). In "Book XIV," often mistakenly included with the original writings of Euclid, Hypsicles improved on Apollonius's approach to problems involving a dodecahedron and an icosahedron inscribed in the same sphere. Hypsicles is also credited with works on polygonal numbers, regular polyhedra, and arithmetic progressions. The latter appears in his *On the Ascension of Stars,* the first astronomical text to divide the Zodiac into 360 degrees.

Li Ch'un-feng. *Ten Classics of Mathematics* (c. A.D. 650). Li Ch'un-feng led a group of scholars who created commentaries on the "Nine Chapters of Mathematical Art," the "Mathematical Classic of the Gnomon of Chou," and other works. These came to be known collectively as the *Ten Classics of Mathematics,* which remained the definitive mathematical text in China for at least four centuries.

Liu Hui. *Commentary on "Nine Chapters of Mathematical Art"* (c. A.D. 263). The "Nine Chapters" was a text of unknown authorship dating back to the first century B.C. The oldest known Chinese mathematical text, it contained 246 problems, which as the title suggested were presented in nine chapters. The first of these concerned arithmetic and the fundamentals of geometry, and included a discussion of the counting rods.

Manava. *Sulbasutras* (c. 750 B.C.). These were early Hindu mathematical texts, of which Manava was the author of one. Mathematics in ancient India primarily served the purposes of priestly rites, and Manava's *Sulbasutra* concerns accurate construction of altars for making sacrifices. Long before Greek mathematicians began trying to square the circle, the Manava *Sulbasutra* offered information on converting squares or rectangles to circles. Among the various values for π given in the text was the figure of 25/8 or 3.125.

Nicomachus. *Arithmetike eisagoge* (Introduction to arithmetic, c. A.D. 100). A highly influential if rather unusual mathematical text. Here, Nicomachus examined odd, even, prime, composite, and perfect numbers. He also presented an interesting theorem in which he showed that by adding consecutive odd numbers, successively including one additional number, it was possible to produce a series of the sums of all cubed numbers. The book also shows the strange, highly unscientific, side of Pythagorean mathematics, which for instance assigned personalities to numbers. Pappus and other mathematicians of the late ancient world despised Nicomachus's book, while Boethius perhaps revealed himself as a true medieval with his admiration of it. He turned it into a schoolbook, and despite—or perhaps because of—its peculiarities, the work became the standard arithmetic text of the Middle Ages. Not until the Crusades (1095-1291), when western Europeans increasingly became exposed to Arab versions of more significant ancient works, was it replaced.

Nicomedes. *On Conchoid Lines* (c. 250 B.C.). This work, the author's most notable—and perhaps only—written work, has been lost. Portions of it that have survived in the writings of others, however, provide historians with knowledge of the conchoid and lemma. The first of these looks rather like what a modern person would describe as a very flat Bell curve, but to the Greeks it appeared like a sea creature; hence the derivation of the name from *konche,* or mussel shell. Below this curve was a line, and beneath that a point parallel to the apogee of the curve. By determining the length of a segment from the apogee to the lower point, it was possible to find segments of equal length—both of which also intersected the curve—on either side of that one. This in turn yielded a trisected angle, providing a solution of sorts to one of the great problems of antiquity. The lemma was a minor theorem he discovered in the course of working on the duplication of the cube problem.

Pappus of Alexandria. *Synagoge* (Collection, fourth century A.D.). While serving as a handbook of geometry, the *Synagoge* also incorporated the work of earlier mathematicians, and in many cases is the only existing source for these ideas. It also contains influential work on astronomical and projective geometry, and both René Descartes and Isaac Newton used Pappus's work.

Panini. *Astadhyayi* (c. 400 B.C.). In this work, Panini gave some 4,000 aphoristic rules for Sanskrit, which—in part thanks to his systematization—remained largely unchanged for the next two millennia. Panini's linguistic formulae have often been likened to mathematical functions, and it has been suggested that the Hindu number system and mathematical reasoning are linked to the structure of the Sanskrit language.

Plato. *The Republic* (c. 380 B.C.). Here Plato wrote that plane and solid geometry were two of the five subjects necessary to the education of the philosopher-king. Mastering geometry trained the student in logical argument and taught him to search for and love truth, which Plato represented with eternal and ideal geometric forms. In other words, knowledge of geometry and logic became prerequisite to the study of philosophy. Plato also recorded the research efforts of Theodorus and Theatetus to geometrically demonstrate the existence of the irrational numbers between $\sqrt{3}$ and $\sqrt{17}$.

Theodosius of Bithynia. *Sphaerics* (second century B.C.). Intended to provide a mathematical grounding for astronomy, the *Sphaerics* expands on the *Elements* of Euclid with regard to spheres.

Wang Hs'iao-t'ung. *Ch'i-ku Suan-ching* (c. A.D. 625). Includes the first known use of cubic equations in a Chinese text. The work offers 20 problems involving mensuration, but does not provide any rule for solving cubic equations. Foremost mathematician in China during the seventh century, Wang Hs'iao-t'ung was regarded as an expert on the calendar.

Won-wang. *I-ching* (c. 1150 B.C.). This work includes some mathematical information. The book discusses the magic square, a matrix in which the sum of all figures added along any straight line is the same. The magic square, which the text attributes to the semi-legendary Emperor Yu, is in turn related to the faces of the dice.

Zeno of Elea. *Epicheiremata* (c. 450 B.C.). Apparently contained the author's famous paradoxes, of which Plato claimed there were 40 or more. Both it and the majority of the paradoxes have long since disappeared, though it is likely that the other problems in principle resembled the four that survive. These four failed in their attempt to prove that motion was impossible, but they did impress philosophers with the importance of logic itself. Through use of logic, Zeno seemingly created a series of statements that could not possibly be true. Thus was born the scientific study of the dialectic, which has captured the imagination of philosophers and mathematicians ever since.

NEIL SCHLAGER

Physical Sciences

Chronology

c. 600 B.C. Thales of Miletus originates both Western philosophy and physics with a treatise in which he postulates that water is the foundational substance of the universe.

570 B.C. Greek philosopher Anaximander recognizes that the heavens revolve around the Pole star, that the sky is a "sphere" rather than an arch over Earth, and that space is three-dimensional.

c. 450 B.C. Leucippus, a Greek philosopher, first states the rule of causality—i.e, that every event has a natural cause.

c. 450 B.C. The Greek philosopher Philolaus is the first to state that Earth moves through space.

c. 425 B.C. Democritus, a Greek philosopher and student of Leucippus, states that all matter consists of tiny, indivisible particles called atoms; it will be some 2,300 years before scientific knowledge catches up to him.

c. 350 B.C. Aristotle states that Earth is constantly changing, and that erosion and silting cause major changes in its physical geography; he also provides observational proof that it is not flat.

c. 300 B.C. Greek physicist Strato is the first to argue that in falling, a body accelerates.

c. 300 B.C. Greek explorer Pytheas is the first to scientifically observe tides, and suggests that they are influenced by the Moon.

c. 260 B.C. Aristarchus, a Greek astronomer, states that the Sun and not Earth is the center of the universe, and that the planets revolve around it; unfortunately, Ptolemy will later reject this heliocentric view in favor of a geocentric universe, a notion only refuted by Copernicus in the 1500s.

c. 240 B.C. Eratosthenes, a Greek astronomer and librarian of Alexandria, makes a remarkably accurate measurement of Earth's size, calculating its circumference at about 25,000 miles.

c. 220 B.C. Archimedes discovers the principle of buoyancy, noting that when an object is placed in water, it loses exactly as much weight as the weight of the water it has displaced.

c. 150 B.C. Greek astronomer Hipparchus, adapting an idea originated some 150 years before by Dicaearchus, develops a system of latitude and longitude; he also creates the first star catalogue, is the first to note the precession of the equinoxes, and accurately calculates the length of a year.

c. A.D. 425 The Byzantine historian Zosimus first notes the electrolytic separation of metals.

A.D. 517 Johannes Philoponus, a Byzantine philosopher, offers a theory of motion which prefigures Newton by stating that a body will continue moving in the absence of friction or opposition from another body.

Overview:
Physical Sciences 2000 B.C. to A.D. 699

When did the study of physical science begin? The Sumerians, who created the first civilization in Mesopotamia around 3200 B.C., were ruled by a relatively complex government, yet considered the natural world to be ruled by a variety of gods. By 2000 B.C., however, Mesopotamia had become Babylonia, and mathematics and astronomy became legitimate fields of study. Egypt, influenced by Babylonian thought, had similar levels of scientific sophistication at the same point in history. Records from India indicate complex astronomical thought before 1500 B.C. In China, long overlooked by historians of science, artifacts from between 1600 and 1400 B.C. indicate observations of comets, nova, and star positions, as well as an elaborate system of mathematics.

Astronomy had its crude beginnings as astrology, when pagan religions sought to interpret the movement of the stars and planets and determine the influences of the heavens on human events. It began in ancient Mesopotamia, and eventually spread to Egypt, Greece, India, and the Orient. This required a keen observation of the stars and planets, and served as a springboard for the science of astronomy and the development of calendars.

Early Greek Philosophy

Eventually astronomy split from astrology, marking the birth of physical science. This was precipitated by the Greeks, beginning with Thales (c. 640-546 B.C.). Later philosophers such as Eudoxus (c. 400-347 B.C.), Apollonius of Perga (c. 240-170 B.C.), and Hipparchus (c. 130 B.C.), studied astronomy by observing the heavens, predicting events, and verifying the results. Two Greek mathematicians, Heraclides of Pontus (387-312 B.C.) and Aristarchus of Samos (c. 310-230 B.C.) proposed a heliocentric universe, with planets revolving around the sun. Greek thought also divided the physical world into two realms, the superior celestial and inferior terrestrial. This division of nature also divided physics into two branches: earth science and astronomy.

This division of nature was further elaborated by Pythagoras (d. c. 497 B.C.) and Plato (c. 427-347 B.C.), who saw perfection in the circular cycles of celestial motion. Terrestrial motion on the other hand, was viewed as rectilinear, im-

perfect, and corruptible, as was everything in that sphere. Aristotle (384-322 B.C.), one of the greatest ancient Greek philosophers, believed that the cosmos was divided into 55 concentric spheres, with Earth at the center.

Aristotle also defined, arranged, and compartmentalized the natural sciences in four separate branches: the celestial (*De caelo et mundo*), the terrestrial, the earth sciences, and chemistry (*Meteorologica*), physics (*De physica*), and organic nature (*De generatione et corruptione*). Like Empedocles (c. 492-c. 432 B.C.), he believed that the four elements of fire, earth, air, and water were the basis of all matter, rejecting Leucippus's (fifth century B.C.) and Democritus's (c. 470-380 B.C.) theory that atoms (from the Greek *atomos,* meaning "indestructible") were the building blocks of nature.

Theophrastus (c. 372-287 B.C.), a Greek philosopher and devoted student of Aristotle, succeeded his teacher as head of the Lyceum in Athens. Although much of his work has been lost, several of his works have survived, either in whole or in part, including his *History of Physics* and nine treatises on physical science, such as *On Stone, On Fire,* and *On Wind.*

Greek cosmology reached its apogee in Ptolemy's (c. 100-170) *He mathematike syntaxis* (The mathematical compilation), or *Almagest* as it came to be known over the centuries. Ptolemy expanded Aristotle's vision, positing that the planets traveled in "epicycles," circular orbits that followed that paths of the spheres; sometimes the epicycles themselves were thought to follow epicycles.

Like the other earth sciences, geodetics (the science of calculating the size and shape of the earth and determining the position of points on its surface) and geography began with the Greeks. Both Plato and Aristotle espoused a spherical earth. Eratosthenes (c. 276-c. 196 B.C.), the first to calculate the circumference of the earth, also wrote a systematic treatise on geography. Ptolemy's second great work, the *Geography,* contained contemporary maps of the ancient world and questioned the possibility of life at the equator and in the southern hemisphere (the antipodes), an idea that was not universally held at the time.

Terrestrial physics were less studied than celestial mechanics, but two areas did receive considerable attention: optics (or perspective) and statics. Euclid (c. 300 B.C.), Aristotle, and Archimedes (c. 287-212 B.C.) all wrote on optics; Archimedes also laid the foundations of hydrostatics, an event immortalized by his apocryphal revelation in the bathtub.

Roman and Medieval Science

As the centuries progressed, Greek science was filtered through Roman practicality of thought, which combined incomplete Greek science with curiosity about nature. Among the more important works of this type are the writings of the poet Marcus Manilius (fl. 40 B.C. to A.D. 20), Ambrosius Theodosius Macrobius (395-423), Martianus Capella (c. 410-c. 429), and Gaius Julius Hyginus, curator of the Palatine Library in Rome (c. 64 B.C.-A.D. 17). A more general science compilation, *Noctes Atticae,* written by Aulus Gellius (c. 123-c. 170) contains fragments from many lost works. Several original works on physical science came from the notable Romans Lucius Annaeus Seneca (c. 4 B.C.-A.D. 65), who wrote *Natural Questions,* and Pliny the Elder (23-79), author of *Natural History,* the first encyclopedic sourcebook of ancient science, which was very influential up through the sixteenth century.

The rise of Christianity in the fourth century heightened suspicions about Greek science and its ties to pagan astrology. By the end of the fifth century, knowledge of the Greek language and science faded away with the Roman Empire. But Christian Neoplatonists provided something of a stopgap in this situation. Representing the last vestiges of seminal Greek philosophy, they borrowed from the spectrum of Greek philosophy, focusing on Aristotle's method and logic. Both Christian and non-Christian schools developed in the second century A.D.. Neoplatonism lent itself well to Christian theology, a preference that would continue through the seventeenth century.

The first important Neoplatonic disciple in the Christian integration of ancient science with Christian theology was a theologian and biblical scholar named Origen (c. 185-c. 254), who was well acquainted with current Greek astronomical knowledge, including precession of the equinoxes and other Greek science. Around 300, a Greek scholar named Chalcidius translated the Platonic dialogue *Timaeus* into Latin; it would be the only one of its kind for the next 800 years. Saint Augustine (354-430), who defended empirical science against biblical articles of literal faith, was a profound Neoplatonic influence during the Middle Ages, paving the way for medieval church acceptance of Greek science, particularly Aristotle.

A few other early century thinkers kept ancient Greek works from disappearing, translating scientific works and compiling encyclopedic commentaries on them. One of the most important of these was the Christian Neoplatonist Anicius Manlius Severinus Boethius (c. 480-524), who attempted to translate all of Aristotle's and Plato's works into Latin. Boethius was followed by Cassiodorus (480-575) and Isidore of Seville (560-636), a monk and archbishop, respecitvely. These men were encyclopedic compilers who preserved many ancient texts at a time when civilization was collapsing around them. Sixth-century Neoplatonists such as Ammonius (fl. 522), Johannes Philoponus (c. 490-570), and Olympiodorus (fl. 530) wrote commentaries on Aristotle's meteorology, astronomy, and physics, displaying their acquaintance with Greek observational sophistication, and adding some much-needed common sense to their commentary and criticism.

Near and Far Eastern Science

Scientific thought appeared much later in the East, around the first century A.D., but fortunately was free of the upheaval that disrupted progress in the West. Chinese science, furthermore, exhibited an orderly regimentation very different from the disputational Greek tradition. An important scientist from this period is Zhang Heng (or Chang Heng or Hong, 78-139) who plotted stars and comets, and constructed what may be the first seismographic instrument. Another important personality is astronomer Tsai Yung (c. 190) who constructed the Chinese calendar. In India astronomy began to flourish around 300 B.C. Hindu astronomers included Aryabhata (476-550) who taught the rotation of the earth, Varahamihira (505-587) who gathered a compendium of Egyptian, Greek, Roman, and Indian astronomy, and Brahmagupta (598-668) who developed algebraic methods for calculating motions and conjunctions of planets and eclipses of the sun and moon.

During nearly three thousand years of development, both West and East, in seminal concepts of physical nature ultimately provided the foundation for all physical science to follow. This was a rich legacy which to this day is marveled over and studied as an epochal continuum of intellectual endeavor unique in the annals of civilization.

WILLIAM MCPEAK

Contributions of the Pre-Socratics

Overview

What we call philosophy began with the *pre-Socratics*, Greek thinkers from 600 to 400 B.C. who preceded Socrates (469-399 B.C.) and who speculated about the origins of things and the order of the universe. Building on and beyond the practical knowledge gained by their Egyptian and Babylonian neighbors to the east, these philosophers rejected the predominate myth of a cosmos ruled by gods and demons in favor of a rational one governed by universal and discoverable laws. The term pre-Socratic is a little misleading, since Socrates was alive until almost 400 B.C. and around to debate pre-Socratic philosophy, which he considered inferior to questions of politics and individual morality. So pre-Socratic is used not in a literal sense, but loosely to indicate a certain way of looking at things. Pre-Socratic philosophers did not themselves leave a written record. What we know of them we know through the writings of philosophers like Aristotle (384-322 B.C.), who followed them.

Background

Thales of Miletus (c. 636-c. 546 B.C.) is the first recorded Western philosopher. His questions regarding the origin and nature of the universe inspired others to think along similar lines. This group was known as the Milesian school, after the prosperous center along the Asiatic coast where they lived. The key question for the pre-Socratics was: what is the universe made of? Thales proposed that a single element lay behind the diversity of nature, and according to him, that element was water. He believed that the Earth was a flat disc floating in an infinite sea, which makes sense given how essential water is to life. An equally important question was the nature of motion. Thales believed that soul was the cause of motion, animating the entire universe, which led him to proclaim that "everything is full of gods."

A contemporary of Thales named Anaximander (c. 611-c. 547 B.C.) defined the primary substance as a formless mass called *apeiron*, from the Greek for "infinite." The basis of his reasoning was that water was already too differentiated (in other words, too specialized) to be able to separate out into all the kinds of things there are in the world. Anaximander further believed that a natural law exerts itself in the world, maintaining a balance between different elements. He

suggested that life arose from water, and that man developed from fish. Anaximenes (c. 570-c. 500 B.C.) was a pupil of Anaximander who believed that air was the fundamental element of the universe. Different objects were simply different degrees of density—for example, condensation or rarefaction—of the one basic element.

Pythagoras (c. 582-c. 507 B.C.) migrated from the Aegean island of Samos to southern Italy in 529 B.C., where he founded a mystical brotherhood. His followers were called Pythagoreans. It was Pythagoras who coined the word *philosophy*. The Pythagoreans taught that souls could be transmigrated to animals and even plants. They also taught that numbers constituted the true nature of things, and consequently that all relationships could be expressed numerically. Pythagorean doctrine was an odd blend of science and mysticism that imposed moral, ascetic, and dietary rules on its members (members of the order were not allowed to eat beans) to purify their souls for the next embodiment. Still, the Pythagoreans were skilled mathematicians. The Pythagorean theorem (which states that the square of the length of the hypotenuse of a right triangle equals the sum of the squares of the lengths of the other two sides) influenced early Euclidean geometry. Pythagoras put too much faith in the power of numbers, however, and at the end of the fifth century B.C., an enraged populace set upon the Pythagoreans for interfering with traditional religious customs, and forced them to flee.

Pythagoras first formulated his theory of numbers after discovering that the relationship between musical notes could be expressed in numerical ratios. This discovery led him to conclude that the universe had a cosmic harmony, whose structure it was the business of philosophy to figure out. Where Pythagoras saw harmony, Heraclitus of Ephesus (c. 535-c. 475 B.C.) saw nothing but flux. Change was the only reality, and permanence an illusion. "Everything flows and nothing stays," he said. Yet Heraclitus also believed that a kind of cosmic justice kept the world in balance, and he called this cosmic justice the *Logos*, or God. For Heraclitus, the primary element of the universe was fire, and he identified life and reason with it.

Parmenides of Elea (born c. 515 B.C.) insisted that it was the changing world that was illusory. Twenty-five years younger than Heraclitus,

Parmenides argued that reason is superior to the evidence of the senses. Existence, he taught, is unchanging, indivisible, and unmoving. Since the world that we see is a changing one, it has no reality and must be an illusion. Parmenides used this same argument to destroy the possibility of generation, destruction, and motion. Zeno of Elea (c. 490-c. 430 B.C.) and Melissus of Samos (fl. 440 B.C.) were followers of Parmenides. Zeno devoted himself to defending Parmenides's ideas by pointing out how absurd the opposite view was in a series of arguments involving tortoises, arrows, and moving blocks. These arguments are known as Zeno's paradoxes.

Change and permanence cannot both be illusory, and Empedocles of Acragas (c. 495-c. 435 B.C.) tried to reconcile the views of Heraclitus and Parmenides. Empedocles proposed that there are four basic elements: fire, air, water, and earth. They are the ultimate roots of things and as such cannot themselves be changed or destroyed. But in combination, they form the changing world of our senses. Because the four basic elements are not self-moving, Empedocles supposed that harmony and discord cause disparate elements to be brought together and to be drawn apart. He was the first of the so-called pluralists.

Anaxagoras (c. 500-c. 428 B.C.), who is credited with transferring the seat of philosophy to Athens, adopted an extreme form of pluralism. His idea was that physical objects result from mixing tiny bits of a wide variety of substances. The bits are shaped by something Anaxagoras called mind or reason, which knows all things and has all power, and is made of a material different from what makes up the rest of the world. In the beginning, the universe was an infinite, undifferentiated mass. From this mass, mind produced all the things that are. Anaxagoras believed that the Sun was a white-hot stone, and the Moon made of earth that reflected the Sun's rays. He was accused of atheism for these beliefs and exiled.

The theory of atomism was first advanced in the fifth century B.C. by Leucippus (about whom we know nothing) and his pupil, Democritus (c. 460-c. 370 B.C.). The atomists posited that the universe consists of many microscopic and indivisible particles of a single kind of matter in constant motion. *Atoma* means uncuttable. These particles are different sizes and shapes, and in space they collide and become entangled. They fall together in a vortex, and from the pattern of falling bodies—larger and heavier to the center, smaller and lighter to the circumference—a cosmos is formed. The early atomists did not sug-

Democritus, Greek philosopher of the late fifth century B.C. *(Bettmann/Corbis. Reproduced with permission.)*

gest a separate cause of motion. Sensory qualities like sight and taste occur when the atoms of external objects and those in our bodies interact.

Impact

Philosophy is the single greatest achievement of the Greeks, and the pre-Socratics gave it birth. Their contribution to Europe, and ultimately the world, was the belief that it was possible to abandon the mythology of ages and to find a coherent and logical explanation for the origin and nature of the universe. Later, Plato (427?-347 B.C.) would apply the rational reflexion begun by the pre-Socratics to the question of how a person should live.

The emergence of philosophy in Greece resulted from the coming together of several factors. Miletus was a bustling, prosperous trading center, and wealthy citizens have time to think. The basic structure of a democracy was in place, the language was suited to precise description, and the Greeks were avid travelers, which facilitated intellectual cross-fertilization. Of course, people did not abandon their gods overnight, and pockets of irrationality and superstition persisted. In fact, in fifth-century Greece, to deny the existence of gods was blasphemy.

The impact of the pre-Socratic philosophers individually and as a group was impressive.

Thales was the first to dispense with mythology in explaining the nature of the physical world. Anaximander not only was the first to attempt to explain all aspects of the world in detail, but centuries later his notion of the indefinite was echoed in the concept of the indestructability of matter, and his view of human development presaged the theory of evolution. Anaximenes anticipated the practice of modern science that seeks to explain qualitative differences quantitatively. And in addition to influencing early Euclidean geometry, the Pythagoreans made important contributions to medicine and astronomy and were the first to teach that Earth was a sphere rotating about a fixed point.

We have Parmenides to thank for demonstrating the power of reasoned proof for assertions, and Zeno for being the first to employ the dialectical method. His paradoxes continue to excite interest among today's philosophers and mathematicians. Empedocles's explanation of the nature of perceived change states a principle that is central to physics. And Democritus's atomic explanation of the physical world was more radical and scientific than any that had gone before.

GISELLE WEISS

Further Reading

Brumbaugh, Robert S. *The Philosophers of Greece*. New York: Thomas T. Crowell, 1964.

Guthrie, W. K. C. *A History of Greek Philosophy*. 2 vols. Cambridge: Cambridge University Press, 1965.

Kirk, G., J. Raven, and M. Schofield. *The Presocratic Philosophers*. Cambridge: Cambridge University Press, 1983.

Early Greek Matter Theories:
The Pre-Socratics to the Stoics

Overview

Between the sixth and the third centuries B.C., the classical Greek philosophers proposed numerous theories regarding the material composition of the universe, with those of the atomists, Aristotle (384-322 B.C.), and the Stoics emerging as the major alternatives. While providing the intellectual foundations for subsequent developments in Western chemistry and physics, the primary purpose of these theories was not to offer a scientific account of nature in the modern sense, but rather to pose and answer philosophical questions regarding the structure of reality, the certainty of knowledge, and the formulation of ethical principles.

Background

Like those of other ancient civilizations, the earliest Greek cosmological theories were mythological accounts of the creation of the universe by the gods. However, in about 600 B.C. a tradition of speculative thought emerged that, while not rejecting religion, sought to offer coherent explanations in nonmythological terms. Unfortunately, the writings of the pre-Socratic philosophers and the later Stoic naturalists have all been lost. While scholars have made speculative reconstructions of their views, based on surviving isolated quotations from their works, and summaries and criticisms of their ideas by Aristotle and other commentators, many basic details remain controversial.

Despite some radical differences of opinion, the early Greek philosophers ("lovers of wisdom") all believed that the universe has a fundamental order (*cosmos* = order), knowable by human minds through the use of reason and sensory perception. They also assumed two initial logical principles: the law of identity ($A = A$) and the law of noncontradiction ($A^1 \neq$ not-A). Together these three principles determined the basic questions these philosophers posed about the constitution and structure of the natural order. What is the cosmos made out of? How is it structured? Is it eternal, or does it have a beginning and an end? How are different objects and events identified, distinguished, and related to one another? Are space, time, and material objects real, or merely subjective perceptions?

These questions sought to address two major problems. The first of these is termed the One and the Many. If the cosmos is One, or a unity, then how do different objects and events arise? But if it consists of many things, how can it have any unity and structure? The second and related problem is that of Continuity and Change. If an object or event is itself, and cannot be something it is not, then why and how do things appear to change into other things? If change is real, then how can anything have a permanent and distinct identity? How can anyone know for sure what a thing is, if it is subject to change? How is reality distinguished from mere appearance, and the permanent from the temporary?

The earliest Greek philosophers were all monists who believed the cosmos to be a unity constituted from a single element; however, they disagreed sharply as to its nature and identity. There is little sound information about Thales of Miletus (624?-547? B.C.), the first Western philosopher known by name, except that he asserted everything is made of water and that "all things are full of gods." He is presumed to have meant that water is a universal elemental principle, observed in gaseous, liquid, and solid forms as steam, water, and ice, whose transformations are evidence of an inherent animating power.

More is known about his successor, Anaximander (c. 610-c. 545 B.C.), though again only a single quotation survives: "Out of those things whence is the generation of existing things, into them also does their destruction take place, as is right and due; for they make retribution and pay the penalty to one another for their injustice." Anaximander held that the universe is constituted out of a single, eternal, divine material principle called the "unbounded" (*apeiron*), which enfolds all things. From the *apeiron* is engendered a primordial pair of opposites, Hot and Cold, whose interaction gives rise to the basic elemental principles (water, air, fire) and opposing powers and qualities (e.g., hot-cold, dry-wet, heavy-light, rough-smooth, bright-dark). The tension between these opposites causes the universe to undergo repeated cycles of generation and destruction, as one extreme or the other gains dominance. Anaximander also used these opposites to explain meteorological phenomena (wind, lightning, thunder), and proposed an elaborate model of the solar system.

The third Milesian philosopher, Anaximenes (fl. c. 545 B.C.), proposed air as the universal element, as shown by his only extant statement: "Just as our soul, being air, holds us together, so do breath and air surround the whole cosmos." Unlike Thales's water and Anaximander's *apeiron,* "air" here denotes not an abstract causal principle but the common physical substance. Air envelops and constitutes all things, even the gods and living souls. Constantly in motion, it

Plato, student of Socrates and teacher of Aristotle. *(Bettmann/Corbis. Reproduced with permission.)*

gives rise to various natural objects and occurrences by successive stages in an ongoing process of rarefaction (generating fire) and condensation (resulting in water and earth).

The next two great monistic thinkers staked out opposing positions that influenced their successors for centuries. For Heraclitus of Ephesus (540?-480? B.C.), from whose writings about 125 cryptic epigrams survive, fire is the universal element and animating divine principle: "All things are an exchange for fire, and fire for all things." The uncreated and eternal cosmic One is process—a cyclic flux of constant change, "an ever-living fire being kindled in measures and being extinguished in measures," governed by a *Logos,* or material principle of Strife between yoked opposites, whose transformations or "turnings" provide continuity, stability, and balance to the whole: "Fire lives the death of earth and air lives the death of fire, water lives the death of air, earth that of water." These four elements correspond very roughly to the modern concept of energy and to the gaseous, fluid, and solid states of matter. The senses provide reliable information, but reason must interpret them, if appearances of permanence are not to deceive: "Upon those who step into the same rivers, different and again different waters flow."

For Parmenides of Elea (515?-450? B.C.), however, the cosmic One—whose nature he did not specify, though it recalls Anaximander's *apeiron*—is not only uncreated, eternal, continuous, and necessary, but also unchanging and unchangeable, indivisible into distinguishable parts, complete and whole in and of itself. From his argument comes the famous principle that "out of nothing, nothing comes to be." All appearances of change, including time and motion, are illusory; commonplace speech and the senses are untrustworthy; reason alone supplies true knowledge. In extant fragments of his poem *On Nature*—which offer the first surviving philosophical argument in Western history—Parmenides distinguishes three paths to knowledge: the right Way of Truth, the false Way of Opinion, and the misleading Way of Inquiry, which nonetheless allows the seeker to find out the Way of Truth. His pupil, Zeno of Elea (495?-430? B.C.), is still famous for his imaginative (if ultimately fallacious) paradoxes, designed to prove the impossibility of plurality, motion, change, or divisions in time and space.

The philosophical impasse created by Heraclitus and Parmenides was broken when subsequent thinkers abandoned monism and embraced pluralism as their fundamental starting principle. By assuming an initial multiplicity of cosmic elements, they obviated the question of the One and the Many by eliminating any need to explain how the latter arose out of the former. This also in part resolved the question of Continuity and Change, by allowing for the reality of both; basic elements could remain unaltered, while their various combinations did not. The most important issues thus became the identities and natures of those elements; the fundamental principles and modes of their interactions; and the nature of matter, time, space, and place. This also coincided with a shift away from models based upon astronomical phenomena toward principles drawn from biology, and a correspondingly increased emphasis upon issues of human ethics and psychology.

Impact

The first pluralist matter theory was that of Empedocles of Acragas (492?-432? B.C.). Drawing upon Heraclitus, Parmenides, and the Pythagorean principle of number as the fundamental reality of all things, Empedocles taught in his poem *On the Nature of Things* (*Peri physikos*) that fire, air, water, and earth are the four eternal and indestructible material elements or "roots" of the cosmos. Swirling in a constantly rotating vortex that constitutes a material plenum without void space, these roots constantly mix, aggregate, and separate to produce various natural objects, each composed of certain elements in characteristic proportions (e.g., bone is composed of earth, water, and fire in a ratio of 2:2:4). Combination and separation of the elements is governed by two primeval forces, Love and Strife, which alternately dominate in a fluctuating cycle of unity and diversity. Empedocles was the first philosopher to distinguish between elements, compounds, and physical mixtures; to assert that compounds are in some sense real and not just phenomenal unities; and to reject strict necessity and introduce "chance" as a causal principle of motion and change. These roots and forces also account for complex biological and psychological factors, and a cosmic religious cycle of sin and redemption.

Empedocles's contemporary Anaxagoras of Clazomenae (500?-428? B.C.) in turn took the concept of multiplicity to the farthest extreme. In his book *On Nature* (*Physika*), he taught that the cosmos is a material plenum, in which there are as many different types of uncreated and eternal original "seeds" (*spermata*) as there are types of "basic matter" (*homoiomere*). This plenum is infinitely divisible—there are no "smallest portions" of any type of matter—with seeds of every type of matter in every object: "in everything there is a portion of everything." Objects are constituted of a particular type of matter because they contain a predominant portion of those seeds in comparison to all the other types of seeds (e.g., a gold bowl is composed primarily of gold seeds). Originally all the seeds and portions were mixed together in an undifferentiated cosmic unity, but subsequently were partially segregated by the rotating cosmic vortex according to the principle of "like attracts like." This cosmic vortex is governed by "mind" (*nous*), an immaterial vital principle that alone is originally unmixed with other things, and which alone knows, rules, orders, and moves all things according to opposing cosmic principles (e.g., hot-cold, dry-wet, heavy-light, rough-smooth, bright-dark). Anaxagoras was the first to distinguish a moving agent from both objects moved and principles of motion.

The third and most important of the pre-Socratic matter theories was that of the atomists Leucippus of Miletus (fifth century B.C.) and his pupil Democritus of Abdera (460?-370? B.C.) The former is a shadowy figure from whom only a single quotation survives: "Nothing happens in vain, but everything from reason and necessity." While over 100 sayings of Democritus are extant, most concern his ethical teachings, with only one directly concerning his theory of atomism: "By custom sweet and by custom bitter, by custom hot, by custom cold, by custom color; but in reality atoms and void." Nevertheless, the content of their theory is well known, both from the extensive criticisms of Aristotle and its later adoption by the Hellenistic philosopher Epicurus of Samos (341-270 B.C.), a few details of which survive in his *Letter to Herodotus.* Subsequently the Roman poet Titus Lucretius Carus (99?-55 B.C.), in his classic work *On the Nature of Things* (*De rerum natura*), preserved many otherwise lost details of Epicurus's doctrine; it is the only surviving intact account from antiquity of a non-Aristotelian matter theory.

According to the early atomists, the infinite cosmos consists of minute, indivisible bits of matter called *atomos* (from the Greek for "indivisible") and void space. Atoms are eternal, uncreated, indestructible, infinite in number, and in constant motion within a vortex through the void; their only properties are size, shape, and solidity. (Whether or not atoms also had weight, and if so in what way, remains controversial.) All atomic motions occur due to strict necessity or fate, not chance or free will. Combinations of atoms may vary in quantity, shape (*H* or *V*), order (*HV* or *VH*), or position (*V* or *L*). All objects and their qualities (color, taste, texture, etc.), and changes in these, are merely phenomenal appearances, resulting from atomic shapes, motions, and interactions due to forces of attraction, repulsion, and cohesion. (For example, pointed atoms produce a bitter taste, and round atoms a sweet one; smooth atoms produce bright colors, rough atoms dark ones.) Objects are perceived because they radiate ephemeral atomic films or "images" (*eidola*) to the observer's eye or field of vision. Since the human soul is also composed of atoms, it disintegrates with death; there is no immortality, and the gods are merely giant *eidola.*

The atomists were the first to posit a strictly mechanical account of motion, and to distinguish between primary and secondary qualities. While atoms are in effect fragments of the Parmenidean One, the atomists broke from Parmenides not only in substituting the Many for the One, but in their crucial distinction between relative material nonexistence, called *ouk on,* (not being), or the void of absolute nonexistence, called *ouden* (nothing). Since the Greeks did not consistently distinguish place as the relative location of objects from space as the extension in place of a given object, or conceive of absolute space as a set of dimensions existing apart from any objects, the void denoted any interval between two atoms, whether separate or conjoined, not the modern concept of a vacuum.

Aristotle's criticisms of the early atomists led Epicurus and Lucretius to revise atomism in several ways. For them weight, defined as downward linear motion toward the center of a spherical cosmos, is a primary atomic property. Atoms move from place to place almost instantaneously at the same speed, rather than varying in speed by size. Space is distinguished from the void as a type of "intangible existence" and hence is also real (though Time remains an accidental property of motion). The complex bodies composed by atomic aggregates, and their qualities, are not merely phenomenal, but also real; thus in their view the senses are more reliable sources of information than for Democritus, and the soul is also a real body. Most important of all was Epicurus's novel idea of an atomic "swerve," or occasional slight random deviation in the motions of atoms. While heavily criticized by other philosophers as illogical, this doctrine sought not only to explain atomic motions and interactions (since atoms falling perpetually in straight lines would never collide), but also to reject the strict causal necessity and fatalism of early atomism and thus preserve a measure of free will and moral responsibility in human actions.

A very different version of atomism was proposed by Socrates's great pupil, Plato (427?-347 B.C.). Ignoring Democritus altogether and drawing for inspiration upon Pythagorean geometry, Plato suggested in his dialogue *Timaeus* that there are five different types of geometrical atoms, corresponding to the five perfect geometrical solids (having sides all of equal length, faces all of equal size and shape, and angles all of equal degree). Four of these solids correspond to the traditional four elements—fire: tetrahedron, earth: cube, air: octahedron, water: icosahedron— and the fifth (the dodecahedron) to the entire cosmic sphere. The faces of the first four solids are in turn readily divisible into either equilateral triangles (for fire, air, and water) or isosceles triangles (for earth). (As the unitary cosmic symbol, the dodecahedron does not require division.) Thus the geometric "atoms" are not indivisible, but are assembled from and disintegrate into their triangular components, as the true, unchanging, indestructible formal elements of the cosmos, with fire, air, and water being interconvertible. Like Democritus, however, Plato attributes secondary qualities to the sizes, shapes, motions, and interactions of his atoms. However, the later Neoplatonists, the most important being Plotinus (A.D. 205-270), demonstrated virtually no interest in scientific questions.

Apart from Aristotle, the most important rival matter theory to atomism in later Western antiquity was that of the Stoics, named for the *stoa* or "porch" that was the original site of their school. Originating with Zeno of Citium (336?-265? B.C.), Stoic philosophy was developed by Chryssipus of Soli (280?-206? B.C.) in a systematic fashion that included cosmological doctrines. Later Stoics of importance included the Roman poet Seneca (4? B.C.-A.D. 65) and the Roman emperor Marcus Aurelius (A.D. 121-180), though their interests lay primarily in religion and ethics.

Inspired by Heraclitus and other pre-Socratics, the Stoics held that the cosmos is a single eternal substance, a kind of universal living being animated by a pervasive "spirit" (*pneuma*) or "soul" (*psyche*), with fire as a divine, active, material principle of creativity. This primeval fire acted as a creative force on a precosmic moisture to generate the cosmos, with "seeds" (*spermata*) of the *pneuma* moisture blend inseminating all things. As a conceptually divisible but actually undivided material plenum, the cosmos is harmoniously unified and strengthened by a universal "tension" (*tonos*), a property resulting from opposing forces acting in the intangible *pneuma* as a material medium. The *tonos* transmits causal "impulses" from one body to another, like the vibrations of a plucked lyre string or the stretching of an athlete's tendons. The entire cosmos is subject to a recurring cycle of conflagration and regeneration, due to an alternating tension between the destructive and creative aspects of the primeval fire.

Stoic matter theory likewise presented a remarkable synthesis of old and new ideas. All real things—even knowledge, virtues, and vices—are material "bodies" (*somata*) with capacities for acting upon, and being acted upon or resisting action by, other bodies. The original combination of *pneuma* with moisture differentiated an initially qualityless primal matter into the traditional four basic elements of ordinary fire, air, water, and earth, by a process of rarefaction and condensation. The elements formed four concentric layers, with fire at the perimeter, followed by air and then water, with earth at the center, subsequently undergoing further intermixture. Rejecting Aristotle's complex scheme of complementary pairs of primary elemental qualities, the Stoics adopted the simpler association of each element with a single primary quality—fire:hot, air:cold, water:wet, earth:dry—made by a pupil of Empedocles, Philistion of Locri.

Particularly controversial was the Stoic doctrine of "total blending" (*synchysis*). The pre-Socratics had struggled with the distinction between chemical combination and physical mixture. Because they denied any possibility of material generation out of or destruction into nothingness, they proposed only theories in which initial components were juxtaposed or mixed in ways that preserved them unaltered in the final product. Aristotle's theories of elemental transformation and combination (*mixis*) allowed for generation and destruction of substances in a restricted but not absolute sense. But the Stoics asserted that because the entire cosmos is a single substance, with all bodies being pervaded by the *pneuma* as a creative and transforming agent, two different bodies can be completely fused and fully interpenetrate one another to form a single new body. A favorite example was the penetration of heated iron by fire. Critics of the Stoics accused them—wrongly, some scholars think—of believing that two bodies can occupy the same place at the same time, a logical impossibility according to the Greek concept of place as a distinct relational location.

The power of Aristotle's own philosophical system, and his penetrating criticisms of the flaws in those of his rivals, ultimately led to the demise of the atomist and Stoic schools during the third century A.D. The rise of Christianity was another major factor, since the basic theism of Aristotle's system proved far more compatible with Christian theology than the atomists' implicit atheism or Stoic pantheism. Atomism as a scientific theory would only be revived in the seventeenth century, and then in a drastically revised fashion. Stoic physics has received renewed attention from historians and philosophers only in recent decades, for ideas in it that faintly foreshadow concepts of modern quantum physics. Nevertheless, all three systems, and those of their pre-Socratic predecessors, contributed profoundly to the intellectual richness and complexity of theories of matter, energy, and physical and chemical interactions in the Western scientific tradition.

JAMES A. ALTENA

Further Reading

Books

Bailey, Cyril. *The Greek Atomists and Epicurus.* Oxford: Clarendon Press, 1928.

Barnes, Jonathan. *The Presocratics*. Rev. ed. 2 vols. London: Routledge and Kegan Paul, 1982.

Furley, David J. *The Greek Cosmologists*. Cambridge: Cambridge University Press, 1987.

Guthrie, William K.C. *A History of Greek Philosophy*. 6 vols. Cambridge: Cambridge University Press, 1962-81.

Hahm, David E. *The Origins of Stoic Cosmology*. Columbus: Ohio State University Press, 1977.

Kahn, Charles H. *Anaximander and the Origins of Greek Cosmology*. New York: Columbia University Press, 1960.

Long, Anthony A. *Hellenistic Philosophy: Stoics, Epicureans, Sceptics*. 2nd rev. ed. Berkeley: University of California Press, 1986.

Lucretius Carus, Titus. *De rerum natura* (On the Nature of Things). Trans. and ed. by Anthony M. Esolen. Baltimore: Johns Hopkins University Press, 1995.

McKirahan, Richard D. *Philosophy Before Socrates: An Introduction with Texts and Commentary*. Indianapolis: Hackett Publishing, 1994.

Sambursky, Samuel. *The Origins of Stoic Physics*. London: Routledge and Kegan Paul, 1959. See Chap. 2.

Sambursky, Samuel. *The Physical World of the Greeks*. London: Routledge and Kegan Paul, 1956.

Schofield, Malcolm. *An Essay on Anaxagoras*. Cambridge: Cambridge University Press, 1980.

Sorabji, Richard R. *Matter, Space, and Motion: Theories in Antiquity and Their Sequel*. Ithaca: Cornell University Press, 1988. See Chap. 2.

Todd, Robert B. *Alexander of Aphrodisias on Stoic Physics: A Study of the "De Mixtione," with Preliminary Essays, Text, Translation and Commentary*. Leiden: E.J. Brill, 1976.

Vlastos, Gregory. *Plato's Universe*. Seattle: University of Washington Press, 1975. See Chap. 3.

Articles

Kerferd, George B. "Anaxagoras and the Concept of Matter before Aristotle." In *The Pre-Socratics: A Collection of Critical Essays,* ed. Alexander P. D. Mourelatos, 489-503. New York: Anchor Press, 1974.

Lloyd, Geoffrey E.R. "Hot and Cold, Dry and Wet in Early Greek Thought." In *Studies in Presocratic Philosophy,* ed. Reginald E. Allen and David J. Furley, Vol. 1, 255-80. London: Routledge and Kegan Paul, 1970.

Longrigg, James. "Elementary Physics in the Lyceum and Stoa." *Isis* 66 (1975): 211-29.

Longrigg, James. "The 'Roots of All Things.'" *Isis* 67 (1976): 420-38.

Strang, Colin. "The Physical Theory of Anaxagoras." In *Studies in Presocratic Philosophy,* ed. Reginald E. Allen and David J. Furley, Vol. 2, 361-80. London: Routledge and Kegan Paul, 1970.

Vlastos, Gregory. "The Physical Theory of Anaxagoras." In *Studies in Presocratic Philosophy,* ed. Reginald E. Allen and David J. Furley, Vol. 2, 323-53. London: Routledge and Kegan Paul, 1970. Reprinted in *The Pre-Socratics: A Collection of Critical Essays,* ed. Alexander P.D. Mourelatos, 459-88. New York: Anchor Press, 1974.

Physical Science in India

Overview

Though there remain essential questions concerning the age of Indian literature, a timeframe of sophisticated ideas from about 1500 B.C. indicates that Indian astronomy and physical thought were on a par with that of Babylonia and Egypt. Some historians have pushed this date back to 6000 B.C. and earlier based on claims that the oral transmission of such ideas far precede written records. Ancient Hindu sources from around 1500 B.C. indicate the existence of advanced concepts of basic astronomical awareness of movement for five planets and the Sun and Moon, and their application to cosmological time cycles, or solar calendars. Other claims suggest an understanding of relative astronomical motions (specifically rotation of the Earth and the Sun as center of the planets), sphericity of Earth, flattened poles of the Earth, and the concept of fundamental matter as atoms.

Some of these ideas require a validated chronology of dating to be considered legitimate. On the other hand, the traditional historical claims that Indian science was dependent on Greek science after 326 B.C. are inaccurate and fail to account for the native evolution of science and mathematics in India for thousands of years B.C. By the fifth century A.D., important Indian astronomers and mathematicians, such as Aryabhata (476-550) and Brahmagupta (598-668), developed advanced physical thought, providing continuity with seminal ancient Indian conceptions in astronomy and other sciences.

Background

Based on interpretation of early Indian Hindu religious literature, which was orally composed well before being committed to writing, some scholars have claimed that a high level of sophistication and innovation existed early in Indian science, perhaps in rudimentary form from 3000 B.C., though more likely 1500 B.C. It remains suspect, however, that oral information alluding to scientific ideas would be exactly transmitted thousands of years later to the written word without cumulative revisions based on improved scientific conceptions and techniques. Early, essentially nonscientific literature has been long accepted as a legitimate source for very early scientific ideas where scientific literature had not yet developed formally. The earliest sources in India are the Hindu *Vedas,* four Vedic texts (*Rig, Yajur, Sama,* and *Atharva*) that are sources of prayers, hymns, magic, and sacrificial formulae with many astronomically advanced suggestions dating from about 1500 B.C., though perhaps earlier.

The *Rig Veda* is referred to the most regarding claims of advanced scientific concepts in India, and these conceptions are largely presented under astrological references that had great significance in Hindu religion. This was also the case with other ancient religions in an effort to understand the heavens and its influence on terrestrial nature. *Veda* concepts entailed the use of a 12-month lunar calendar, knowledge of the precession of the equinoxes, a solar day ritually divided into 3, 4, 5, or 15 equal divisions, and a rather modest 27-star catalog of names.

Another assemblage of early literature, based on the text fragments and oral traditions of the Jain (a religion independent from Hinduism), were recomposed much later (probably about 350 B.C.). *Ardha-Magadhi Prakrit,* consisting of approximately 50 texts that entail more specific mathematical and astronomical information. These texts also imply that the scientific concepts included were updated over time. Of these writings, two of the books known as the *Angas* deal with astronomy and mathematics. In the set known as the *Upangas* there are references to astronomy and the concept of time, *asankhyata,* meaning "inscrutable infinitesimal time," and *sirsaprahelika,* meaning "millions of years." The *Culikasutra* is a treatise on astronomy and mathematics. A post-Jain work from the second century A.D. called *Tattvarthadhigma Sutra* was composed by the astronomer Umavati (185-219) and deals with astronomy and Indian cosmology. Additionally, there are other works on Jain theoretical and observational astronomy written up until the seventh century.

Jain cosmology featured the conception of Mount Meru in India as the central axis of the Earth, a motionless planet, surrounded by the Sun, Moon, planets, and constellations of the stars. The Jain cosmographical diagram showed Mount Meru at the center (the pole star above it) with the 12 months, planetary cycles, and the movements of the Sun and Moon wrapped congruently around it. A major city or area considered the center of the Earth was a typical conception of most ancient cultures. The Indians, like other early cultures, were also interested in understanding worldly phenomena such as the terrestrial water cycle, the tides, and the change of seasons.

Claims that the Indian conception of the sphericity of the Earth is earlier and significantly different than that of the ancient Greeks remain suspect, as they are dependent on the validation of unrevised-by-time passages in ancient Indian literature, such as the *Rig Veda,* to predate that of Greece. Though some claim that the term "bowl," which appears in early Vedic texts and the Puranic works *Markandeya Purana* and *Vishnu Purana,* alludes to the sphericity of the Earth, it was actually quite common among ancient cultures to envision the cosmos as a flat or hollow bowl-shaped Earth, or sometimes an inverted bowl, with the heavens stretched over it. These are typical ancient interpretations based on the perspective of one standing on Earth and observing and interpreting what he or she saw at face value around them. But these texts do have interesting anticipations of later physical thought that suggest the cause of twilight, the blue of the sky, the phases of the Moon, allusions to some notion of gravity, and perhaps the Sun as center of the Solar System. A group of texts called the *Vaiseshika Sutras* (dating from as early as the second century to as late as the fifth century) entail an elaborate analysis of matter as atoms, even in combination (the idea of molecules), under the influence of time and direction.

Impact

Aside from considerations of the religious and astrological origins of Indian cosmology and astronomical thinking, the more historically precise textual proof of the advances of Indian astronomy and its ancient foundation begins with the fifth century, particularly the astronomer Aryabhata (476-550). His *Aryabhatiya,* a multi-disciplinary work of mathematics and astronomy, was composed in about 498 and is an enlightening source for observing the advance of Indian science from its foundation in ancient times. It contains a discussion of spherical astronomy (an application for astronomy) along with calculations of planet mean positions and rules for calculating solar and lunar eclipses. Most importantly, it contains his view that Earth rotated on its axis, though it is uncertain whether or not this concept was the result of Greek influence on Indian astronomy.

Aryabhata was perhaps the earliest astronomer to begin a continuous counting of solar days by revolutions of the Earth (rather than the usual conception of orbits of the Sun around the Earth)—that is, determining the length of the year. He also used this data with that on orbits of the Moon (the ratio of the former to the latter) to provide a mathematical rendering of the month, and thus an early astronomical ratio. His methods of calculating planetary motions would be adopted by astronomers in his native Kerala, a state in southern India, who by 683 agreed to modify this method with a revision to be known as the Parahita system.

The *Siddhantas* texts of the sixth and seventh centuries reveal more sophisticated astronomical calculations, indicating an advance from Aryabhata. And indeed a few of the these are in fact technical in nature, one being written by astronomer and mathematician Varahamihira (505-587), who came from the area of Ujjain, where a famous astronomical observatory would arise. As a philosopher as well, Varahamihira made a thorough study of Western and Middle Eastern astronomical rudiments, including those of the Egyptians, Greeks, Romans, and Indians, and compiled a comparative and comprehensive work called *Five Treatises*. Though an example of intellectual exchange, Varahamihira's work might have provided a source for the inaccurate traditional Western view that Indian astronomy was dependent on other cultures.

The Ujjain school of astronomy produced another significant astronomer in the person of Brahmagupta (598-668), who wrote a comprehensive revision of an old text of astronomy, the *Brahma Siddhanta*, called the *Brahma Sphuta Siddhanta*. Brahmagupta's revision included pure math and astronomical applications. Essentially, he provided a systematized rendering of rules for algebra, geometry, other mathematics, and astronomy. Brahmagupta may have been the first astronomer to apply algebraic techniques to astronomical problems and, evidently, also came up with the concept of using zero and a solution to the indeterminate equation, a significant advance in numerical theory before the eighteenth century. His methods of astronomical calculations were thorough and included discussion of the motions, positions, rising and setting, and conjunctions of

planets, and eclipses of the Sun and Moon. He believed the heavens and Earth were round or spherical, but he did not believe that Earth rotated or moved. He also took a hand at geodetics, resulting in a fairly good approximation of the circumference of the Earth as 5,000 yojanas (about 4.5 miles [7.2 km] per the ancient unit of yojanas or 22,369 miles [36,000 km]).

Another area in which some historians theorize that Indian scientists made very early innovations concerns the understanding of gravity and heliocentrism, as understood by the major Indian astronomers already noted. Varahamihira might have been the first Indian thinker to suggest a force holding the Earth and the celestial bodies in place. Brahmagupta had said: "Bodies fall toward the earth as it is in the nature of the earth to attract bodies, just as it is the nature of water to flow." Yet earlier claims point to some Sanskrit words found in Vedic literature that are interpreted as dealing with attraction in the gravitational sense, suggesting to some that some idea of gravity was known at an early date. And again, the Vedic literature is held up as the earliest source, as well as an ultimate reference, to the central place of the Sun as gravitational source and center of the universe—thus, its heliocentric implication. However, heliocentrism was arrived at later in Indian astronomy, first by Aryabhata, as already noted.

Some historians assert this early Indian concept of heliocentrism, based on the Vedic sources, predates the Greek concept of heliocentrism that appeared in about the mid-fourth century B.C. However, this claim is highly interpretative since, after all, the Sun's importance as a light and heat source, whether as a deity or natural influence, is primary in all ancient civilizations, thus casting doubt on such a purported gravitational or heliocentric view. Additionally, words relating to the prominence of the Sun would be subjective in meaning and could be interpreted as anticipating such scientific ideas in other ancient cultures as well. At the same time, Aryabhata's applications of heliocentric astronomy in his calculations would logically point to perhaps some earlier tradition of heliocentrism as a foundation for his ideas. But a more exacting, objective analysis of scientific meaning in Indian literature will have to validate what would otherwise be interpreted as only subjective claims.

Nonetheless, the credit due to Indian scientific thinkers from 1500 B.C. to the end of the seventh century A.D. has been long in coming.

Indian historians and other interdisciplinary international historians have proceeded with important research in Indian source translation and commentary that is necessary to properly interpret India's scientific past and its place in the larger framework of intellectual history.

Certainly, in the more than 2,000 years leading up to the seventh century Indian thinkers provided a progression from sophisticated religious refinement of the ordering of the physical cosmos to important practical developmental concepts in observational astronomy, mathematical application to astronomy, and various physical theories centering on the perception of the Sun and planets. As with other ancient peoples, India holds a unique place by virtue of its cultural characteristics and innovative intellectual endeavors, as well as its contri-butions to foundational thought and legacy of physical scientific knowledge.

WILLIAM J. MCPEAK

Further Reading

Aryabhata. *Aryabhatiya of Aryabhata,* edited and translated by K.V. Sarma and K.S. Shukla. New Delhi: Indian National Science Academy, 1976.

Bose, D.M., et al. *A Concise History of Science in India.* New Delhi: Indian National Science Academy, 1971.

Kay, G.R. *Hindu Astronomy, Ancient Science of the Hindus.* New Delhi: Cosmo Publications, 1981.

Sarma, K.V. *A History of the Kerala School of Hindu Astronomy.* Hoshiarpur, India: Vishveshvaranand Institute, 1972.

Sens, S.N., and K.S. Shukla. *History of Astronomy in India.* New Delhi: Indian National Science Academy, 1985.

Astrology and Astronomy in the Ancient World

Overview

The first records of systematic astronomical or astrological observation and interpretation lie in the scattered remains of ancient Egyptian and Babylonian civilizations. The earliest evidence of the development of astronomy and astrology—in the modern world distinctive representatives of science and pseudo-science, respectively—establish that they share a common origin grounded in mankind's need and quest to understand the movements of the celestial sphere. Moreover, evidence suggests an early and strong desire to relate earthly everyday existence to the stars and to develop a cosmology (an understanding of the origin, structure, and evolution of the universe) that intimately bound human society to a coherent and knowable universe.

Background

The most primitive origins of both astrology and astronomy predate the human written record. There is abundant archaeological and artistic evidence that long before there were the stirrings of true civilization in ancient Egypt or Babylonia, humans constructed elaborate myths and folk tales to explain the wanderings of the Sun, Moon, and stars across the celestial sphere.

In ancient Egypt, priests became the first practicing practical astrologers by linking religious beliefs to the apparent movements of celestial objects. The type of observations and predictions made by the earliest Egyptian astrologers were, however, fundamentally different than those of the later dynastic periods when the cosmology of ancient Egypt yielded to the influence of Greece. Although the point is hotly debated by many archaeoastronomers (scientists who study the relationships between archaeology and ancient astronomy), there is little convincing evidence that the ancient Egyptians constructed anything approaching the horoscope of modern astrology. Nor is there clear evidence that the constellations of the zodiac were especially important in ancient Egyptian society.

The emphasis on a zodiacal division of the constellations was first advanced in Babylonia and in other branches of Mesopotamian civilization. These fundamental divisions and groupings of the constellations lying along the plane of the ecliptic—the plane of Earth's orbit and the Sun's annual path—subsequently influenced the development of Greek cosmology, and via that culture, subsequently altered the course of Egyptian cosmology. Only in the architectural remains from the later stages of an

Egyptian civilization dominated by Greece are zodiacal signs evident.

Impact

The extant record indicates that astrological interpretations of celestial patterns date to ancient Mesopotamia. Astrology evolved from simple celestial observation, onto which was laid a theological base of interpretation. The movements of celestial objects were used as portents of the future—a methodology to predict the rise of kings, the fate of empires, and other issues critical to the continuation of power by the ruling priestly class.

Aside from a desire to elevate mankind's terrestrial existence to an astral plane, the development of astrology in Babylonian society provides evidence that in the development of Babylonian cosmology, the universe was thought to be a vital (living) entity. This societal worldview and quest for the heavens is also strongly reflected in the construction of Mesopotamian ziggurats (tiered towers with temples).

The experience of Babylonia was repeated in the rise of astrology in India, China, and among the Mayan civilizations in Central America.

However errant by the standards of modern science, the development of a zodiacal-based cosmology in ancient Babylonia signaled an attempt by early man to rely on something fixed and objective as a determinant force in human affairs. Prior to the development of ancient astrology, the tide of events was left more to the whimsy of widely varying bias toward dreams and visions as portents of future events.

The accurate prediction of the movements of the Sun, Moon, and celestial sphere took on an enormous practical importance to stable and successful agricultural development. In a very real sense, the rise of ancient astrology in Babylonia was an outgrowth of continual refinements to ancient calendars that were themselves predictors of the ebb and flow of the seasons. Accordingly, it may be fairly argued that this desire for prediction underpinning astrology also spurred the rise of real astronomical science as a more mundane cyclical predictor of celestial and seasonal occurrences. There was, for example, a chain understanding of terrestrial seasons and events derived from the regularity of variation in the location of the rising and setting Sun.

Over time, the regularity of observation first emphasized by Babylonian astrologers made the accurate prediction of the flooding of the Nile River a practical benefit of later Egyptian astronomy. Regardless of the initial religious importance of the movements of bright star Sirius, eventually the location of its rise on the horizon of the Nile plain became an accurate predictor of annual Nile flooding.

Although a proper understanding of the celestial mechanics associated with solar and lunar eclipses would await the Copernican revolution more than a millennia distant, the regularity of such occurrences was noted in the religious practices associated with these phenomena. Indeed, the need to develop increasingly accurate calendars was often driven by priestly desire to make timely predictions of celestial events that could be interpreted, with due variance to local need and custom, as messages from the gods.

The emphasis on the supernatural qualities of astrology continued to develop and influence the affairs of society. At the same time, astrology became fused with astronomical precision. Thus, only with the accurate measurement of the celestial sphere could there be accurate prediction.

Following the death of Alexander the Great (356-323 B.C.), who spread the Greek philosophical tradition and intellectual culture across much of the known world, astrology began to take on an emphasis in Greek society that soon overshadowed pure astronomical observation. Influenced by Eastern traditions, a more mundane form of everyday astrology became commonplace in Greek society, and later in Roman civilization. No longer regulated to the prediction of grand affairs of state or religion, astrology became used by Stoics as a practical medicinal art. Good evidence of this everyday application of astrology is found in surviving Greek poems and plays that show that the position of the planets was used as a guide to ordinary affairs.

Although there is often an emphasis on the influence of the supernatural upon ancient society, this masks real achievements that resulted from an increased emphasis upon astronomical observations. Notable among such observations and calculations are Aristotle's (384-322 B.C.) observations of eclipses that argued for a spherical Earth, Aristarchus of Samos's (310-230 B.C.) heliocentric model that proposed that Earth rotated around the Sun, and Eratosthenes of Cyrene's (276-194 B.C.) accurate measure of the circumference of Earth. Stimulated by astrological mythology, in 370 B.C. Euxodus of Cnidus (c. 400-c. 350 B.C.) developed a geocentric-based (Earth-centered) mechanical system that set out to explain the observed motions of the stars and planets. Moreover, these advances in astronomy laid a foundational base for the scientific development of astronomy. Hipparchus's (fl. 146-127 B.C.) classifications of magnitude of brightness, for example, are still a part of the modern astronomical lexicon.

Later, Greek astronomer Ptolemy's (fl. second century A.D.) *Algamest* became the most influential work of the scientific astrology produced in the ancient and classical world. Although his models of an Earth-centered universe composed of concentric crystalline spheres were incorrect, they dominated the Western intellectual tradition for more than a millennium.

During the decline of the Roman Empire, the tenuous place of scientific astronomy was completely overwhelmed by either a renewed emphasis on astrology, or upon an avoidance of both astronomy and astrology as contrary to the tenets of a growing Christian civilization.

The lure of astrological explanations in ancient Babylonian civilization evolved into a desire among the philosopher-scientists of Greece and Rome to define the essential elements of life—and of the forces that influence these elements. In addition, early astrology provided a coherent worldview that reconciled astronomical science with myth and religion, thus providing social stability. The development of a stable civilization and society was enhanced by astrological interpretations that provided a sense of divine control and immutable fate to human affairs.

K. LEE LERNER

Further Reading

Bronowski, J. *The Ascent of Man.* Boston: Little, Brown, 1973.

Deason, G. B. "Reformation Theology and the Mechanistic Conception of Nature." In *God and Nature: Historical Essays on the Encounter between Christianity and Science,* ed. by David C. Lindberg and Ronald L. Numbers. Berkeley: University of California Press, 1986.

Harrold , Frances B. and Raymond A. Eve., eds. *Cult Archaeology and Creationism: Understanding Pseudoscientific Beliefs about the Past.* Ames: University of Iowa Press, 1987.

Ancient Scientists Learn about the Planets

Overview

Today we know that the planets are nine large spherical objects that rotate around the Sun and reflect light from the Sun and that Earth is a planet. This knowledge has come from observations and theories made and refined over the centuries and is quite different from the concept of planets held by the ancients. Indeed, some of the planets we know today were not known in ancient times because they are not visible without the use of a telescope.

The so-called fixed stars hold constant positions relative to each other and all move together in a regular manner. During each night they appear to travel together eastward across the sky. Each morning at sunrise their positions have shifted slightly eastward in the sky relative to the day before so that a star's sunrise position seems to make a complete circle of Earth each year. The Moon, the Sun and the five planets visible to the naked eye—and therefore visible to the ancients—behave differently. Their positions shift daily relative to each other and to the regular positions of the stars. The ancients, therefore, identified as planets all bright objects in the heavens that moved in a manner different from the stars. The Greek name for these objects, *planetes,* means "wanderers," and recognizes their somewhat irregular motion. Originally, the ancients recognized seven planets. They included the Sun and Moon along with the five visible planets: Jupiter, Mars, Mercury, Saturn, and Venus.

Background

In ancient times, the sky was a much more dominant feature of life than it is today. Since they spent more time outside and had no bright electric lights, early humans were more aware of celestial phenomena. Very early, seemingly long before recorded history, humans began to try to understand the world around them by attempting to find some order in the apparently random chaos of the objects in the sky. They noticed the constant positions of the stars relative to each other and picked out patterns of stars, grouping them together and assigning names to them. We call these patterns constellations. They also noticed the regularity of some celestial events and began to correlate this regularity with occurrences in the natural world around them. This was particularly true of the activity of the Sun and Moon.

The Sun shifts eastward almost one degree each day relative to the fixed stars. In one year (365 days) it moves through 360 degrees. Its height in the sky varies through the year. This motion was clearly related to the changing seasons, to warmth and cold, to the proper times for planting and harvesting. The Moon moves steadily along, close to the same path in the sky as the Sun (called the ecliptic or zodiac), overtaking the Sun every 29.5 days, and waxing and waning through phases throughout this time. It was observed that such things as the tides and the menstrual cycles of women were somehow correlated with these regular changes in the Moon.

Because of the regularity observed for the Sun and Moon and their apparent effect on earthly activities, it was assumed that the other five planets would also demonstrate similar regularity and effects. However, the five planets behaved somewhat differently from the Sun and Moon. Even though they generally move in an eastward direction across the sky on paths similar to those of the Sun and Moon (the ecliptic), they sometimes reverse themselves and move westward (termed retrograde motion) for a time before continuing eastward. The name planets (wanderers) seemed to apply especially well to them. Although the evidence that the behavior of these planets directly affected their well being could not have been as clear as the effects exhibited by the Sun and Moon, the ancients were convinced of such correlations and studied the movements of the planets in an attempt to understand and predict their behavior.

The pseudo-science of astrology grew out of these efforts to predict the effect of the planets on events on Earth. Since much of process of astrology depended on predicting the positions of the wandering planets among the constellations along the zodiac, astrologers made accurate observations of planetary positions and attempted to develop mathematical methods for such predictions. As a result, significant advances were made in astronomy, such as the scientific observation and knowledge of celestial phenomena. These advances also included the development of methods for defining time and preparing calendars, for defining direction and improving techniques for locating oneself during travel over long distances (navigation), and for predicting the reoccurrence of seasonal events and thus improving agriculture.

Impact

Because the behavior of the seven planets (the five visible planets plus the Sun and Moon) was so different from that of the fixed stars, and because their behavior seemed to affect the world and their people's lives, the ancients attributed godlike characters to the planets. This attribution was not limited to the people of one geographical area but was worldwide. In the New World, there is evidence that various prehistoric native cultures, including the Maya and the southwest American Indians, regarded the planets and other celestial objects as having the power of gods. Prehistoric petroglyphs, stone circles, and aligned standing stones in Europe, Great Britain, and other areas attest to the religious identification of the heavenly objects.

The Egyptians made observations and drew constellation maps. They seem to have regarded the planet we call Venus as a god, and astrological horoscopes that mention planets, particularly in relation to the Dog Star Sothis (Sirius) have been found which date to 221 B.C. There does not appear, however, to have been any systematic effort to develop numerical methods for predicting the motion of the planets. Similarly, the ancient Chinese, Persians, and Indians were aware of the planets and their behavior but made little effort to make accurate observations or develop quantitative methods.

The Babylonians led the development of astrology and, therefore, of astronomy. Beginning around 1800 B.C., they undertook systematic observations, developed theories, and made accurate predictions in regard to planetary, as well as solar and lunar, motion. Archeologists have found cuneiform tablets containing detailed calculations and lists of star and planet positions, as well as evidence that this information was used for making astrological predictions. Sargon of Akkad, who ruled in Mesopotamia about 2000 B.C., employed astrologers to choose auspicious times for his important activities, and Esarhaddon, who ruled from 681 to 668 B.C., also depended heavily on astrologers. The first surviving horoscope that refers to the location of the planets in the zodiac constellations dates from 410 B.C.. Chaldea, a region of Babylonia in Mesopotamia, became known for its astrologers, and the term *Chaldean* is still synonymous with *astrologer*. The services of the Chaldean astrologers were sought throughout the ancient world, and they spread their astrological and astronomical understanding, practices, and theories to Egypt and to Greece.

The Greeks, like the Babylonians and other groups, related the planets to their mythological gods. They gave the name of their principal god Zeus to the planet the Babylonians called Neberu. Later the Romans gave this planet the name of their principal god Jupiter, the name that is still used today. Similarly, the Babylonian Delebat became the Greek Aphrodite and the Roman Venus, goddess of love and beauty. Sihtu became the Greek Hermes and was then named after the Roman messenger god Mercury. Kayamanu became known by the Greeks as Cronus and by the Romans as Saturn, god of agriculture. And the Babylonian planet Salbatanu was related to the Greek god Ares and the Roman god of war Mars.

The days of the week were named by the Babylonians after the seven planets, reflecting the importance placed on the planets and their use in the definition of time. The Romans followed this example, and the current English names were derived from the Anglo-Saxon names of the gods of Teutonic mythology. Sunday is the day of the Sun; Monday, the day of the Moon. Tuesday is named for Tiu (or Tiw), the Teutonic god of war (equivalent to Mars of the Romans). Wednesday comes from Woden, the principal Teutonic god (Mercury). Thursday is named for Thor, god of thunder (Jupiter). And Friday is derived from Freya, goddess of love and beauty (Venus).

Like the Babylonians, the Greeks did not distinguish between astrology and astronomy. In fact, astrology remained the prime motivation for the study of the heavens until the end of the Middle Ages. The Greeks, however, applied their philosophical methods of rational thought to celestial phenomena and made significant contributions to the understanding of the planets. The Greeks believed that the circle played a particularly important role in the universe, and much of the their understanding of planetary motion was based on circular orbits of spherical bodies.

Eudoxus of Cnidus (c. 408-c. 355 B.C.) is credited with the first theory of planetary motion. He proposed a set of nested crystalline spheres to which the stars, Sun, Moon, and planets were attached. These spheres were centered on Earth and moved in various ways, accounting for the motion observed on Earth. Aristotle (384-322 B.C.) adapted and extended the theory of Eudoxus. Aristarchus of Samos (third century B.C.) proposed that the observed celestial phenomena could be explained better if Earth rotated on its axis every 24 hours and revolved with the other planets around the Sun. This model was rejected, however.

Greek astronomy, including the geocentric model of planetary motion, was compiled and refined by Ptolemy (c. 100-165 B.C.) in the *Almagest*. Ptolemy replaced Eudoxus's nested crystalline spheres with circular orbits centered on Earth, around which each planet moved at its own characteristic speed. In order to account for observed variations (including retrograde motion), he proposed that each planet moves at a uniform speed around a small circle (epicycle) whose center is located on the circular orbit (deferent) whose center is Earth. The location of Earth was also shifted slightly from the center of the planetary deferents (large circles centered on Earth) which then became eccentric circles.

The geocentric Ptolemaic model of our planetary system remained the accepted one until it was successfully challenged by Copernicus (1473-1543), Kepler (1571-1630), and Galileo (1564-1642) who reintroduced a model of planetary motion in which the planets, including Earth, rotate in elliptical orbits around the Sun. The successful mathematical treatment of this model by Newton (1642-1727) led to the Scientific Revolution.

J. WILLIAM MONCRIEF

Further Reading

Evans, James. *History and Practice of Ancient Astronomy*. New York: Oxford University Press, 1998.

Hoskin, Michael A. *Cambridge Illustrated History of Astronomy*. Cambridge: Cambridge University Press, 1996.

Rey, H. A. *The Stars, A New Way to See Them*. Boston: Houghton Mifflin Company, 1975.

Development of Calendars

Overview

Keeping track of the passage of time has been a human preoccupation since the dawn of history. Calendars helped societies to understand and track the changing seasons and to mark the elapse of time between important human events such as religious holidays. But producing accurate calendars was an enormous challenge, and proved to be a great stimulus for the advance of astronomy and mathematics from ancient Mesopotamia through Renaissance Europe.

Background

Some of the very earliest evidence of recorded history reveals the importance to all civilizations of calendars. Traces of calendars from prehistoric civilizations have been found at Stonehenge as well as at sites in the Americas, but among the earliest calendars about which historians have significant and detailed evidence are those of the Egyptians and Babylonians. The Egyptian calendar relied on practical principles rather than elaborate astronomical investigations; that is, it was a calendar intended to facilitate civic life rather than one produced by or aimed at supporting astronomical observations. A year consisted of 12 months of 30 days each, with 5 additional, or intercalary, days at the end of the year. Paradoxically, this calendar, with its reliable simplicity, became a widely used reference standard for astronomical calculations that was still in use at the time of Nicolaus Copernicus (1473-1543). While the Egyptians did not make use of careful observation or mathematics to produce their calendar, it nevertheless played an important role in the history of science.

The Babylonians and Greeks both developed far more complex calendar systems. The Babylonians were sophisticated astronomers. Their calendar was based primarily on observations of the Moon—a new Babylonian month began with the evening when the crescent moon was first visible at twilight, and days were counted from sunset to sunset. (Different cultures maintained different definitions of a day, starting variously at times from dawn to midnight.) Although the new moon seems like an easy and intuitive way to demarcate the calendar, in fact it leads to problems of great complexity because of the slight irregularities in reconciling each lunar and solar cycle. Their powerful interest in astro-

logical and astronomical observations, coupled with their complicated calendar, helped lead to advanced development of mathematical techniques by the Babylonians. The Jewish calendar, dating from around the ninth century A.D. and still in use for religious purposes in the twenty-first century, was derived from the Babylonian calendar in combination with Biblical calculations to place the date of creation at approximately 3761 B.C.

Like the Babylonians, Greek observers as early as the fifth century B.C. constructed astronomical calendars. They too wrestled with the challenge of reconciling lunar and solar observations: because a solar year is not an integral multiple of a lunar month, one cannot create a simple calendar that matches months to seasons and to years. One early strategy tried by Greek astronomers was to note that 19 solar years are very nearly equal to 235 lunar months, and to design a calendar that included 12 years of 12 months and 7 years of 13 months in every 19-year cycle. Hipparchus refined this calendar in the second century B.C. to take into account the precession of the equinoxes. This enabled him to give a value for the length of a year that is very close to modern calculations. This sophisticated calendar was used for astronomical investigations but had little impact on civil life, as politicians in the various Greek city-states chose instead to use lunar calendars, which they adjusted more or less at will to suit their needs.

Elsewhere in the world, societies struggled with the same needs and scientific challenges in calendar design. The oldest known Hindu calendar, based primarily on lunar revolutions with necessary adjustments for solar reckoning, dates back to 1000 B.C. Evidence from ancient bone inscriptions show that the Chinese had already established the length of the solar year at $365\frac{1}{4}$ days and the lunar month at $29\frac{1}{2}$ days as early as the fourteenth century B.C. The Chinese also made use of the relationship of solar years to lunar months discovered by the Greeks—in fact, they may have discovered this relationship sometime earlier than the Greeks. In the Americas, civilizations including the Maya made use of a civil calendar of 260 days as well as a solar calendar of 365 days, taken together to form a longer cycle of 52 years known as the Calendar Round.

Impact

The early Roman republic relied upon a lunar calendar that became frustratingly out of phase with the seasons as years passed. Julius Caesar (102-44 B.C.), in the middle of the first century B.C., invited an Alexandrian astronomer to reform the Roman calendar. This spurred the adoption of an essentially solar calendar with a year of 365¼ days; it was decided that each year would have 365 days with every fourth year containing an extra day. This "Julian" calendar required subsequent modifications, as errors were made in assigning dates and calculating changes. But it proved to be quite robust, and remained the dominant calendar in Europe until the sixteenth century.

For Christian Europe, the greatest concern surrounding the calendar was to assure the proper identification of feast days, especially Easter. In the early centuries of Christianity, different sects chose to celebrate Easter on different days, based in part on conflicting testimony from the four Gospel texts. Justifying the selection of a particular date for Easter was thus of great importance in helping to support the ultimate authority of each church. The pursuit of accurate astronomical data to support the proper identification of Easter was a great stimulus to the collection of solar and lunar information by the Church. The date of Easter depended upon both the vernal equinox and the phases of the Moon, and so required a reconciliation of lunar and solar calendars—the perennial challenge of calendar design. In addition, the date of Easter also depended upon the date of Passover in the Jewish calendar, in which days were calculated to begin at sunset and were therefore confusing to reconcile with other calendars. Medieval churchmen relied upon a system of patterns of dates to pick Easter Sunday in a given year. Their method was ultimately deemed unsatisfactory, as it was often out of cycle with the astronomical full moon.

By the sixteenth century, efforts to maintain the date of Easter were in crisis. Calendrical errors had compounded to the point that the vernal equinox had moved 10 days from its true date. The pope, under orders from the ecclesiastical Council of Trent, began in 1545 to try to correct the situation. No solution was forthcoming until 1582, when a series of reforms suggested by astronomers was finally put into place. In addition to laying down new rules for the calculation of the date of Easter, the new "Gregorian" calendar adjusted the number of leap years and changed the length of the year slightly. Adoption of the Gregorian calendar was not uniform throughout Europe, and it took more than three centuries before all of Europe finally adopted the "New Style" calendar. In fact, Greece only adopted this calendar in 1923.

The construction and maintenance of calendars through history has been a project informed both by science and civic life. The collection of accurate astronomical information was recognized by all as very valuable to calendar design, and throughout history the political desire to have useful and practical calendars stimulated and supported astronomical research. The Catholic Church illustrated this most dramatically during the sixteenth through eighteenth centuries. At the same time that the Church was condemning astronomers such as Galileo (1564-1642) for advancing Copernicus's hypothesis that Earth revolved around the Sun, Church-sponsored astronomers made use of Copernicus's models to get more accurate data to establish the necessary dates for Easter and the other Christian feast days. Several major cathedrals were in fact designed for use as solar observatories to further facilitate these investigations.

As the world became more interconnected, the need for shared calendars or at least reliable means of translating dates from one calendar to another grew. Scientists and engineers also came to require more and more accurate means of timekeeping, which in turn produced ever more subtle adjustments to the practical calendars in use during the twentieth and twenty-first centuries. Calendars are completely man-made guides to the predictable but complex patterns of nature and the universe. Many of the difficulties in calendar creation were recognized during ancient times, and astronomical and calendrical sophistication supported one another's development for many centuries. Calendar creation and improvement is perhaps the earliest example of political, civic, and religious leaders turning for answers to scientific experts to solve their practical problems. That relationship of science to power has over the centuries grown to include fields from munitions to medicine, but none has exceeded mundane calendar creation in its basic importance to civic life.

LOREN BUTLER FEFFER

Further Reading

Berry, Arthur. *A Short History of Astronomy from Earliest Times through the Nineteenth Century*. New York: Dover, 1961.

Dreyer, J.L.E. *A History of Astronomy from Thales to Kepler*. New York: Dover Press, 1953.

Evans, James. *The History and Practice of Ancient Astronomy*. New York/Oxford: Oxford University Press, 1998.

Heilbron, J.L. *The Sun in the Church: Cathedrals as Solar Observatories*. Cambridge: Harvard University Press, 1999.

Lindberg, David. *The Beginnings of Western Science*. Chicago: University of Chicago Press, 1992.

Lloyd, G.E.R. *Early Greek Science: Thales to Aristotle*. Cambridge: Cambridge University Press, 1970.

Lloyd, G.E.R. *Greek Science after Aristotle*. Cambridge: Cambridge University Press, 1973.

Neugebauer, Otto. *The Exact Sciences in Antiquity*. Princeton: Princeton University Press, 1952.

Richards, E.G. *Mapping Time: The Calendar and its History*. Oxford: Oxford University Press, 1998.

The Importance of the Eclipse in Ancient Society

Overview

Humans have always sought to bring order and stability to their lives. From the earliest times, ancient people were drawn to the majesty and permanence of the heavens. Over time, early societies attached religious and spiritual significance to the regular motions of the celestial realm. When an astronomical event such as an eclipse took place, most ancient people believed it was the action of a divine being. When the rays of the Sun or the glow of the Moon were extinguished, early societies believed this was a terrible omen and that some type of disaster was imminent.

Background

Life for early man was short and traumatic, and he longed for permanence in a world that was governed by change. Humans were able to succeed in this highly competitive environment because they had important advantages over animals. Specifically, their large brains allowed them to develop the critical thinking skills that eventually evolved into their most powerful tool. This reasoning capacity was expanded into methods of practical application because their hands were free from the task of locomotion and their opposable thumbs gave them the dexterity to create and use tools and weapons.

Early survival of humans was linked to this very powerful combination of reason and physical skill. Most importantly, human ancestors developed the intellectual skill of recognizing cause and effect relationships. For instance, this powerful habit of mind allowed them to make the connection between water holes, animal herds, and successful hunting. In time, the perception that all reality revolved around this cause and effect intellectual model stimulated them to inquire into the ultimate cause of all natural phenomena. Eventually, they gave spiritual significance to the forces they could not control. Wind, rain, thunder, drought—indeed, severe natural occurrences of any kind—were believed to be the actions of very powerful spiritual beings. These ideas eventually evolved into a primitive theology that gave these powerful entities control over every aspect of existence.

The belief that gods controlled the forces of nature also compelled these early ancestors to attempt to develop techniques of divination. This intellectual heritage drove them to interpret natural phenomena as a sign of the will and intention of divine beings.

This quest for understanding eventually was focused toward the heavens. Because change, sometimes in violent form, was a constant in the lives of Paleolithic societies, they were moved to find some stability in their existence. Close observation of the heavens gave them the vision of a celestial realm of consistency and permanence. Coupled with the observation that most dramatic natural events originated in the sky above them, Paleolithic society reasoned that the heavens were the domain of these powerful beings that had such a profound impact on their daily lives. This naturally increased the importance of astronomical events, and from the very beginning of the systematic observation, these celestial occurrences took on great religious significance.

The intellectual capacity of humankind increased with the onset of the Neolithic Revolution. Sedentary agriculture and urbanization dramatically restructured human society and accelerated the invention of new skills. The most important individual advance of the Neolithic period was the invention of writing. This revolutionary tool evolved from early attempts to keep track of food surpluses to the eventual creation of a system of self-expression where everything from listing basic statistical data to the discussion of complex ideas could be recorded, transmitted, and preserved.

This created the greatest increase in knowledge and creativity in human history, because writing permitted a more elaborate intellectual life. In the area of scientific inquiry, it proved to be a revolutionary skill. The ability to keep extensive data increased the accuracy of scientific investigation. This was especially true in the collection of astronomical data, which was so important in the creation of the first calendars. Neolithic society relied heavily on the accuracy of these calendars for the construction and operation of their irrigation systems, which were the basis of the Neolithic agricultural revolution. This coincided with the development of more abstract religious thought and the creation of early epic literature, which was used to convey new value systems.

An intellectual connection developed between the newly organized religious beliefs and the science of astronomy. The Paleolithic concept of a heavenly realm continued into the Neolithic Revolution. The importance of this connection grew as the necessity of accurate astronomical data increased. Celestial events took on added importance because they connected the actions of the gods with the success of the great agricultural enterprise of the Neolithic Revolution. Events such as eclipses portrayed the disruption of both the natural and religious order of the universe.

Very few astronomical events have the impact of a solar or lunar eclipse. The darkening of the sky, whether it is day or night, was looked upon by these societies as a sign that the natural order of the cosmos was disrupted.

Impact

The first Neolithic societies to develop the scientific foundation of astronomy were the early people of Mesopotamia. Because the weather in that area was so harsh and unpredictable, it was paramount that accurate predictions concerning the onset of heavy rains be made. This concern focused many Mesopotamian intellectuals on the science of astronomy. These early sky watchers kept very detailed records, and the first account of an eclipse in this region occurred in 1375 B.C. Beginning at least with the reign of King Nabonassar (r. 747-734 B.C.), ancient Babylon-

ian society kept a detailed log of astronomical events, including both solar and lunar eclipses. Eventually, Babylonian astronomers were able to determine the intervals between such eclipses, intervals that broadly repeat every 18 years—a cycle called the *saros*. Among the early Babylonian astronomers who made important discoveries regarding eclipses were Nabu-rimanni (fl. c. 490 B.C.) and Kiddinu (fl. c. 350 B.C.).

The Sumerians and Babylonians both developed religions based upon a pantheon of anthropomorphic gods. These powerful beings were given exaggerated human characteristics. It was widely accepted by both societies that these deities controlled natural phenomena. History's first great intellectual class was the priests of ancient Mesopotamia, who were both scientists and theologians. They had the important assignment of performing the rituals necessary to placate the gods and at the same time make the astronomical calculations needed to run their irrigation systems. Eventually, religion and astronomy merged, and an important celestial event such as an eclipse was perceived to have a potential impact on everyone in Mesopotamia.

Ancient China was another example of the merging of the science and culture. In this case, it was the relationship among government, political philosophy, and science. By the time of the Shang Dynasty (1600-1050 B.C.), professional astronomy would be part of the government bureaucracy. These intellectuals were charged with keeping track of the movement of the Sun and Moon in relation to Earth. The earliest record of a solar eclipse occurs in Shang literature in 2134 B.C. It was during the rule of this dynasty that it became widely accepted that all astronomical events reflected the will of the gods.

The importance of astronomy, and in particular an eclipse, rose to unprecedented heights following the rise of the Zhou dynasty (1027-221 B.C.). The Duke of Zhou led a successful military campaign against the Shang Dynasty. The new emperor declared his cause was just because of the corruption of the Shang Dynasty, and he set about creating an entirely new social and political paradigm. China's major deity was now referred to as "Heaven." The new ruler was now considered to be the son of Heaven and his divine right to rule was referred to as the Mandate of Heaven. This new political philosophy was based upon a hierarchical structure with the power flowing from God to the monarch. The emperor only maintained his power as long as he was in the good graces of the chief deity. The ruler maintained his positive relationship with Heaven as long as Chinese society was economically and politically stable. As soon as he forgot about or violated this responsibility, he would lose the mandate and another contender had the right to overthrow his government. The astronomical bureaucracy now took on added importance, because it played an important role in the enforcement of the Mandate of Heaven. Unscheduled important celestial events were regarded as a sign of Heaven's displeasure. If the government made an accurate announcement about an eclipse, this would be accepted as the affirmation of the emperor's performance. When in fact the astronomical bureau missed an eclipse, the political and social consequences were perceived to be so grave that the leader of the bureau would be executed by decapitation.

The importance of an eclipse changed radically with the development of ancient Greek culture. The classical Greeks were the first people to separate science from religion. This was the result of their rejection of the traditional religious explanation for natural phenomena. The Greeks believed that the natural world was governed by universal laws and not by the whim of supernatural beings. Through the development and use of reason, the Greeks believed that these natural laws could be discovered, understood, and used for the benefit of the human community. Their natural philosophers were the first to develop the astronomical subdiscipline of cosmology, which sought to discover the origin, foundation, and laws of the universe. Through their investiga-

tions, the Greeks tried to ascertain the relationship between these natural laws and the natural order of the universe. By discarding the theological implications of natural events, the Greeks were able to focus strictly on the basic results of their observations and data. From observing the interaction of the Sun, Moon, and Earth, these early scientists began to develop the first natural models of the universe. By observing an eclipse from a purely natural vantage point, the Greeks correctly described Earth as a sphere and the cause of an eclipse as the interaction of the movements of the Sun, Moon, and Earth. They debated whether the Sun or Earth was at the center of this system and eventually—incorrectly—chose Earth. Nevertheless, the Greeks were perhaps the first people to ask purely scientific questions and then to seek to answer these inquiries through scientific observation. This movement away from theological speculation toward scientific observation would set the stage for the great scientific discoveries of both the Islamic and Western civilizations.

RICHARD D. FITZGERALD

Further Reading

Krupp, E.C. *Echoes of the Ancient Skies: Astronomy of Lost Civilizations*. New York: Oxford University Press, 1983.

Temple, Robert. *The Genius of China: 3,000 Years of Science, Discovery and Invention*. New York: Simon and Schuster, 1986.

White, K.D. *Greek and Roman Technology*. Ithaca: Cornell University Press, 1984.

Cosmology in the Ancient World

Overview

Attempts to make sense of the universe have been made by human societies since prehistoric times. These efforts, taken as an effort to understand the universe as a whole, are called cosmologies. While the earliest known cosmologies explained phenomena in terms of specific mythological accounts, Greek philosophers in the sixth century B.C. first sought to find naturalistic explanations for the motions of the stars and planets. By the time of Ptolemy in the second century A.D., a cosmology and a mathematically complex model of the universe had been constructed that would remain essentially unchallenged until the sixteenth century.

Background

Evidence of Babylonian and Egyptian achievements in mathematics, astronomy, and astrology date back as far as 3000 B.C. The Babylonians eventually achieved very accurate knowledge of the motions of the Sun, Moon, and planets. Their astronomical knowledge made use of mathematical theory, not merely observations, and was motivated by a desire to record and understand periodic phenomena (including eclipses) for religious and astrological purposes, as well as to support their lunar calendar and agricultural activities. But this fairly sophisticated mathematical astronomy made no claims about the causes of the motion of the heavenly bodies, or about their nature or origins. To answer these questions, the Babylonians turned to mythological accounts of the behavior of their gods. For example, one Babylonian account attributes the origin of the world to a sexual liaison between the god of the waters, Enki, and the goddess of the soil, Ninhursag. The Egyptians did not achieve the same high level of mathematical sophistication in their astronomy, but their cosmology was similarly limited to mythical explanations rooted in the actions of specific gods.

The conceptual and cultural leap from explaining the universe by reference to capricious acts of single deities to seeking explanation based on universal principles or laws cleared perhaps the greatest hurdle in the history of ancient science. In making this leap, gods and myths were not abandoned. But their role in describing the origins and nature of the physical world was diminished as people sought explanations of phenomena that were regular, harmonious, and above all natural. While these early naturalistic accounts may still sound fantastic and even mythical to modern ears, they nevertheless represented a major shift in worldview for the people who believed them. The Milesians philosophers, including Thales (c. 624-c. 547 B.C.) and Anaximander (c. 610-c. 546 B.C.), are considered to be among the first to speculate about the world around them in a manner that sought natural causes and employed critical analysis in their study. The Milesians produced naturalistic accounts of such phenomena as earthquakes, thunderbolts, and the origin of animals, as well as a general cosmology that suggested that the universe grew from a seed made of a primary, undifferentiated substance. A dual process of condensation and evaporation explained change of all kinds.

The Pythagoreans, a group of Greek thinkers devoted to mathematics that flourished in the fifth century B.C., were the first to postulate a law-based, quantitative cosmology. The Pythagoreans believed that everything in the universe was made up of numbers and could be analyzed through study of proportions. They had discovered that musical harmonies could be expressed as numerical ratios, and they applied this idea to the motion of the heavens. They argued that the movements of the stars and planets made music, music that we cannot hear because we have been perpetually exposed to it since birth. The Pythagoreans believed that Earth and all the other bodies revolved around an invisible central fire, and suggested that an invisible "counter-earth" also circled the central fire. This latter, rather fantastic, idea was postulated to account for the frequency of lunar eclipses.

Another group of early Greek philosophers is known to us as the atomists, and included Leucippus (fl. fifth cent. B.C.) and Democritus (c. 460-c. 370 B.C.). The basic premise of atomism was that atoms and the void between them are the fundamental elements of the universe. Differences among physical entities were caused by differences in the shape, position, and arrangement of the constituent atoms. These atoms are considered to be in constant motion, with collisions among them responsible for change of all kinds. While it is a mistake to identify these ideas too closely with the mechanistic science of the seventeenth century and beyond, the ancient atomists did take the important step of conjuring a worldview in which nothing depended on gods or human minds.

One of the most important cosmologies that have survived from ancient Greece is Plato's famous dialogue, the *Timaeus*. This work continued to be influential well into the Middle Ages. Plato (c. 427-347 B.C.) rejected the idea that the universe could be reduced to mere matter and motion. He contended that it was instead the handiwork of a divine craftsman that he called the Demiurge. The Demiurge constructed the universe on rational geometric principles, associating the elements of earth, air, water, and fire with four of the five regular geometric solids. The fifth solid, the dodecahedron, was associated with the cosmos as a whole. As with the cosmology described by the Pythagoreans, this was an early effort to describe the universe in wholly mathematical terms. Although the *Timaeus* was especially important to scholars of the early Middle Ages, by the thirteenth century—as more sources were being recovered from Arabic translations—Plato's influence was eclipsed by that of Aristotle (384-322 B.C.).

Impact

No Greek thinker had more influence on Western thought than Aristotle. Aristotle's cosmos—which he held to be eternal, denying the possibility of a creation or an end—was entirely spherical. A lower region, demarcated by the orbit of the Moon, included Earth at its center and allowed the possibility of changes such as birth, death, and ordinary motion. The upper celestial region consisted of eternally unchanging bodies that moved about in an ether—there was no void or empty space in Aristotle's universe. Aristotle assigned qualities of hot, cold, wet, dry, heavy, and light to the terrestrial elements, and used these qualities to explain various observed phenomena. In the celestial region, the heavenly bodies were understood to be attached to spheres made of an incorruptible crystalline substance. All motion in the celestial region was spherical, a condition that led to complex arrangements of spherical patterns to explain the observed motions of the planets. The mathematical astronomer Ptolemy (c. A.D. 100-c. 170), who worked in the second century A.D. in the late years of the Hellenistic period, synthesized several centuries of advances in mathematics and astronomy to produce remarkably sophisticated mathematical models of the motion of the stars and planets. Ptolemy's mathematical astronomy, based on Aristotelian cosmology, remained predominant in Western thought until the work of Nicolaus Copernicus (1473-1543) and Galileo (1564-1642) many centuries later.

Perhaps the greatest intellectual project of the Middle Ages was the effort to reconcile Aristotelian cosmology with Christian theology. Issues arising from conflicts between the two resonated throughout the schools and churches, and several times were the subject of papal decrees. For most of the Middle Ages, scholars were primarily concerned with explaining Aristotelian principles such as the crystalline spheres in a way that seemed compatible with Christian beliefs. But as observational and mathematical astronomy gained in sophistication, scholars began to find problems with Ptolemy's mathematical models as well. These challenges culminated in the famous conflicts of the period known as the Scientific Revolution, when Copernicus, Galileo, and others broke with Ptolemaic tradition and argued that Earth was not at the center of the universe after all. Because of the tight connection that had been forged between Aristotelian cosmology and Christianity, it was difficult for the Church to accept any such challenges. Rather than merely intellectual speculation, the suggestion that Earth moved about the Sun was considered blasphemous, and its proponents suffered various well-known consequences.

Since ancient times, cosmology has marked the boundary between science as a way of explaining the universe and the larger worldview that gives it meaning. Cosmology addresses grand philosophical questions such as what is the nature of change and how did the universe come to exist; it also generates specific, testable questions about motion and matter. Between the speculation necessary to address the largest questions, and the detailed analysis required to address the more specific, the pursuit of understanding the universe has shaped both science and religion. The most important cosmologies of the ancient Greeks influenced Western and Eastern thought for 1600 years, and thus formed the context for the emergence of the modern sciences of astronomy and astrophysics.

LOREN BUTLER FEFFER

Further Reading

Berry, Arthur. *A Short History of Astronomy from Earliest Times through the Nineteenth Century.* New York: Dover, 1961.

Dreyer, J.L.E. *A History of Astronomy from Thales to Kepler.* New York: Dover Press, 1953.

Evans, James. *The History and Practice of Ancient Astronomy.* New York/Oxford: Oxford University Press, 1998.

Furley, David. *The Greek Cosmologists.* Cambridge: Cambridge University Press, 1987.

Lindberg, David. *The Beginnings of Western Science.* Chicago: University of Chicago Press, 1992.

Lloyd, G.E.R. *Early Greek Science: Thales to Aristotle.* Cambridge: Cambridge University Press, 1970.

Lloyd, G.E.R. *Greek Science after Aristotle.* Cambridge: Cambridge University Press, 1973.

Neugebauer, Otto. *The Exact Sciences in Antiquity.* Princeton: Princeton University Press, 1952.

Geocentrism vs. Heliocentrism:
Ancient Disputes

Overview

During the second century A.D., Greek-Egyptian astronomer and mathematician Ptolemy (100-170) summarized eight centuries of Greek geocentric (earth-centered) thought about the nature of the cosmos. Despite the heliocentric (sun-centered) theories of Aristarchus of Samos (320?-250? B.C.) and a few others, Ptolemaic geocentrism dominated Western astronomy until Nicolaus Copernicus (1473-1543) proposed his heliocentric theory in the sixteenth century.

Background

In the sixth century B.C. the philosopher Pythagoras (580?-500 B.C.) founded a school of thought that concentrated on order, harmony, permanence, rationality, and regularity. His ideals were music and mathematics. Music was viewed as the origin and expression of harmony, and mathematics the rational explanation of music. Pythagoras believed that everything could be understood in terms of number, and therefore that everything is accessible to the mind, since the concept of number is intelligible. He posited a geocentric universe in which the Moon, Sun, and the five known planets (Mercury, Venus, Mars, Jupiter, and Saturn) all moved in perfect geometrical order by virtue of their natural and eternal mathematical relationships. He saw the geometry of space as "the music of the spheres," the ultimate harmony. He clearly recognized that the Earth is a sphere.

The most productive periods in the history of science are Pythagorean periods, that is, periods in which number and quantity are given preeminent roles in scientific investigation. Especially in the ancient world, nearly all progress in science was influenced in some way by Pythagoras.

In the fifth century B.C. the Pythagorean astronomer Philolaus departed from the geocentric model. He suggested that the Earth revolved not around the Sun, but around a cosmic central fire, around which the Sun also revolved. To explain why this central fire was never seen from Earth, Philolaus imagined the existence of an "anti-earth" always between the Earth and the fire.

The universe propounded by Plato (427?-347 B.C.) in the early fourth century B.C. was essentially Pythagorean. He emphasized the perfection, divinity, and eternity of spheres and circles, but disdained empirical observations of the heavens. Even though Platonic cosmology was grounded in questionable science, its influence on Western theology, philosophy, and culture persists even into the twenty-first century.

Sophisticated geocentric theory began with Plato's contemporary, Eudoxus of Cnidus (400?-350? B.C.), who proposed an onion-like arrangement of 27 concentric spheres, with the Earth innermost and the fixed stars outermost. Each planet needed four spheres to explain its observed motion, the Sun and Moon each needed three, but the fixed stars only needed one. Later in the fourth century B.C. this system was made more complicated by Callipus, and even more complicated by Aristotle (384-322 B.C.), who posited 55 spheres, all moved by an eternal "unmoved mover" or *primum mobile* outside the outermost sphere. The advantage of Eudoxus, Callipus, and Aristotle over Plato consisted in their use of observation as well as speculation. As a result, the influence of Aristotle's cosmology on Western theology, philosophy, and culture has been even greater than that of Plato, especially through the legacy of Thomas Aquinas (1225-1274). Aristotle's successors continued to modify geocentric theory until the second century A.D.

One reason that the ancients, especially the Greeks, favored a geocentric model of the universe is that the Earth obviously had great weight, whereas the Moon, planets, and stars were believed to be light and airy or fiery. Weight would naturally determine the center of the celestial spheres, i.e., whichever body weighed the most would be central.

In Alexandria, Egypt, in the third century B.C., Aristarchus of Samos, a student of the Aristotelian Strato of Lampsacus (?-270? B.C.), calculated the relative distances of the Sun, Earth, and Moon from one another by measuring the Moon-Earth-Sun angle as it changed through the various phases of the Moon. His observations and calculations showed that that Sun was about 20 times the moon's distance from the Earth, and that the Sun was much larger than the Earth or Moon. His method was sound, but his primitive instruments impaired his results. In fact, the Sun is about 390 times the Moon's distance from the Earth. He further reasoned that since the sun is so large and so far away, it must also be proportionately greater in weight.

Aristarchus's correct conclusions that the Sun is larger and more massive than the Earth, and so distant, led him to suppose that the universe is heliocentric. He was the first major thinker to suggest such a theory and to support it with empirical data. His book, *On the Sizes and Distances of the Sun and Moon*, survives, but all his writings on heliocentrism are lost. Thus his heliocentric theory is not known in detail, and is only known at all through reports by Archimedes (287?-212 B.C.) in the third century B.C. and Plutarch in the second century A.D.

The Greeks generally rejected Aristarchus's heliocentric theory, but he had a few supporters. Seleucus, for example, as reported by Plutarch, defended heliocentrism in the second century B.C. The observations of Timocharis and Aristyllus, contemporaries of Aristarchus, may have been intended to support his heliocentrism.

Around the beginning of the second century A.D., Menelaus of Alexandria, a Hellenistic Egyptian mathematician and astronomer, invented spherical geometry. This innovation was of tremendous importance for astronomy, because the treatment of concentric arcs in the geometry of spheres is analogous to the treatment of lines in the Euclidian geometry of planes. Although Menelaus's original Greek book on spherical geometry, *Sphaerica,* is lost, its content survives in Arabic translation.

About a generation later, Ptolemy, another Hellenistic Egyptian mathematician and astronomer, achieved a mathematically complete geocentric system and published it in a large book best known as the *Almagest.* Ptolemy's mathematics was able to explain, albeit in a very complicated way, all the apparent retrograde motions of the planets. He relied heavily upon the mathematics of epicycles developed by Apollonius of Perga (262?-190? B.C.) in the third century B.C. and upon the trigonometry developed by Hipparchus of Nicaea in the second century B.C. Ptolemy's work soon became the definitive geocentric system.

Impact

Perhaps the clearest example of the extent to which Ptolemaic geocentrism determined Christian cosmology throughout the Middle Ages is Dante Alighieri's *Divine Comedy,* written in about 1310. Dante depicted the Earth as a sphere, with Jerusalem or Zion diametrically opposite the mountain of Purgatory. Hell was inside the Earth, and the Garden of Eden was atop Mount Purgatory. The Earth was surrounded by a sphere of fire, beyond which was Heaven or Paradise, consisting of 10 concentric spheres: 1) the sphere of the Moon; 2) the sphere of Mercury; 3) the sphere of Venus; 4) the sphere of the Sun; 5) the sphere of Mars; 6) the sphere of Jupiter; 7) the sphere of Saturn; 8) the fixed stars and the zodiac; 9) the Aristotelian *primum mobile;* and 10) the Empyrean sphere, i.e., the sphere of pure light, beyond which was only God. Ten is a perfect number in Pythagorean numerology. As Dante, on his journey through the Christian theological universe, emerged from the Garden of Eden and first perceived Paradise, he heard the Pythagorean music of the spheres.

No serious thinker since the time of Ptolemy has believed that the world was flat. Contrary to popular mythology, Christopher Columbus (1451-1506) did not sail west to prove to Europeans that the world was round. All educated people of his time already knew it was round. Columbus sailed to prove that he could navigate safely to Asia westward across open ocean, out of sight of land, thus avoiding the two disadvantageous eastward routes, either the perilous trek overland or the long voyage around Africa, keeping land in sight all the way.

Geocentric cosmology dominated Western thought until the early modern era. Copernicus developed a plausible heliocentric theory in about 1512, but circulated it privately and secretly because he was afraid of the possible reaction against him. The Roman Catholic Church vigorously opposed heliocentric cosmology and persecuted the thinkers who believed in it. Copernicus finally published his heliocentric conclusions in *De revolutionibus orbium coelestium* (On the revolutions of the heavenly spheres) in 1543, the year of his death.

Heliocentrism is much simpler than geocentrism because it has no need for elaborate mathematical stratagems to account for retrograde motions. This fact was very attractive to Copernicus and his successors.

During the 1633 Inquisition, the fearful Galileo (1564-1642), recalling that another defender of Copernicus, Giordano Bruno (1548-1600), had been burned at the stake by the Church in 1600, publicly recanted his own findings and affirmed the Church's official view that the Sun orbited a stationary Earth. But in an aside, in accordance with his 1632 book, *Dialogo dei due massimi sistemi del mondo* (Dialogue on the two main world systems), which argued that the Earth orbited the Sun, he is supposed to have muttered under his breath, "And yet it moves." The Inquisition convicted Galileo, and he spent the rest of his life under house arrest and close scrutiny.

ERIC V.D. LUFT

Further Reading

Brecher, Kenneth, and Michael Feirtag, eds. *Astronomy of the Ancients.* Cambridge, MA: MIT Press, 1979.

Britton, John Phillips. *Models and Precision: The Quality of Ptolemy's Observations and Parameters.* New York: Garland, 1992.

Evans, James. *The History and Practice of Ancient Astronomy.* New York: Oxford University Press, 1998.

Gingerich, Owen. *The Eye of Heaven: Ptolemy, Copernicus, Kepler.* New York: American Institute of Physics, 1993.

Goldstein, Bernard R. *Theory and Observation in Ancient and Medieval Astronomy.* London: Variorum, 1985.

Hadingham, Evan. *Early Man and the Cosmos.* New York: Walker, 1984.

Hetherington, Norriss S. *Ancient Astronomy and Civilization.* Tucson, AZ: Pachart, 1987.

Krupp, Edwin C., ed. *Archaeoastronomy and the Roots of Science.* Boulder, CO: Westview Press for the American Association for the Advancement of Science, 1984.

Krupp, Edwin C. *Echoes of the Ancient Skies: The Astronomy of Lost Civilizations.* New York: Oxford University Press, 1994.

Neugebauer, Otto. *A History of Ancient Mathematical Astronomy.* Berlin: Springer, 1975.

Taub, Liba Chaia. *Ptolemy's Universe: The Natural Philosophical and Ethical Foundations of Ptolemy's Astronomy.* Chicago: Open Court, 1993.

Thurston, Hugh. *Early Astronomy.* New York: Springer, 1994.

Aristotle's Chemical Theory of Elements and Substances

Overview

Aristotle (384-322 B.C.) remains the single most influential philosopher in Western history. His theories of substance and accidents, the four elements, and of elemental transformations and combinations, dominated Western matter theory, alchemy, and early chemistry for some 2,000 years. Although neither the theories themselves nor their specific details are accepted today, the issues he framed and questions he posed in offering them remain vital to both modern philosophical and scientific thought. Because of the extraordinary richness and complexity of his ideas, many points of interpretation remain controversial, and in recent years new explanations have challenged traditional views, significantly revising how Aristotle is being read and understood.

Background

Like other philosophers before and after him, Aristotle's primary concern was to explain the fundamental principles of being, existence, and physical reality. His main account of these is set forth in the *Metaphysics,* an extraordinarily difficult but seminal philosophical work. In it, Aristotle employs a highly technical vocabulary to present a complex theory of substance and nature; form and matter; actuality and potentiality; essences, properties, and accidents; and causes and changes. For Aristotle, being, existence, and reality are defined by "substance" (*ousia*). The technical, logical definition of a substance is a subject of predication, but is itself not a predicate of anything else. (For example, in the statement "the blind old dog," "blind" and "old" are

predicates attributed to "dog," but not the reverse.) In more concrete terms, a physical substance is a unitary entity having a particular "nature" (*physis*), or innate principle that describes and directs all of its activities. Each physical substance is a composite—a complex unity of definition or "form" (*eidos*) and content or "matter" (*hyle*), which respectively manifest its "actuality" (*energeia*), or present reality and pattern of activity, and "potentiality" (*dynamis*), or latent capacity and power for alternative activities.

The form–matter relation is thus not just one of a determinate container to its indeterminate contents (e.g., of a glass to water), but rather one of defining characteristics for patterns of action (e.g., of the actual programming of a given computer to other possible ways of programming it). Together, the form and nature express the essence, called *to ti en einai* (literally, "what it is to be") or the necessary, primary defining features of a substance that determine its characteristic patterns of activity. All the activities of a substance are directed toward the ultimate purpose or goal (*telos*) of complete and perfect realization (*entelecheia*) of its essence. The essence and nature of a substance also determine which "properties" (*pathe*), or permanent characteristics, and which "accidents" (*symbebekota*) or incidental characteristics, it may possess.

Since essence primarily defines substance, and Aristotle frequently identifies essence with substantial form, there is a sense in which form itself is "primary substance," as antecedent to form–matter composites, while matter is only substance "in potentiality." However, Aristotle generally seems to consider form–matter composites to be the truest substances, devoting the overwhelming bulk of his analysis to these. (Whether an Aristotelian substance can have only one unique substantial form, or whether it may have two such forms, a generic one that is purely formal and a specific one as an individual form–matter composite, is presently a topic of scholarly debate.) As complex form–matter unities of essences, natures, properties, and accidents, with distinctive actualities and potentialities, composite substances may be either inorganic materials or organic beings. These range in complexity from the four terrestrial elements, through simple uniform materials (*homoiomere*) such as iron and blood, and biological limbs and organs, to complete biological organisms, which for Aristotle are substances par excellence.

Having accounted for existence or being, Aristotle next turns his attention to explaining changes in being or "becoming," a main topic of his *Physics*. Substances can undergo two types of change, essential (substantial) or accidental (nonsubstantial). In essential change, one substance undergoes "destruction" (*phthora*) while another undergoes "generation" (*genesis*)—e.g., a caterpillar becomes a butterfly. In accidental change or "motion" (*kinesis*), a substance alters in size, place, or quality, but its essence remains the same—e.g., a man grows thinner, sits down instead of standing, or has hair that changes color from black to gray, but he still remains a man. In both types of change, however, there are three principles involved: two "contrarieties" (*enantion*) which are the initial and final entities or states that respectively are lost and acquired, and the continuant or "underlying subject" (*hypokeimenon*) that undergoes the change by moving from "privation" (*steresis*) to "possession" (*hexis*). As the innate directive principle of substantial activity, a *physis* determines and restricts what kinds of changes a substance may undergo. Thus, whereas the early pre-Socratics denied any possibility of essential change, according to the principle that "nothing can come to be out of nothing," Aristotle circumvented this difficulty by distinguishing between absolute and relative nonbeing—i.e., "X does not exist" vs. "X is not Y"—and by making change a reciprocal relation between two existing things, rather than between existence and nonexistence.

Aristotle also distinguished four modes of change—natural, unnatural, spontaneous, and chance. Natural and unnatural changes take place in accordance with or contrary to a *physis*, the latter occurring by "art" (*techne*) or forcible intervention of an outside agent—e.g., a fish swimming in water vs. one carried through the air by a seagull. "Spontaneous" (*automata*) and "chance" (*tyche*) changes result from intentional acts and unintentional coincidences that are not directed toward realization of a substantial *telos*. Finally, all change is to be accounted for in terms of four "causes" (*aitia*) or "principles" (*archai*)—formal, material, efficient, and final—which correspond to the form, matter, active agent, and goal of a change. These causes explain not just how a substance changes, but also why, supplying the process with an origin, content, mechanism, and purpose.

Impact

Aristotle's elemental matter theory, to which he applies the foregoing analyses of substantial being and change, exists in two versions. (Some scholars have therefore suggested that Aristotle's elements are conceptual abstractions rather than real physical entities.) In Books III and IV of *De caelo* (On the heavens), he presents a model of a finite spherical cosmos, divided into two realms. In the outer superlunary sphere, the Sun, stars and planets move in unvarying perfect circular orbits through an inert celestial element of *ether*, initially set into and perpetually kept in motion by the Prime Mover, an impersonal divine being and original source of all being and becoming. In the inner sublunary sphere, the Earth and its atmosphere consist of the four traditional terrestrial elements (*stoicheia*) of fire, air, water, and earth. These elements are the simplest, most basic composite substances, composed of a completely indeterminate, qualityless substrate of "prime matter" and relative cosmic place as their respective "forms"—fire as absolutely light, air as relatively light, water as relatively heavy, and earth as absolutely heavy. Left undisturbed, the four elements would completely segregate into four concentric layers, with earth at the center, then water, then air, with fire bounded at the outer perimeter by the *ether*. However, the motions of the superlunary bodies transmit perturbative effects to the sublunary sphere that impose unnatural motion upon the elements, keeping them mixed in a constant state of agitation.

In his *De generatione et corruptione* (On generation and destruction), however, Aristotle's account of the four elements is very different. Here, the four terrestrial elements are composite substances, whose forms are not relative places but rather complementary pairs of four primary qualities—hot, cold, dry, wet—that "inform" the indeterminate prime matter. In this scheme, fire

= hot + dry, air = hot + wet, water = cold + wet, earth = cold + dry. (Some scholars contend that these qualities, not the terrestrial "so-called elements," are Aristotle's true sublunary elements.) Any single element can be transformed into another element by a change of one or both of its primary qualities into its opposite, which is a process of substantial generation and destruction. When two different elements come into direct contact without external interference, the less preponderant element spontaneously changes into the more preponderant one. By making the elements composites of complementary pairs of primary qualities with prime matter, rather than simply identifying each element with a single quality as some pre-Socratics did, Aristotle provided a basic mechanism to account for every type of physical change.

The next stage in Aristotle's elemental theory is his account of the *homoiomere,* or basic tangible physical substances, and of *mixis* or chemical combination between them. The individual elements are imperceptible to the senses; the most basic perceptible kinds of cosmic matter are the *homoiomere* (e.g., granite, iron, blood, bone). Each type of *homoiomere* is materially uniform in composition, being constituted out of all four elements in certain characteristic proportions. As constituents of *homoiomere* rather than autonomous substances, the elements have only potential rather than actual existence. Similarly, the primary qualities become merely relative rather than absolute contraries, as nonessential substantial properties rather than essential parts of substantial forms, which mediate and temper each other rather than transforming into one another. The *homoiomere* are thus infinitely divisible, with every part being identical to another. Compatible *homoiomere* can be physically combined to constitute *anhomoiomere,* or complex, higher-order parts and organisms (e.g., leaves, hands, plants, human beings), which are also substances.

Two *homoiomere* also can be chemically combined. For the pre-Socratics, only physical mixture (synthesis) or conjunction between different types of matter was possible, because these could not be created, destroyed, or transformed into one another. But for Aristotle, a truly complete combination or *mixis* of two *homoiomere* is possible, because their constituent elements have merely potential rather than actual existence. In order for two *homoiomere* to undergo *mixis,* three conditions must exist: they must be related as different species of the same material genus; they must be easily divisible in order to facilitate the combination of infinitely

small parts; and they must be present in balanced amounts, as otherwise the dominant substance will merely transform the other completely into itself (e.g., a drop of wine added to 10,000 gallons of water is converted into water). When these conditions are present, the constituent elements and qualities of the two initial *homoiomere* undergo mutual tempering and blending (*krasis*), and are then united by "concoction" (*pepsis*), or the action of heat upon moisture, to constitute a uniform material product. (Scholars disagree over whether *mixis* is a genesis or generation of a new substance, an *alloiosis* or qualitative alteration as a type of *kinesis,* or a *pathos* or material property that merely describes the final product.)

In Book IV of the *Meteorologica,* Aristotle lists and discusses 18 pairs of changes in material substances due to changes in one or more of the primary qualities: susceptibility or resistance to solidifying, melting, softening by heat, softening by water, bending, breaking, fragmenting, impressing, molding, or squeezing, and being or not being tactile, malleable, fissible, cuttable, viscous vs. friable, compressible, combustible, or fumigible. While this work has sometimes been termed "Aristotle's chemical treatise," this is a misnomer. Most of these examples involve simple changes in physical states or single properties, by an increase or decrease of the proportional presence of one primary quality; none involves *mixis,* and few, possibly none, entails an elemental transformation.

In recent years, the standard interpretation of Aristotle's matter theory has been disputed at two major points—the existence of prime matter and the status of the elements as true substances. The newer interpretation holds that prime matter, which is never explicitly mentioned in Aristotle's writings, is an incorrect inference read back into his system by commentators in late antiquity who wished to reconcile Aristotle with Plato (427?-347 B.C.). Instead, the elements are simply their pairs of primary qualities; as such, these pairs are not substantial forms, and the elements are not composite form–matter unities and thus not substances, but mere "heaps" (*soros*), as Aristotle calls them in the *Metaphysics.* References to the elements as substances in the *Physics* and *De caelo* are explained as being informal, nontechnical uses of that term. Accordingly, the elements are considered collectively to constitute the lowest or primary matter of the sublunary sphere, and the *homoiomere* and not the elements are the lowest order of substance.

Due to its far greater sophistication and explanatory power, Aristotle's metaphysical theory of substance and qualitative theory of elemental matter quickly eclipsed the rival theories of the atomists and Stoics, and remained unchallenged in ancient and medieval Western and Islamic culture for almost 2,000 years. While the details of his chemical theory have now been discarded, his concept of substance still remains an indispensable starting point for most philosophical analyses of the nature of physical reality. His concept of being as a dynamic network of activities, rather than a static set of properties, has proved to be extraordinarily farsighted. Recent years have seen a renewed appreciation for applications of the actuality–potentiality distinction to problems in quantum physics, physical chemistry, genetics, and developmental psychology. In many ways, as much as ever, Aristotle still remains a dominant intellectual figure in Western history, "the master of those who know."

JAMES A. ALTENA

Further Reading

Books

Aristotle. *The Complete Works of Aristotle: The Revised Oxford Translation.* 2 vols. Rev. and ed. by Jonathan Barnes. Princeton: Princeton University Press, 1985.

Anton, John P. *Aristotle's Theory of Contrariety.* London: Routledge and Kegan Paul, 1957. See Chaps. 1-5.

Cohen, Sheldon M. *Aristotle on Nature and Incomplete Substance.* Cambridge: Cambridge University Press, 1996. See Chaps. 2-3.

Düring, Ingemar. *Aristotle's Chemical Treatise: Meteorologica, Book IV; With Introduction and Commentary.* New York: Garland, 1980.

Gill, Mary L. *Aristotle on Substance: The Paradox of Unity.* Princeton: Princeton University Press, 1989. See Chaps. 2 and 7.

Solmsen, Friedrich. *Aristotle's System of the Physical World.* Ithaca: Cornell University Press, 1960. See Chaps. 11-21.

Sorabji, Richard R. *Matter, Space, and Motion: Theories in Antiquity and Their Sequel.* Ithaca: Cornell University Press, 1988. See Chap. 2.

Williams, Christopher F.J. *Aristotle's De generatione et corruptione.* Oxford: Clarendon Press, 1982.

Periodical Articles

Bogaard, Paul A. "Heaps or Wholes: Aristotle's Explanation of Compound Bodies." *Isis* 70 (1979): 11-29.

Charlton, William. "Prime Matter: A Rejoinder." *Phronesis* 28 (1983): 197-211.

Joachim, Harold H. "Aristotle's Conception of Chemical Combination." *Journal of Philology* 29 (1904): 72-86.

Sokolowski, Robert. "Matter, Elements and Substance in Aristotle." *Journal of the History of Philosophy* 8 (1970): 263-88.

Ancient Views on Earth's Geography

Overview

Early in the sixth century B.C., Greek philosophy began to develop, and philosophers of this period undertook a serious inquiry into the nature and organization of the world. They explored questions involving the natural sciences, including questions about the substances that constitute Earth and the sky above it. They also began to ponder geographical questions about the shape of Earth and the nature of its origins. Initially, the endeavors of these philosophers were steeped in the pervasive myths of their day. As philosophers began to pose new questions and come up with new answers, however, they began the gradual move away from using mythology to explain natural phenomena. No longer using the arbitrary acts of the gods to explain natural phenomena, philosophers began to believe that Earth was actually an orderly and predictable world that was governed by universal principles. They argued that the causes of lightning, volcanic eruptions, and earthquakes were the same throughout the world and should not be attributed to the acts of various gods. Just what these principles were that governed such natural phenomena proved to be harder to figure out. Firm in their belief, however, Greek philosophers began to apply the term *kosmos* (cosmos) to the orderly world. As Greek philosophers began to understand more about the manner in which the world operated, they began to explore physical aspects of the world—including the study of geodesy (the size and shape of Earth).

Background

Early Greek philosophy increasingly began to concern itself with physics, or natural explanations of phenomena. Most of the work in this area was done in the Greek colony of Ionia, in the southwestern area of modern Turkey, among a group of philosophers now known as the Milesian philosophers (named after Miletus, an Ionian

mary substance that composed Earth and the rest of the universe was a material substance and not an ethereal and unquantifiable substance. Second, the Milesians were monists, meaning that they believed that the primary substance was one and the same throughout the universe.

The Milesians rejected the idea that the world could have been formed out of nothing. Instead, they believed that the world was formed out of a simple material. They believed that this material was finite, meaning that the world, in turn, also had to be finite. In addition to being finite, the Milesians believed that the world was more or less circular, though not necessarily spherical. Milesians believed that the sky, like Earth, was finite, meaning that it too had specific boundaries. Milesian philosophers also believed that there was an immortal source of energy that guided all actions on Earth and in the heavens. The belief in an immortal energy force acting on the universe shows that the Milesians maintained some aspects of their belief in the gods. The Milesian philosophers simply chose to call this force by another name.

The Milesian ideas of materialism and monism greatly influenced future Greek natural philosophers, including the atomists. The atomists, of which Leucippus of Miletus (fl. c. 440 B.C.) and Democritus of Abdera (fl. c. 420 B.C.) were the most influential, flourished in the second half of the fifth century B.C. Leucippus and Democritus argued that the world was composed of atoms, which were tiny particles far too small to be seen with the naked eye. Leucippus and Democritus also argued that these atoms were all made of the same primary substance. The atomists asserted that the motion and configuration of atoms led to the diverse range of objects that exists in the universe.

THE CATACLYSM OF A.D. 535

During the late twentieth century, a number of theories came to the forefront regarding catastrophic geological change with implications for Earth's life forms. Among the most prominent of these was the idea that a comet had wiped out the dinosaurs millions of years ago. But no one was around to see such an event; whereas if dendrochronologist Mike Baillie and amateur archaeologist David Keys are correct, plenty of people saw the catastrophe of A.D. 535. In studying tree rings, Baillie discovered a pattern of severely curtailed growth for the period 535-541. He later published his findings in *Exodus to Arthur,* while Keys laid out his own theory in *Catastrophe*. The latter cited a number of historical texts, including writings by Byzantine, Chinese, and Anglo-Saxon scholars, suggesting that something cataclysmic had happened in 535.

Some geologists maintain that it was an eruption of Krakatoa, the Indonesian volcano that famously exploded in 1883, which would have hurled enough dust into the atmosphere to cause an artificial winter. Whatever the cause, it appears that as Procopius (d. 565) wrote, "The sun gave forth its light without brightness ... for the whole year." Soon afterward, the Byzantine Empire from which Procopius came was struck by the first in a series of plagues. The latter may perhaps be attributed to an imbalance between predators and disease-carrying rats, an imbalance caused by environmental changes.

Those same changes may have also prompted a food shortage on the steppes of central Asia, which in turn spawned a new wave of European invasions, this time led by the Avars. It also appears that the plague, combined with the Avar invasion, forced the Byzantines to give up the attempted reconquest of the Western Roman Empire under Justinian I (r. 527-565). Thereafter western Europe was plunged into a dark age from which it would not soon recover—and it is possible that the cause lies with a volcano on the other side of the world.

JUDSON KNIGHT

Impact

Despite the best efforts of early Greek philosophers, many of them still found it difficult to separate their ideas from the notion that the gods played at least some role in the formation of Earth and in the natural phenomena that occurred on Earth and in the heavens. One exception to this philosophical view was exemplified in the arguments of Xenophanes. Working in the late sixth and early fifth centuries B.C., Xenophanes developed radical ideas on natural philosophy and geography. In a major departure from the Milesian philosophers, Xenophanes asserted that Earth stretched on infinitely in length and width, as well as in depth below its surface. Likewise, the heav-

city). Although only fragmentary and fairly unreliable evidence about the Milesians has passed down to modern scholars, two ideas of the Milesians can be stated authoritatively. First, the Milesians were materialists. They believed that the pri-

ens stretched on infinitely above the surface of Earth. According to these arguments based upon the concept of infinite extent, Xenophanes was forced to rule out the possibility that the Sun and the stars made regular visits to observers on Earth. Therefore, Xenophanes concluded that the rising and setting of the Sun and stars was an optical illusion to observers. He claimed that rising vapors from clouds became incandescent clouds once they had risen to a sufficient height. These incandescent clouds formed the Sun, Moon, and stars with precise regularity. Accordingly, each day brought forth the rise of a new sun, separate and distinct from the one that had risen the day before.

Xenophanes asserted that this infinity of the universe existed with creation by or help from the gods of the Greeks. Xenophanes further asserted that Greek gods merely followed convention and that every culture modeled their gods on themselves and the problems of nature that were particularly relevant to their society. Accordingly, Xenophanes's god—in a foreshadowing of Aristotle's (384-322 B.C.) idea of the Prime Mover—moved the cosmos with the power of his thought.

Although modern observers often criticize Xenophanes, who was one of the most independent thinkers of his day, Xenophanes introduced two radical ideas that had an immeasurable influence on the natural philosophers that followed. First, he was the first philosopher to free himself from the shackles of the myth-based models that governed the endeavors of previous philosophers. Second, he was one of the first natural philosophers to adamantly advocate using observation of that which could be seen to explain that which could not be seen. Although he did not come to the correct conclusions, Xenophanes changed the dialogue among natural philosophers and forcing them to consider new questions and different approaches in their attempts to better understand the world around them.

Another Greek natural philosopher who did not share the monist or materialist view of the Milesians was Empedocles (c. 490-430 B.C.). A rough contemporary of Leucippus and Democritus, Empedocles believed that four distinct elements comprised all objects in the universe: earth, air, fire, and water. The belief that these elements combined in various combinations to form all observable objects was adopted and popularized about 150 years later by Aristotle. Perhaps more important than Empedocles's assertion that these four materials formed all objects was his introduction of immaterial principles governing the manner in which elements joined together. Empedocles believed that love led the four elements to join together in certain combinations and strife led them to separate.

Like Xenophanes, Empedocles believed that cyclical changes occurred in the cosmos. Although Xenophanes saw these cycles occurring every day with the rising and setting of the Sun, Moon, and stars, Empedocles asserted that cyclical changes involving the creation and destruction of the entire universe occurred over vast periods of time. Empedocles held that initially all four of the elements (earth, air, fire, and water) were blended together into a homogenous sphere, held together by the love of the gods. Eventually, strife began to enter this sphere, breaking it apart. Empedocles viewed the world as being in a state in which this process was being carried out, with air separated from earth and water, and earth and water separated to form land and the oceans. Eventually, all of the elements would separate completely, forming four spheres, each representing one of the elements. Love would then enter the equation, bringing these four spheres back together again.

The influence of both Xenophanes and Empedocles can be found in the works of Aristotle. In accord with Xenophanes, Aristotle firmly believed that observation was the key by which one could better understand terrestrial events that could not otherwise be explained. Aristotle's observations led him to two important conclusions concerning Earth and the heavens: 1) Earth is a realm of constant change in which birth, growth, and death occur everyday; 2) in contrast, the heavens are an area comprised of bodies of uniform circular motion and unwavering change. In his work *On the Heavens,* Aristotle argued, "In the whole range of times past, so far as our inherited records reach, no change appears to have taken place either in the whole scheme of the outermost heaven or in any of its proper parts." Therefore, Aristotle concluded that the heavenly bodies were composed of a fifth element, after earth, air, fire, and water. Accordingly, Aristotle called this fifth, incorruptible element "quintessence" or "ether."

Aristotle used his skills of observation and analysis in making what has to be viewed as one of the most significant contributions to earth science up to the time: proving that Earth was indeed spherical. Aristotle's teacher, Plato (c. 428-c. 348 B.C.), had argued that since a sphere is perfect, then Earth in its perfection must be spherical. Merely assuming that Earth was a sphere was, however, not good enough for Aristotle's inquiring mind, so he set out to prove (or disprove) this belief. Rather than looking to Earth for an answer to this question, Aristotle turned his eyes to the heavens. During a lunar eclipse, where, in modern understanding the orbital dynamics cause Earth to align between the Sun and the Moon and thereby cast an observable Earth shadow upon the Moon, Aristotle observed that the shadow cast onto the Moon by Earth has a rounded edge, indicating that Earth is a sphere. As Aristotle noted, "The sphericity of the earth is proved by the evidence of our senses, for otherwise lunar eclipses would not take such forms; for ... in eclipses the dividing line is always rounded. Consequently, if the eclipse is due to the interposition of the earth, the rounded line results from a spherical shape."

Armed with the knowledge that Earth was spherical, mathematicians and natural philosophers in ancient Greece now had to consider the size of Earth. This problem was not solved for over a century until Eratosthenes of Cyrene (276-194 B.C.), head of the great Library of Alexandria in Egypt, solved this dilemma. Eratosthenes learned of a well near present-day Aswan, south of Alexandria, which at noon on the day of the summer solstice had sunlight hit the bottom of the well. Eratosthenes realized that by measuring the length of a shadow cast in Alexandria at noon on the day of the summer solstice and using the measured distance from Alexandria to Aswan, he could determine the circumference of Earth. Depending on the assignment of value to some of the ancient units used by Eratosthenes, his ingenious calculations produced a remarkably accurate result. The circumference of Earth as calculated by Eratosthenes remained the closest measurement for over 1,500 years. Although Eratosthenes's calculations were disputed in his own time, they allowed the subsequent development of maps and globes that remained among the most accurate produced in the ancient and classical world. His calculations, based upon the spherical nature of Earth, also allowed for calculable antipodes and the development of an early theory of climatic zones. In his work *Geography,* Eratosthenes was the first to use the term "geography" to describe the study of Earth.

Based on the value for the circumference of Earth that he had worked out, Eratosthenes became the first to endeavor to make a map using lines of latitude (north-south position) and longitude (east-west position). Working nearly a century after Eratosthenes, Hipparchus (fl. 146-127 B.C.), an astronomer, became the first person to try to specify the location of places using latitudinal and longitudinal coordinates. Hipparchus accomplished this by using work that he had performed on spherical geometry. He was not entirely successful in his task for reasons that will be discussed later.

Although the work of Eratosthenes was never lost, it was ignored by many in future generations. This was due in large part to work done by Ptolemy (c. 100-175) in mid-second century A.D. Ptolemy, like Eratosthenes, worked in Alexandria, and was aware of Eratosthenes's calculations regarding the circumference of Earth. For an unknown reason, however, Ptolemy concurred with Posidonius (c. 135-c. 51 B.C.), a geographer from northern Africa who argued for a smaller value for the circumference of Earth. Posidonius and Ptolemy placed the size of Earth at what would now be approximately 18,000 miles (29,000 km), or 50 miles (80.5 km) per degree of longitude at the Equator. This estimate was 28% too small, making it not nearly as accurate as the estimate of nearly 70 miles (112.6 km) per degree as put forth by Eratosthenes. Ptolemy was highly revered in his day and in the centuries to follow, so his figure became the accepted measure.

In his influential 8-volume work *Hyphegesis Geographike,* or "Guide to Geography," Ptolemy showed how to make maps and listed the latitude and longitude for numerous cities. Like Hipparchus, Ptolemy had no trouble determining the latitude of various cities. Both men understood that by measuring the angle of the North Star (Polaris) above the horizon, one could easily calculate latitude. Longitude, however, proved to be a problem for Hipparchus, Ptolemy, and astronomers and geographers for centuries to come.

Ptolemy and other Greek geographers and astronomers knew that every 24 hours the Sun lay in exactly the same position in the sky. Since the Sun had to travel 360° around the Earth (so they thought) in order to accomplish this feat, then the Sun moved 15° every hour (360° ÷ 24 = 15°). Based on this information, when it is high noon in a city exactly an hour after high noon in a previous city, then the two cities lie exactly 15° apart. While this might sound easy enough to modern observers, the ancient Greeks did not have instruments that could accurately measure the time in faraway lands. Ptolemy, therefore, had to base longitudinal measures on the reports of travelers, resulting in highly inaccurate longitudinal coordinates. The problem with longitude proved to be such a monumental problem that it was not fully solved until the British developed a way to keep longitude on long sea voyages with precision timepieces in the eighteenth century—nearly 1,600 years after Ptolemy. However, the early attempts that Eratosthenes, Hipparchus, and Ptolemy made to determine longitude were so far-reaching that latitude and longitude, though revised, are still used today.

Another major contribution that Ptolemy made to geography was his rebuttal of the idea that a vast ocean surrounded the world that was known to the peoples of the Mediterranean. Ptolemy instead theorized that out in the ocean lay *terra australis incognita*, or literally, "unknown southern land." Ptolemy, like several philosophers before him, argued that there had to be a continent in the Southern Hemisphere in order to keep the world from becoming too top-heavy and tipping over due to the weight of the land in the Northern Hemisphere. Europeans adopted this idea in the fifteenth century, when Ptolemy's *Geography* was translated into Latin and disseminated throughout Europe. The idea gained further credence following the circumnavigation of the globe by Ferdinand Magellan (c. 1480-1521). This voyage added to the speculation that *terra australis incognita* had to exist. Therefore, when Captain James Cook (1728-1779) was commissioned by England to circumnavigate the Earth in the eighteenth century, he was instructed to find and claim the "Land, Southern, Unknown" that Ptolemy had mentioned.

While modern observers might consider the work of Ptolemy and his predecessors to have little or no impact on them except for their historical and cultural significance, two points must be kept in mind. First, Ptolemy had a major impact on cartographers, or mapmakers, for centuries after his death. Most medieval European cartographers—who were not familiar with Ptolemy's work—made maps with east at the top of the map, probably to correspond with the rising Sun. After the translation of *Geography* into Latin, however, most Europeans began to copy Ptolemy's placement of north at the top of the map. Ptolemy did this because the Mediterranean world that was known to him was twice as wide (east to west) as it was tall (north to south). Therefore, Ptolemy found it easier to make maps on the rolled scrolls of his day with north on the top. Subsequent cartography in medieval Europe also suffered greatly before the translation of *Geography* because mapmakers of the period had difficulty accurately portraying a three-dimensional sphere on a flat piece of paper. *Geography* taught European mapmakers the mathematical techniques by which they could more accurately project representations of spherical bodies on paper. Secondly, and perhaps of greater consequence, some scholars argue that Ptolemy's mistake in accepting Posidonius's circumference of Earth over Eratosthenes's value forever changed the course of world history. This is because Ptolemy's smaller, less accurate, value for the circumference of Earth became widely accepted in Europe and ultimately led Christopher Columbus (1451-1506) to believe that he could reach Asia by sailing west.

JOSEPH P. HYDER

Further Reading

Boardman, John, et al, eds. *The Oxford History of the Classical World*. Oxford: Oxford University Press, 1986.

Clagett, Marshall. *Greek Science in Antiquity*. London: Abelard-Schuman, 1957.

Lindberg, David C. *The Beginnings of Western Science*. Chicago: Chicago University Press, 1992.

Lloyd, G.E.R. *Magic, Reason, and Experience: Studies in the Origins and Development of Greek Science*. Cambridge: Cambridge University Press, 1979.

Seismology in Ancient China

Overview

Located in a seismically active part of the world, China is beset by earthquakes, earthquakes that have routinely caused thousands of deaths in one of the most densely populated nations on Earth. This has been the case throughout recorded history, and likely much longer. Little wonder, then, that scientists in ancient China were the first to develop a seismograph to monitor for earthquakes in their nation. The ability to quickly identify the occurrence and general direction of an earthquake could help the government muster assistance more rapidly. Over time, the original Chinese invention has become an important diagnostic tool, too, providing views of the deep structure of the Earth as well as giving us the ability to monitor for surreptitious nuclear weapons testing.

Background

In the Chinese capital city of Sian, in a room in the Imperial Chancellery for Astronomical and Calendrical Sciences, stood a bronze pot, about 6 feet (1.8 m) in diameter. Around the periphery of the pot were eight dragon heads, each holding a ball in its slightly open mouth. Directly beneath each dragon stood a bronze toad, looking expectantly upward with open mouth. In about A.D. 132, a large earthquake struck a city about 400 miles (644 km) northwest of Sian. It was announced in Sian by the loud ringing of a bronze ball falling from the northwest dragon's mouth into that of a toad.

Skeptical observers questioned the event, suspicious that it was simply a false alarm. Several days later, a messenger arrived, announcing the earthquake and banishing all remaining doubts. This event was recorded by the official historian:

> On one occasion one of the dragons let fall a ball from its mouth though no perceptible shock could be felt. All the scholars at the capital were astonished at this strange effect occurring without any evidence of an earthquake to cause it. But several days later a messenger arrived bringing news of an earthquake in Lung-His. Upon this everyone admitted the mysterious power of the instrument. Thenceforward it became the duty of the officials of the Bureau of Astronomical and Calendar to record the directions from which earthquakes came.

Earthquakes were capable of major disruptions in ancient China, just as they are in the modern world. In recorded history, food riots and rebellions followed in the civil unrest that accompanied them. In addition, earthquakes interrupted normal trade, including the shipments of food upon which the cities depended, and which constituted tax payments from many of the provinces. It was also important to send both food and the military to care for the citizenry as well as to quell any thoughts of rebellion. For these reasons, learning of earthquakes as quickly as possible was considered vitally important to Chinese emperors of this time, although how to accomplish this was not known.

Sometime around the year A.D. 130 the brilliant Chinese scientist Chang Heng (78-139) began to see an approach to solving this problem. Although his thoughts have not been recorded, reasonably informed speculation is possible. It is likely that Chang Heng, realizing that the shaking of the earth diminished with distance from the site of an earthquake, also realized that this shaking might continue for very great distances. It would be easy to consider that, at some point, the shaking might still exist, albeit too weakly for humans to feel it. Following this line of thought, it would seem possible to construct a device more sensitive than human perception that might be able to detect earthquakes, even at great distances. The challenge, of course, would be to construct such a device, and to construct it in such a way that it would announce earthquakes in an unmistakable fashion.

The exterior of the device Chang Heng finally developed is described above. The interior, however, is what made it work. Although at least two different models have been proposed, it seems likely that this device consisted of a rod with a weight at the top. This rod was carefully balanced on a base, and the whole structure rested securely on the ground. At the top were eight horizontal rods, each resting gently against the brass balls nestled in the dragons' mouths.

The rod, heavier at the top, was balanced in what physicists call "unstable equilibrium." It is akin to balancing atop a ball; as long as you don't move, you remain balanced. However, as soon as you move a little bit, you continue moving in that direction because of the shape of the ball. The further you move from the point of stability, the more unstable you become. Stable equilibrium is like putting a ball in the center of a bowl; if you move it a little in any direction, the shape of the bowl returns the ball to the bowl's center. In this case, moving further from the center increases the forces that tend to return you to the center.

In this case, an earthquake causes waves to ripple through the earth's crust. These waves would upset the unstable equilibrium of the rod, causing it to fall in the direction from which the waves originated. The rod would then fall against one of the horizontal rods, pushing it into the ball and causing the ball to fall into the bronze toad's mouth with a resulting loud noise. Thus, the device registered both the time and the direction of earthquakes. This is the first known seismograph.

Impact

Opinions vary as to whether Heng's "earthquake weathercock" was sufficiently sensitive to have accomplished much. On the one hand, there is little doubt that it detected the occurrence and direction of at least one earthquake at a great distance. On the other hand, some modern seismologists question whether the device itself might not have had too much internal friction to have had much more sensitivity than an attentive person. Unfortunately, no detailed description of the internal workings has survived, and no working models have either. Therefore, it is likely we will never know for certain whether this device was accurate or simply lucky. In either event, it was forgotten by China for well over a millennia, until after modern seismographs were developed in the eighteenth century.

In spite of this, or perhaps because of it, it seems most appropriate to discuss the impact of the modern science of seismology on society, because the earthquake weathercock was the most distant ancestor of today's devices, even though it vanished for so long. Some of these impacts would have been felt by ancient Chinese society, too. Specifically, they are the impact of seismology on earthquake warning and preparedness and the scientific advances stemming from seismology.

Chang Heng built his device almost solely to help detect earthquakes for the purpose of hurrying assistance to the scene as rapidly as possible. The ability to do so significantly helps to mitigate the damage caused by a serious earthquake, a fact to which inhabitants of San Francisco, Kobe, and other cities can amply attest. In fact, the most common refrain from earthquake survivors is that help took too long to arrive, leading to hunger, suffering, and death that could otherwise have been avoided.

In second-century China, of course, the only thing that could be done was to rush aid to a stricken city as quickly as possible. Toward this, Heng's seismograph helped because these preparations took time. By starting to assemble food, troops, and emergency supplies as soon as the ball dropped, they could simply await the arrival of a messenger to tell them which city was affected. Although not perfectly responsive, the government could at least be ready to travel immediately upon hearing this news.

Today, of course, we have many more options, including a limited ability to detect small tremors that often precede larger earthquakes. In addition, modern seismic networks integrate information from hundreds of seismographs spread throughout the world. This lets us instantly identify the exact location, time, and magnitude of events anywhere in the world, and lets governments immediately know that assistance may be required. While a telephone call or e-mail can also notify the government of an earthquake, it is not uncommon for phone lines to be disabled during natural disasters. Thus, the seismograph acts as a reliable backup for today's nearly instantaneous communications.

In addition to the civil disaster and public policy aspects of seismology, it has also given us some handy tools for mapping the Earth's interior. One of the first observations made was that virtually every earthquake on Earth is near the boundaries of tectonic plates. This, along with other conclusive evidence, made the case for plate tectonics overwhelmingly convincing to all but a few die-hard skeptics. Seismology has also helped us map plate boundaries, which is important because these are the areas most subject to both volcanoes and earthquakes.

Finally, seismology has given us a wonderful way to look inside the Earth. Seismology has shown us that the outer core is liquid, for example, and has helped to map chemical and physical variations in rocks of the deep mantle that are otherwise out of our instruments' reach. It has also shown us the topography within the Earth, and much more. It is safe to say that our modern conception of the Earth's structure is shaped almost entirely by our interpretation of seismological information, all stemming from remote descendents of Chang Heng's machine.

Did Heng conceive of all these uses when he ordered the bronze cast for his first device? Probably not. Most likely, he was simply trying to find a slightly better way to help his fellow Chinese. However, the modern descendents of his machine are giving us far more.

P. ANDREW KARAM

Further Reading

Bolt, Bruce. *Earthquakes and Geological Discovery*. New York: Scientific American Library, 1993.

Temple, Robert. *The Genius of China: 3000 Years of Science, Discovery, and Invention*. New York: Simon & Schuster, 1986.

Aristotelian Physics

Overview

Aristotle's (384-322 B.C.) *Physics* was one of the most influential pieces of writing in science. It defined the field of physics for centuries after it was collated from Aristotle's notes by one of his students. It became the cornerstone of Western science as allowed by the Church. While many of the claims within the *Physics* are incorrect, it represents one of the first attempts to offer a coherent, logical, and natural explanation of motion and change within the physical world.

Background

Greek philosophy and speculation on the world has a long history before Aristotle. His writings build from, or criticize, earlier philosophical ideas about the world. One of the main directions in early Greek philosophy was to speculate on whether the world was made from many substances or simply one in many forms. For example, Anaximenes of Miletus (c. 545 B.C.) suggested air was the base of everything, with water and earth being condensed air, and fire rarefied air.

Indeed, the concept of "The One" versus "The Many" was a common theme in Greek speculation. Parmenides (515?-445? B.C.) said that while things in the world seem to change and move, in reality they are unchanging and at rest. For Parmenides, a being cannot be created or grow, it simply is, and has no past or future. While this may seem ridiculous to us, Parmenides's pupil, Zeno of Elea (c. 495?-425? B.C.) showed that the ideas of motion and change we take for granted from our senses lead to some puzzling paradoxes. The most famous of these paradoxes is Achilles and the tortoise, where Zeno proves that even though Achilles is 100 times faster than the tortoise, he can never catch it. By the time Achilles reaches the point the tortoise started from, it will have moved a hundredth of that distance. By the time he then reaches that new starting point, the tortoise will have again moved on one hundredth of that distance, and so on to infinity. Such ideas concerned Greek thinkers greatly, and indeed, Zeno's paradoxes provoke argument and discussion even today.

Another popular philosophy was that of atomists such as Democritus (460?-370? B.C.) and Leucippus (fifth century B.C.). Atomists believed that everything was made from tiny components, which they called atoms. Some atomists believed there were an infinite amount of atoms. The notion of infinity also concerned many Greek philosophers.

Socrates (470-399 B.C.) tried to steer philosophy away from worrying about paradoxes and from esoteric speculation about the composition of the universe. He tried to answer questions that related to the human condition, such as "What is justice?" and "What is courage?" After Socrates's death, one of his pupils, Plato (427-384 B.C.), established the Academy at Athens to promote the learning of philosophical values. Aristotle arrived as a student at the Academy in 367. He was impressed by the mathematical study there, and in particular by the method of logical deductions from a small set of assumed truths (axioms). Plato suggested that the whole of science might be derived from axioms, an idea that inspired Aristotle. In Aristotle's *Prior Analytics,* he gave this example of the axiomatic method—if every Greek is a person, and every person is mortal, then every Greek is mortal.

Aristotle was profoundly influenced by Plato's philosophy but often disagreed with his teacher. Aristotle specifically rejected Plato's idea of another realm of ideal forms and his concept of the soul as a unifying principle. Aristotle wanted to deal only with the real world, not with abstract concepts, although he was not completely successful in this aim.

Impact

Aristotle's *Physics* is a work in eight sections, each divided into a number of subsections. The work is not just about the subject known today as physics, it also attempts to define all natural sciences of the physical world. However, the focus of *Physics* is on motion and change, and most of the examples refer to bodies in motion, so it came to define the scope of physics for centuries to come.

Aristotle thought that the terrestrial world was characterized by change and decay, whereas everything above the moon was eternal, perfect, and changeless, so the main concern of *Physics* is with everything below the orbit of the moon. This division of the heavens and the sublunar realm was to last until the sixteenth century.

In *Physics,* Aristotle made a distinction between natural objects and created artifacts. Natural objects have an intrinsic principle of motion and rest, as well as growth and propagation. For instance, if you were to plant a wooden bed, suggested Aristotle, you would not expect it to grow baby beds. If anything a tree would sprout from the wood, for a tree is wood's natural form.

Physics then went on to define luck and chance. Aristotle recognized that some events occur due to luck or accident, and he argued for a purpose in Nature, but he did not go as far as other Greek philosophers who argued for a world consciousness.

Much of *Physics* deals with problems raised by earlier philosohers, such as Zeno's paradoxes, and the ideas of the atomists. It contains an extended discussion of terms that perplexed earlier philosophers, such as infinite, void, and time. Aristotle rejected the notion of infinity as suggested by some of the atomists. He argued that even mathematics does not need infinity. This rejection helped him overcome Zeno's "Achilles and the tortoise" paradox, although many later commentators have noted that his argument is not convincing. However, Aristotle's rejection of infinity was only partial, as, for example, he also tried to say that the world has always existed and will not end. His arguments against infinity led later atomists to either abandon the notion or try to disprove Aristotle's arguments.

Another reason why Aristotle rejected infinity was directly related to his concept of motion. He believed that all things were made from four elements (air, water, earth, and fire), and they all tried to move to their natural places in the universe. The heavy elements want to move to the center of Earth, and the light to the edge of the heavens. Because he believed this, Aristotle could not accept the notion of an infinite universe—it must have an edge.

Aristotle believed that all objects have weight and lightness, which are two separate qualities, and that weight and lightness determine an object's natural motion. He stated that heavy objects fall faster than light objects, a claim that caused problems for later scientists whose experiments showed otherwise.

Aristotle tried to keep his analysis of motion simple, allowing only straight or curved motion. Curved motion, such as that of the planets, can be eternal. However, straight motion stops and starts. This simplistic analysis

caused some complications for Aristotle's view, so he devoted a large section to dealing with the end points of motion.

Aristotle related the speed of a moving object to its weight and the density of the

JOHN PHILOPONUS CHALLENGES ARISTOTLE'S THEORY OF MOTION AND LAYS THE GROUNDWORK FOR THE DEVELOPMENT OF THE CONCEPT OF INERTIA

Aristotle distinguished between natural and violent motion. Natural motions are those a body exhibits when unimpeded. Violent motions occur when a body is displaced from its natural resting place. Aristotle maintained that the force responsible for violent motion must be in constant physical contact with a moving body. The originating motive impulse for a shooting arrow is the bow string. Aristotle argued that the original mover not only sets the arrow in motion but also activates the surrounding medium—in this case air. The air parts before the arrow and circles back to maintain a continuous force behind it. This force gradually diminishes due to the resistance of the medium. After completely dissipating, the arrow falls downward according to its natural motion.

John Philoponus (fl. c. A.D. 540) argued convincingly that the medium is not the agent or cause of continued violent motion. Otherwise, it would be possible to move an object by simply agitating the air behind it. This, however, is contrary to experience. John concluded that violent motion occurs by the mover transferring to the object of motion an incorporeal kinetic power—later known as impetus.

Impetus theory was developed most comprehensively by Jean Buridan (c. 1295-1358). He ascribed permanence to impetus, claiming it endured indefinitely unless diminished by external resistance. This implied that without resistance a body in motion would continue in motion indefinitely at a constant speed. Buridan's formulation bears a striking resemblance to the concept of inertia and helped prepare the way for its development.

STEPHEN D. NORTON

medium it is moving through. For this reason, Aristotle denied that there could ever be a vacuum in nature, as that implied a zero density, and so an object moving in a vacuum would have an infinite speed, which would be impossible.

Aristotle claimed that everything that moves is moved by something. This led him to conclude that there must be some original source of motion, a "changeless source of change." This "Prime Mover" was taken by some readers to mean a god, and was later interpreted by Christian writers to mean the Creator God.

Aristotle's writings dominated subsequent Greek philosophy and the studies of Byzantine and Arabic scholars. *Physics* became the key, and for many the only, text on the subject of motion and change. However, the writings of Aristotle were reinterpreted in a variety of ways by later copyists. Christian scholars in Europe, such as Thomas Aquinas (1225-1274), were particularly adept at rewriting Aristotle's words to support their religious beliefs. Christians rejected Aristotle's statement that the world was eternal, as the Bible stated it had both a beginning and an end. They blended Aristotle's ideas with the works of other Greek thinkers to form a coherent philosophy of the world that agreed with the Bible. Later commentators even attempted to reconcile the ideas of Aristotle with those of his teacher Plato, in spite of the fact that Aristotle had specifically opposed many of Plato's doctrines.

After some initial difficulty, Aristotle's works were accepted as the basis of science as taught in European universities from the thirteenth to the seventeenth centuries. Aristotle's *Physics* defined the field. Students committed his works to memory and considered every word he had written to be true. Indeed, when Nicolaus Copernicus (1473-1543), Galileo Galilei (1564-1642), and others began to challenge the concepts of Aristotle's *Physics* through observation and experiment, they found themselves accused of challenging the theological basis of Christianity, the two concepts had become so intertwined. Only when Isaac Newton (1642-1727) published his *Principia Mathematica,* was Aristotle's *Physics* supplanted as the most widely read and influential natural philosophy text.

Aristotle wrote a number of other works on science that developed his overall worldview of change and cause, covering such diverse subject areas as zoology, astronomy, chemistry, geography, meteorology, and psychology. He also wrote on law, constitutional history, ethics, the arts, logic, language, and other topics.

Many of his logical steps are questionable, and his conclusions often incorrect, but in his *Physics,* Aristotle defined the subject and offered a coherent explanation for the observed motion of objects in the real world without resorting to supernatural or abstract explanations. The success of *Physics* was also its failure, for it was so influential that later thinkers treated it as gospel and did not even consider testing or challenging Aristotle's conclusions for many centuries.

DAVID TULLOCH

Further Reading
Books

Barnes, Jonathan. *Aristotle.* Oxford: Oxford University Press, 1982.

Barnes, Jonathan, ed. *The Cambridge Companion to Aristotle.* Cambridge: Cambridge University Press, 1995

Barnes, Jonathan, ed. *The Complete Works of Aristotle.* 2 vols. Princeton, NJ: Princeton University Press, 1984.

Lang, Helen S. *Aristotle's Physics and Its Medieval Varieties.* New York: State University of New York Press, 1992.

Internet Sites
http://classics.mit.edu/Aristotle/physics.html

Biographical Sketches

Anaxagoras of Clazomenae
c. 500-c. 428 B.C.
Greek Philosopher and Astronomer

The first professional philosopher to teach in Athens, Anaxagoras introduced Ionian physical speculation to mainland Greece. He correctly explained the phases of the Moon as well as the eclipses of both the Sun and Moon. He was also the first to clearly distinguish between mind and matter.

According to the most commonly adopted chronology of his life, Anaxagoras was born in Clazomenae around 500 B.C. This Greek colony in Asia Minor was 75 miles (121 km) north of Miletus—home of Thales (c. 625-c. 547 B.C.) and Anaximander (c. 610-c. 546 B.C.). Born to a wealthy family, Anaxagoras devoted himself to

Anaxagoras of Clazomenae, seated at right, listening to Pericles. *(Corbis Corporation. Reproduced with permission.)*

the study of natural philosophy. In either 480 or 456 B.C. he settled in Athens and established a school. He was a member of the enlightened and skeptical circle that gathered around Pericles (d. 429 B.C.). He was later prosecuted for impiety by enemies of Pericles and exiled to Lampsacus on the Hellespont. There, he founded another school shortly before his death.

Anaxagoras accepted the Parmenidean dictum that nothing comes into being and nothing perishes. However, unlike Parmenides (b. c. 515 B.C.), he rejected the idea that reality is a unity and motion impossible. Anaxagoras affirmed as real the multiplicity of forms and change we perceive about us and sought to reconcile this with Parmenidean logic. To this end, he postulated a plurality of primary elements. Infinite in number, these were understood to be ungenerated and indestructible.

Empedocles (c. 492-c. 432 B.C.) argued that the characteristics of different substances were determined by the relative mixtures of four primary elements—earth, air, water, and fire. Thus, a gold cup consisted of the four elements mixed in the appropriate proportions. Anaxagoras took exception to this because it seemed to contradict Parmenidean logic; specifically, it appeared to require something, in this case gold, coming into existence. As an alternative, Anaxagoras formulated his principle of homoemereity, which states that all perceptible bodies or natural substances

are composed of an infinite number of infinitely divisible smaller parts. Furthermore, each part retains its characteristic features upon division. Thus, unlike Empedocles, he argued that a gold cup is composed of smaller parts, each of which is made of gold and nothing else.

Interpreting the principle of homoemereity satisfactorily has been confounded by Anaxagoras's further statement that "there is a portion of everything in everything." This would imply that every part of a gold cup contains a mixture of everything else, including flesh, wood, wine, etc. On the face of it, this contradicts his claim that the parts of the cup are composed solely of gold. Various suggestions have been put forward to resolve the conflict. One solution has it that Anaxagoras simply meant that gold predominates in each part. An objection leveled against this interpretation is that at some point in the process of subdivision there will be parts no longer dominated by gold. However, if the principle of homoemereity is only applied to the predominant ingredient, then inconsistency can be avoided. This may have been what Anaxagoras had in mind.

Anaxagoras taught that *noûs*, or mind, rules the world and brings order to it. He maintained that the universe originated as a homogenous, motionless mixture that noûs operated on by initiating a vortex. This gradually caused the dense, cold, and wet matter to concentrate at the center of the mixture and form into a disk-like Earth. The rare, hot, and dry matter floated free and supported Earth. The Sun, Moon, planets, and stars were torn from Earth by the continued vortical action and ignited by friction. In this cosmology, noûs is conceived of as distinct from that which is moved. Anaxagoras thus made the first clear distinction between mind and matter.

STEPHEN D. NORTON

Anaximander of Miletus
c. 610-c. 547 B.C.
Greek Philosopher

Anaximander is famous for introducing the concept of the *apeiron*—the first use of an unobservable entity to explain empirical phenomena. He also developed the first geometrical model of the universe, drew the first Greek map of the inhabited world, produced the first Greek star-map and celestial globe, and adapted the gnomon sundial for measuring the hours of the day and annual variations in the Sun's path. His now lost work, *On the Nature of Things*, is believed to be the first scientific treatise.

Anaximander was born around 610 B.C. in Miletus, which at the time was the most powerful Greek city in Asia Minor. Little is known of his life. Tradition has it that he was a younger friend of Thales (c. 624-c. 546 B.C.), possibly his student. Regardless, Anaximander was clearly influenced by Thales. The only other detail we have concerning his life is the unsubstantiated report that he led a Milesian colony to Apollonia on the Black Sea.

Thales was the first thinker to describe the world in terms of the modifications of nature itself. He taught that the apparent chaos of the cosmos conceals an underlying permanence rooted in the substance of which the world is composed. He believed this primary substance, or *arche*, to be water (*hydor*). Thales was certainly influenced by myth, but his determination to fit his observations and beliefs into a rational scheme represented a radical break with previous modes of thought. This critical spirit and emphasis on natural causes was the hallmark of Milesian physical speculation about the origins of the world.

Anaximander agreed with Thales that a single substance underlies the plurality and apparent chaos of the cosmos but objected to his choice of water. According to Anaximander, the cosmos is composed of a warring concourse of opposites. There are two sets of opposites in continual conflict: (1) hot and cold, and (2) wet and dry. Observations revealed to him that when one element is in abundance, it has a tendency to consume its opposite. Thus, a particular element, such as water (wet), could not be the *arche* of all things because it would initially have to exist in sufficient quantity to give rise to the world. In such a state, it would enjoy permanent dominance over its potential opposite (dry), thereby preventing it from ever coming into existence.

To explain the origin and nature of the cosmos, Anaximander postulated as *arche* an undifferentiated mass of enormous extent that he called the *apeiron* or the "boundless." He understood the *apeiron* to be both spatially and temporally unbounded and without internal distinctions. The primary opposites were conceived of as potentialities of the *apeiron* that emerged by virtue of its eternal motions. Though unobservable and in all ways transcending the processes of this world, the *apeiron* circumscribes and governs all natural phenomena. Thus, with the introduction of the *apeiron*, Anaximander became the first to attempt an explanation of the perceptible in terms of the imperceptible.

In *On the Nature of Things* Anaximander described the universe as spherical with Earth occupying a stable position at its center. He argued that Earth had no reason to fall in particular any direction since it was equidistant from every point on the periphery of the celestial sphere. This symmetry argument was the first to reject the idea of a physical support for Earth.

Anaximander created the first mechanical planetary theory, imagining heavenly bodies to be fire-filled wheels rotating about Earth. He also realized Earth's surface must be curved to account for the changing position of stars. Thus, he conceived of Earth as a convex cylinder whose height was one-third its width.

STEPHEN D. NORTON

Archimedes
c. 287-212 B.C.
Greek Mathematician and Engineer

Heralded as one of the foremost mathematical geniuses of all time, Archimedes made major contributions to the fields of geometry and mechanics, and laid the groundwork for the much later development of logarithms and calculus. Some of his most famous work included the relationships between the volumes and surface areas of spheres, cones, and cylinders that share the same dimensions of base and height. His work influenced scientists for years to come, including Galileo Galilei (1564-1642) and Isaac Newton (1642-1727).

Born around 287 B.C., Archimedes was the son of the astronomer Phidias and was possibly related to Sicily's King Hieron II. Archimedes temporarily left his childhood home in the Sicilian port of Syracuse to take up studies in Alexandria, the cultural hub of Greece, at the school established by Euclid (c. 300 B.C.) a few decades earlier.

The young man quickly became enthralled by mathematics and maintained his enthusiasm for the field throughout his life. Over the years, he produced many mathematical treatises, including *On the Equilibrium of Planes, On the Sphere and Cylinder* and *On Floating Bodies.* In the first, he considered the mechanics of levers and the importance of the center of gravity in balancing equal weights.

In *On the Sphere and Cylinder,* Archimedes built on the previous work of Euclid to reach conclusions about spheres, cones, and cylinders. As described in *The Scientific 100* (Simmons 1996): "He showed that if these figures have the same base and height-imagine a cone inscribed in a hemisphere which itself is inscribed within a cylinder-the ratio of their volumes will be 1:2:3. In addition, the surface of the sphere is equivalent to two-thirds of the surface of the cylinder which encloses it." Archimedes was immensely pleased with this discovery, even requesting that his family have a sphere and cylinder engraved on his tombstone.

Archimedes used *On Floating Bodies* to recount his theory regarding water displacement and help found the science of hydrostatics. In this book, he demonstrated that when an object of any shape and weight is floated in water, its vertical, buoyant force is equal to the weight of the water it displaces. One legend of Archimedes holds that he first understood this connection between the weight of a floating object and the resulting increase in water level while watching bath water rise as he sunk his body into a tub. He was said to have been so excited by this insight that he jumped from the tub and ran stark naked through the streets proclaiming his discovery.

Although mathematical theory was his passion, Archimedes often put his ideas to practical use. For example, he used his theory of water displacement to confirm King Hieron's suspicion that a golden wreath (some say it was a crown)

was not actually pure gold. He likewise utilized his insights on levers and constructed a device to launch a particularly large ship. One of his inventions was the Archimedean screw, which helped raise water from underground.

Archimedes also helped defend Syracuse from the invading armies of Roman general Marcellus by creating mechanical devices to heave stones and beams at the attacking troops, and to damage enemy ships. Archimedes efforts to protect his homeland were not sufficient, however, and after eight months, Marcellus's armies were victorious. Archimedes, then 75 years old, died at the hands of the Roman soldiers. Some reports state that his preoccupation with mathematics played a role in his demise. As the story goes, Archimedes was too wrapped up in thought to pay attention to the demands of a soldier, who killed him for his insubordination.

As Archimedes wished, his grave was marked with the geometrical figures of a cylinder and sphere.

LESLIE A. MERTZ

Aristarchus of Samos
c. 310-c. 230 B.C.
Greek Astronomer and Mathematician

Aristarchus is famous for developing the first heliocentric planetary theory. For this, he has come to be known as the "Copernicus of antiquity." He also made the first rational estimates of the distance to the Sun and Moon as well as the size of those bodies.

Very little is known of Aristarchus's personal life. He was born on the Aegean island of Samos sometime around 310 B.C. He made his way to Alexandria sometime before 287 B.C. There he studied under Strato of Lampsacus (d. c. 270 B.C.). His only surviving work is *On the Size and Distances of the Sun and Moon.* The details of his heliocentric theory were preserved by Archimedes (287-212 B.C.) in *The Sand-Reckoner.*

Aristarchus was first to attempt a determination of astronomical distances and dimensions by geometrical analysis. The basis of his method was the realization that at the Moon's quadrature— when exactly half-illuminated by sunlight—the Sun (S), Moon (M), and Earth (E) occupy the apices of a right triangle. Angle SME is a right angle, and Aristarchus believed angle MES could be observationally determined. From this information, angle ESM could then be deduced as well as the ratios of the solar and lunar distances.

Though Aristarchus's mathematical reasoning was flawless, the necessary observational techniques did not exist. First, he had no way of determining the precise moment of quadrature. Second, no instrument was capable of measuring angle MES with sufficient accuracy. Small errors in either value would result in seriously inaccurate results. In fact, his conclusion that the solar distance was between 18 and 20 times greater than the lunar distance was approximately 200 times too small.

In *On the Size and Distances* Aristarchus also attempted to determine the diameters of the Sun and Moon. By noting the size of Earth's shadow cast during an eclipse of the Moon, he determined the lunar diameter to be one-third that of Earth. Though his geometrical argument was again sound, inaccurate measurements meant this estimate was slightly too large. However, his estimate that the Sun's diameter is seven times that of Earth's was grossly in error—the actual value is closer to 100 times. Nevertheless, the fact that the Sun was larger than Earth may have suggested to him the possibility that Earth traveled about the Sun.

The groundwork for such an idea had been prepared by Pythagorean philosophers. Philolaus of Crotona (fl. 440 B.C.) postulated a universe of concentric spheres at the center of which was a central fire. Earth, an anti-Earth, and the other heavenly bodies, including the Sun, all moved in circular orbits about this central fire. Furthermore, Hicetas of Syracuse (fl. fifth century B.C.) attributed an axial rotation to Earth.

Aristarchus combined these ideas into a true heliocentric model. His universe was spherical with a stationary Sun at its center and the stars fixed at the periphery. Following Hicetas, he had Earth rotate about its axis. He then introduced the revolutionary concept of Earth traveling in a circular orbit about the Sun.

Earth's orbital motion implied solar and stellar parallax. Aristarchus argued, respectively, that Earth's orbital radius was so small in comparison with the Sun's distance and the distance of the stars so great that neither effect was large enough to observe. Though indeed prescient, Aristarchus's theory failed to explain the inequality of the seasons and other phenomena better handled with epicycles in a geocentric model. Thus, the heliocentric hypothesis attracted little attention, that is, until Nicolaus Copernicus (1473-1543) reinvented it 18 centuries later.

Vitruvius (fl. c. 25 B.C.) credited Aristarchus with inventing the widely used *skaphe* sundial. Aristarchus also developed the first geometric procedure for approximating the sine of small angles.

STEPHEN D. NORTON

Saint Bede the Venerable
c. 672-735
British Theologian, Historian, and Writer

Saint Bede the Venerable is regarded as the most learned man of the seventh and eighth centuries. He was convinced that the Christian Church could bring order and culture out of the violence and ignorance of the Dark Ages that followed the downfall of the Roman Empire. Acting on this conviction, he remained personally committed to cultural progress throughout his life. He wrote on virtually all areas of knowledge of interest in his time, including natural science.

Nothing is known of Bede's family and birth. At the age of seven, his family left him at a Benedictine monastery in Wearmouth, Northumbria, in northern England. He later became a monk and resided in the monastery all of his life, making few trips into the outside world. There he studied and wrote, and his work made Northumbria a center of the revival of ancient learning, influencing scholarship in Britain and on the continent. The British scholar Alcuin (735-804) used Bede's scholarship as a basis for the instruction in the cathedral schools that he established for Charlemagne (742-814), and thus insured Bede's influence on the Carolingian Renaissance.

As a churchman, Bede regarded the Bible not only as a source of literal truth but also containing the rich symbolic meaning of allegory. As a result he was more open to scientific observation and explanation than many of his contemporaries.

Bede wrote between 40 and 60 books on a broad range of subjects, including science, history, biography, scriptural commentary, and grammar. He was also a poet. His best known work dealt with the history of the Christian Church and with the setting of dates (chronology). In his *Historia ecclesiastica gentis Anglorum* (Ecclesiastical history of the English people) Bede covered the history of the conversion of the Anglo-Saxons in Britain to Christianity from the Roman invasion (55-54 B.C.) to the coming of Saint Augustine of Canterbury (A.D. 597).

Certainly one of his most important contributions to chronology was the introduction of

Saint Bede, "the Venerable." *(The British Library. Reproduced with permission.)*

attempting to extract general laws that were internally consistent and that agreed with observational evidence. His own original applications of scientific knowledge and thought dealt with such practical matters as the tides, the calendar, and problems in arithmetic. Bede's principal works that dealt with scientific matters were *De natura rerum, De temporibus,* and *De temporium ratione.*

J. WILLIAM MONCRIEF

Brahmagupta
598-665?
Indian Astronomer and Mathematician

Brahmagupta, a Hindu astronomer and mathematician, is best known for work performed while head of the Ujjain astronomical observatory, one of the leading centers for astronomical and mathematical research in ancient India. While there, he developed a number of important mathematical concepts, investigated the motions of planets and other celestial bodies, and arrived at a fairly accurate estimate of the length of the terrestrial year. His most important contributions are the introduction of the number zero to mathematics and his masterpiece, *Brahma Sphuta Siddhanta* (The opening of the universe), in which many of his mathematical and astronomical discoveries were set out in verse form.

Brahmagupta was likely born in northwestern India, and spent the majority of his life in what is now the city of Bhinmal, in the Indian state of Rajasthan. Little else is known of his early life.

By the age of 30, Brahmagupta had completed a near total revision of an old work of astronomy, the *Brahma Siddhanta.* In this work, he devoted several chapters to mathematics, including the first mathematical treatment of the number zero and the mathematical rules surrounding its use. Brahmagupta also included over four chapters on pure mathematics and several additional chapters on applied mathematics. Among the topics he addressed (many of which were new to science) were methods for solving quadratic equations, arithmetic progressions, and mathematical methods for approaching astronomical problems.

This last topic may be Brahmagupta's most significant contribution to science, because he was the first person to consider using mathematical (particularly algebraic) techniques to predict astronomical phenomena such as planetary motions, solar and lunar eclipses, and so forth. In particular, he was able to describe the manner in

the practice of dating historical events from the birth of Christ, using the terminology *anno Domini* (in the year of our Lord) or A.D. He also developed a systematic method for calculating the date of Easter, simplifying the previous complicated procedures that were a consequence of combining the Roman solar calendar with the Hebrew lunar calendar. His approaches, contained in *On the Reckoning of Time,* were eventually adopted throughout the Western world.

Bede's writings on scientific subjects are of two types: summaries of natural science as it was then understood and more original applications of scientific thought to practical problems. His works are excellent compilations and commentaries on the state of understanding of the natural world in the period prior to the translation of Plato, Aristotle, and other Greek philosophers. Bede's knowledge was based on the encyclopedia of Roman writer Pliny the Elder (23-79), on the cosmology of the Christian Fathers—such as Saint Ambrose (340-397), Saint Augustine of Hippo (354-430), Saint Basil (329-379), and Saint Gregory (540-604)—and on the encyclopedic works of Isidore of Seville (570-636). He did, however, add elements of his own in his distillation of their writings. Bede's writings contain the basic elements of modern science. He downplayed the mystical in the natural world and sought explanations in terms of cause and effect,

which the rising and setting of the planets could be calculated well in advance, as well as their predicted positions in the sky. Seemingly simple today, this was a major advance at that time, and involved many laborious calculations.

Brahmagupta was also involved in the debate over the shape of Earth and the universe. He disagreed with the school of thought that asserted that Earth was flat or concave (like the inside of a bowl). Instead, he thought it most likely that both Earth and the universe were round. However, he also felt certain that Earth was stationary in the universe, a view that is now known to be incorrect.

One of Brahmagupta's later accomplishments was calculating the length of the solar year. In early work, he determined this to be 365 days, 6 hours, 5 minutes, and 19 seconds. This figure was later revised to 365 days, 6 hours, 12 minutes, and 36 seconds. However, some suspect this latter value to have been taken from the work of Aryabhata (476-550?), from which Brahmagupta's value differs by only a few seconds. In any event, both values are very close to today's accepted value of about 365 days, 5 hours, 48 minutes, and represent a remarkably accurate calculation based on information obtained by very primitive instruments.

Perhaps the greatest tribute to Brahmagupta's mathematical talents is that his book was used to introduce the basic ideas of algebra to Islamic mathematicians, who were later given credit for its invention. His work was expanded upon to form the basis of what is now taught in schools, and the Islamic mathematicians must be given credit for adding their original contributions to Brahmagupta's work, but their credit must also be shared with him.

Brahmagupta died between 660 and 670. At the time of his death he was widely acknowledged as the greatest mathematician of this period of Indian history, and one of the greatest Indian mathematicians of any time. This is reflected in one of his honorary titles, "Ganita Chakra Chudamani," bestowed by a fellow scientist and translated as "the gem of the circle of mathematicians."

P. ANDREW KARAM

Callipus

c. 370-c. 300 B.C.

Greek Astronomer and Mathematician

Callipus is famous for refining the planetary theory of Eudoxus by adding additional

spheres. He also made accurate determinations of the lengths of the seasons and constructed a 76-year period to more accurately align the solar and lunar cycles. This Callipic cycle remained the standard for dating and correcting astronomical observations for many centuries.

One of the greatest astronomers of ancient Greece, Callipus was born sometime around 370 B.C. in Cyzicus, located in Hellespontine Phrygia on the southern shores of the Propontis (known today as the Sea of Marmara). According to Simplicius, he studied with Polemarchus (fl. c. 340 B.C.), a former student of Eudoxus (c. 408-c. 355 B.C.). Callipus followed Polemarchus to Athens. He eventually came to live with Aristotle (384-322 B.C.), who encouraged him to devote his efforts to improving the Eudoxean system of concentric spheres.

Plato (c. 428-347 B.C.) first challenged astronomers to explain the apparently irregular movements of celestial bodies in terms of uniform circular motions. Eudoxus accepted this challenge "to save the phenomena" and developed a system of concentric spheres with Earth as their common center. Each planet, as well as the Sun and Moon, was attached to a single sphere. This, in turn, was part of a set of interconnected spheres, each of which rotated about its own axis at a different rate and orientation. The combined motions were then adjusted to approximate the observed movements of the body in question. Eudoxus employed 27 spheres: three each for the Sun and Moon, four each for the five planets, and one for the fixed stars.

Callipus realized that the Eudoxus's system required the Sun to move with an apparent constant velocity against the background of the fixed stars. This implied that the seasons were of equal length, which was contrary to common knowledge. Based on his own careful observations, Callipus accurately determined the lengths of the seasons. To account for his results, he found it necessary to refine the Eudoxean model by adding two more spheres for each of the lunar and solar models and one additional sphere each for the mechanisms of Mercury, Venus, and Mars. This brought the total number of spheres to 34.

Aristotle further modified this system, but unlike Eudoxus and Callipus, he maintained the spheres were material bodies. Accordingly, certain presuppositions of Aristotelian physics needed to be satisfied. This required 22 additional spheres, for a total of 56. Unfortunately, all concentric-sphere models were incapable of explaining or reproducing certain phenomena, specifically the variation in the apparent diameters of the Sun and Moon and the requirement that the hippopede (horseshoe-curve) of retrograde motion repeat itself exactly from one orbit to the next. Nevertheless, the Aristotelian version of Eudoxus's system survived for many centuries and greatly influenced Hellenistic (Greek) astronomy.

By accurately determining the lengths of the seasons (94, 92, 89, and 90 days respectively from the time of the vernal equinox), Callipus reconciled the lunar and solar calendars. The Athenian astronomer Meton (fl. fifth century B.C.) previously established a 19-year luni-solar calendric cycle. Callipus showed that this Metonic cycle was slightly too long. To bring the calendars into alignment, he combined four 19-year Metonic cycles, dropping one day from each.

The resulting 76-year Callipic cycle provided a much more accurate measure for the year. It also became the reference standard by which later astronomers recorded their observations. The existence of this calendric standard made it possible to correct and correlate observations much more accurately. This in turn greatly contributed to the development of future astronomical theories.

STEPHEN D. NORTON

Confucius (K'ung fu tzu)
551-479 B.C.
Chinese Philosopher

Confucius, the Latinized name of K'ung fu tzu (which means "Master K'ung"), was one of world's great philosophers and the inspiration for one of the world's great religions. Although he was apparently not directly responsible for any significant scientific breakthroughs, his philosophies and beliefs were important factors that helped to stir some of the early Chinese mathematical and scientific traditions.

Like so many ancient historical figures, very little is known of Confucius's youth other than that he was apparently born into a noble family in the China of the Chou dynasty. According to traditional stories, he quickly rose in the esteem of his superiors until, at the prompting of jealous counselors, his prince gradually turned against him. By the age of 40 he had embarked on the life of an itinerant philosopher and scholar, traveling from town to town as various noble families forced him onward.

In reality, this is probably far from the truth, and it appears as though he actually spent much of his life as a retainer of the same duke and his successor. Along the way, particularly towards the end of his life, Confucius began to attract students who came to study with the master. His students listened and recorded many of his most important thoughts, publishing the *Analects,* a collection of Confucius's teachings, after his death in 479 B.C. Ironically, Confucius never considered himself a religious figure and died disappointed, convinced his teachings would die with him.

"Confucius said . . . ," often heard in the modern West as the comic cliché was for over two millennia equivalent to the Christian invocation, "Jesus said . . . ." Confucius was not only a great philosopher, but was also the father of Confucianism, a religion with more followers than virtually any other religion in history.

Confucius's impact on Chinese thought and culture certainly did not die with him and. In fact, it had a profound impact on China for over 2,000 years. One of the impacts of Confucianism's success was in its encouragement of scholarship and study. This helped inspire many of the ancient Chinese scientists and mathematicians, with the result that many of the West's technological innovations were either borrowed from the Chinese or had already been independently developed by the Chinese centuries earlier.

Confucius's life was nearly contemporaneous with that of Socrates (470?-399 B.C.), but their lives could hardly have ended differently. Socrates, celebrated until he was condemned to death for not believing in the gods he tried to understand, left a lasting impression on Western thought, but little else. Confucius, on the other hand, was largely ignored (with the exception of his students) and died certain he would be quickly forgotten. Not only was he remembered, but he was revered as a god himself by future generations.

P. ANDREW KARAM

Democritus
c. 460-c. 370 B.C.
Greek Philosopher

Democritus embraced the atomism of Leucippus and worked out its detailed application. His atomic theory was the basis for Epicurean philosophy and all later ancient materialistic schools of thought.

Democritus was born at Abdera, in Thrace, although the exact date is unknown. One chronology maintains he was born soon after

Confucius. *(Bettmann/Corbis. Reproduced with permission.)*

500 B.C. and died about 404 B.C., while another places his birth around 460 B.C. Tradition has it that he lived to great age; so, if the latter chronology is correct, he lived well into the fourth century B.C. In *Little World-system* he described himself as a young man when Anaxagoras (c. 500-c. 428 B.C.) was old. This lends support to the latter chronology, which is widely accepted by modern scholars.

Apparently wealthy, Democritus traveled to Egypt, Chaldea, and the Red Sea. His literary efforts extended over a considerable period and generated over 60 works, including treatises on astronomy, mathematics, music, physics, biology, medicine, and ethics. Known as the "laughing philosopher"—because of his amusement with the follies of man—he left many students. One of these, Nausiphanes, introduced Epicurus (c. 341-270 B.C.) to Democritean atomism.

The only known teacher of Democritus was Leucippus (fl. c. 450-420 B.C.), from whom he learned the essentials of atomism. Leucippus formulated his atomic theory to escape certain difficulties associated with Parmenidean metaphysics. According to Parmenides (b. c. 515 B.C.), nothing can come into being from that which is not, and anything that exists cannot be altered. However, earlier thinkers had argued that all things were derived from a single primary substance: Thales (c. 625-c. 547 B.C.) be-

lieved it was water, Anaximander (c. 610-c. 547 B.C.) the *Apieron*, Anaximenes (fl. 545 B.C.) air, and Heraclitus (fl. 500 B.C.) fire. Unfortunately, each of these theories required an alteration of the primary element.

To avoid this difficulty, Leucippus postulated the existence of an infinite number of unalterable atoms whose combinations give rise to the sensible properties of bodies. Furthermore, he agreed with Parmenides that without a void, motion was impossible. Nevertheless, Leucippus accepted the existence of empty space, since it was that which separated and through which atoms moved. The ideas of Leucippus were highly speculative. Democritus must be given credit for their detailed development.

The atoms of Democritus's system were solid corporeal bodies, both infinite in number and shape as well as of different sizes (though not so big as to be visible). They were devoid of any perceptible qualities and scattered throughout the void, through which they moved perpetually according to unalterable laws of nature. Democritean atoms were certainly extended in space, but they were physically indivisible and otherwise indestructible. Whether or not he thought they were conceptually divisible remains a matter of dispute.

Democritus undertook a detailed exposition of the relationship between specific atomic configurations and the sensible qualities they give rise to. Atomic motions generate collisions that either result in their deflection or adhesion. When hooked or barbed atoms collide, they adhere to one another, forming compound bodies. The various qualities a body possesses, such as color, taste, and temperature, are the consequence of the total number present, their particular arrangement, and the different shapes of the individual atoms. For example, an object's weight varies according to the number of atoms and amount of void contained therein. Additionally, all change was conceived of as a result of atomic combinations, recombinations, and dissociations.

Democritus also taught that all bodies emit thin films of atoms that interact with the organs of sense perception to generate our sensation of objects. Likewise, he taught that thought was the result of interactions with the atoms of the soul. Democritus advocated critical reflection on the evidence of the senses as the best means of obtaining truth.

STEPHEN D. NORTON

Empedocles. *(Archivo Iconigrafico, S.A./Corbis. Reproduced with permission.)*

Empedocles of Acragas
c. 492-c. 432 B.C.

Sicilian Philosopher, Poet, and Physician

Empedocles is considered, perhaps incorrectly, the originator of the four-element theory of matter that dominated natural philosophy and greatly influenced Western medical thought until the time of the Renaissance.

The facts of Empedocles's life are obscure. He was born at Acragas (later known as Agrigentum, modern Agrigento), on Sicily's southern coast, around 492 B.C. Though born into a wealthy aristocratic family, he championed democratic principles and instigated the overthrow of the tyrannical oligarchy of Acragas known as "the thousand." He was offered the kingship but refused it, preferring instead to continue his study of nature and philosophy. While traveling abroad, his enemies at home raised sufficient support to oppose his return. He spent his remaining years in exile, dying on the Peloponnese around 432 B.C.

Empedocles was greatly influenced by Parmenides of Elea (b. 515 B.C.). He accepted the Parmenidean dictum that nothing can come from nothing and that what exists cannot perish. He also affirmed the Parmenidean denial of a void; but unlike Parmenides, he rejected the further conclusions that reality is a unity and motion impossible. Empedocles's modification of Parmenidean metaphysics and synthesis with previous views and observations generated his now-famous four element theory.

Empedocles maintained that there exists a plurality of primary elements or *archai*. These were understood to be ungenerated and indestructible as well as qualitatively unalterable and homogeneous. Because matter appears in four forms: vapors (gases), liquids, solids, and fire (associated with aethereal matter), Empedocles thought it plausible that the elements were four in number. Thales (c. 625-c. 547 B.C.) had previously argued that all things were composed of water, Anaximenes (fl. 545 B.C.) air, and Heraclitus (fl. 500 B.C.) fire. To these, Empedocles added earth. He considered them together as the "root" of all things. Accordingly, he argued the characteristics of different substances were determined by the relative mixtures of the four elements.

This plurality allowed Empedocles to explain motion without postulating a void. He taught that the elements continually replace one another, much as one object in a series slides into the place occupied by the object directly before it. Such motion requires no empty spaces.

All modes of change, including generation, corruption, and local motion, were the result of mixing, unmixing, and remixing of the four elements. But because the four were passive,

Empedocles believed they could interact only under the influence of Love and Strife—Love being the power of aggregation (attraction) and Strife of separation (repulsion). The same powers that fill men's hearts, Empedocles argued that Love and Strife operate simultaneously but have opposite effects.

The cosmology of Empedocles was cyclical: (1) Under the influence of Love, the four elements fuse into a homogenous sphere; (2) with the gradual ascendancy of Strife, a process of increasing differentiation begins; (3) ultimately, the *archai* completely separate from one another as the influence of Love wanes; and (4) as Strife then begins to wane and Love increases, a period of progressive integration occurs. Presumably, the cosmos, as we know it, can only exist in stages (2) and (4).

Associated with the homogenous sphere of his cosmic cycle is Empedocles's picture of the universe as a spherical crystalline plenum enclosing Earth. The fixed stars and planets were pockets of fire embedded in this rotating sphere. It is doubtful that he considered Earth itself a sphere. He also correctly explained solar eclipses.

Empedocles was also concerned with zoology and botany. Invoking chance and natural selection, he applied his principles of elemental mixture to the emergence of life. He described earlier life forms and how those best adapted to their environment are able to survive and reproduce. Unlike the later Darwinian theory of evolution, Empedocles's evolutionary mechanism ceases to function precisely when heredity becomes important.

STEPHEN D. NORTON

Epicurus
c. 341-270 B.C.
Greek Philosopher

Epicurus founded the Epicurean school of philosophy, which sought attainment of happiness through simple living. His importance to science lies in modifying and promulgating Democritean atomism.

Epicurus was born at the Athenian colony of Samos around 341 B.C. His father, Neocles, was a schoolmaster who had emigrated from Athens. Epicurus traveled to Athens in 323 B.C. to complete his mandatory military service. He later rejoined his family at Colophon, on the coast of Asia Minor. There he studied philoso-

phy with Nausiphanes, a former student of Democritus (c. 460-c. 370 B.C.) . He established a school at Mytilene, on the island of Lesbos, and another at Lampsacus, on the Hellespont. In 307 or 306 B.C. he returned to Athens and established an Epicurean community known as the Garden, which admitted men, women, and slaves alike. This remained the center of his activities until his death in 270 B.C.

As Alexander the Great (356-323 B.C.) swept away the last vestiges of Greek democracy and substituted for it monarchical authoritarianism, there developed in the Hellenistic consciousness a growing sense of the impotence of the individual. Gone also was the sense of community and civic duty so characteristic of the Greek city-state. This gave rise to a new philosophical attitude. Philosophy was no longer primarily thought of as an intellectual activity to be pursued in its own right. Rather, it was viewed as a potential refuge from the despair and vicissitudes of life. Epicurus developed his philosophy in this atmosphere. His primary goal was to teach men how to cultivate an attitude toward life that guaranteed happiness. The result was a moderate hedonism.

Central to Epicurean teaching was the emphasis on peace of mind as of way attaining happiness. Epicurus believed this was threatened by ignorance of the natural world that generated widespread belief in supernatural powers and doubts about potential rewards or punishments in the afterlife. He found the atomism of Leucippus (fl. c. 450-420 B.C.) and Democritus congenial to his needs and adapted it accordingly.

In *De natura*, he developed a mechanistic explanation of the world grounded in Democritean atomic theory. He accepted that all natural phenomena are generated by atoms and the void. The atoms are imperceptibly small, of different shapes and sizes but composed of the same substance, infinite in number, and free to move through the void. Sensory qualities of physical bodies, such as taste, color, and weight, vary according to the number of component atoms, their arrangement, and the presence of empty space. This picture left no room for supernatural forces.

The Epicurean atom was physically indivisible just as was the Democritean atom. However, unlike Democritus, Epicurus claimed the atom was composed of minimal parts that could be conceptually divided. Democritean atomism also seemed to make no allowance for human volition since the motions of the atoms composing the mind were completely determined by their

observations during his life, and may have been one of the first to suggest that some of the other planets (although probably not Earth) revolve around the Sun. This was a major departure from the thinking of the day and, in fact, it would not be until the time of Nicolaus Copernicus (1473-1543) that this would again be suggested.

Like so many others born over two millennia in the past, virtually nothing is known of Heraklides's early life. In fact, very little is known about his life at all, with the exception of some of his cosmological observations.

Apparently, in addition to his musings on philosophy, Heraklides spent a great deal of time watching the skies and plotting the positions of the planets in the heavens. Cosmology, the study of the origins and structure of the universe, was an important field of philosophical speculation to the ancient Greeks, and Heraklides tried to make his observations fit into the day's cosmological framework—that Earth was the center of the universe and everything rotated around Earth.

Today, this cosmology, called geocentric (or Earth-centered) seems silly. However, we make such an assessment with the perspective of several centuries of scientific observation with instruments and theories of which Heraklides could not even conceive of, let alone use. The ancient Greeks could hardly be faulted for devising a cosmology based on observations with their most sophisticated scientific instrument—the naked eye. And what the eye sees is an endless cycle of stars and planets rising in the east, passing overhead, and setting in the west. No wonder so many believed for so long that the universe rotated around us, especially because, not understanding gravity, many were convinced that a spinning Earth would fling everyone off into space.

However, Heraklides seems to have been interested in the fact that the planets moved against the backdrop of the fixed stars. Consider that over 2,000 stars are clearly visible at any one time in a dark sky, and about 6,000 or so stars are clearly visible during the course of the year. All of these stars appear fixed to the sky; over a human lifetime (over several lifetimes) they remain unmoving with respect to each other. Of all these stars, only five are ever seen to move: Mercury, Venus, Mars, Jupiter, and Saturn.

The Greeks were seemingly born to speculate, and it is no surprise that they would try to find a reason why only five "stars" were allowed to move. Over the following centuries, this ques-

Epicurus. *(Bettmann/Corbis. Reproduced with permission.)*

previous motions and interactions with environment. To avoid the undesirable consequence of pure determinism, Epicurus introduced his famous atomic "swerve," which was a spontaneous deviation from natural atomic motions.

These occasional deviations served another purpose. In Democritean atomism, the natural motion of atoms was left as an undefined axiom of the theory. Epicurus found this unsatisfactory and argued that they have a natural tendency to fall "downwards" due to their weight. But, he realized atoms would merely fall through the void without ever interacting unless there existed some mechanism to alter these motions. The atomic "swerve" explained how atomic paths crossed.

The Epicurean "canonic," or theory of knowledge, held that all knowledge comes from the senses. Furthermore, anything not in conflict with experience could be regarded as true.

STEPHEN D. NORTON

Heraklides of Pontus
388-315 B.C.
Greek Philosopher and Astronomer

Heraklides of Pontus was one of the many early scientists to come from ancient Greece. An astronomer, Heraklides made many important

tion plagued many of the early scientists, and ever-more elaborate theories were constructed to try to explain this phenomenon. Heraklides was the first to suggest that, instead of orbiting Earth, Venus and Mercury actually orbited the Sun. In so doing, he made an intellectual leap of astounding proportions, because it was so different from what the eye saw.

Heraklides appears to have limited his theorizing to the planets closer to the Sun than Earth, and their motions are certainly easier to plot. Both of these planets appear only in the morning or the evening, and both are usually close to the Sun in the sky. Their motions across the sky are by far more simple than those of the outer planets. Nonetheless, Heraklides's suggestion was revolutionary, so much so that it was not to be taken up again for nearly 2,000 years.

Heraklides died in about 315 B.C., and his ideas do not seem to have taken hold during his lifetime. However, they did inspire Copernicus who, in his book introducing the Sun-centered (heliocentric) universe, cited Heraklides as one whose work supported his own. Few scientists can ask for more.

P. ANDREW KARAM

Hipparchus

180-c. 126 B.C.

Greek Astronomer, Geographer, and Mathematician

Often described as the greatest astronomer of antiquity, Hipparchus (also known as Hipparchos, Hipparchus of Nicaea, or of Rhodes) revealed the precession of the equinoxes, determined the lengths of the four seasons on Earth, studied annual solar movements, and considered the distances of the Sun and Moon from Earth. He was also the first person to use longitude and latitude in an attempt to pinpoint locations on Earth. In addition, he indexed the latitude, longitude, and brightness of approximately 850 stars, creating the most complete astral catalog ever before assembled.

Born around 180 B.C. in Nicaea, Bithynia (in Anatolia), Hipparchus spent much of his life as an astronomer in Rhodes (one of the Greek islands), although he may also have spent some time at Alexandria, Egypt. Although only one of his 14 books remains, a review of Aratus's *Phaenomena*, his contributions to mathematics, astronomy, and geography were described in the famed work by Ptolemy known as the *Algamest*.

Hipparchus's principal scientific contribution was his detection of the precession of the equinoxes. By contrasting his observations of the Sun's path through Earth's sky against similar observations made in 281 and 432 B.C., he found that the path shifts from year to year, intersecting the celestial equator in slightly different positions. This shift is known as the precession of the equinoxes. From this data, Hipparchus was also able to infer the length of the year and was correct to within 6.5 minutes.

For much of his work, Hipparchus relied on a combination of traditional mathematical calculations and a "table of chords" that he developed. This table, a forerunner of the sine used in modern trigonometry, was critical in his estimations of stellar and planetary positions. His observational skills also played an important part in his studies. Using his own and historical observations of lunar eclipses, Hipparchus mathematically explained the apparent movement of the Moon as viewed from Earth. To resolve the Sun's apparent movements, he measured the span from vernal equinox to summer solstice, and from summer solstice to autumnal equinox. This work was particularly useful during the period when astronomers viewed Earth as the center of a spherical universe. In addition, Hipparchus employed his mathematical skills to determine the approximate size of the Sun and Moon, along with their relative distances from Earth. While he greatly underestimated the Sun's actual size, his assessment of the Moon's diameter was off by fewer than 500 miles (800 km).

In his other work, Hipparchus wished to extend the application of latitude and longitude so that both stars and terrestrial positions could be mapped. He put together a catalog charting the locations of some 850 stars and indicating their brightness using a scale of magnitude similar to the scale used today. The catalog became the standard reference. Back on Earth, however, Hipparchus was less successful. Although he calculated a value for the approximate size of one longitudinal or latitudinal degree, he was unable to develop a practical terrestrial system.

To future scientists, the work of Hipparchus became a testament to the importance of careful astronomical observation, a stepping stone in the development trigonometry, and a demonstration of how mathematics and observation could combine to give rise to important discoveries.

LESLIE A. MERTZ

Saint Isidore of Seville

560-636

Spanish Roman Encyclopedist

As the title of his *Etymologies* suggests, Isidore of Seville intended his monumental work as an exploration of word origins. What resulted was both something less and something more than he intended. On the one hand, the book has its ludicrous moments, particularly when the author's slavish commitment to his original purpose forced him to stretch credibility beyond the breaking point. On the other hand, the *Etymologies* comprises an encyclopedia of all that was known at the time concerning science and technology, and provides a panoramic view of ideas on education, theology, and other subjects in late antiquity— or more properly, the early medieval period.

Isidore's family were native Spaniards trained in Roman traditions, but the Western Roman Empire had long since fallen, and Spain had belonged to the Visigoths for more than a century when he was born. His father served the Visigoth rulers, and after his death, Isidore—still a small child—went to live with his older brother, Leander.

Leander later became archbishop of Seville, and in about 599 Isidore, by then nearing the age of 40, replaced him in that position. During this time, the Visigoths converted from the Arian heresy to Catholicism, which had long been the religion of Spanish Romans such as Isidore. This placed him in a uniquely powerful position to inspire unity, and to provide spiritual and intellectual leadership. Taking advantage of this situation, Isidore set out to educate his flock with a series of works on theology, liturgy, and religious controversies. By far the most important writing to emerge from Isidore during this time, however, was his overview of all that was known to Western Europe at that time, the *Etymologies*.

The *Etymologies*, which occupied him from 622 to 633, just three years before his death, consists of 20 books. The first three concern the seven subjects passed down from the Roman educational system: the trivium (grammar, rhetoric, and dialectic or logic) and the quadrivium (arithmetic, geometry, astronomy, and music.) The fourth book addresses medicine, and the fifth a history of the world from the Creation to A.D. 627. The next three books involve theological subjects: sacred writings and church offices (Book VI); God, the angels, and members of the Church (Book VII); and the Church and its heretical opponents (Book VIII).

With Book IX, Isidore turned to subjects of geopolitics, including language, states, and peoples. Book X consists of a dictionary, and Book XI discusses the human race. The next three books address scientific subjects: zoology (XII); cosmography, or a general discussion of the world and universe (XIII); and geography (XIV). After a book on monuments and means of communication (XV), the text returns to topics of direct interest to the scientist, including petrography and mineralogy (XVI), and agriculture and horticulture (XVII). The remainder of the book is taken up with more or less science-related subjects: the military arts and sports (XVIII); ships, housing, and apparel (XIX); and, finally, food, agriculture, and what a later age would call home economics.

The organizational scheme of this work may strike a modern reader as odd: Isidore seemingly wandered from topic to topic, and lumped together topics that do not have an immediately apparent connection. Part of this involves differences in mindset between his time and the present, but it also owes something to Isidore's assumption that the name of a subject is a key to understanding the subject itself. Thus he took a formal approach, one governed by a preoccupation with ideas rather than objects.

Such a viewpoint, of course, leads to many forced or strained connections, and this is further compounded by Isidore's attempts to reconcile the Bible with classical philosophy and science. Thus he identified the Book of Genesis, for instance, with the science of physics—an early example of attempts to treat the Bible as a scientific, rather than as a theological, poetic, and historical, work.

Yet in other parts of his *Etymologies*, Isidore showed himself willing to separate scientific from religious approaches. Thus in his cosmographical discussion (Book XIII), he presented the atomic theories of Epicurus (341-270 B.C.) and his predecessors. This was an idea he had attacked in his earlier book on the Church (VIII), maintaining that atomism conflicts with the notion of an all-providing Creator; but in the scientific passage, he did not repeat this attack, and instead presented atomism as a viable theory.

The *Etymologies* represents the culmination of a lifetime of reading on the part of a highly educated figure—perhaps the most educated man of his day. That it strikes the modern reader as disjointed and unwieldy says more about the time than it does about its author: inasmuch as it was possible to make sense of the Western world in the seventh century, Isidore did so, and his work remained influential for centuries after his death.

JUDSON KNIGHT

Leucippus
480?-420? B.C.
Greek Philosopher

It may surprise many modern people to learn that the concept of the atom is an ancient one, but in fact it had its roots with the Greek philosopher Leucippus and his student Democritus (460?-370? B.C.). It is perhaps ironic that these two should have been responsible for the idea that all of nature can be broken down into infinitesimal and indivisible parts: likewise it is difficult to separate the contributions of Leucippus from those of his more famous pupil in forming the theory of atomism.

Details of any kind about Leucippus's career are scant, and this is particularly so with regard to his early life. It is possible he was born in Miletus, a town in Asia Minor (now Turkey) that was home to numerous philosophers and scientists. He may have founded the school of philosophy at the city of Abdera (now Advira) on the Thracian coast of Greece, and it seems fairly well established that he as least lived in Abdera.

Yet Epicurus (341-270 B.C.), writing less that two centuries later, maintained that Leucippus was only a legendary figure, since so little was known of his life. Even Aristotle (384-322 B.C.), who along with his own pupil Theophrastus (372?-287 B.C.) credited Leucippus with establishing the theory of atomism, seems to have been uncertain about Leucippus's role in developing atomic theory, at times suggesting that Democritus worked alone.

Some scholars have maintained that Leucippus studied under Zeno of Elea (495?-430?). It is more likely, however, that he was simply influenced by Zeno, as well as by Zeno's mentor Parmenides (b. 515? B.C.). Implicit in the paradoxes of Zeno is the idea that matter can be divided endlessly; but Leucippus maintained that the quality of divisibility does not continue endlessly. Rather, in Leucippus's view, at the smallest level of existence were indivisible monads too tiny to see.

In using the term *atom,* it should be stressed that Leucippus's understanding of atoms was different not only in degree, but in kind, from the understanding of atomic particles by physicists today. He had no concept of elements—not even the primitive idea of four elements, such as fire and water and so on. To him and to all thinkers of his time, all matter consisted of the same substance: individual atoms differed only in position and perhaps shape.

Nonetheless, Leucippus broke new ground by finding divisibility in nature, where up to that time scientists saw only an undifferentiated mass. His achievement becomes particularly impressive when viewed in light of the Greeks' tendency to regard empty space as nothingness, devoid of content. It was for this reason, for instance, that Greek architecture avoided the arch, which seemed to place an emphasis on the void rather than on the physical substance of the stone framing the void. Yet according to Theophrastus, it was Leucippus's position that both matter and void had an existence.

Leucippus was said to have been responsible for two texts, *The Great World System* and *On the Mind.* In fact only fragments of either survive, and then primarily through the works of others. In commenting on Leucippus, Diogenes Laertius (third century A.D.) indicated that *The Great World System* put forth a cosmology regarding the creation of worlds. It was Leucippus's view, Diogenes wrote, that worlds were created by the agglomeration of large atoms at the center—an explanation that sounds rather like a precursor to the law of gravity.

Along with Democritus, Leucippus has long enjoyed a reputation for astounding prescience, anticipating the subjects of twentieth century physics by a good 2,500 years. Perhaps even if he could see all that would unfold, Leucippus would not have regarded later discoveries as a surprise: in the single surviving direct quote from the teacher, he maintained that "Nothing happens in vain, but everything from reason and of necessity."

JUDSON KNIGHT

Martianus Capella
Fourth century-fifth century? A.D.
Roman North African Scholar

Martianus Capella is remembered almost solely for *De nuptiis philologiae et Mercurii* (the marriage of Mercury and philology), also known variously as the *Satyricon* and *Disciplinae.* An allegory concerning the arts and sciences, the book was destined to have an enormous impact on medieval learning.

He was a native of Madaura, Numidia (now in Algeria), the same town where the philosopher Lucius Apuelius (124?-170?) had been born three centuries earlier. After moving to Carthage (now in Tunisia), Martianus worked as a solicitor or attorney, and married. He later had a son named Marianius, to whom he dedicated *De nuptiis*.

In Martianus's time, North Africa had been overrun by the Vandals, and he and his family apparently struggled to maintain some semblance of a normal life in Carthage. Thus it is all the more interesting and significant that he decided to compose an encyclopedic overview of his era's liberal culture—no doubt in part to preserve the learning that had lately come under great threat.

Writing in a combination of prose and verse, Martianus took for his conceit a story in which the god Mercury weds a maiden named Philologia (the study of literature.) At her wedding she is presented with seven slaves, each of which stand for one of the liberal arts: Grammar, Dialectic, Rhetoric, Geometry, Arithmetic, Astronomy, and Harmony. In earlier times, the scope of learning had been broader, but by Martianus's age it had narrowed to just these seven. These would—thanks in part to his influence—remain the principal topics of study throughout most of the medieval period.

As for its scientific significance, *De nuptiis* discussed not only astronomy but cosmology, geography, and astrology. It also examined a number of mathematical areas: dialectic or logic and harmony (which involved the mathematical study of music), as well as geometry and arithmetic. There was little in the way of original thinking presented: instead, Martianus summed up the knowledge that had come down to his own time from the less enlightened past.

In composing his allegory, which proceeded from the scene of the marriage to explication of the various disciplines, Martianus sought to imitate Apuelius, but perhaps unwittingly created a new type of literature. Apuelius had written allegories, or stories in which characters represent ideas, but those characters possessed a degree of life that made them more akin to the figures represented in purer forms of literature. Martianus, by contrast, presented his seven slaves as thoroughly lifeless abstractions, thus providing a model for numerous medieval allegories. (This form would reach its apotheosis with the *Pilgrim's Progress* of John Bunyan, written more than a millennium after Martianus.)

Despite its unabashedly pagan character—an unusual quality, given the fact that it was written long after Christianity had triumphed over the Roman religion—*De nuptiis* virtually became required reading in the schools of Western Christendom, and remained so from the sixth century until practically the dawn of the modern age. During that time, its presentation of the seven arts became so imprinted on Western consciousness that Martianus's description of their physical embodiments provided the model for statues representing the arts at churches around Europe. Numerous figures of some importance to scientific history, among them John Scotus Erigena (810?-877?), Alexander Neckam (1157-1217), and others, wrote commentaries on *De nuptiis*.

JUDSON KNIGHT

Oenopides
490?-420? B.C.
Greek Philosopher

One of the many Greek natural philosophers who benefited from the flowering of culture that accompanied Pericles's reign, Oenopides is perhaps best known for his contributions to mathematical astronomy. Among his accomplishments are determining the angle of Earth's axis relative to the plane of the ecliptic, accurately reporting the length of the lunar month, and calculating the length of the "Great Year," during which the Sun, Moon, and planets all return to the same relative locations in the sky.

Almost nothing is known about Oenopides's early life except that he was born on the island of Chios. He may have visited Athens at some point, but even this is not certain. However, knowledge of his scientific contributions is more certain.

Perhaps Oenopides's best-known contribution to science is his estimate of the angle at which Earth's axis is tilted with respect to the plane of the solar system, also called the ecliptic. Simply looking at the night sky, it is not obvious that all the planets orbit the Sun in a common plane. In fact, this is the case, and this plane is roughly the same as the plane of the Sun's equator, projected into space. Early astronomers tracked the positions of the planets in the sky for centuries, and they noted that the planets always appeared in a relatively narrow swathe of sky. It is this swathe that holds all the constellations of the zodiac and, indeed, these constellations were given special meaning because they are the only ones in which the planets appear.

Looking into the sky (or at a star chart), one can see that these constellations make an arc across the sky. As mentioned above, astronomers determined early on that the planets appeared only along that arc. It wasn't really until Oenopides that the significance of this arc was understood. The arc showed the plane in which the planets were confined, and its elevation in the sky showed how much that plane was tilted with respect to Earth. In fact, Oenopides determined this angle to be about 24°, only slightly more than today's accepted value of 23.5°.

Two of Oenopides's other contributions to science—determining the exact length of the lunar month and the length of the Great Year—are somewhat linked. There is still some debate over whether Oenopides intended his Great Year to refer to the time in which *all* celestial bodies would return to the same relative positions, or if his observations were limited to the Sun, Moon, and the inner planets. In any event, he was able to show that a Great Year lasted 59 years, and the lunar month was 29.53013 days long (as compared to a modern value of 29.53059 days). With respect to Oenopides's meaning of a Great Year, it certainly included the motions of the Sun, Moon, Venus, and Mercury. Mars and Jupiter also return to nearly the same position, with only Saturn out of place, though by only a few degrees. However, although the planets all have the same positions relative to each other every 59 years, they do not return to the exact same position in the zodiac, suggesting that Oenopides did not intend to chart their motions. In any case, his calculations for the lunar month and the length of the Great Year are impressively accurate.

Oenopides was also important in determining the rules under which the Greeks approached some types of mathematical problems, particularly those involving ruler and compass constructions. He also is known to have espoused the concept of Earth as a living being, with God as its soul. He died around 420 B.C., and was later mentioned in some of the works of Plato (427?-347 B.C.) and other philosophers.

P. ANDREW KARAM

Parmenides

b. 515? B.C.

Greek Philosopher

In a fragmentary poetic text, Parmenides outlined what he described as "The Way of Truth"—that is, the way of intellect, which penetrates the unchanging nature of true being—and "The Way of Opinion," or of the senses. Not only does this allegory represent one of the first attempts at philosophical discourse, but it had numerous (mostly unintended) implications for the development of scientific thought.

Born in Elea in southern Italy, Parmenides may have traveled to Athens in about 450 B.C.; this, at least, is the account provided by Socrates (470?-399 B.C.) in Plato's (427?-347 B.C.) *Thaetetus*. Aside from this detail, little else is known about his life, but from his philosophy—not to mention the fact that he came from an area where Pythagorean thought prevailed—it appears that he was heavily influenced by the ideas of Pythagoras (580?-500? B.C.) and his school.

What remains of Parmenides's thought can be found in two poems, "The Way of Truth" and "The Way of Opinion." Most of the latter has been lost, but a substantial part of the former still exists, including the stirring prologue in which Parmenides describes a journey "on the mystic way / That takes the man who knows through all the towns of men." Entering an ethereal bastion of truth through "the gates to the ways of Night and Day," he is greeted by an unnamed goddess who tells him, "No evil fate has sent thee on this way / (Though it lies far from travelled paths of men) / But divine will and justice. / It is fitting that you shall learn all things, / Both the constant heart of encircling Truth, / And also mortals' thoughts, where not one true belief lies. . . ."

What follows is a discussion of what Parmenides presents as the true nature of being: timeless, motionless, and changeless. According to Parmenides, nonbeing cannot exist, and this leads to a rejection of the principle put forth by Heraclitus (540?-480? B.C.) that being and nonbeing can exist simultaneously.

In order to assert that change is impossible, it is necessary to refute the validity of sense-impressions—which Parmenides seeks to do in "The Way of Opinion." To Parmenides, the physical objects of perception (though of course he would reject that very word) exist purely in the intellect, a position that helps explain his later appeal to Plato.

It is ironic that the man who went to such great pains to deny the existence of the physical world was in fact interested in a number of the sciences, most notably astronomy and biology. Parmenides's surviving work contains detailed descriptions of physical phenomena, a fact that has long puzzled scholars, some of whom have suggested that he was in fact attempting to summarize existing thought patterns, or "The Way of Opinion."

Yet he went so far as to develop something that amounted to a form of cosmology, though the fragments of Parmenides's writing are too scanty to draw conclusions regarding its overall schema. Enough material does exist, however, to conclude that Parmenides was a discerning astronomical observer capable of insightful commentary on the subject. Also interesting, inasmuch as it marks another milestone toward the understanding of matter on the part of the ancients, was his theory that all physical substance (something of which he denied the existence!) is made up of Fire and Darkness, two opposing elements that appear in varying degrees within all forms of life.

By far the greatest impact Parmenides exerted on thought, however, lay in the combination of logic and mysticism that underlay his philosophical principles. Here was the first rigorous application of the logical method that would later be formalized by Aristotle (384-322 B.C.), and indeed Parmenides was the first to suggest something like the law of the excluded middle: that a thing cannot be both *A* and not-*A* at the same time.

And in his somewhat mystical attempt to prove that change does not exist, Parmenides perhaps inadvertently opened new paths of scientific investigation. Up to that point, practitioners of the new discipline of philosophy—Thales (625?-547? B.C.) was the first philosopher in the Western sense of critically approaching abstractions—had maintained that the entire universe consisted of one substance. Thales had called this water, Heraclitus fire. Parmenides had easily accepted the one-substance notion because it fit with his belief that change was impossible: if everything is one substance, then change really does not occur. Thus in order to maintain that change *is* possible, as scientists almost universally did, it became necessary to view the world in terms of multiple substances and elements.

Finally, Parmenides exerted an impact through his student Zeno, whose famous paradoxes examine the nature of space and motion and the continuity of reality. Like his teacher, Zeno in the end stumbled upon a set of questions greater than the answers he originally intended to offer.

JUDSON KNIGHT

Philon of Byzantium
280?-220? B.C.
Greek Physical Scientist

What little is known of Philon suggests that he was a relatively wealthy man who helped design and construct a large number of machines. Most of these machines seem to have been designed to help fight and win battles, and many of his designs are described in his treatise, *Mechanics.*

Virtually nothing is known of Philon's life and, in particular, his early life is apparently lost to history. However, the devices he helped to invent or to make useful were considered sufficiently important that, even a few centuries later, Roman architect and engineer Vitruvius (first century B.C.) mentioned him as one of the great inventors, and Heron of Alexandria (first century A.D.) also notes some of Philon's writings in his own work.

Philon's lived after the great flowering of Greek science and philosophy that reached its peak during the fourth and fifth centuries B.C.. Between the reign of Pericles and Philon's birth, Athens lost the Peloponnesian War to Sparta, Alexander the Great (356-323 B.C.) conquered the known world, and Alexander's empire disintegrated following his death. During Philon's lifetime, the city of Rome was winning the wars that would give it one of history's largest empires, though the Romans had not yet conquered Greece, and the Persian empire was still a concern to the East. It is no wonder that some of the best and the brightest minds of Greece concerned themselves with helping to develop military technology. Even Archimedes (287?-212 B.C.), one of the greatest thinkers of the day, was as well known for his military devices as for his brilliant discoveries in mathematics and physics. It is against this backdrop that Philon's inventions must be viewed.

Philon is thought to have been wealthy enough to have been well traveled in the ancient world. During his travels, he was able to meet many inventors, and he was inventive enough to not only remember what he saw, but to use many of these ideas as inspiration for his own inventions. One example is the catapult, which was invented by Ctesibius (second century B.C.). Shortly after learning about this device, Philon wrote of it in his masterwork, *Mechanics,* as well as discussing its military applications with the rulers of Alexandria and other Greek cities.

Other military devices Philon wrote about were siege engines and fortresses and the art of defending and besieging towns.

His *Mechanics* consists of a total of nine books which, taken as a whole, summarize much of the world's knowledge about a variety of devices and technologies during his time. Though not all of his writings have survived, Philon repeats and refers to previous works with sufficient frequency that most of his ideas, if not his exact words, remain.

In addition to his technological accomplishments, Philon did have one contribution to mathematics: his description of a method to duplicate the cube. By this, he meant to construct a cube with exactly twice the volume of a given cube. In reality, this accomplishment was not driven by intellectual interest so much as a practical application for designing catapults, but the method of solving the problem did help to advance mathematics and deserves recognition.

Although Philon is known to have written other works, they have not survived. However, *Mechanics* seems to be his most important work, and it is certainly the only one to be referenced after his death at the age of about 60.

P. ANDREW KARAM

John Philoponus
490?-570
Byzantine Scholar

Known also as Johannes Philoponus and John the Grammarian, the Byzantine scholar John Philoponus wrote on a variety of subjects, from theology to physics. Most notable among his writings was his critique of Aristotle's ideas regarding motion: whereas his great predecessor had incorrectly maintained that a body in motion requires a continued application of force to remain in motion, Philoponus held that a body will keep moving in the absence of friction or opposition.

Philoponus grew up in Alexandria, Egypt, where he studied under Ammonius Hermiae (fl. c. 550), a well-known commentator on the ideas of Aristotle (384-322 B.C.). As a student, Philoponus was one of the first to attempt a synthesis between Aristotelian thought and Western spiritual beliefs. Later other thinkers—including the Christian Thomas Aquinas (1225?-1274), and before him the Muslim Ibn Rushd (Averroës; 1126-1198) and the Jew Moses Maimonides (1135-1204)—would attempt a similar synthesis, but Philoponus was centuries ahead of them. In particular, he was one of the very first to identify Aristotle's idea of the First Cause with the Christian God, a notion that would become highly influential when Aquinas expressed it more than half a millennium later.

Much of Philoponus's work concerned Aristotle: thus he produced commentaries on the *Physics, Metaphysics, Organon, De anima* (On the soul), and *De generatione animalium* (On the generation of animals). The subjects of these works varied between issues of science and theology, a range reflected in the career of Philoponus himself. Though his greatest historical importance today lies in his scientific discussion of the *Physics*, his most significant work in his lifetime was *Diaitetes e peri henoseos* (Mediator; or concerning union), a discussion of the nature of Christ and the Trinity.

In the latter he maintained that every being has a singular nature, and thus Christ could only be divine, not human as well. This placed him dangerously close to the heresy known as Monophysitism, and though Philoponus attempted to exonerate himself with logical acrobatics, the Third Council of Constantinople censured him in 681. By then, however, Philoponus was long dead, and thus did not suffer as a result of his flirtation with heretical thought.

Just as he proved forward-thinking in his attempts to reconcile Aristotle with the Bible, Philoponus was even more ahead of his time in his critique of Aristotelian views on motion. According to Aristotle's kinetic theory, an object does not move unless an external force acts on it, and it can only remain moving as long as that force continues to be applied. According to this explanation, wind and air themselves act as a form of propellant. Philoponus, on the other hand, held that velocity is proportional to the positive difference between force and resistance—i.e., the force must be greater than the resistance—and that the body will remain in motion as long as the former exceeds the latter.

One can see this principle in action by observing a ball in motion. As Aristotle had maintained, the ball cannot move unless an external force acts on it: for instance, a person must pick up the ball and roll it across the floor. (Aristotle's theory, of course, makes no sense when discussing an animate being such as a person or animal.) Initially the ball has considerably more force than the resistance of the floor, which will be lesser or greater depending on the floor's covering: in other words, marble would provide a much easier rolling surface than carpet. As the ball rolls, however, the resistance becomes increasingly greater than the force (actually, modern physicists would call this kinetic energy) of the rolling ball, and eventually it comes to a stop.

The preceding is an illustration of the first law of motion formulated by Isaac Newton (1642-1727). Before Newton, however, there was Jean Buridan (1300-1358), who held that one object imparts to another a certain amount of power, in proportion to its velocity and mass, that causes the second object to move a certain distance. This was an accurate and prescient observation, as was Buridan's position that air resistance slows an object in motion.

But even before Buridan there was Peter John Olivi (1248-1298), credited as the first Western scholar to challenge Aristotle's erroneous notions on kinetics; and before Olivi came was Ibn Sina (Avicenna; 980-1037), like Philoponus a devotee of Aristotle who nonetheless broke with the latter's teachings on matter in motion. Yet long before all these was Philoponus himself, who also challenged the Aristotelian (and indeed Greco-Roman) view that all persons reflect a universal mind. It was Philoponus's position, as would become the view of modern man, that each person is an individual possessed of a separate intellect.

JUDSON KNIGHT

Plato
427-348? B.C.
Greek Philosopher

Although Plato himself did not contribute substantial works directly to science and mathematics, his philosophy and methods of education heavily influenced developments in these fields for many centuries. Many of his ideas were taken from earlier Greek philosophers, especially Socrates (470-399 B.C.). However, Plato was the first to produce a large body of writing that covered the major aspects of philosophy discussed today.

Plato was born into an upper-class family in Athens at a time when the power of his home city was in decline. He fought in the Peloponnesian War (409-404 B.C.) against Sparta, and came to respect the Spartan system of rigid government and strict social rules which appeared to give them the power to defeat Athens. Plato seemed destined for a political career, but the corrupt politics of the age and the treatment given to his

teacher Socrates led Plato to pursue philosophy instead. Socrates had been found guilty of corrupting Athenian youth and questioning the gods, for which crimes he was condemned to death. Disillusioned, Plato traveled to Egypt, Sicily, and Italy, where he learnt mathematics from the Pythagoreans, aristocratic fraternities whose main achievements lay in the fields of music, geometry, and astronomy. He returned to Athens around 387 B.C. and founded the Academy, a place of higher learning designed to instill in the elite youth of Athens the moral values Plato believed would make them better leaders.

Plato wanted to develop in philosophical subjects the certainty he found in mathematics. He hoped that the whole of science could be deduced from a few basic assumed truths, or axioms. With the ideas of Socrates as a starting point, Plato used the method of written dialogues to pursue answers to questions such as "What is courage?" and "What is justice?" Plato tried to explain the relationship between abstract ideas and their representations in the real world. For example, a line is a length with no width, but any line drawn will always have width. Plato imagined a realm of abstract ideas where the perfect, eternal, forms of all things existed. He used the analogy of being chained to the back of a cave, facing the wall, when all you can see of the objects in the cave are the shadows the objects make on the wall. Plato believed that the world was like a shadow of the perfect realm, which would contain items such as the perfect circle and the perfect dodecahedron (geometrical figure with 12 faces), but also the perfect dog, horse, man, and the perfect courage and justice.

Plato's ideas suffered from the reinterpretations of later writers. For example, Plotinus (A.D. 204?-270) altered the central ideas of Plato's ideas to suit his own beliefs, creating a new philosophy later called Neoplatonism. Plato's pupil Aristotle (384-322 B.C.) was to have a greater impact on science than his teacher. Aristotle's work survived the alteration caused by copying, translation, and reinterpretation better than Plato's. Arab thinkers, medieval scholars, and theologians had ready access to Aristotle and found his ideas fitted their own biases, or could be modified to do so. Plato's works were rediscovered during the Renaissance in Europe, when Platonic and Neoplatonic ideas influenced many thinkers such as Johannes Kepler (1571-1630), Isaac Newton (1642-1727), and others. In the seventeenth century, so many scholars of Cambridge University were so influenced by Plato's ideas that they became known as the Cambridge Platonists.

Plato wrote about art, music, poetry, drama, dance, architecture, ethics, metaphysics, the ideal form of government, and the nature of reality. He lived to be about 80, and the Academy he founded in Athens continued until A.D. 529. Plato contributed little to what we would call scientific subjects, but his ideas on education and what constituted knowledge inspired his followers to explore the world in new ways. His stress on mathematics and philosophy, and his insistence on defining terms rather than trusting intuition, influenced many later thinkers.

DAVID TULLOCH

Pliny the Elder
23-79
Roman Scholar

Although only one of his works, *Natural History*, has survived, Pliny the Elder became famous for that work. *Natural History* was a 37-volume encyclopedia that covered topics ranging from anthropology, astronomy, and mineralogy to geography, botany, and zoology. Although the encyclopedia mixes fact and fiction, it nonetheless provides a view of the state of science in antiquity.

Pliny the Elder (full name: Gaius Plinius Secundus) was born at Novum Comum, which is now known as Como, Italy, but spent most of his early life in Rome where he received his education. In his early 20s, he served in Germany in the Roman cavalry, a typical choice for a young man who was born to a prosperous family. After a decade in the military, Pliny briefly shifted his focus to law before settling into his career as a scholar.

His roles as writer and scholar suited his personality well. Apparently a perpetually curious man, Pliny spent years collecting information from many sources and on many subjects, and wrote more than 100 volumes describing this wealth of material. Some of these works included such diverse subjects as grammar usage, the fine arts, oration, military and Roman history, and even the use of javelins as weapons.

Although Pliny bequeathed all his manuscripts to his nephew Pliny the Younger, only one remains—*Natural History* (*Historia Naturalis*). A monumental effort, *Natural History*

summarized much of the material Pliny collected over his lifetime. In the dedication, he claimed that it contained 20,000 pieces of information that he gathered from reviews of 2,000 works by more than 100 different authors. Of the 37 volumes, Pliny devoted five books to astronomy and geology, five to zoology, eight to botany, 13 to medicine and drugs, and five to mineralogy. The first volume was basically a table of contents and list of references.

Although *Natural History* contained a great deal of information, critics have found that it also holds many errors of translation, as well as false statements caused by inadequate fact-checking. For example, the zoology books rely heavily on the scientific work of Aristotle (384-322 B.C.), but also include Pliny's descriptions of legendary animals and folklore. In the astronomical and geological books, the mathematical and technical portions are often incorrectly translated or lack critical details. In spite of these mistakes and the merging of fact with fiction, *Natural History* represented the first truly comprehensive reference work, and its influence continued well into the fifteenth century.

Pliny finished writing *Natural History* around 77 and published 10 of the 37 volumes before he accepted an official position as commander of a fleet in the Bay of Naples in 79. Although his charge was to employ the fleet in the suppression of piracy, Pliny became sidetracked by the eruption of Mount Vesuvius. According to his nephew, Pliny led his fleet ashore to aid in a rescue, but he was soon overcome by the fumes from the erupting volcano. Pliny the Younger supervised the publication of the remaining 27 volumes of *Natural History*.

LESLIE A. MERTZ

Ptolemy
c. 100-c. 170
Greek Astronomer and Geographer

Ptolemy is known historically through his written works. His earliest and most noted treatise is the 13-volume set commonly known as the *Almagest,* which he probably wrote around 150. In these books, he pinpointed the location of more than 1,000 stars, identified the so-called "classical 48" constellations, explained how to calculate latitude and longitude, and predicted solar and lunar eclipses. He also used often complicated mathematical models to help explain the movements of the various celestial bodies. The complexity, in part, derived from his belief that Earth was at the center of the universe, and all stars and planets revolved around Earth.

For example, Ptolemy developed an interesting system to explain why the planets usually, but not always, appear to move forward in their paths across the night sky. Under the erroneous assumption that the planets revolve around Earth, Ptolemy resorted to planetary movements known as *epicycles* to explain the path abnormalities.

Ptolemy's epicycle hypothesis persisted for well more than 1,000 years. Eventually, however, astronomers understood that the planets only appear to backtrack when viewed from Earth. This illusion results because the planets revolve around the Sun in different orbits and at different speeds. Earth passes a planet in an outer orbit much like a race car passes another in an outer lane. To bystanders in the stands surrounding the race track, it is clear that both cars are moving forward in their paths around the track. The view from the inner car is different. If a video camera were mounted on the car's fender, it would record the outer car apparently slowing as the inner car approached. As the inner car reached and passed the outer car, the video would show the outer car momentarily stopping before beginning to go backward. As the distance between the two cars increased, the camera would eventually see the outer car apparently cease its reversal and begin again to gain forward speed. Likewise, when viewed from Earth, other planets move across the sky at a relatively constant speeds most of the time, but occasionally appear to slow down, remain still, and then backtrack before regaining a forward trajectory.

While the underlying notion of Earth as the middle of the universe was incorrect, Ptolemy's intricate mathematical models were very precise in predicting celestial movements as viewed from this planet.

Ptolemy was also very interested in astrology and the impact of planetary position on human society. His four-volume *Apotelesmatica* became a primary reference for horoscope readers.

More notable from a scientific perspective, however, were Ptolemy's contributions to geography. His *Geography,* an eight-volume set, listed the latitude and longitude for many major localities, included a wealth of regional cultural information, and also presented mathematical models describing how to depict the spherical Earth on a two-dimensional map.

Ptolemy also used mathematics in music theory and optics, but the most influential by far were his contributions to astronomy and geography. Ptolemy's work left an imprint on these fields for hundreds of years. His conclusions prevailed well into the sixteenth and seventeenth centuries, when—largely through the studies of Nicolaus Copernicus—the scientific community finally stripped Earth of its central placement in the universe and modern astronomy began.

LESLIE A. MERTZ

Thales of Miletus
630?-547 B.C.
Greek Engineer

Thales is remembered as a founder of rational scientific inquiry and one of the first in a long and distinguished series of Greek scientists in the ancient world. Interested in virtually everything, he was perhaps the first person to try to postulate rational reasons for the phenomena he saw in the world, rather than relying on superstition or religion to explain everyday occurrences. In so doing, he helped set the stage for the great flowering of Greek science and philosophy that was to come, and he left his imprint on many of the Greek scientists who were to follow his example.

Thales was the son of Examyes and Cleobuline, themselves members of a distinguished family. Although Thales was likely born in the Greek city of Miletus, his parents may have been Phoenicians who lived in Miletus. However, Thales was Greek by upbringing if not by ethnicity, and he spent most of his life in Greece or its territories.

Among other things, Thales is credited with making significant contributions to geometry. For example, he was one of the first to accurately measure the height of the Egyptian pyramids, and he is frequently cited as one of the first to develop some of geometry's most basic theorems. In addition, many claim that he was the first to arrive at the concept of a logical mathematical proof, which forms the basis of much of mathematics and geometry today. However, for each such claim, there is a counter-claim that Thales was an empirical scientist who made use of a number of rules of thumb and observations, but did not necessarily understand why they worked.

These claims are difficult to sort out today, partly because none of Thales own writings have

Thales of Miletus. *(Corbis Corporation. Reproduced with permission.)*

ever been found and partly because of the tendency of the ancient Greeks to attribute much more to him than he ever could have actually accomplished. One example of this is his purported prediction of a solar eclipse in 585 B.C. Given the fact that solar eclipses affect only a very small part of Earth, it is now considered highly unlikely that one could have been predicted so long ago, especially in light of that time's astronomical knowledge. It is thought more probable that Thales's contemporaries reasoned that so intelligent a man must have foreseen the eclipse, and never told anyone. Over the years, this transformed into the belief that he, in actuality, had predicted the eclipse.

However great such doubts, it is certain that Thales was highly respected during his time and for many centuries after his death. There can also be no doubt that this respect was deserved. Perhaps the best example of Thales's contribution to Greek science is his own particular cosmology. He felt that Earth consisted of a disk floating in water, and that some aspects of this setup also could be used to explain earthquakes. Of course, this explanation is wrong, but that is not important. What is important is that this marked the first time in recorded history that a person tried to explain Earth using a rational physical explanation rather than resorting to superstition. In this, Thales made a huge conceptual step, believ-

ing that a fallible human could explain concepts that were formerly the sole domain of the gods.

On a somewhat lighter side, Thales may also have given rise to a new form of humor. Walking along one evening, staring at the stars, Thales fell into a ditch. When he was helped out by an attractive servant girl, she is said to have asked him how he could possibly understand the heavens if he couldn't even see what was at his feet. This could well be the first absent-minded professor joke ever told.

P. ANDREW KARAM

Xenophanes
570-475 B.C.
Greek Philosopher

Xenophanes was one of the first Greek philosophers to question the existence of multiple gods, and among the first to deny the possibility that any sort of knowledge could be either absolute or objective. In the second of these, he anticipated questions and philosophies that would arise again in future centuries and, in so doing, was influential in the philosophy of science.

Xenophanes was born in Colophon, Greece. He became a wandering poet in his mid-20s, and made his living in this manner until at least the age of 92. He mixed his poetry with philosophy and, in many of his recitations, he raised serious questions about the multiple deities of the Greek pantheon. This line of thought also led him to question what we know, how we know it, and whether anyone other than God could actually have objective and complete knowledge of anything. It also led him to adamantly deny that gods would have the human characteristics of those in Greek mythology.

Traditional Greek mythology described gods with supernatural powers who, nonetheless, acted very much like people. To Xenophanes, this seemed unlikely. Instead, he first questioned whether or not such powerful beings would indeed act in any recognizable manner. To him, it seemed more likely that such exceptionally powerful beings would act in a manner very unlike mortal and limited people.

Xenophanes then started speculating about whether or not it made sense for multiple gods to exist. He was able to dismiss most of the acts of the gods as natural phenomena, such as weather. Further thought led him to the conclusion that there could be only a single god, and that this god would more likely resemble the universe rather than a person. In his thinking, possibly influenced by the cosmology of Anaximander (610-547? B.C.), God was spherical, incorporeal (that is, without a physical body), and eternal.

Central to Xenophanes's cosmology was the concept of eternity. He felt that, for anything to exist now, it must have always existed. Or, put another way, if there was ever a time when nothing existed, then matter could not have been created from this nothingness. Because of that, he felt that the universe must be eternal. Xenophanes also felt that the universe must be infinite in extent and, on the larger scale, should also be homogeneous. Finally, he concluded that there could be only one god who, like the universe, must resemble it. His description of god was of a being who has no discrete sense organs but, instead, "sees all over, thinks all over, hears all over," who exists everywhere simultaneously, and who exists "without toil."

Xenophanes's contributions to the philosophy of science involved questioning whether it was possible to have complete or absolute knowledge of any aspect of the universe. His feelings were that the universe is so complex that it is impossible to completely understand it. He further suggested that, even if we did know the whole truth about anything, we would be unaware of this and would be unable to communicate this truth to anyone else. His contention was that only educated guessing was possible for humans. Or, in his words, "Here then let these Opinions stand; in resemblance to the reality."

P. ANDREW KARAM

Zhang Heng
78-139
Chinese Astronomer and Mathematician

Zhang Heng was a Chinese mathematician and scientist, born in the first century A.D. In addition to being one of the emperor's astrologers, Heng made a number of interesting astronomical discoveries. However, his most important invention was the seismograph, used to detect earthquakes at a great distance.

Virtually nothing is known of Heng's early life or, indeed, of his nonprofessional life. What is known is that he was a multitalented scientist who became a minister in the Chinese government during the rule of Emperor An'ti. During this time, he was the chief astrologer as well as a respected mathematician and scientist.

Heng is known to have made a number of important contributions during his life. As an astronomer, he made a very good attempt to bring some order to the stars, describing the sky as being composed of "124 groups which are always brightly shining. Three hundred twenty stars can be named. There are in all 2,500, not including those which the sailors observe. Of the very small stars there are 11,520." Presumably, Heng's reference to the sailors is acknowledging the fact that the skies at sea are darker and more clear than those ashore, allowing more stars to be seen.

Another of Heng's accomplishments was that, in A.D. 123, he revised the Chinese calendar in an effort to bring agreement between the calendar and the seasons. This is similar to corrections made in Europe, including the introduction of the leap year, and the addition of 11 days to the calendar during the eighteenth century. Heng's solution was not a permanent fix, but it did serve its purpose in reconciling the calendar and seasons for some time.

It is likely, however, that Heng's most significant contribution was the seismograph. Today, seismographs are extraordinarily sensitive instruments that not only alert us of earthquakes anywhere on Earth, but also serve notice of nuclear weapons testing. Tracking the reverberations of earthquakes, too, has given scientists an unparalleled tool to look inside Earth, tracking how the waves from earthquakes propagate through the planet and what this tells us about Earth's interior.

Heng's instrument was not, of course, nearly so sophisticated. Nonetheless, it was a significant accomplishment that, in and of itself, was sufficient to assure his reputation as a scientist. What most amazed his contemporaries were those instances in which an earthquake was detected so far away that a messenger would take several days to reach the imperial capital with the news. His instrument consisted of four dragon's heads, each holding a ball. When an earthquake struck, the ball was knocked loose, falling into a metal bowl. The direction in which the earthquake occurred could also be determined by seeing which ball of the four had fallen.

In spite of his relatively long and productive life, little else is known of Heng than this. However, his invention of the seismograph alone would be sufficient to assure him a place in the annals of science. Heng died seven years after first demonstrating his seismograph, at the age of 61.

P. ANDREW KARAM

Biographical Mentions

Anaximenes of Miletus
570?-500?B.C.

Greek philosopher who thought the source of all things to be infinite and that air was the basic element from which all others came. According to Anaximenes, air, when evenly distributed, is invisible; when condensed, it becomes water; and if further condensed, becomes earth. Fire was regarded as rarefied air. Anaximenes sought the natural causes of phenomena such as rainbows, rather than accepting supernatural explanations. He may have been a student of Anaximander, or at least had read his work.

Andronikos (Andronicos) of Kyrrhestes
First century B.C.

Greek clockmaker who built the Tower of the Winds in Athens, a water clock combined with sundials and wind indicators in an octagonal building. The tower also displayed the seasons of the years and astrological dates. The building still stands and is one of the few examples of Corinthian architecture from ancient Greece. In later times, engravings were added to the building. The clock mechanism, however, did not survive.

Aristotle
384-322 B.C.

Greek philosopher who is considered the most influential ancient philosopher of the sciences. Aristotle wrote founding texts in physics, astronomy, meteorology, psychology, and biology. A student of Plato and a member of the Academy, his writings often contradicted his teacher's ideas, and he founded a rival center of learning, the Lyceum. Later thinkers venerated Aristotle's words, and they were copied, revised, Christianized, twisted, and eventually criticized and overthrown in the late Renaissance. He also wrote on numerous other topics, including rhetoric, politics, and ethics.

Aryabhata the Elder
476-550

Indian astronomer and mathematician whose *Aryabhatiya* helped usher in a period of increased scientific curiosity in his country. Aryabhata suggested that the reason the stars and planets seem to move around Earth is that Earth is in fact rotating on its axis—an idea literally a thousand years ahead of its time. He also maintained that the Moon and planets reflect light rather than generating it, and that the planets'

orbits describe ellipses rather than circles. In addition, Aryabhata provided an accurate explanation of the causes for solar and lunar eclipses, supplanting a prevailing superstition that these were caused by a demon named Rahu.

Saint Augustine of Hippo
354-430

Tunisian orator and Christian bishop whose religious and philosophical views dominated Medieval thought and greatly influenced the development of Western science. Augustine maintained that the universe was formed according to order, form, and number by an intelligent Creator and was thus intelligible. This reinforced the neo-Platonic belief in the mathematization of nature that later proved central to the seventeenth-century Scientific Revolution. Augustine's emphasis on the meaning and direction of human history prepared the way for developmental thinking—understanding things in terms of their origins and causal antecedents.

Aulus Gellius
123?-170

Roman writer and lawyer who spent a year in Athens compiling information for the composition of *Noctes Atticae* (Attic nights), a compendium on ancient culture and knowledge. It is comprised of 20 books, of which all but part of one survive, and is an example of the Roman commonplace book of general antiquarian knowledge. Noted for its compilation of quotations from lost ancient works, *Noctes Atticae* also contains a fairly important survey of physical science knowledge, including physics, natural phenomena, and astronomy.

Bhaskara I
c. 600-c. 680

Indian mathematician and astronomer who wrote a number of texts on topics relating to his two disciplines. His *Mahabhaskariya* concerns planetary longitudes and conjunctions, solar and lunar eclipses, risings, and settings, and the lunar crescent. *Aryabhatiyabhasya* (629) is a commentary on the *Aryabhatiya* of Aryabhata (476-550), of whom Bhaskara was a follower.

Anicius Manlius Severinus Boethius
480?-525?

Roman mathematician and philosopher who translated many Greek texts into Latin in large encyclopedic collections. These collections were almost all that survived of ancient Greek learning in Europe until Arabic sources were eventually translated into Latin centuries later. Boethius helped define the divisions in education (the trivium and quadrivium) that were to last throughout the medieval period. A Roman aristocrat who probably studied in Alexandria or Athens, he was condemned to death after becoming embroiled in court politics.

Chalcidius (or Calcidius)
Fourth century A.D.

Roman scholar of Greek language who was one of the few to translate Greek into Latin during the third and fourth centuries A.D., providing a portion of the natural science corpus of Greek knowledge to the West. He translated most of the *Timaeus* (Plato's dialogue on his cosmology with its incipient theism) into Latin and added to it the first commentary on the *Timaeus*. It would remain the only Latin translation of a Platonic dialogue for eight centuries and an important source of Greek cosmology in the Middle Ages and Renaissance.

Chen Zhuo
fl. c. 300

Chinese astronomer noted for his star map. In about A.D. 300 Chen Zhuo created a chart that brought together three different star maps created by Chinese astronomers of the fourth century B.C.: Shih Shen, Gan De (Kan Te), and Wu Xien (Wu Hsien).

Chrysippus of Soli
c. 279-207 B.C.

Greek Stoic philosopher who conceived of sound as a wave motion in air. Chrysippus succeeded Cleanthes as the third head of the Stoa in Athens and is considered its second founder. According to Diogenes Laertius, he wrote 705 books, about half of which dealt with logic and language. Chrysippus maintained that knowledge of the world is obtained through the sense organs and true perceptions can be distinguished from illusory ones through deliberation.

Cleanthes of Assos
c. 331-232 B.C.

Greek Stoic philosopher who instigated campaigns of popular resentment against Heraklides of Pontus (c. 390-c. 310) and Aristarchus of Samos (c. 310-c. 230). Cleanthes attacked Heracleides for claiming that Earth rotated about its own axis; and he charged Aristarchus with sacrilege for proposing a heliocentric cosmology in which the world was displaced from its rightful place at the center of the universe. Cleanthes's main contribution to Stoic physics was the introduction of the concept of *tonos* (tension).

Dicaearchus
c. 355-c. 285 B.C.

Greek philosopher who first established latitude as a scientific concept. In about 300 B.C. Dicaearchus noted that at all points along a straight line from east to west on any given day, the noonday Sun appeared at an equal angle from the zenith. A student of Aristotle (384-322 B.C.), Dicaeaarchus wrote about a number of subjects, including the history of Greece up to his own time, and his work had an influence on such later figures as Cicero (106-43 B.C.) and Plutarch (A.D. 46?-c. 120).

Diocles
fl. second century B.C.

Greek mathematician who, according to Arabic tradition, discovered the parabolic burning mirror. Diocles discusses the theory of spherical and parabolic burning mirrors in his only extant work, *On Burning Mirrors*. However, this treatise is primarily a collection of theorems in higher geometry. Among those mentioned are two solutions to the problem of doubling of the cube: (1) using the intersection of two parabolas, and (2) using a spherical curve known as the cissoid.

Eratosthenes of Cyrene
276-194 B.C.

Greek astronomer and mathematician who wrote on geography, mathematics, philosophy, geology, chronology, and literary criticism, though only fragments of his texts remain. Eratosthenes studied at Athens and became head of the library of the Museum of Alexandria. He founded a system of chronology based on the fall of Troy, was an authority on Old Comedy, and introduced mathematics into the field of geography. He estimated the size of the Earth from shadows cast in bowls, and invented a simple method for finding prime numbers (Eratosthene's sieve).

Eudoxus of Cnidus
400?-347? B.C.

Greek astronomer and mathematician who wrote on astronomy, mathematics, geography, philosophy, and drafted laws for his native town. He attended some of Plato's lectures and established his own school, where he lectured on astronomy, theology, and meteorology. Only fragments of his writings survive. His lost books include *On Velocities,* which influenced the planetary theory of Aristotle, and *Tour of the Earth,* which was quoted heavily by later geographers.

Ezekiel
fl. sixth century B.C.

Israelite prophet and priest who described (Ezekiel 1:4-28) a vision of a wheel spinning in the air that has been interpreted as a prediction of the internal combustion engine. This vision has also been interpreted as a description of a UFO. He was a prophet of doom before the capture of Jerusalem by Nebuchadnezzar II, the spiritual leader of the Jews during their deportation to Babylon, and lawgiver after Jerusalem's independence was restored.

Gan De (Kan Te)
fl. fourth century B.C.

Chinese astronomer known for his star catalogue and his studies of sunspots. Along with Shih Shen and Wu Xien (Wu Hsien), Gan De was one of three scholars who independently developed a star catalogue that would, centuries later, be included in a single star map. As for his studies of sunspots, which he described as eclipses that began at the center of the Sun and moved outward, Gan De's work in this regard was about 12 centuries ahead of the West: only in A.D. 807 did a European scientist mention sunspots.

Geminus of Rhodes
fl. 70 B.C.

Greek astronomer and mathematician who authored *Introduction to Astronomy* and *Theory of Mathematics*. The former is an elementary astronomy textbook based on the theories of Hipparchus (c. 170-c. 120 B.C.). The latter was more substantial, providing an exposition and classification of the principles of the mathematical sciences. In it, Geminus carefully defined mathematical terms and concepts including "hypothesis," "postulate," "axiom," "line," "surface," and "angle" as well as criticizing Euclid's (c. 330-c. 260 B.C.) parallel postulate.

Heraclitus of Ephesus
fl. c. 500 B.C.

Greek philosopher to whom is attributed the doctrine that "all things are in flux and nothing is stable"—meaning the world is composed of opposites whose dynamic and constant tension gives rise to the apparent stability around us. Though often interpreted to mean everything is continually changing and therefore unknowable, Heraclitus did not deny the possibility of obtaining knowledge from sensory experience. On the contrary, he maintained that sensory experience, guided by proper understanding, was necessary to discovering the "Logos" that underlies and explains all things.

Ken Shou-ch'ang
fl. c. 52 B.C.

Chinese astronomer who built an early armillary sphere or equatorial ring to show the positions of important circles in the heavens. Some 130 years later, in A.D. 84, Fu An and Chia Kuei added a second ring to show the ecliptic, and in 125 Change Heng added two more rings for the meridian and horizon, thus constituting the first complete armillary sphere.

Kiddinu
fl. c. 350 B.C.

Babylonian astronomer who may have developed the 19-year Babylonian calendar, and who was among the first to recognize the irregular velocity of the Sun and planets. Director of the astronomical school at Sippar, Kiddinu probably created the system by which the Babylonians aligned their lunar and solar calendars by including intercalated months at assigned intervals over a 19-year-period. He also developed what later came to be called System B, the Babylonian method of describing the irregular velocities of heavenly bodies. In addition, Kiddinu calculated the length of the synodic month at 29.530614 days, almost exactly the correct value.

Lucian of Samosata
120-c. 180

Syrian-born Roman satirist who wrote the earliest stories about voyages to other planets. In both his *Icaromenippus* and *True History*, Lucian depicted trips to the Moon and outer space. He also wrote a number of works critical of impostors, including a magician and "wonder-worker" named Alexander, who he attacked in *Alexander the Paphalagonian*.

Ambrosius Theodosius Macrobius
396-423

Roman Neoplatonist philosopher of Greek parentage who was best known for his *Saturnalia* and a commentary on the *Insomnium Scipionis* (Dream of Scipio) in Cicero's *De republica*. His commentary was essentially an encyclopedic rendering of the Platonic interpretation of terrestrial and celestial science. Among many topics, the subject of the possible celestial nature of the Milky Way is an apt example with a lengthy discussion noting the ideas of Democritus, the Stoic Posidonius, and Aristotle's student Theophrastus. It was an important source of Platonic scientific thought and opinion to well into the Middle Ages.

Marcus Manilius
48? B.C.-A.D. 20

Roman scholar who researched the foremost authors of astrology and astronomy in the composition of his long poem "Astronomicon." As such, it is a compendium of ancient astrological and astronomical conceptions, with details including a discussion of the nature of the Milky Way. It was the first Roman work on the subject of astrology and was a popular focus of translation for astrological commentators late into the seventeenth century.

Julius Firmius Maternus
Fourth century A.D.

Roman notary under Constantine the Great and astrological scholar who compiled an astrological work entitled *Mathematics, or the power and the influence of the stars composed of eight books.* The work provided a comprehensive reference to astrology and became a standard authoritative source on the subject until the sixteenth century.

Pomponius Mela
fl. c. 44

Roman geographer whose system of five temperature zones remains in use today. In A.D. 44 Mela introduced his system in *De situ orbis,* a geographical work destined to have enormous impact. The book, which divided Earth into north frigid, north temperate, torrid, south temperate, and south frigid zones, later influenced the work of Pliny the Elder (c. 23-79) and others. Unlike many works of antiquity, it has remained influential well into modern times.

Melissus of Samos
Fifth century B.C.

Greek philosopher of the Eleatic school of Parmenides (and most likely his student) whose work *About Nature and Reality* has survived only as a fragment. The extant text reflects Melissus's attempt to integrate the ideas of Parmenides with the earliest Greek philosophy of the Ionian school. He criticized sense knowledge, noting the concepts of change, motion, and the multiplicity in nature as illusions. He defined "infinity" as being that which has neither "body" or "beginning or end."

Menelaus of Alexandria
c. 70-c. 130

Greco-Roman astronomer and mathematician who wrote a number of works, of which only his *Sphaerica* survives. The latter dealt with spherical triangles, and like most of his writings involved the application of mathematics to astronomy. The titles of selected other works by Menelaus have been translated as "On the Knowledge of the Weights and Distribution of Different Bodies," "The Book on the Triangle," and "Three Books on the Elements of Geometry." Ptolemy (c. 100- 170) recorded astronomical observations made by Menelaus in Rome on January 14, 98.

Meton
fl. fifth century B.C.

Athenian astronomer who introduced the 19-year luni-solar calendric cycle known today as the Metonic cycle—the period after which the Moon's phases recur on the same day of the year. Possibly influenced by a similar Babylonian cycle, Meton's purpose was to provide a fixed calendrical system for recording astronomical observations. His observations, conducted with Euktemon around 433 B.C., are the earliest known serious astronomical observations by Greeks. They also established the solstitial points.

Nabu-rimanni
fl. c. 490 B.C.

Babylonian astronomer credited with devising what came to be known as System A, a set of tables accounting for the positions of the Sun, Moon, and planets at any particular time. As a result of inaccuracies in this system, Kiddinu (fl. c. 350 B.C.) would later correct the earlier method with what historians refer to as System B. Nabu-rimanni calculated the synodic month (i.e., from new Moon to new Moon) as 29.530614 days, a figure that according to modern calculations was correct to the third decimal place.

Olympiodorus of Alexandria
Sixth century A.D.

Greek Neoplatonist philosopher of the Athenian school who, along with John Philoponus, was a student of Ammonius Hermiae. Olympiodorus joined them in performing perhaps the first critical commentaries on Aristotelian science. He wrote an extensive commentary on the *Meteorologica,* wherein he disagreed with Aristotle's terrestrial Milky Way in relation to observations by parallax (a realization by Greek astronomical observation) with which he was acquainted. He also showed comprehension of varied optical topics, such as rules of refraction, and the importance of cloud drops in discussing Aristotle's rainbow theory.

Origen (Oregenes Adamantius)
185?-254

Alexandrian philosopher and Christian theologian who was one of the earliest intellectual fathers of the Greek Church. Origen sought to amalgamate a comprehensive rendering of Hellenic science in a Christian interpretation. He defined Christian cosmology from Plato's theistic cosmological origins, with God creating an infinite series of finite and comprehensive worlds, mutually alternative, with the visible world being stages toward an eternal cosmic process. He was well acquainted with Greek astronomical theory, which he integrated with commentary on *Genesis*.

Philolaus of Crotona
fl. fifth century B.C.

Pythagorean philosopher whose idea of a spherical universe exerted considerable influence on later Greek astronomy. According to Philolaus, the universe consisted of concentric, or nested, spheres at the center of which was a central fire that was never visible to man because Earth is always turned away from it. Earth, an anti-Earth, and the other heavenly bodies, including the Sun, all moved in circular orbits about this central fire.

Plotinus
204?-270

Greek philosopher whose writings were posthumously published in six books called the *Enneads*. Plotinus studied at Alexandria and went on to teach in Rome. His writings reinterpreted the ideas of Plato, which led many later thinkers to misunderstand the original works. The ideas of Plotinus and his followers became known as Neoplatonism, and the early Christian Church adopted some of their ideas. He attempted, unsuccessfully, to found a model city to be called Platonopolis.

Posidonius of Rhodes
c. 135-c. 51 B.C.

Stoic philosopher and astronomer who improved on Hipparchus's (c. 170-c. 120 B.C.) estimate of Earth's distance to the Sun. He also disagreed with Eratosthenes's (c. 276-c. 294 B.C.) excellent calculation of 252,000 stades for Earth's meridian circumference. Posidonius's poorer and much smaller value of 180,000 stades encouraged Christopher Columbus to undertake his voyage of discovery. Posidonius was also the first to draw attention to the spring and neap tides. On a more philosophical note, he emphasized the hypothetical nature of scientific knowledge.

Protagoras of Abdera
c. 490-c. 421 B.C.

Greek sophist who said "man is the measure of all things," meaning all perceptions are true and only individuals can judge the quality of their own sensations. According to Protagoras, the phenomenal world is composed of contradictory qualities. When one experiences something as cool while another experiences it as warm, neither is wrong because both qualities are in the thing. The qualitative experience each has derives from their selective perception of qualities that coexist in matter.

Pseudo-Dionysius the Areopagite
fl. c. 500

Philosopher and theologian, probably a Syrian monk, known as Pseudo-(false) Dionysius the Areopagite, because he wrote under the name of the companion of St. Paul mentioned in the Bible (Acts 17:34). Dionysius wrote a series of Greek treatises and letters to unite Neoplatonic philosophy, which believes there is a single source from which all forms of existence come, and that the soul seeks a mystical union with this source—with Christian theology and mystical experience. His works such as *The Divine Names* and *The Mystical Theology* had a vast influence on medieval thought.

Seleucus of Seleucia
d. c. 150 B.C.

Chaldean astronomer who speculated that the tides were caused by the motions of the Moon. Born in the Babylonian town of Seleucia, he is today primarily remembered as the only known supporter of Aristarchus's heliocentric theory, maintaining that it accurately described the physical structure of the universe. In particular, he accepted Aristarchus's conjectures concurring the centrality of the Sun, Earth's diurnal rotation, and the infinity of the universe.

Lucius Annaeus Seneca
4? B.C.-A.D. 65

Roman Stoic philosopher and playwright who wrote on topics pertaining to natural science. Educated in a philosophy that integrated Stoicism and neo-Pythagoreanism, he probably studied natural science and composed his *Natural Questions* during his banishment from Rome (A.D. 41). This work poses insightful queries dealing with natural science, particularly terrestrial phenomena, revealing Seneca's curiosity about meteorological phenomena (rainbow, thunder, lightning, etc.), comets, and earthquakes. It was a popular source during the Middle Ages and Renaissance among scholars interested in these phenomena. Seneca looked at scientific study as something of a religious exercise and a basis for moral philosophy, as in his relating physics to ethics.

Shih Shen
fl. c. 350 B.C.

Chinese astronomer often credited with creating the first star catalog. In about 350 B.C. Shih Shen developed a map showing some 800 stars, and shortly thereafter Gan De (Kan Te) and Wu Xien (Wu Hsien) created their own star catalogues. More than six centuries later, Chen Zhuo (fl. c. A.D. 300) would bring togetherthese three catalogues in a single star map.

Socrates
470?-399 B.C.

Greek philosopher who exerted a profound influence on subsequent critical thought and intellectual development in ancient Greece. Socrates left no writings of his own; it is only through his pupils, mainly Plato, that we know of his ideas. While not directly concerned with what may be regarded as science, his philosophy shaped later developments. His emphasis on human life, as opposed to the heavens, led to new study of the physical world. His philosophical method—starting with a likely hypothesis, then considering the consequences—is still influential. Found guilty of corrupting the Athenian youth with his ideas, he committed suicide.

Sosigenes
First century B.C.

Greek astronomer who was among those consulted by Julius Caesar on calendar reform. Sources suggest that Sosigenes used Egyptian astronomical calculations, and was the person who convinced Caesar to use a 365-day year with a leap year every four years. He may have also been the one who suggested making the year 45 B.C. 445 days long to bring the seasons back in line with the calendar. He possibly wrote three books, but none survive.

Themistius
c. 317-388

Greco-Roman educator important for his role in preserving writings by Plato (427-347 B.C.) and Aristotle (384-322 B.C.). Themistius produced a number of works in which he paraphrased the earlier philosophers as a means of making their thought accessible to a larger audience. Many of these versions would later reappear in the Arab world, where they exerted enormous influence during the Middle Ages.

Wang Ch'ung
27-c. 100

Chinese philosopher who commented on a number of scientific subjects. In his *Lun-heng*, Wang Ch'ung, a former official who had left office in order to pursue a career as a philosopher, criticized superstitious thinking. Departing from orthodoxy, he attacked not only Taoism, but Confucianism—by then a well-established doctrine—from a position that oriented toward rationalism and naturalism. He also described the human psyche as a purely mechanical, rather than a spiritual, construct.

Wu Xien (Wu Hsien)
fl. fourth century B.C.

Chinese astronomer who created one of the first star catalogues. Wu Xien was one of three scholars, the other two being Gan De and Shih Shen, who independently created star catalogues or star maps in the fourth century B.C. Some 650 years later, Chen Zhuo (fl. c. A.D. 300) would combine these three in a single star map.

Zeno of Elea
490?-425? B.C.

Greek philosopher and student of Parmenides who is remembered chiefly for his paradoxes, which illustrate the nature of motion and time. Among the most famous of these is "Achilles and the tortoise," which posits that if Achilles and a tortoise were to race, with the much slower tortoise given a lead, Achilles could never overcome the tortoise, who would continue to outdistance Achilles, however marginally, as long as it continues to move. Another paradox, "the flying arrow," suggests that motion is impossible. Many thinkers since have been fascinated and frustrated by Zeno's paradoxes.

Zou Yan
fl. c. 270 B.C.

Chinese philosopher who developed a system of five elements: earth, water, fire, metal, and wood. The system of elements, which was associated with the idea of the opposing yin and yang forces, became the basis of the Chinese Naturalist school. These concepts later spread to Korea and other parts of East Asia.

Bibliography of Primary Sources

Anaximander. *On the Nature of Things* (c. 550 B.C.). Now lost, this work is believed to be the first scientific treatise. In *On the Nature of Things* Anaximander described the universe as spherical with Earth occupying a stable position at its center. He argued that

Earth had no reason to fall in particular any direction since it was equidistant from every point on the periphery of the celestial sphere. This symmetry argument was the first to reject the idea of a physical support for Earth.

Aristarchus of Samos. *On the Size and Distances of the Sun and Moon* (c. 270 B.C.). The author's only surviving work, *On the Size and Distances* contains Aristarchus's attempt to determine the diameters of the Sun and Moon. By noting the size of Earth's shadow cast during an eclipse of the Moon, he determined the lunar diameter to be one-third that of Earth. Though his geometrical argument was sound, inaccurate measurements meant this estimate was slightly too large. However, his estimate that the Sun's diameter is seven times that of Earth's was grossly in error—the actual value is closer to 100 times. Nevertheless, the fact that the Sun was larger than Earth may have suggested to him the possibility that Earth traveled about the Sun, a revolutionary concept.

Aryabhata the Elder. *Aryabhatiya* (c. 525 B.C.). This work, written in versed couplets, helped usher in a period of increased scientific curiosity in India. Aryabhata suggested that the reason the stars and planets seem to move around Earth is that Earth is in fact rotating on its axis—an idea literally a thousand years ahead of its time. He also maintained that the Moon and planets reflect light rather than generating it, and that the planets' orbits describe ellipses rather than circles. In addition, Aryabhata provided an accurate explanation of the causes for solar and lunar eclipses, supplanting a prevailing superstition that these were caused by a demon named Rahu.

Aulus Gellius. *Noctes Atticae* (Attic nights, second century A.D.). A compendium on ancient culture and knowledge, comprised of 20 books, of which all but part of one survive. As an example of a Roman commonplace book of general antiquarian knowledge, it is noted for its compilation of quotations from lost ancient works and also contains a fairly important survey of physical science knowledge, including physics, natural phenomena, and astronomy.

Bhaskara I. *Mahabhaskariya* (c. A.D. 650). This work concerns planetary longitudes and conjunctions, solar and lunar eclipses, risings, and settings, and the lunar crescent.

Capella, Martianus. *De nuptiis philologiae et Mercurii* (The marriage of Mercury and philology, c. fourth century A.D.). Also known variously as the *Satyricon* and *Disciplinae*. An allegory concerning the arts and sciences that provided a model for numerous medieval allegories. It became required reading in the schools of Western Christendom, and remained so from the sixth century until practically the dawn of the modern age. During that time, its presentation of the seven arts became so imprinted on Western consciousness that Martianus's description of their physical embodiments provided the model for statues representing the arts at churches around Europe. *De nuptiis* also discussed astronomy, cosmology, geography, and astrology, and a number of mathematical areas: dialectic or logic and harmony (which involved the mathematical study of music), as well as geometry and arithmetic. There was little in the way of original thinking presented: instead, Martianus summed up the knowledge that had come down to his own time from the less enlightened past.

Epicurus. *De natura* (c. 300 B.C.). In this work the author developed a mechanistic explanation of the world grounded in Democritean atomic theory. He accepted that all natural phenomena are generated by atoms and the void. The atoms are imperceptibly small, of different shapes and sizes but composed of the same substance, infinite in number, and free to move through the void. Sensory qualities of physical bodies, such as taste, color, and weight, vary according to the number of component atoms, their arrangement, and the presence of empty space. This picture left no room for supernatural forces.

Geminus of Rhodes. *Introduction to Astronomy* (c. 70 B.C.). An elementary astronomy textbook based on the theories of Hipparchus.

Isidore of Seville. *Etymologies* (c. A.D. 600). A monumental encyclopedia of all that was known at the time concerning science and technology, as well as providing a panoramic view of ideas on education, theology, and other subjects in early medieval Europe. Ostensibly concerned with word origins, as the title suggests, the *Etymologies* includes coverage of the Roman educational system, medicine, world history, sacred writings, religious figures and controversies, geopolitics, lexicography, cosmography, geography, communication, petrography and mineralogy, agriculture and horticulture, the military arts and sports, ships, housing, apparel, and food.

Lucian of Samosata. *Icaromenippus* and *True History* (c. A.D. 170). Lucian depicted trips to the Moon and outer space in these two books, which contain perhaps the earliest stories about voyages to other planets.

Manilius, Marcus. *Astronomicon* (c. first century A.D.). An extended poem that represents a compendium of ancient astrological and astronomical conceptions, with details including a discussion of the nature of the Milky Way. It was the first Roman work on the subject of astrology and was a popular focus of translation for astrological commentators late into the seventeenth century.

Maternus, Julius Firmius. *Mathematics, or the power and the influence of the stars composed of eight books* (fourth century A.D.). A comprehensive reference on astrology that served as a standard authoritative source on the subject until the sixteenth century.

Mela, Pomponius. *De situ orbis* (A.D. 44). In this work Mela introduced a system of five temperature zones that remains in use today. The book, which divided Earth into north frigid, north temperate, torrid, south temperate, and south frigid zones, later influenced the work of Pliny the Elder and others. Unlike many works of antiquity, it has remained influential well into modern times.

Melissus of Samos. *About Nature and Reality* (fifth century B.C.). Exists as only as a fragment, but the extant text reflects Melissus's attempt to integrate the ideas of Parmenides with the earliest Greek philosophy of the Ionian school. He criticized sense knowledge, noting the concepts of change, motion, and the multiplicity in nature as illusions. He defined "infinity" as being that which has neither "body" or "beginning or end."

Menelaus of Alexandria. *Sphaerica* (first century A.D.). The author's only surviving work. It dealt with spherical triangles, and like most of his writings involved the application of mathematics to astronomy.

Parmenides. *The Way of Truth* and *The Way of Opinion* (c. 550 B.C.). A fragmentary poetic text consisting of an allegory and detailed descriptions of physical phenomena, representing one of the first attempts at philosophical discourse in the West. His theory that all physical substance is made up of Fire and Darkness, two opposing elements that appear in varying degrees within all forms of life, also marks a milestone toward the understanding of matter on the part of the ancients.

Plato. *Timaeus* (c. 350 B.C.). Ignoring Democritus altogether and drawing for inspiration upon Pythagorean geometry, Plato suggested in this dialogue that there are five different types of geometrical atoms, corresponding to the five perfect geometrical solids (having sides all of equal length, faces all of equal size and shape, and angles all of equal degree). Four of these solids correspond to the traditional four elements—fire: tetrahedron, earth: cube, air: octahedron, water: icosahedron—and the fifth (the dodecahedron) to the entire cosmic sphere. The faces of the first four solids are in turn readily divisible into either equilateral triangles (for fire, air, and water) or isosceles triangles (for earth). (As the unitary cosmic symbol, the dodecahedron does not require division.) Thus the geometric "atoms" are not indivisible, but are assembled from and disintegrate into their triangular components, as the true, unchanging, indestructible formal elements of the cosmos, with fire, air, and water being interconvertible. Like Democritus, however, Plato attributes secondary qualities to the sizes, shapes, motions, and interactions of his atoms.

Pliny the Elder. *Natural History* (c. A.D. 77). A 37-volume encyclopedia that covers topics ranging from anthro-pology, astronomy, and mineralogy to geography, botany, and zoology. Although the encyclopedia mixes fact and fiction, it nonetheless provides a view of the state of science in antiquity.

Ptolemy. *Almagest* (c. A.D. 150). In this 13-volume set, Ptolemy pinpointed the location of more than 1,000 stars, identified the so-called "classical" constellations, explained how to calculate latitude and longitude, and predicted solar and lunar eclipses. He also used often-complicated mathematical models to help explain the movements of the various celestial bodies.

Ptolemy. *Apotelesmatica* (c. A.D. 150). Four-volume work that became a primary reference for horoscope readers.

Ptolemy. *Geography* (c. A.D. 150). Eight-volume set that listed the latitude and longitude for many major localities, included a wealth of regional cultural information, and also presented mathematical models describing how to depict the spherical Earth on a two-dimensional map.

Seneca, Lucius Annaeus. *Natural Questions* (c. A.D. 41). This work poses insightful queries dealing with natural science, particularly terrestrial phenomena, revealing Seneca's curiosity about meteorological phenomena (rainbow, thunder, lightning, etc.), comets, and earthquakes. It was a popular source during the Middle Ages and Renaissance among scholars interested in these phenomena. Seneca looked at scientific study as something of a religious exercise and a basis for moral philosophy, as in his relating physics to ethics.

Wang Ch'ung. *Lun-heng* (c. A.D. 75). Departing from orthodoxy, the author in this work attacked not only Taoism, but Confucianism—by then a well-established doctrine—from a position that oriented toward rationalism and naturalism. He also described the human psyche as a purely mechanical, rather than a spiritual, construct.

NEIL SCHLAGER

Technology and Invention

Chronology

c. 3500 B.C. The wheel is invented in Sumer.

c. 3000 B.C. First instances of weaving (cotton fabric, in the Indus Valley) and cultivation of vegetables (potatoes, in the Andes).

c. 2650 B.C. Imhotep designs the Step Pyramid of Saqqara, the world's first large stone structure; within a century, the Great Pyramid of Cheops is built.

c. 2600 B.C. Peoples of the Near East harness oxen for plowing, the first significant use of domesticated work animals.

c. 2500 B.C. Beginnings of the Iron Age in the Near East; development of papyrus in Egypt; and establishment of the first major cities, Harappa and Mohenjo-Daro (which feature urban planning and drainage systems) in India.

c. 2400 B.C. The abacus makes its first appearance in Babylon.

c. 400 B.C. The wheelbarrow is developed in China.

c. 300 B.C. The Romans develop sophisticated techniques of road-building and construct aqueducts many miles long.

c. 300 B.C. Gears, or toothed wheels, are developed in Alexandria.

c. 250 B.C. Archimedes invents a number of useful mechanisms, including the Archimedes screw, a device for raising water that is still in use in parts of the world today.

c. 100 B.C. The Romans introduce chain mail armor, which will remain in use until the fourteenth century.

A.D. 105 Chinese inventor Ts'ai Lun perfects a method for making paper from tree bark, rags, and hemp.

c. A.D. 450 Stirrups, brought westward by invading nomadic tribes, make their first appearance in Europe; in the view of many historians, this is one of the most important inventions in history since it makes warfare on horseback effective, and thus opens the way for knights and feudalism.

c. A.D. 600 Block printing makes its first appearance in China, where the first viable form of ink—made from lamp-black and later incorrectly dubbed "India ink"—was developed two centuries earlier.

Overview: Technology and Invention
2000 B.C. to A.D. 699

Technology in the ancient and classical worlds reached impressive levels of achievement. The use of simple tools, skilled management of large numbers of workers (many of them slaves), and the absence of time pressure allowed these societies to create both productive farms and thriving cities. Many achievements of this era reflect the ingenuity and skill of these early engineers.

Agriculture

Agriculture was the foundation of preindustrial societies from ancient Egypt to early medieval Europe. The simple scratch plow and use of domesticated animals, especially in the ancient Mediterranean world, grew the grain that allowed civilizations to emerge. In places like Egypt, where regular river flooding required the storage and control of water, irrigation ditches and canals were essential for society's survival. Water control was a central function of society, and social and political structures emerged to develop and maintain it. These "hydraulic cultures," like many preindustrial societies, valued stability and discouraged technological innovation; the pace of change in them was often very slow.

Although technical change occurred slowly, it gradually diffused throughout the known world. Many advances came from the East, including the use of natural fertilizers, such as manure and lime. In addition, the Roman empire spread techniques such as crop rotation, grafting, and mariculture to much of its empire, giving the world enough food to allow civilization to flourish. Between the fifth and eighth centuries, three key developments would revolutionize early medieval agriculture and society: the stirrup, the horseshoe, and the padded horse collar.

Urban Technology

This agricultural base allowed cities to develop, and their needs fueled still further technological developments. Water supply and sewage systems, monumental structures, bridges, roads, arenas, public buildings, central heating systems, defensive city walls, urban planning, and record keeping all stemmed from the needs of cities. As urban centers grew, so did their buildings. The brick and stone of early construction eventually gave way to hydraulic cement, which

allowed the construction of bigger and stronger structures. The Romans also introduced a new motif in building architecture with the use of the semicircular arch, which they used in bridges and aqueducts.

Cities in this era were ceremonial, commercial, political, and trading centers. Each had its special technological needs, including the need to manage large labor forces. Despite their limitations, early societies produced impressive results, from Mayan temples to Roman aqueducts. Just as agrarian hydraulic societies relied on a highly ordered social system, the urban segment depended on a well managed and controlled population for the construction and maintenance of its technology. Those technological features, when well maintained, provided decades or even centuries of use, a testament to the quality and design prowess of ancient engineers.

Mechanics and Metallurgy

Not all ancient technology was purely functional; many inventions were intended for amusement and decoration as well. For instance, the ancient Greeks used their mechanical ingenuity to produce a series of technological toys or gadgets, using their sophisticated understanding of hydraulics, pneumatics, and mechanics to produce steam-powered toys, water organs, water clocks, and pumps. Unfortunately, with an army of slaves to do the heavy lifting, there was little incentive to transform these "toys" into labor-saving devices. Even so, the Greeks' ability to apply a theoretical understanding of simple machines such as the lever, screw, wedge, and pulley led to several devices like the olive press, crane, and water screw. This fundamental understanding of mechanics and the resultant engineering of such men as Archimedes and Ctesibius provided a foundation for later developments in the Roman and medieval eras.

The decorative arts were also the foundation of metallurgy in the ancient world. The desire to produce beautiful pieces for the nobility or religious hierarchy created a core of skilled artisans adept at working with gold, bronze, copper, silver, and iron. Ancient Egypt and China, for example, had sophisticated metal-working techniques in place centuries before this kind of metallurgy became a hallmark of industrialism.

The Written Record

Ancient and classical cultures learned to write and store information to record their achievements. Stone tablets, sheets of papyrus, scrolls of parchment, and pages bound in book form were among the many ways to store words and symbols. The Romans, for example, developed a set of rectangular single sheets bound together in a form that became the prototype of the printed book.

With these documents came the need for libraries. The largest and most influential of these, at Alexandria, Egypt, housed as many as 500,000 scrolls containing the work of the ancient and classical world's significant scientists and philosophers. Unfortunately, in the late third century A.D., the library was destroyed and much ancient knowledge was lost.

Empirical Technology

Throughout this era, almost all technological developments were empirical: brought about by the direct observation of a need. Invention based on theory or scientific hypothesis was rare in the ancient world. Although astronomy and mathematics, especially geometry, were used to site structures or to operate simple machines such as the screw or lever, science played a minor role in technology. This trial-and-error method, while it hindered rapid progress, did allow engineers to learn by doing. Its benefits are obvious in the era's durable and practical construction. Indeed, the Romans were renowned for their ability to organize and complete large-scale projects.

A lack of theoretical understanding in engineering efforts extracted a price as well. Often the projects created were heavier than necessary and overdesigned or overengineered. Relatively cheap labor and materials removed the need to worry about efficiency or economy, so the use of extra materials or labor was commonplace. Massive, solid projects with a high margin of safety marked the technology of the time. A greater theoretical analysis would have given engineers more efficient results using less material with the same margin of safety.

Conclusion

Early technological achievements prove that simple tools, well-managed labor forces, and practical knowledge, when used by skilled and experienced craftsmen can produce impressive results. Early innovations created a successful agriculture that eventually supported the development of larger civilizations. The pace of change was slow—because many cultures sought stability over transformation—yet their technology was well learned, durable, and skillful, features desirable for any era.

H. J. EISENMAN

Early Agriculture and the Rise of Civilization

Overview

People began farming at different times in different parts of the world. Around 8500 B.C. hunter-gatherers in the area of southwest Asia known as the Fertile Crescent began to cultivate wild grains and domesticate animals. One thousand years later, people in northern and southern China were growing rice and millet and raising pigs. Archeological evidence shows that crops were planted in Central America as early as 7000 B.C., and around 3500 B.C. in the Andes mountains and Amazon river basin of South America. Farmers in Africa began growing crops around 5000 B.C. Three thousand years later, native Americans in the eastern United States planted a few crops, but still depended on hunting and gathering. As agriculture evolved in these locations, so did the social, economic, and cultural practices that led to what is known as civilization.

Background

The shift from hunting and gathering to farming was a gradual process that happened 10,000 years ago in some parts of the world, 5,000 years ago in others, and never in still others. These disparities are due not to human differences from place to place, but to differences in indigenous plants and animals and local climate and geography.

For the thousands of years before plants and animals were domesticated, people roved in small bands, foraging for enough food to stay alive. Because of the abundance of wild foods in the Fertile Crescent, hunter-gatherers settled there permanently. They progressed from gathering wild grains to planting them, choosing seeds from plants with the most desirable characteristics. Their first crops were emmer wheat and barley, which were high in protein and easy to domesticate compared to plants native to other parts of the world. Cultivated emmer wheat, for example, is very similar to its wild ancestor, while it took thousands of years for modern corn to evolve from its half-inch-long ancestor.

The climate and geography of the Fertile Crescent were varied, ranging from valleys to mountains and from deserts to riverbeds. In addition to supporting a variety of plant life, this diversity supported a variety of mammals. Surrounded by an abundance of natural resources, it is not surprising that the people of the Fertile Crescent were the first farmers. But agriculture appeared in less fertile parts of the world as well. The first crop grown along the Yellow River in China was millet, followed by rice and soybeans, significant sources of protein. In Central American, the earliest cultivated foods are still characteristic of that region: squash, beans, tomatoes, avocados, chocolate, corn, and chili peppers. Farther south, on the Pacific coast of modern Panama, archeologists have found traces of manioc, yams, arrowroot, and corn on ancient grinding stones. Central Americans also domesticated wild turkeys. The first farmers in the Andes mountains of South America raised sweet potatoes, manioc, peanuts, and quinoa (a grain), as well as llamas, alpacas, and guinea pigs. The only crops domesticated in the eastern United States were squash and a few seed plants. Sorghum and millet were cultivated in sub-Saharan Africa, tropical West Africa, and Ethiopia, but archeologists are unsure whether agriculture arose there independently or in response to outside influences.

Impact

In the case of agriculture, necessity was not the mother of invention. It was hunter-gatherers who already had enough to eat that made the shift to farming. Permanent homes and stockpiles of wild cereals gave them enough time and energy to experiment with cultivating seeds and breeding animals without the risk of starvation. As food was grown and stored more efficiently,

populations increased and settlements grew larger, creating both the incentive and the means to produce even more food on more land.

Agriculture spread at different rates, depending on climate and geography. From the Fertile Crescent, it moved west through Europe and Egypt and east through Iran and India, reaching the Atlantic Coast of Ireland and the Pacific Coast of Japan by the beginning of the Christian era. From its origins in China, agriculture moved south, eventually spreading across the Polynesian islands. In contrast, agriculture passed either slowly or not at all through the tropical and desert climates surrounding early agricultural sites in Egypt, sub-Saharan Africa, Central America, and the Andes. Domesticated animals did not reach South Africa until around A.D. 200, the same time corn reached the eastern United States. It was therefore the plants, animals, and farm-related technologies of the Fertile Crescent and China that had the greatest impact on future civilizations.

The hunter-gatherers of the Fertile Crescent and China had been making tools from stone, wood, bone, and woven grass for thousands of years. Once farming took hold, people improved their tools so they could plant, harvest, and store crops more efficiently. One of the earliest tools was a pointed digging stick, used to scratch furrows into the soil. Eventually handles were attached to make a simple plow, sometimes known as an ard. Around 3000 B.C. Sumerian farmers yoked oxen to plows, wagons, and sledges, a practice that spread through Asia, India, Egypt, and Europe. After iron metallurgy was invented in the Fertile Crescent around 900 B.C., iron tips and blades were added to farming implements. The combination of iron-tipped plows and animals to pull them opened previously unusable land to cultivation. Although seeds were most often simply thrown into furrows, some farmers in Egypt and Babylonia dropped seeds through a funnel attached to the end of the plow. The seeds were then trampled into the ground by a person or a herd of sheep or pigs. Grains were harvested with wooden-handled sickles, with either stone or iron blades.

The evolution of agriculture can also be traced through the evolution of containers, essential for storing surplus harvests. Nomads favored portable leather or straw baskets and also dug underground storage pits. When people began to live in permanent settlements, they built heavier but more functional storage containers from clay that they dried in the sun. They also lined underground ovens with clay. Not only did the ovens bake the ground-wheat batter, they also hardened the clay into pottery. Experience with high-temperature underground ovens later proved crucial to the invention of bronze and iron metallurgy.

The domestication of animals was a gradual process. Some animals were easy to tame and breed; others impossible. The most docile or productive animals were allowed to reproduce while the least were slaughtered. Animals also evolved in response to their new environments, some becoming larger and others smaller. The first domesticated animal was the dog, which was bred for hunting and food in several places around the world. Other small animals and birds, like guinea pigs in South America, turkeys in Central America, ducks and geese in Eurasia, and chickens in China, provided food, eggs, and feathers. But it was the five mammals found in the Fertile Crescent—sheep, goats, cattle, pigs, and horses—that had the greatest impact on food productivity. The power of oxen and horses was harnessed to pull plows and wagons, grind grain, and build irrigation projects. Grazing animals also fertilized fields with their manure and cleared them of undergrowth. In contrast, the rest of the globe had either no large mammals (as with North America, Australia, and sub-Saharan Africa) or only one (the ancestor of the alpaca and llama in South America). The result was both short- and long-term advantages for civilizations with domesticated animals. They had more food, larger populations, and land transportation, so were able to move into neighboring territories and ultimately conquer other continents, as the Spanish did in Central and South America.

The domestication of animals also influenced the rise of epidemic diseases like smallpox, influenza, and measles. Using manure and human waste as fertilizer infected people with harmful bacteria. Once people started to live in close contact with animals, they were exposed to animal viruses that over time mutated into new ones causing human epidemics. When carriers of these diseases invaded unexposed populations—again, as the Spanish did in Central and South America—the result was devastating. For example, the natives of Hispaniola were entirely wiped out by germs carried by Christopher Columbus (1451-1506) and his sailors. The same process of virus mutations in farm animals is believed to occur today in southern China, where certain influenza viruses periodically shift to new forms that require new vaccines.

Early farmers along the banks of the Tigris and Euphrates rivers of Mesopotamia used three methods of water regulation. They dug shallow canals through high riverbanks to allow water to flow into nearby fields. Because the population continued to grow and crops had to be cultivated farther away from rivers, the canals were extended. Farmers also built embankments around their fields to protect them from too much water. Another method of water regulation was dams, constructed upstream of fields before spring flooding. These techniques allowed a smaller number of people to farm larger areas of land, which no longer had to be naturally productive. Water regulation not only developed engineering and construction skills, but also had important social effects. A stratified society of laborers, supervisors, and administrators was necessary for planning, building, and maintaining large-scale dams and canals. The intensive farming made possible by irrigation and embankments also led to social stratification since productive land became much more profitable. Some acquired more wealth and power than others, and it did not take long for societies to be divided into royalty, peasants, and slaves, as happened in Sumeria, Egypt, and China.

In a two-way circle of growth, populations and food production kept increasing and so did technology. Since every moment no longer had to be spent on food, people could specialize in occupations, such as that of the potter, baker, metallurgist, and engineer, that supported agriculture. These new occupations, in turn, increased population density and food production even more. A counting system based on clay tokens was devised in the Fertile Crescent to keep track of crop yields and animals. The first writing system, Sumerian cuneiform, joined pictures of objects to numerals scratched on flat clay tablets, giving birth to another specialist, the scribe. Other writing systems from China, Egypt, and Mexico also spread and evolved, providing a tool for governance. As societies learned to produce, store, and distribute food, they developed the characteristics of modern civilizations: densely populated cities, centralized government, organized religion, private property, specialized occupations, public works, taxation, technology, and science. People lived as hunter-gatherers for tens of thousands of years before they began to plant crops and domesticate animals. Once this happened, however, the transition to modern civilization was rapid and fundamental.

LINDSAY EVANS

Further Reading

Cowan, C. Wesley, and Patty Jo Watson, eds. *The Origins of Agriculture.* Washington, DC: Smithsonian Institution Press, 1992.

Diamond, Jared. *Guns, Germs, and Steel. The Fates of Human Societies.* New York: W. W. Norton, 1997.

Heiser, Charles B., Jr. *Seed to Civilization.* San Francisco: W. H. Freeman, 1973.

MacNeish, Richard S. *The Origins of Agriculture and Settled Life.* Norman: University of Oklahoma Press, 1992.

Rahn, Joan Elma. *Plants that Changed History.* New York: Atheneum, 1982.

Smith, Bruce D. *The Emergence of Agriculture.* New York: Scientific American Library, 1995.

Struever, Stuart, ed. *Prehistoric Agriculture.* Garden City, NY: Natural History Press, 1971.

The Domestication of the Horse

Overview

The horse is best described as a single species (*Equus*) that consists of various breeds. It is a hoofed, herbivorous mammal that has had a significant impact throughout the history of man. In fact, the horse has been referred to as "the proudest conquest of man," by the French zoologist Georges Buffon (1707-1788). Its importance to the course of human history cannot be understated.

The relationship between horse and human is quite unique. Unlike most other large animals, the horse has been likened to a partner and friend. From its earliest history, the horse has served as a food source, aided agriculture by plowing fields and bringing in the harvest, helped to track and hunt down game, served as an important transport vehicle for goods and people, as well as an integral part of war and conquest, and has even provided many forms of recreational activities.

The earliest form of the horse, known as *Hyrocotherium,* or *eohippus* (dawn horse) is believed to have first appeared on Earth approximately 50 million years ago. It is characterized as a being a timid creature about the size of a dog. It lived in swampy areas and its range extended over much of the world. As the habitat of the early horse changed from swamp to forest to grasslands, it went through many evolutionary changes until it reached its present form, *Equus,* that is seen today. It should be noted that although the range of the horse initially spread throughout the world, it became extinct in the Western world some 8,000 years ago and was brought into Western hemisphere by the Spanish about 500 years ago.

Initially, the wild horse was used as an important food source. This use is depicted in Stone Age cave paintings, such as those seen at Lascaux, France. These are believed to be nearly 20,000 years old and show the horse as an object of human prey. In order to make better use of the horse, it was domesticated. There is little agreement as to when this actually happened, but it is likely that other animals such as dogs and cattle were domesticated first. Some authorities argue that early Cro-Magnon farmers were nomadic and domesticated the horse for use as a pack animal. Others make the case that domestication took place 6,000 years ago by native tribes living on the steppes adjacent to the Black Sea. After this time, horses quickly became indispensable to the people who used them.

Background

After domestication, horses were still used as a ready food source for their meat and mare's milk. But there is recent evidence that humans began to ride horses soon after domestication. The minimum amount of equipment needed in order to control the horse would be some sort of bit or mouthpiece attached to reins. Archeological evidence shows horse teeth with bit wear dated as early as 4000 B.C. The early forms of bits were probably just ropes around the jaw, but more sophisticated bridle setups using antlers coupled with some other pliable material have been uncovered. This type of system would give the rider superior control over the horse and indicates that these early nomads were probably accomplished horsemen.

Early humans had seen the power of draft animals and the yoke was believed to be invented around 5000 B.C. to harness this source of power. In the Near East, cattle were used to drive sledges, but wheels were added in the next millennium. As horses were brought from Asia between 2000 and 3000 B.C., it became apparent that these animals were much faster than cattle

Merchants carrying their goods on horseback. *(Archivo Iconigrafico, S.A./Corbis. Reproduced with permission.)*

and they quickly became the draft animal of choice. The yoke collar needed to be modified for use with horses because it restricted their breathing. Breast straps and other types of collars were invented in order to circumvent this problem. By the fifteenth century B.C., Egyptians developed a wishbone-shaped yoke, joined by straps, that enabled the horse to breath without restriction while being used as a draft animal.

The use of the horse for war and hunting was not overlooked at this time either. An early form of a chariot pulled by horses was developed for these purposes. It was a light, responsive vehicle that gave the driver much control. All metal bits were introduced for use on chariots around 1500 B.C. These made the use of the chariot in warfare more effective because they were stronger and gave greater control over the team of horses. At that time, warfare was still restricted to large chariot forces requiring disciplined horses that were in good condition. It would be another 500 years before men got directly on the horse and began forming cavalry units.

Speculation exists as to the reason why early humans generally preferred driving horses rather than riding them. One reason commonly given was the small stature of early breeds. However, given the fact that small horses are used today as effective mounts, this assumption is questionable. It may have been sociological

factors that influenced this, as it seems the upper class had a great aversion to horse sweat, so it may have been undignified to come into direct contact with the horse.

The first group to make wide use of the horse for successful warfare was a united group of nomadic tribes from the Russian steppe, collectively called the Scythians, in 800 B.C. They mastered the technique of horseback archery and measured their wealth in terms of horses. As feared warriors, they influenced future generations with their demonstration of the horse's great power and advantage. By 700 B.C. riding horseback was much preferred over a chariot for both hunting and war.

Native tribes of the steppes made most of the technological advances involving the horse, and these advances were brought west as they tribes invaded neighboring lands. The first horseshoes are no exception. Horseshoes of various types were used by these tribes, but it was the Romans who made wide use of them. Horseshoes were used by the Romans to reduce wear of the hooves, especially on paved roads. These iron shoes were known as "hipposandals" and were tied to hooves. The modern iron horseshoe, which is nailed to the hoof, first appeared around the fifth century A.D.

The advent of the saddle, a seat for the rider, was an important development. The leather saddle, developed in around the third century B.C., greatly improved the utility of the horse. In conjunction with this, the stirrup—used to support the rider's feet—was used to help in the control of the horse as well. Attila the Hun (406?-453) is credited with introducing the stirrup to Europe, giving rider's in the West much more maneuverability and control. Subsequent advances in both the saddle and stirrup occurred throughout early history, with modern stirrups and saddles bearing a striking resemblance to those of the Middle Ages.

Impact

Throughout human history the horse has been extremely beneficial and has helped to shape society into what it is today. The horse has been an important component in recreation, travel, labor, and especially war. First and foremost, its domestication was absolutely essential for society to realize how useful and valuable the animal could be. Secondly, the invention of important technological advances, such as the invention of the collar, bit, bridle, stirrup, saddle, and horseshoe, allowed humans to make the horse much more functional, which was important in its rise to significance.

The first domesticated horses were used as food and their hide, but subsequent generations began to use the animal for other things. It seems likely that the next major step was to use the horse to lessen the work done by man. As the horse was harnessed as a power source, the need for human power was reduced. This had the effect of reducing human slavery, since slaves were used primarily as a major source of power.

The legacy of the importance of the horse in antiquity would not be complete unless its significance in waging war was discussed. Armies that used horses for conquest had distinct advantages over those that did not. During antiquity the world was overrun by various ethnic groups that relied heavily upon horses during battle. The initial preference was for two-man chariots, that consisted of a driver and an archer positioned to shoot arrows at the enemy. Later, men directly mounted horses for combat purposes. This gave well-trained armies using the horse a significant advantage and proved to be a key component of conquest for thousands of years. The horse figured significantly in the course of human history and those that could best harness its natural power played more dominant roles until the technological advances of machines made the horse relatively obsolete in modern countries.

JAMES J. HOFFMANN

Further Reading

Caras, Roger A. *A Perfect Harmony: The Intertwining Lives of Animals and Humans Throughout History.* New York: Simon & Schuster, 1996.

Clutton-Brock, Juliet. *A Natural History of Domesticated Mammals.* New York: Cambridge University Press, 1999.

Facklam, Margery. *Who Harnessed the Horse? The Story of Animal Domestication.* New York: Little Brown & Co., 1992.

The Domestication of Wheat and Other Crops

Overview

Although it may sound exaggerated, civilization would not have advanced without the domestication of plants and animals. Since the domestication of wheat and other crops is so important to the development of civilization, its origins must be studied to understand the links between farming and other innovations that form an advanced society.

Background

Although domestication of plants and crops cultivated for consumption has been carried on for

11,000 years, this figure pales in comparison with the seven million years humans fed themselves by hunting wild animals and eating wild plants. Without the transition, however, mankind could not have completed its social and cultural evolution. Cultivation of cereals played a major part in the shift from hunting and gathering to plant and animal husbandry.

Several regions were first movers in developing independent domestication. Bread wheat, barley, oats, and rye in the Middle East; rice and millet in Southeast Asia; and corn, beans, and squash in Central America (Mesoamerica) all supported the rise of civilizations. Scientists and archaeologists have pinpointed these areas, along with the Andes and the Amazon basin of South America (potato, manioc) and the eastern United States (sunflower, goosefoot) as the main regions in which food production arose independently.

In total, only a few areas of the world developed food production independently. Often people in neighboring areas adopted food production techniques, yet some continued on as hunter-gatherers. The regions that were early food producers gained a foothold in the evolutionary cycle that ultimately led toward firepower, political systems, and metalwork. Like all evolutions, though, the move from hunter to producer took thousands of years.

The earliest known emmer wheat dates back to 8500 B.C. and came from a region in the near Middle East, called the "Fertile Crescent." After its domestication there, it spread further west, to Greece in 6500 B.C. and Germany in 5000 B.C. Perhaps the most widely used wheat, bread wheat (dated to 6000 B.C.), is strictly a domesticated species. It emerged accidentally in the near Middle East when different species of wheat were grown together. Domesticated barley appeared in the Fertile Crescent around 7000 B.C.

The Fertile Crescent is a hotbed of research activity linked to the rise of civilization. Although the development of cities, empires, and writing happened there, food production outdated them all. Thus, the area is studied to find out how domestication gave the region such an enormous head start. Because people in the Fertile Crescent were the first to develop concentrated food production techniques and animal husbandry, they experienced dense concentrations of human population, which in turn, enabled them to advance rapidly into technology, education, political systems, and even disease. Disease played a role in warding off potential enemies and thinning populations that had yet to build up immunity.

The domestication of rice dates to about 4000 B.C. in mainland Southeast Asia and China. Cultivation of this species usually involves flooded conditions in paddies, although it is also grown in upland regions. Today, nearly half the world's rice farming takes place in China and India and less than 1% in the United States. A comparatively small number of rice farmers basically supply the principle dietary staple for more than half the world.

Corn (also called "maize") was first grown in the highlands of Mesoamerica (Mexico) about 6000 to 5000 B.C. The development of corn is noticeably different from the cereals grown in the Fertile Crescent. Unlike the crops grown there, corn is much larger and adapts to warm seasons.

In terms of world production, four of the best known crops are members of the grass family: sugarcane, wheat, rice, and corn. Domestication of sugarcane occurred in Southeast Asia. Early cultivators discovered that the stem was a rich source of sugar and high in caloric value.

Bamboo is an overlooked but remarkably useful crop. Some analysts suggest that the tree grasses (or bamboos) have more uses than any other plant on Earth. Young shoots of several species are important vegetables in the daily diet in Japan, China, and Taiwan. Other cultures view bamboo as a gourmet item. In China, Southeast Asia, and Brazil, bamboos have been used in papermaking. In India the majority of the pulp for paper production comes from bamboos. The amazing strength and lightness of bamboo stems make them an excellent building material in the construction of houses and temples. Bamboo has also been used for centuries to make woven mats and a number of containers, including bowls and trays.

Impact

Food production is a prerequisite for the development of guns, germs, and steel, which function as the building blocks of world history. In hindsight, the cycle of evolution seems remarkably simple: the availability of more calories per person from farming leads to denser populations. The resulting crops feed geometrically more people than hunter-gatherer communities, which leads to greater numbers of people in the farming-based communities.

Once a food stockpile is acquired, then a political elite develops to control the surplus, which includes taxation. Some of the taxes are used to build and maintain armies, other money is used for building public works and cities. Since there are more food producers, they have a decided edge in military battles. The more complex a political system becomes, the better a society can mount a sustained war of conquest. In the meantime, an entire set of cultural, educational, and artistic endeavors arise because people have more free time and are healthier due to farming.

As food plants, cereals have many advantages, including a high yield per acre of growth. They are also wonderful sources of carbohydrates, fats, proteins, minerals, and vitamins. From this standpoint, the cereals can be considered the staff of life. People also developed innovative uses for plants and cereals. Alcoholic beverages were distilled from other crop grasses: barley provides beer malt, rice is used in the production of sake, and corn for bourbon. Wheat, rye, corn, and barley contribute to the making of whiskeys and vodka.

Most of the crops that were domesticated were also used for natural fibers to make clothing, blankets, nets, and rope. Fiber crops included cotton, flax, and hemp. Grasses were also used for livestock feed, erosion control, and turf.

Crop cultivation aided in the development of civilized society by forcing farmers to live in fixed areas. The crops had to be sown and reaped at specific times of the year, thus the population was forced to stay in the area. Inevitably, down time was filled with education, which leads to further progress.

In today's society, although the majority of farm production is derived from agri-business conglomerates, the economic importance of grasses remains firm. Nearly 70% of the world's agricultural land is used for crop grasses and more than 50% of the world's calories come from grasses, particularly the cereals.

The gradual development of farming communities over thousands of years led to the first hint of civilization. The domestication of plants and animals provided the instrument for mankind to cultivate the social, cultural, and political ideologies associated with civil society. It is no wonder that Charles Darwin (1809-1882) begins *On the Origin of Species*, his seminal work describing evolution, with an account of how the domestication of plants and animals took place through artificial selection by humans. For his subject, Darwin chose how farmers develop varieties of gooseberries and weed out the best variety, continually improving the quality by sowing seeds from the best plants.

BOB BATCHELOR

Further Reading

Baker, Herbert G. *Plants and Civilization*. Belmont, CA: Wadsworth, 1970.

Diamond, Jared. *Guns, Germs, and Steel: The Fates of Human Societies*. New York: W. W. Norton, 1999.

Dodge, Bertha S. *It Started in Eden: How the Plant-Hunters and the Plants They Found Changed the Course of History*. New York: McGraw-Hill, 1979.

Zohary, Daniel, and Maria Hopf. *Domestication of Plants in the Old World*. Oxford: Oxford University Press, 1993.

The Pyramids of Ancient Egypt

Overview

One of the most enduring accomplishments of early civilizations was the creation of monumental architecture in the form of pyramids, a shape that has fascinated humans ever since. The Greeks and Romans considered the Great Pyramid at Giza in Egypt as the most impressive of the seven wonders of the world. Even today, a pyramid is printed on the back of American dollar bills, and in the 1980s, a glass and steel pyramid was built in the courtyard of the Louvre in Paris as a new entrance to the museum. Although pyramids were constructed in the ancient world in areas as far apart as Peru, Central America, Mesopotamia, and Indonesia, those built in the Old Kingdom of Egypt have drawn the most attention in the Western world.

Background

Although the pyramids of ancient civilizations were massive structures, their most impressive aspect is not their size, but rather the fact that they were built at all. Judged by modern standards, they performed no pragmatic, functional purpose. They were not storehouses for food, provided no protection from invaders, nor did they offer shelter from the elements. Instead, they were built for religious purposes by peoples willing to expend extraordinary amounts of labor and economic resources in the construction of these structures, which served as burial sites or as monumental platforms for temples.

The earliest of these monuments were built in Peru about 3500 B.C., in the form of truncated mounds of earth and stone rubble, topped with a temple. Much later, pyramids were constructed in Central America (termed Mesoamerica by archaeologists), particularly in present-day Mexico. The Olmec peoples built tamped-earth pyramid mounds beginning around 1000 B.C. Near what is now Mexico City, an unknown people constructed the massive Pyramid of the

Sun in about A.D. 100, and both the Maya and Aztecs built huge pyramid structures. Elsewhere, pyramids were being built in Egypt by 2700 B.C.; in nearby Mesopotamia, pyramids called ziggurats were constructed by the Sumerians from 2100 B.C. on. Considered by some archaeologists to be pyramids, dome-shaped towers (stupas) were built by Buddhists in ancient India and Indonesia. While not all ancient civilizations constructed pyramids, their form seemed to satisfy a religious need in various polytheistic cultures.

The most common form of these ancient monuments was the step pyramid, a structure built in five or six stages, or steps, each rectangular stage smaller in area than the one below it. Outside staircases rose to the top platform, which held a temple or shrine. It was built in a heavily populated area and was intended to provide a dramatic setting for religious ceremonies. The "true" pyramids of Egypt were different in both form and function. They had a square ground plan with the four walls of triangular shape meeting at an apex above the center of the base; they supported no architectural or decorative features. Built outside cities on the edge of the desert, they served as burial tombs not accessible to the general population. But whichever form a pyramid took, it served as an ideological icon, as a stairway or mountain reaching to the heavens.

The Egyptian pyramids have received the most attention from historians, archaeologists, and the general public. Aesthetically, their closed geometric form probably functions as a subliminal metaphor for perfection. In addition, the Mesoamerican pyramids were far more inaccessible to scholars and tourists than were those of Egypt, which were easily reached by travel up the Nile River. Also, Sumerian ziggurats and many Central American pyramids were built of mud bricks and other less stable materials that have crumbled into chaotic mounds. Many of the stone pyramids of Egypt, while damaged, are still close enough to their original shape to be impressive. Another factor explaining the popularity of the Egyptian pyramids is the writings of the so-called *pyramidiots* who, for over a century and a half, have published scores of books revealing the "secrets" of the pyramids. Some have claimed they are divine prophecies in stone predicting future events; others view them as proof that Earth was once visited by aliens. Theories such as these, coupled with myths of the "curse" of the pharaohs and the mummy's "revenge," have made the Egyptian pyramids irresistible.

Impact

The age of the pyramids in Egypt began about 2780 B.C. and ended around 1550 B.C., although it should be noted that scholars do not agree upon the dates (or even the chronology) of

BUILDING THE GREAT PYRAMID OF CHEOPS

When Herodotus saw the Great Pyramid of Cheops and the other monumental structures of ancient Egypt, they were nearly as ancient to him as he is to the modern observer. In fact it is likely that modern scholars know more about Herodotus than he did about the pyramids or the almost unbelievably ancient civilization that produced them. He speculated that the Egyptians had used giant cranes to build them, and elsewhere wrote that Cheops was a cruel king who compelled some 100,000 slaves to toil in the construction of his pyramid.

The myth of slave labor building the pyramids has continued throughout history, as reflected in the 1956 film *The Ten Commandments*. While certainly great entertainment, the movie compounds its flawed portrait of history by depicting the Israelites' captivity as coinciding with the building of the pyramids—which is a bit like portraying John F. Kennedy and Charlemagne as contemporaries. In fact the workers who built the pyramids were Egyptians, and they seem to have done so voluntarily, believing their labor an act of service to the gods. They even left behind graffiti suggesting the pride they took in their toil—the various work gangs touting themselves, for instance, as "Vigorous Gang" or "Enduring Gang."

Nor were the gangs who built the Great Pyramid nearly as large as Herodotus envisioned: just some 4,000 men, working during a period of little more than 20 years. The idea that they could build a perfectly proportioned structure—and do so without the wheel, draft animals, or iron tools—has long been perplexing, and has led to much speculation. In fact historians believe that, though an impressive feat by any standards, the building of the pyramids could be (and was) accomplished without recourse to giant cranes, slave-labor gangs, extraterrestrial intelligence, or any other outside help.

JUDSON KNIGHT

Egyptian history. Over 70 royal pyramids are known to have been built in this period; it is impossible to estimate how many more lie hidden under the desert sands or have crumbled, leaving no traces. Much later, kings in Nubia (pre-

The pyramid of Zoser, constructed during the third dynasty. (*Charles & Josette Lenars/Corbis. Reproduced with permission.*)

sent-day northern Sudan) built about 180 smaller, inferior pyramids during the period 720 B.C. to A.D. 350. The golden age of the Egyptian pyramids was in the Old Kingdom, particularly during the fourth dynasty (2575-2465 B.C.) when, in a single century, the greatest of these monuments were constructed. Subsequent Egyptian pyramid builders looked back to this period for inspiration and models.

Ancient Egypt, which stretched along the Nile River, was unified in about 3000 B.C. With unification, the kings (later called pharaohs) and their officials learned to organize large numbers of Egyptians to control the annual flooding of the Nile and to irrigate the fields as the waters slowly receded. Such organizational skills were essential for successful completion of the pyramids and the many buildings and walls in the pyramid complex. Thousands of conscripted peasants (not slaves) had to be organized into teams to transport and then raise into place the huge stones used in the pyramids. Thousands of others were needed to work in the quarries and at the pyramid site as skilled laborers. Farms had to be established to feed these workers. Only a strong central government could provide the necessary organizational expertise and finances for such an effort.

The Egyptians regarded their king as an incarnation of the god Osiris and the son of Re,

the sun god. After the king died, they believed that he joined the gods and hence could intercede on their behalf with the divine powers. But this did not automatically occur. Egyptians viewed death as a continuation of life, and were convinced that the survival of the soul (the *ka*) in the afterlife depended on its rejoining the body's manifestation (the *ba*) after death. This would not happen if the body was decaying. Hence they mummified and embalmed the body of the deceased, removing all the viscera to prevent decay. They also believed that all the material needs for the *ba*, such as food, must be supplied for eternity or the *ka* would perish. Thus it was imperative that they build a tomb for the king that would be ready at his death, would protect his remains, would provide the necessities for his eternal survival, and would remind him of the loyalty of his subjects.

Initially, the kings were buried in mastabas, large flat rectangles of mud brick, under which lay a burial chamber and storerooms for the necessities of the *ka*. In this early period, the belief developed that any object or drawing placed in the tomb would magically provide for the needs of the deceased in the afterlife. So while boats were buried in pits outside the mastabas to provide the king with transportation for his heavenly journeys, model boats and drawings of boats were placed in the tomb as further insurance. However, mastabas were not secure from robbers seeking the treasures buried with the king. A ruler named Zoser ordered his architect, Imhotep, to remedy this. Imhotep made two crucial innovations in the tomb he built at Saqqara (c. 2780 B.C.). He used six mastabas to cover the burial chamber, placing one upon another, each smaller than the one beneath it. The result was a step pyramid, the first in Egypt. Equally important, Imhotep built the structure entirely of stone, the first burial monument so constructed.

Around 2570 B.C. King Sneferu tried to build a smooth-sided pyramid at Dahshur. However, for some reason the builders changed the angle of the sides, resulting in the so-called Bent Pyramid. It was Sneferu's successor, Khufu (Cheops, 2551-2528 B.C.), who built the first true pyramid. Known as the Great Pyramid of Giza, it contains about 2.3 million blocks of stone, averaging about two and a half tons each, with some weighing as much as 15 tons. Each side measured 756 feet (230 m) and was aligned almost exactly with true north and south or east and west. Standing 481 feet (147 m) high and covered with a smooth casing of white limestone, it marked the pinnacle of pyramid building. Even today, with its casing gone and its crude inner courses of stone revealed, it still awes onlookers.

The Great Pyramid of Giza, and all subsequent pyramids, contained a hidden burial chamber to protect the king's body. Carved reliefs and paintings in the chamber guaranteed the necessities for a comfortable afterlife. Also buried with the king were valuable objects such as jewelry and, in later pyramids, copies of the so-called Pyramid Text containing rituals and magic spells to insure an untroubled afterlife. All pyramids were built on the edge of the desert on the west side on the Nile, where the setting sun symbolized death. The pyramids themselves were part of a large complex that included a chapel where priests brought daily offerings of food. The complex also contained a mortuary temple, a small ritual pyramid (of unclear purpose), and a number of boat pits. A covered and walled causeway led down to a valley temple that was located on a canal connecting the complex to the Nile. After the king's mummy was placed in the burial chamber, the pyramid entrance was sealed and hidden behind one of the casing stones. Kings endowed their monuments with large estates to support the priests, workers, and guards who maintained the complex.

Much about the pyramids puzzle scholars, such as how they were aligned so perfectly with certain stars. Their biggest mystery is how they were built; that is, how such huge blocks of stone were raised into position. Most archaeologists and historians believe that some system involving ramps was employed, but there is no agreement as to how they were arranged. Nor is there agreement on the purpose of the smaller, subsidiary pyramids in the pyramid complex; although it is clear some were tombs for members of the king's family, others seem to have no obvious purpose.

Built to last through eternity, the pyramids suffered from two problems. As centuries passed, subsequent builders removed their limestone casings for other uses, such as constructing new pyramids. Once the casing was removed, the inner core deteriorated. The second problem greatly concerned later kings. The pyramids were very conspicuous and in times of anarchy such as in the First and Second Intermediate periods, the burial chambers were all broken into and looted by robbers. Since the pyramids failed to protect the remains of the king against such desecration, newer burial methods had to be developed. Ahmose I (c. 1550 B.C.) seems to have been the last major pyramid builder in Egypt. Instead of building massive monuments, the royal tombs were now hidden in the cliffs of inaccessible valleys across the Nile from Thebes. Thutmose I (c. 1500 B.C.) was probably the first of the many kings to be buried in the isolated valley known today as the Valley of the Kings.

ROBERT HENDRICK

Further Reading

Books

Andreu, Guillemette. *Egypt in the Age of the Pyramids.* Trans. by David Lorton. Ithaca: Cornell University Press, 1997.

Bierbrier, Morris. *The Tomb-Builders of the Pharaohs.* New York: Charles Scribner's Sons, 1984.

Clancy, Flora Simmons. *Pyramids.* Montreal: St. Remy Press, 1994.

Edwards, I. E. S. *The Pyramids of Egypt.* Rev. ed. Baltimore: Penguin Books, 1961.

Fakhry, Ahmed. *The Pyramids.* 2nd ed. Chicago: University of Chicago Press, 1969.

Lehner, Mark. *The Complete Pyramids.* London: Thames and Hudson, 1997.

Periodical Articles

Spence, Kate. "Ancient Egyptian Chronology and the Astronomical Orientation of Pyramids." *Nature* 408 (November 16, 2000): 320-24.

Wilford, John Noble. "Early Pharaohs' Ghostly Fleet." *New York Times* (October 31, 2000): F1, F4.

The Rise of Cities

Overview

In early farming societies, improved farming methods and a reliable food supply led to permanent settlements. Surplus food not only caused an increase in population. Freed from the all-consuming quest for food, laborers could begin to specialize in activities such as metalworking and weaving, and some jobs became more important than others. A social hierarchy took shape. Writing enabled administrative record keeping and, later, literature.

Background

Technological advance, including writing, beyond the folk society was a necessary condition for cities to take root. But it was not sufficient. Two additional considerations came into play. One was the ability to collect, store, and distribute the agricultural surplus, with all the social organization that implied. The second was a favorable environment in the form of fertile soil for growing crops and a water supply capable of supporting the needs of both agriculture and urban consumption. These are the conditions that describe the river valleys in which the world's earliest cities arose.

All early cities shared common organizational forms. They were theocracies (the government was divinely guided), and the elite and their entourages lived in the city center. This arrangement both facilitated interaction among the elite and offered protection from outside attack. The shops and dwellings of artisans such as carpenters and jewelers were located at a distance from the city center, and the poorest urbanites and the farmers lived at the periphery.

The very first cities arose in the Fertile Crescent, a crossroads that includes Mesopotamia, around 3500 B.C. The quality of the soil and the exposure over millennia to people of different cultures were an added stimulus to the evolution of cities in the area. These cities had similar technological bases and power structures, and by 2000 B.C. probably housed up to 10,000 people. By that time, urban communities had also been flourishing in the Nile delta for over a thousand years. Whether these communities arose as the result of Mesopotamian influence or whether they began independently is a matter of debate.

By the third and second millenniums B.C., cities were already widespread. By 2500 B.C., the Indus Valley civilization had produced the urban centers of Mohenjo-Daro and Harappa in what is now Pakistan. In another thousand years, urban settlements would spring up along the Yellow River in China.

Mohenjo-Daro and Harappa are a special case. These centers, which flourished from c. 2500 B.C. to c. 1700 B.C., were a miracle of city planning. Well-fortified citadels, with integrated palaces, granaries, and baths, protected towns laid out in a rigorously rectangular pattern. Drainage systems ran from houses, which had bathrooms, to brick-lined sewers. The economy was based on agriculture and trade. Unlike the civilizations of Egypt and Mesopotamia, the Indus people left very little written information behind, only brief inscriptions on seals that have yet to be decoded. The strict uniformity of many features of so-called Harappa culture, from the planning of the streets to weights and measures, suggests a high level of organization and achievement. Yet the culture never adopted the technical advances of Mesopotamia, with which they traded. The civilization died out mysteriously.

In the fourth century B.C. the city of Pataliputra, in the east of India, controlled the entire basin of the Ganges River and was run by an administrative board of 30 members. It was there that Emperor Asoka had his palace, a brick and wood construction that reveals considerable technical expertise.

In Mesoamerica, the Maya, Zapotecs, Mixtecs, and Aztecs developed their own substantial cities, though how and why these cities arose remains unknown. Teotihuacán, in the region of modern Mexico City, boasted 100,000 inhabitants during the first millennium A.D. Here, as in the East, writing and literacy seem to have been essential in the formation of urban centers and the development of science. But urbanization in Mesomerica also contradicts assumptions about the origin of cities: cities managed to develop there without technologies such as animal husbandry and the wheel, and minus abundant water and fertile soil. In the Andean region, on the other hand, where writing was unknown, even sophisticated engineering and a complex social structure were not enough to catalyze the emergence of cities.

Between the sixth and third centuries B.C., Persia, India, and China all experienced waves of urban expansion that extended the previous reach of their empires. Toward the end of the second millennium B.C., the shipbuilding Phoenicians had begun a westward spread that gave them dominion over the Mediterranean. The Greeks followed a similar course several centuries later, and Greek city-states (cities plus their surrounding countryside) became established along the Mediterranean shore from Asia Minor to Spain and France.

In building cities to maintain military supremacy in the numerous regions it conquered, Rome did more than any other empire to diffuse city life into previously nonurban regions. At its largest, the city of Rome itself may have had 300,000 inhabitants. The first Italian city-states were Greek colonies. Later, native city-states formed, including Rome, which was founded by a foreign power in 753 B.C. In 500 B.C. the Romans overthrew their foreign rulers and declared a republic that governed for the next 400 years. Rome became an empire, and the exposure to new perspectives that goes with empire building proved a fortuitous stimulus to literature and the arts.

Overattention to conquest, however, caused trouble at home. Periods of civil strife alternated with periods of peace, during which a superb system of roads and a postal system were developed. A fire in A.D. 64 had the happy result of shifting the population to the right bank of the Tiber, where broad streets and impressive buildings were built. In the course of the next few centuries the empire fell into disarray, and in 395 Rome lost its political importance. Little was left of the original city, yet what did remain included baths, temples, theaters, and 19 aqueducts. The city had enjoyed the protection of police, sanitation, and a seven-brigade fire-fighting force. Central heating and running water existed in some quarters.

Paris was founded in the third century B.C. by a Celtic tribe called the Parisii, who set up huts on what is now known as the Ile de la Cité. When Julius Caesar (100-44 B.C.) conquered Paris in 52 B.C., it was a fishing village called Lutetia Parisiorum. During the reign of Augustus (63 B.C.-A.D. 14), the town spread to the left bank of the Seine River. The Roman occupation was an era of peaceful expansion, during which a Roman network of roads built originally for the military enabled new trade routes and the development of new cities. During the prosperous second century A.D., the city went into a fever of building. The catacombs under Montparnasse and the baths still remain from Roman times.

Impact

One practical consequence of the rise of cities was to separate people from the resources they needed to survive. Economic, administrative, and social systems arose to deal with those needs. For example, because city dwellers were no longer living close to sources of sustenance, food had to be grown outside the city and transported in. Water management became an issue, not just making drinking water available to residents but also getting rid of wastewater. The Indus Valley and Roman civilizations provide examples of drainage and irrigation systems, sewers, baths, and aqueducts. Public works that required large numbers of people became imaginable owing to available manpower. City planning evolved to enable cities to grow in an organized way.

The specialization of workers that occurred once surplus food could be produced contributed to an urban revolution of major proportions. Cities were located along major transportation routes, which enabled a continuous influx and outflow of ideas and inventions. Moreover, the mere concentration of so many experts encouraged innovation. Technical advances encouraged cities to expand further, which meant that inhabitants had access to human and material resources over a far wider region. The increased access fostered the growth of even more cities, and so on.

The entire ecology of groups changed. Scourges like smallpox and childhood diseases only persist in large populations, so city dwellers lived under the perpetual threat of infection from microbes and from poor sanitation and hygiene. Of course, infectious disease also became a kind of biological weapon. A conquering people used to living with disease could easily introduce it into naïve populations. People so stricken quickly succumbed, facilitating territorial expansion.

Nor was disease the only threat to inhabitants of cities. Breakdowns in transportation and crop failures could both spell famine in short order. Thus, cities began to depend on rural cultivators to produce surpluses of both food and workers to compensate for shortfalls. So dependent, in fact, were cities on this supply line, that civilizations could not have progressed without it. But the flow of people did not always go toward the city. Sometimes, particularly in prosperous times, it happened that surplus workers were forced back out to the countryside. Revolt, civil wars, and wars of conquest were a kind of self-correcting means of reducing surplus population from the countryside.

In all the early civilizations, cities were centers of innovation and development. The efflorescence of culture that went hand in hand with the rise of cities became a tool for remaking the world. With the fall of the Roman Empire, cities in western Europe went into eclipse, and their function as centers of learning and art passed to the monasteries. In the East, failed cities were the exception. With time, urban Europe came to

life again, a result of the cycle of technology transfer and human migration that helps to keep cities going.

GISELLE WEISS

Further Reading

Ancient Cities. Scientific American Special Issue. New York: Scientific American, 1994.

McNeill, William H. *Plagues and Peoples.* New York: Anchor Books, 1976.

Stambaugh, J. E. *The Ancient Roman City.* Baltimore: Johns Hopkins University Press, 1988.

Tomlinson, R. *From Mycenae to Constantinople: The Evolution of the Ancient City.* London: Routledge, 1992.

Triumphs of Ancient Architecture and Art: The Seven Wonders and the Parthenon

Overview

The expression "Seven Wonders of the World" has become commonplace, so much so that people often speak of things or even people as "the eighth wonder of the world." There are lists of seven modern wonders, seven natural wonders, seven medieval wonders, and so on; but all these have their roots in the concept of the ancient world's seven wonders. Whereas Roman engineering represents an ancient triumph in the practical application of technology, these structures stand for something else: aesthetic beauty, physical magnificence (either in size, detail, or both), and the political unity of the peoples who built them.

The Seven Wonders of the Ancient World include two tombs, two statues, a temple, a lighthouse, and a set of hanging gardens. Included are one each from Egyptian and Babylonian civilizations, as well as five structures from the Hellenic and Hellenistic civilization scattered over three continents. Indeed, one notable feature of the list is the preponderance of Greek structures included, a fact that results partly from the fact that the idea of such a grouping comes from the Greek historian Herodotus (484?-420? B.C.).

But there is more to it than that. Influenced by the Egyptians, builders of the oldest Wonder, the Greeks created new standards of beauty and durability that created lasting definitions of architectural and artistic achievement. It is for this rea-

son that the buildings of the Athenian Acropolis, and particularly the Parthenon, deserve to be discussed alongside the structures included on the canonical, or accepted, list of Seven Wonders.

Background

In fact, Herodotus could only have known three of the Seven Wonders, and perhaps only two, since some historians question the existence of Babylon's Hanging Gardens as anything other than a figment of Greek writers' imaginations. His list included the Pyramids and Hanging Gardens, as well as the Walls of Babylon, and—though the latter did not make the final roster—it appears on the first definitive version of the list, attributed to the historian Antipater of Sidon in the second century B.C. Only in the early modern era did the final list take form, thanks to Dutch painter Maarten van Heemskerck (1498-1574), who created depictions of each.

In Heemskerck's time, the number of existing Ancient Wonders was the same as it is today: one. That one is the Great Pyramid of Giza, the first of the Wonders to be built and already some 2,000 years old before the erection of the second Wonder. The last of the Wonders, the Lighthouse at Alexandria, was also the last destroyed, when just a few years before Heemskerck's birth it ceased to function.

There is of course a certain symmetry in the fact the first and last Wonders were both in

The Parthenon, built in the mid-fifth century B.C. on the Acropolis. *(Susan D. Rock. Reproduced with permission.)*

Egypt, but the Egypt of the pyramid-builders was quite different from the one that saw the creation of the Alexandrine lighthouse 23 centuries later. By that time Greek civilization had spread from its homeland in southeastern Europe to permeate the entire Mediterranean world: hence the distinction in terminology between Hellenic (Greek) and Hellenistic (Greek-influenced). Alexandria became home not only to the lighthouse but to another wonder of sorts, the city's library, whose librarian Callimachus of Cyrene (305-240 B.C.) composed "A Collection of Wonders Around the World." The particular buildings on his list will never be known, because his manuscript—like the library—has been lost.

This sad record of loss and destruction only serves to highlight the world's great fortune that at least one of the monuments survives, and the fact that the Great Pyramid is also the most massive and, in some ways, the most perfectly constructed only enhances its value. Many things about this structure are compelling, not least of which is its antiquity: built between about 2550 and 2530 B.C., the pyramid was as ancient to Antipater of Sidon as he is to the modern observer.

All the great pyramids of Egypt were built during the Old Kingdom (2650- 2150 B.C.), but there is only one Great Pyramid with a capital "G," erected for the pharaoh Cheops (reigned 2551-2528 B.C.). Originally 481 feet (147 m) tall, it is now 33 feet (10 m) shorter due to the removal of its finished exterior by Egypt's Arab rulers in medieval times, but it remains a structure of staggering proportions. Its height makes it equivalent to that of a 50-story building, and it would be almost 4,000 years before there was a taller structure, the thirteenth-century cathedral of Cologne, Germany. At 755 feet (230 m) along each side, St. Peter's of Rome, Westminster Abbey, and St. Paul's Cathedral in London, as well as the cathedrals of Florence and Milan, could all fit comfortably on its base. Or, to put it in modern American terms, the pyramid occupies an area equivalent to 10 football fields.

Two million blocks of stone, each weighing more than 2.5 tons (2,270 kg), comprise the Great Pyramid. It has been estimated that there is enough stone in the structure to build a wall 10 feet (3 m) high and 1 foot (0.3 m) thick all the way around the nation of France. Yet as is well known, the interior stones fit together so well, and without mortar, that it is impossible to slip even a card between them. The pharaoh's interior burial chamber bears an incredible burden of stone, and does so through an ingeniously designed triangular roof that deflects the weight of the structure atop it. Other corridors and escape shafts run through the pyramid as well. Most impressive of all is the fact that the maximum difference between the lengths of the sides on this massive structure is less than 0.1%, making it much more square than the average home built in America today.

It is amazing, then, to consider that the people who built the pyramid apparently had no knowledge of iron tools or the wheel. These facts, combined with the structure's precise orientation—each side faces one of the cardinal points of the compass—has led in modern times to pseudo- or plainly unscientific suggestions that extraterrestrial beings were responsible for its construction. Much more interesting, however, are the more realistic theories concerning how the pyramid was built—theories that, rather than calling upon magic from the sky, highlight the resourcefulness of the civilization that built this great structure.

In fact modern archaeologists generally believe that with the use of the Egyptians' large work gangs (whose members, contrary to popular belief, were not slaves), it would have been possible to raise the blocks of the pyramid by use of sloping earthen ramps. These would have been built up alongside the pyramid itself, then removed when the structure was completed. It is possible, also, that the Egyptians used long levers, one of the earliest forms of machine. As for the moving of the blocks from stone quarries up the Nile in Aswan, this too could have been achieved by use of the barges available to the Egyptians at the time.

Impact

Compared to the Great Pyramid, the Hanging Gardens of Babylon seem relatively recent, yet so little remains of them that they have sometimes been dismissed as a fiction. Legend holds that the great Babylonian king Nebuchadnezzar II (630?-562 B.C.) wanted to build a structure to make his wife happy: she came from the mountainous land of Media, and was homesick in the flatlands of Mesopotamia. Therefore he ordered the building of a huge set of terraced structures containing all manner of lush plants watered by a complex irrigation network.

One Greek writer described the Hanging Gardens as "a work of art of royal luxury," but it is likely he was drawing on imagination rather than firsthand experience. Babylonian records from Nebuchadnezzar's time contain no mention of the Hanging Gardens, and some historians maintain that the very idea of the gardens came from tall tales related by soldiers of Alexander the Great (356-323 B.C.) after they saw the great

city many years later. In the late twentieth century, however, archaeologists began to find remains in Babylon, by then a ruin in southern Iraq, including massive walls that may have been stepped to create terraces.

Historians know far more about the Temple of Artemis at Ephesus in Asia Minor, built in about 550 B.C. by Croesus (reigned 560?-546 B.C.), king of Lydia. Though Croesus preceded the golden age of Greek civilization, the temple is linked with Hellenic culture because over the centuries it acquired increasing enhancements from great Greek sculptors such as Phidias (490?-430? B.C.) and Polyclitus (fifth century B.C.). The deity worshipped was closely related to the Greeks' own Diana, but the temple also bore elements of other cultures, reflected in decorations brought from Persia and India. Adding to its international character is the fact that the temple, or at least the cult of Artemis, is mentioned in the Bible by the Apostle Paul (A.D. 5?-67?), who preached to the Ephesians against idol-worship.

The temple, which earned its place on the list of Seven Wonders purely on the basis of its aesthetic qualities rather than any particular feat of engineering it represented, went through several incarnations. On July 21, 356 B.C., a certain Herostratus burned the temple to the ground. The date was significant because it also marked the birth of Alexander, who helped rebuild the temple. In A.D. 262 an invasion by the Goths destroyed the temple, but the Ephesians rebuilt it again—only to see it destroyed for good in A.D. 401. by the zealous Christian leader St. John Chrysostom (347-407).

Next among the Seven Wonders was the Statue of Zeus at Olympia by Phidias, but that great sculpture must be viewed in light of the larger spectacle of the Greek Golden Age, which lasted from the victory over the Persians at Salamis in 479 B.C. to the defeat of Athens by the Spartans in the Peloponnesian War 73 years later. It was an era when Athens, led by Pericles (495?-429 B.C.), ruled the Greek world—an age that in its great intellectual and cultural advancements rightly deserves the appellation of "golden."

Symbolic of Athenian self-confidence was the great program of rebuilding undertaken by Pericles, who intended not only to restore the city, destroyed by the Persians, but also to erect lasting monuments to the democratic state he and others had helped create and sustain. Like other Greek cities, Athens had an acropolis, an elevated fortress, and it was there Pericles arranged the construction of several great structures: the temple of Athena Nike; the Erechtheion, famous for its Porch of the Caryatids; the monumental entryway called the Propylaia; and a temple to Athena, the maiden or *parthenos* for whom Athens was named.

In its damaged beauty—it suffered an explosion during a war in A.D. 1687 between Venice and the Turks, when it was used as a munitions warehouse—the Parthenon seems to whisper of humankind's deepest aspirations, symbolized both by its physical power and the delicate simplicity of its design. By the time it was built, two distinctive styles of Greek architecture had developed, the Doric and Ionic. (A third style, the Corinthian, had also emerged, but it was actually a version of the Ionic, distinguished primarily by a difference in the design of the capital, or the decorative top of the column.) Both the simple Doric and less formal Ionic showed the influence of the Egyptians, who drew upon natural forms in the design of columns, and used these to strike a powerful emotional response in the viewer.

The Parthenon itself was of Doric design, but with Ionic features such as a continuous frieze, or relief sculpture, which occupied the pediment, or triangular roof gable. The roof was supported by a single row of columns, eight at each end and 17 along a side. The architects, Ictinus and Callicrates, clearly possessed an advanced understanding of how humans see objects in space. In a number of places where lines appear straight, these are actually curved, because a truly straight line would have *seemed* curved. Likewise the columns bulge slightly at the center in order to look as though they taper gently from bottom to top.

Historians have a less clear idea about the interior of the Parthenon as it was in classical Greece. At the center was a huge statue of Athena, designed by Phidias, that was made of gold and ivory. It probably showed Athena with a shield and weapons of war; in any case, few Athenians ever got to see it, because the Parthenon's inner chambers were the reserve of priests. The statue of Athena is also tied with the decline of Pericles's influence: his enemies charged that Phidias had stolen part of the gold for the statue, and carved pictures of himself and Pericles on Athena's shield—claims which, if true, would have constituted a serious offense to the goddess. They managed to make the charges stick, and Phidias was forced to leave Athens.

Yet Phidias's most renowned work lay in the future, when the city of Olympia, home of the Olympic Games, commissioned him to create a statue of Zeus for their temple. Covered in gold and ivory, the seated god was 40 feet (13 m) tall, and the proportions of the building were such that if he had stood up, his head would have gone through the roof. Some regarded this as a design flaw, but in fact it was a hallmark of the statue's brilliance, emphasizing the greatness of the deity's size.

As with the temple of Artemis, this one contained gifts from other parts of the world, befitting its status as a gathering place for the city-states of Greece. The glory of the temple, however, faded with that of Greece itself, and after Rome's emperor Theodosius I (347-395) banned the Olympics as a pagan spectacle, the site fell into disrepair. Earthquakes, landslides, and floods destroyed much of the temple, and a fire in A.D. 462 left it a ruin.

Fifth among the Seven Wonders was the Mausoleum at Halicarnassus (Herodotus's hometown) in Asia Minor. It was built for Mausollos (reigned 377?-353? B.C.), who ruled a vassal kingdom of the Persian Empire. As with the nearby temple of Artemis, a number of Greek sculptors contributed their efforts, and the tomb was decorated with all manner of statues either life-size or larger, depicting humans, lions, horses, and other animals. Indeed, one of its distinctive features was the fact that its artwork illustrated natural forms, rather than gods and heroes as in most Greek decorative arts. Thanks to this structure—destroyed when the Knights of St. John of Malta used its stone to build a fortress in the late fifteenth century—aboveground tombs today are still called mausoleums.

As the mausoleum was the second tomb, so the Colossus of Rhodes was the second statue on the list, and like the Athenian Acropolis it represented the self-confidence of a people who had just won a victory against outsiders. In 305 B.C. the Antigonids of Macedonia attacked the city-states of Rhodes, which had united a century earlier. After a long siege, Rhodian resistance forced the Antigonids to withdraw. The invaders had left a large quantity of weapons, which the Rhodians sold to raise money for the building of a stature in honor of the sun god Helios or Apollo.

Completed in 282 B.C., the Colossus stood about 110 feet (33 m) high. Though in a sense it "guarded the harbor," it did not straddle the harbor's entrance, as numerous illustrations mistakenly suggest. For it to have done so would have required an even more massive statue than the incredibly impressive Colossus; and furthermore, when an earthquake broke the statue at its knee just 54 years after its construction (making it the shortest-lived of all the Wonders), the broken statue would have blocked the harbor.

As it was, Pliny the Elder (A.D. 23?-79) wrote that "few people can make their arms meet round the thumb" of the fallen giant. The ruins stayed where they were, and when the Arabs invaded Rhodes in A.D. 654, they dismantled the remains and sold them to a Jewish merchant from Syria, who reportedly required 900 camels to transport the pieces. In later centuries, the Colossus served as model for numerous giant statues depicting human figures, most notably the Statue of Liberty in the United States.

Finally, there was the Lighthouse at Alexandria, the only one of the Seven Wonders that served a practical purpose. It was built on the island of Pharos (and thus is sometimes called the Pharos Lighthouse), beginning in about 290 B.C., and its architect was named Sostratus. The coastline near Alexandria was a dangerous one, and thus the lighthouse, which stood as tall as a 40-story building, constituted a highly advantageous addition to the region. The core of the structure was a shaft with pulleys for raising fuel, which lit fires that burned by night. During the day, a gigantic mirror reflected the Sun's rays and provided a signal to ships as far away as 35 miles (50 km) out at sea. The mirror also reportedly served a defensive function, using sunlight to burn enemy vessels at a distance.

Like many of the other Wonders, earthquakes—which in this case occurred in 1303 and 1323—helped bring the lighthouse to an end. When the Mameluk sultan Qait Bay needed to fortify the city's defenses in 1480, he used the fallen stone and marble from the lighthouse to build a fortress on the spot where it had stood. Then, more than half a millennium later, in 1996, divers off the Egyptian coast discovered the remains of the famed lighthouse, and in 1998 a French company announced plans to rebuild the great structure on the spot where it had originally stood.

The latter illustrates the continuing allure of the Seven Wonders, though it should be pointed out that any number of other ancient structures— the Great Wall of China, for instance, or the Pyramid of the Sun in the Mesoamerican city of Teotihuacan—certainly deserve the appellation of "ancient wonder." The structures on the canonical list, however, along with the Parthenon and the other great buildings of classical Greece, exert a special pull on Western civilization. Together they symbolize the greatest of human aspirations, and the continuing allure that lies in the application of intellect to mastery of the physical world.

Further Reading

Bowra, C. M. *Classical Greece.* New York: Time-Life Books, 1965.

Lullies, Reinhard, and Max Hirmer. *Greek Sculpture.* New York: H. N. Abrams, 1957.

Robertson, Donald Struan. *A Handbook of Greek and Roman Architecture.* New York: Cambridge University Press, 1959.

Scarre, Christopher. *The Seventy Wonders of the Ancient World: The Great Monuments and How They Were Built.* New York: Thames & Hudson, 1999.

JUDSON KNIGHT

Building an Empire and a Legacy:
Roman Engineering

Overview

The engineers of ancient Rome designed and constructed many projects to serve the needs of an urban and an imperial nation. With their use of the semicircular arch, the barrel vault, and hydraulic cement, they transformed architecture and construction in the ancient world. The results were impressive in scale and practicality and influential in their shaping of a timeless architectural style.

Background

Roman civil engineers and architects created a series of structures intended to address the multifarious needs of Roman society. From the religious to the secular, from the recreational to the utilitarian, and from the military to the domestic, they engaged in a wide range of projects. Perfecting techniques inherited from earlier cultures such as Egypt and Greece, they added their own special contributions that came to mark a structure or design as Roman. Relying on a variety of materials such as clay, brick and mortar, limestone, marble, and tufa (a form of volcanic mud), they addressed the needs of an urban-centered society that expanded its reach and influence into the known Western world.

Among those many needs were the requirements of a hydraulic culture in which supplying and controlling water dominated the activity of the societies. The Romans responded with aqueducts, tunnels, siphons, dams, and sewers. They built massive aqueduct systems of brick and stone that brought water from the mountains into urban centers. Using the gravity feed principle allowed Roman engineers to transfer water in some aqueduct systems for as much as 40 to 50 miles (64 to 80 km).

Using arches to span valleys, these engineers produced elegant and effective water channels that delivered a volume of water equivalent to the basic needs of many twentieth-century European cities. For example, the Pont du Gard, in the south of France, spanned the river Gard with a three-tiered arch bridge reaching a height of 160 feet (49 m) and brought water to the city of Nimes. To supplement the expensive aqueduct system, Roman builders also used tunnels and siphons to move water as well. As a complement to the aqueduct system, dams composed of rubble, brick, and stone along with reservoirs provided a store of water for domestic use or to power water mills, especially those for grinding grain. Romans also used their hydraulic engineering skills to supply water to various public baths, to provide water for domestic needs, and to remove wastes through an extensive sewer system.

To meet the many challenges of their urban needs, the Romans relied greatly on a water-proof material, hydraulic cement, available to them because they had access to vast quantities of pozzolana sand or ash generated by volcanoes. This special substance gave them a material that could be used under water for bridge piers, was fire-resistant, and could withstand the effects of the weather. It also added to the durability of mortar used to secure bricks or stone in place in many structures. The full utilization of this material allowed Roman engineers to construct durable buildings and bridges and other projects on a massive scale.

The extended arch or barrel vault gave the Romans a new technique for enclosing space. In buildings such as theaters, amphitheaters, circuses, public baths, and basilicas, this new architectural element appeared repeatedly. The arch itself became a defining element in various stadia such as the Colosseum in Rome, functioning as entrances and exits and in tiered form to add to the structure's height. Used as a vault, it defined passageways, ceilings, and graceful interiors that created vastly larger spans than had been present in the ancient world. The Colosseum itself, with its many arched openings, gave the impression of vast open interior space when, in fact, a massive hidden structure provided the support for this interior, which would hold between 45,000 and 50,000 spectators. Eighty arches on the exterior wall allowed for easy entry and exit and provided ready access to the whole building. This successful design was also combined with the post and lintel style of many columns in symmetrical array so that the arch and vault created a durable classical style of architecture that permeated the Roman world.

The arch and barrel vault were combined in one of the most impressive basilicas in ancient Rome, the Basilica of Maxentius. With its 260-foot (79-m) length and 80-foot (24-m) vaults, this structure dominated the Roman Forum. Three main vaults with coffered ceilings, a common decorative feature of Roman interiors, dominated the building. The result was the largest hall built in antiquity.

Elaborating on the arch and vault, Roman architects also pioneered the use of the circular dome. Unlike the Greeks, the Romans built enclosed space and focused on interiors. The most spectacular result of dome construction was the Pantheon in Rome, a temple whose dome, complete with a 25-foot-wide (7.6-m) oculus, was 142 feet (43 m) in both diameter and height, the largest dome in the West until St. Peter's was built in Rome in the sixteenth century. Like many other Roman interior spaces, the dome had a coffered ceiling with the simple geometry of nested squares; the resulting rotunda interior created an overwhelming sense of uninterrupted space, of a dome the floated in space, in a simple design unequalled in the ancient world.

An outgrowth of Roman experience with the arch, the Pantheon's dome was constructed as a series of linked segmental arches or vaults, a technique widely used by Roman engineers. Step rings, a solid foundation, and the building itself support this very large heavy dome, which has survived for centuries and, as one of several extant original Roman buildings, remains a testament to Roman engineering ingenuity and as an example of the best of Rome's concrete vaulted architecture.

The Romans used their skill as civil engineers on roads as well as buildings. Treating a road surface like a wall buried in the earth, they created a series of primary and secondary roads that together covered almost 200,000 miles (321,900 km). Built to last for a century, these roads shared the characteristics of a straight path, gradual gradients, curved surfaces for water run-off, curbs, and gutters. Often 6 feet (1.8 m) thick, the primary roads consisted of a series of rock, stone, and gravel layers covered with paving stones. Because they were a means of moving men and materiel as well as an effective means of communication, they were as essential to the successful operation of the nation and empire as the methods of water control and distribution so characteristic of Roman achievements.

This extensive road system benefited from Roman bridge-building techniques. The semicircular arch was the basic motif of Roman bridges, with a range of graceful spans from a single stone arch to multiple arches covering larger areas. The use of hydraulic cement allowed the builders to erect durable stone bridges that have stood and been used for centuries. Dispersed throughout the Roman world, these bridges became hallmarks of the classical architectural style as elegant ways to span space.

As a major force in the classical world, Rome required buildings and structures to serve its military needs. Massive stone walls, forts, and watch towers emerged over a period of years to protect and expand the Roman Empire. In an era when stone walls surrounded towns and cities in order to protect those within from attack, the Romans built impressive stone barriers, often with several gates and towers, to control access to urban centers. These defensive perimeters evolved into an elaborate system of walls, watch towers, forts, castles, and turrets that, in some cases, successfully served as defensive perimeters for more than 1,000 years.

Another Roman architectural legacy is the triumphal arch. Varying in style from one to four archways, these monuments celebrated leaders of the empire, military figures or victories, towns or cities, and various religious figures. As decorative features in an urban setting, these arches often served as a focal point in a city center or in defining a major gateway.

Roman domestic architecture produced a spectrum of housing from luxury villas to apartment houses in cities. Usually constructed of brick or stone, houses often were built around an atrium and, if space was available, included a garden. Rooms were arranged so that inhabitants could move from place to place within to seek or to avoid the sun depending on the climate or the season. Various dwellings also incorporated a means of coping with inclement temperatures. During cool or cold days, a hypocaust or central heating system provided warmth within homes and other buildings such as public baths. Thick ceramic tile floors were supported by regularly placed pillars; the resultant underground chambers created by these pillars allowed the heat from a charcoal or wood fire to permeate the space and to radiate from the thick floor tiles into the spaces above. For Mediterranean and other moderate climates, this system performed well in providing heat for inhabitants of the Roman world.

The extensive building and construction projects of Rome led to urban planning. Their carefully planned cities consisted of a regular grid of streets intersecting at right angles. Main roads were flanked by sidewalks alongside domestic and commercial buildings, with open squares often covered with decorative mosaics. Commercial centers such as the forum, positioned near or at the city and placed near the intersection of two main streets, served as focal points, with an array of domestic, governmental, religious, and recreational buildings filling out the cityscape. These buildings usually shared some common architectural element or design so there was a uniformity to the various districts that reinforced the visual message that these were planned communities.

Impact

The effects of Roman architecture and building were both immediate and enduring. Without the urban-oriented technology that dominated much of Roman engineering, the culture known as ancient Rome would not have flourished. The many roads, bridges, stadia, public buildings, and water supply systems produced in the era aided in the operation and survival of the Roman world. Further, these engineering successes allowed the Roman Empire to expand and to dominate much of the known world in the years from 200 B.C. to A.D. 400

Roman engineers demonstrated the potential of simple technology matched to the astute management of workers, slave or free. Relying on a commitment from the society, these ancient engineers constructed projects that were built to last. For example, most major Roman roads were designed to be in service for a century,

A model of ancient Rome showing the Colosseum and the complex of Ludus Magnus. *(Araldo de Luca/Corbis. Reproduced with permission.)*

compared with the modern world's goal of 20-40 years. To this day, many amphitheaters, public baths, aqueducts, and bridges remain intact and in use throughout Europe and other areas that previously were part of the Roman Empire, from Britain to Asia Minor. The Roman installation at Bath, England, and the extensive ruins of Ephesus, Turkey, attest to the durability of Roman engineering. Because these skilled engineers were so successful in completing massive projects, the term "Roman project" has come to mean a viable large-scale engineering endeavor.

Because they were a pragmatic people, the Romans profited from diffusion and stimulus technology. Borrowing heavily from earlier civilizations, especially ancient Egypt and Greece, Roman engineers were able to perfect known techniques. In doing so, they replicated the style and refined the classical motif of colonnaded buildings, extended city planning, and introduced their own variation of the style with the use of the arch. Without a strong theoretical grounding to their work, these engineers sometimes produced "over-engineered" structures. These products of empiricism with their high margins of safety usually contained far more material than necessary for structural integrity and occasionally resulted in an overly heavy and cumbersome design. Although durable, these projects left a legacy that empiricism alone does not always produce the most elegant results.

The practical bent of Roman engineers manifested itself in another way as well. Aware that the products of their engineering ingenuity would require upkeep and continual attention, Roman designers provided means for maintenance technology in many of their structures. Projecting stones and cavities in walls as a permanent feature in things such as buildings and bridges made it easy to install scaffolding for the repair and maintenance of these items. Likewise, roads were monitored carefully for any problems that might cause structural weaknesses or deterioration so they could be repaired in a timely manner. This approach to maintenance technology carried over into the Medieval era, when cathedral builders incorporated features such as hidden stairways, exterior walkways, and passages to all sections of a building from the foundation to the spire to aid in the monitoring and maintenance of these stone churches.

Roman success with the technology of building and construction influenced the architectural style of several later eras. The basic basilica design, a rectangular building with carefully placed columns, was a prototype for many Renaissance churches and public buildings. The Romanesque style, incorporating semicircular arches and barrel vaults, also became a favorite for Mediterranean churches in the era. The success of the Romanesque design, along with the incorporation of the Roman dome, allowed the classical style to dominate much architecture in Renaissance Europe and in the emerging American republic. With their ordered columns, arches, vaults, and domes, many public buildings such as libraries, museums, city halls, state capitols, stadia, and monuments were copies of Roman designs. The graceful lines of a typical Roman semicircular arch bridge have made that style a favorite for many communities, including America's capital city of Washington, D.C., with its many bridges across the Potomac River. Indeed, the new American nation looked to classical Rome for many of its symbols and styles, from the eagle to the United States Supreme Court and the Capitol buildings. In addition, Thomas Jefferson, with his design of the library at the University of Virginia, took his inspirations from the Pantheon in Rome. The Roman classical style is so deeply ingrained into Western culture that for centuries many public buildings across the Western world were built with that architectural design.

Modern architects as well have embraced the arch and barrel vault as a motif for their buildings. H. Richardson, a renowned late-nineteenth-century architect, transformed American architecture with his neo-Romanesque style, which relied heavily on arches, stone exteriors, turrets, and vaulted spaces. Richardson's influence touched a wide range of projects, from warehouses to train stations, libraries, and churches throughout America. In the twentieth century, Louis Kahn incorporated the semicircle and vault in many of his designs, including the graceful Kimball Art Museum in Fort Worth, Texas, which is considered one of the most elegant buildings of its kind.

Beyond specific applications, the Roman record in technology left a legacy of the efficacy of empirical methods. Having both material and human resources, determination, ingenuity, and the ability to learn by using technology, the Romans achieved magnificent results. Relying on highly skilled craftsmen and artisans, the Romans, like many preindustrial societies, were able to produce massive, durable projects using relatively simple tools. Their ability to organize and manage large numbers of workers aided in their quest to provide a technology to serve both the urban and imperial Roman worlds.

Those talents were nurtured by a nation that used its architecture and engineering to expand the empire and to make a statement about the power of that empire. The impressive scale of many of Rome's monumental buildings were tangible reminders of the strength and ambitions of ancient Rome. In this way, engineering served the state in both functional and symbolic manners. The *Pax Romana*, that era in the which Rome dominated much of the Western world, was due in large measure to the ancient technol-

Roman aqueducts in Spain. *(Archive Photos. Reproduced with permission.)*

ogy Romans borrowed and perfected. The extensive road and bridge systems, arenas and stadia, public baths and other civic buildings, aqueducts, fortifications, and monuments served to unify the disparate elements of Rome's world. In the process, Romans spread and adopted various technological methods so that even the far reaches of the empire mirrored the life style and physical artifacts of the society, much as the influence of the American dominance in the last half of the twentieth century spread throughout the globe.

The legacy of Roman architecture and engineering is enduring. Beginning with the heritage of earlier Greek classical designs and methods, Roman engineers gradually modified, refined, and improved on these inherited styles. Especially in their use of the arch and of concrete, the Romans created their own unique architecture, which played a key role in serving the needs of an urban culture and an empire. The vastness of their projects, from aqueducts to arenas, and the ingenuity of their methods earn them an impressive reputation as very successful engineers. The durability of that technology and the degree to which it spread through the Roman Empire attest to its usefulness and functional design. Because they used empirical methods, Roman engineers demonstrated the value of that kind of technology; anyone who experiences an original Roman structure is deeply impressed with the craftsmanship, artistry, and architectural insight that made it possible. The hallmarks of Roman architecture and engineering impressed people of the classical era and of many later periods, including Renaissance Europe and the new nation of America. The achievements of Rome remind us that through determination, dedication, skill, simple technology, and the prudent management of large labor forces, this society created magnificent results. They also provided the basis for future urban planning, for urban-based technology, for the creation of large-scale projects, and for a distinctive and widely copied architectural style. Those achievements continue to generate awe and admiration and stand as one of the most significant engineering triumphs of the preindustrial world.

H. J. EISENMAN

Further Reading

Barton, Ian M., ed. *Roman Domestic Buildings*. Exeter: University of Exeter Press, 1996.

Sear, Frank. *Roman Architecture*. Ithaca, NY: Cornell University Press, 1982.

MacDonald, William. *The Architecture of the Roman Empire*. Vol. 1, rev. ed. New Haven, CT: Yale University Press, 1982; Vol. II, 1986.

Ward-Perkins, John B. *Roman Architecture*. New York: Harry N. Abrams, Inc., 1977.

White, K. D. *Greek and Roman Technology*. Ithaca, NY: Cornell University Press, 1984.

Water Management in the Ancient World

Overview

Water, one of the basic necessities for human life, was the lifeblood of early civilization. Indeed, the ability of ancient societies to harness the power of water facilitated the rise of agriculture and the first urban centers. So important was water to these early people that historians refer to these first societies as "River Valley Civilizations."

Background

The primary characteristic of these advanced societies was their reliance on sedentary agriculture in which people farmed the same land for generations. The combination of rich soil, mild climate, and a reliable source of water gave ancient people the ability to create crop surpluses. This movement away from nomadic wandering to a more localized existence is known as the Neolithic Revolution. It began about 8,000 years ago in the lush, fertile land surrounding the Tigris, Euphrates, Nile, Indus, and Huang He rivers. In these "Cradlelands of Civilization" dramatic events took place that would forever change the human condition.

The control and successful management of water had an important impact on early society. Sedentary agriculture created the world's first urban environment. Humankind had to develop ways of dealing with an entirely new social structure. A rigid new class system developed from the necessity of controlling large populations and the requirements of constructing and maintaining extensive public civil engineering projects.

The success of these early civilizations was based upon the movement of water into their agricultural fields. Great irrigation projects directed the substantial amounts of water necessary for agricultural success. Projects of this magnitude needed extensive planning and supervision. An intellectual elite arose to deal with the construction and operation of these systems. These individuals were the first highly skilled engineers in history. Eventually, they also developed the foundation for the discipline of mathematics and the science of astronomy. Mathematical concepts, especially geometry, were developed to deal with the challenges of the construction of the canals, dams, and dikes that controlled the flow of water. The continued success of these great agricultural civilizations also depended upon the accurate prediction of when to plant their crops so as to take advantage of the seasonal rains. These early engineers also had to develop maintenance schedules to repair any structural problems in the irrigation system. Water levels had to be low in order to successfully carry out these projects, so it was of the utmost importance that accurate weather predictions were made.

The same held true for planting. If the seeds were planted before the final onset of heavy rain, it would be washed away. The population would then be faced with potential famine and the political and social consequences that would follow from such a catastrophe. This required the development of highly accurate calendars. The recorded movements of the Moon and Sun formed the foundation for the development of the first calendars, thus many historians of science now believe that astronomy was humanity's first theoretical science. Detailed astronomical records allowed these early civilizations to accurately predict the seasonal changes so important to their survival. The most important aspect of these weather changes dealt with the impact of rainwater on their irrigation systems.

As societies became more accomplished they experienced a population exposition. New ruling elites developed to control both the distribution of food and the ever-increasing urban population. In time, written codes of law were produced to help create an orderly society.

Impact

The first successful application of water management was in Mesopotamia. The ancient people known as the Sumerians conquered and occupied the area bordering the Tigris and Euphrates rivers. Historians refer to this highly productive land as the "Fertile Crescent." It is in this area, in about 4000 B.C., that the first successful irrigation projects were created. This land became so productive that Biblical scholars now believe that the Fertile Crescent is the area referred to as the "Garden of Eden" in the "Book of Genesis." The weather in Mesopotamia is very harsh and unpredictable. Flash floods are prevalent; therefore, scholars also believe that the story of the "Great Flood" is a myth whose foundation is based on the severe flooding that occurs in this

region. This unpredictability necessitated the construction of an intricate irrigation system consisting of canals, dams, and dikes to control, store, and direct the water for use in the fields.

The result of this successful management was the development of the world's first civilization. During this time a system of record keeping was developed to keep track of the food stored in various warehouses throughout Mesopotamia. This system of record keeping evolved into the first written language, known as "cuneiform." The Sumerians used wedge-style ideographs on wet clay tablets to express complex ideas. Over time, a Sumerian literature developed that resulted in the first recorded myth, the "Epic of Gilgamesh." This story centers around the impact of irrigation and sedentary agriculture on human society. It details the new problems faced by the human community as the result of this agricultural explosion. It also describes the struggle between the new emerging urban civilization and the traditional nomadic pastoral peoples. These monumental changes took place as the result of the successful use of water.

The most dominant people in ancient Mesopotamia were the Babylonians, who developed a flourishing civilization around 1800 B.C. King Hammurabi unified Mesopotamia and constructed an extensive irrigation system. He was so successful that the population of the area grew to unprecedented levels. He developed his famous legal code to ensure the proper regulation of his society. So important was the movement and management of water that a section of Hammurabi's Code deals with the regulation of construction guidelines for his irrigation system.

The reign of King Nebuchadnazzar (605-562 B.C.) marked the high point of the sophisticated utilization of water under Babylonian rule. His palace had an extensive water system, which consisted of both a private shower and toilets. His administrative buildings also used an advanced system. Water was so highly prized in Babylonia that a ritual evolved around washing one's hands before a royal audience as a sign of great respect. There was also an extensive drainage system under the palace complex to dispose of the waste from the sacrifice of animals. The most important and elegant water system of the time was Nebuchadnezzar's fabled "Hanging Gardens." They were constructed as a gift to his wife, in which he attempted to recreate the beautiful mountainous green landscape of her native land. The structure consisted of multiple levels of stone covered with rich soil and irrigated by an underground system of pipes. So magnificent was this structure that it would be designated one of the "Seven Wonders of the World."

Water was also very important in ancient Egypt and would have a profound impact on both Egyptian culture and its economics. The Nile River Valley was very fertile and did not require an extensive irrigation system. Instead, a highly accurate calendar was needed to predict the annual flooding, when the river would deposit both fertile soil and life giving water upon their agricultural lands. This abundant existence led to the development of a belief in the afterlife; it was the hope of every Egyptian that the fruits of this earthly existence would be extended into eternity. An elaborate system of funeral rituals developed and was designed to ensure a successful transition to the afterlife. The body was prepared for the journey through the process of mummification. An extensive system of water-based waste management was developed in the Egyptian mortuaries to aid in the disposal of the byproducts of mummification. The Egyptians also constructed pyramids to house people on their journey to the next world. Since it was believed that basic human needs would remain the same in the next life, the pyramids were built with sophisticated water systems that included individual bathrooms.

The cities of Harappa and Mohenjo-Daro in the Indus River Valley were similar to Egypt in that the water flowed freely into the agricultural lands. Fed by the melting snows of the Himalayas, the Indus provided a steady source of water for its civilization. This resulted in a highly productive agricultural sector that formed the economic base for the two great cities. Each was laid out on a square grid pattern that was subdivided into streets. Numerous public baths were supplied with water by a citywide water system. Archeologists in Mohenjo-Daro unearthed the largest public bath in the ancient world. Known, as the "Great Bath," it measured approximately 39.4 feet (12 m) by 23 feet (7 m) and it was 7.9 feet (2.4 m) deep. The private homes of wealthy aristocrats and merchants also had intricate water systems that included indoor plumbing.

The most powerful civilization in the classical period, Rome, made great advances in the movement and use of water. Rome was the most populous city in the world and maintaining its supply of fresh water was a constant challenge to the city's authorities. This problem was overcome by a system of aqueducts that transported fresh water from the countryside into the urban areas.

Roman cities used water in unprecedented amounts. Every Roman city had a system of public baths that were the envy of the classical world. They all used underground pipes, and many of them had heated floors and hot and cold running water. These baths were the great gathering places in the Roman cities, and both men and women used these facilities. The private homes of well-to-do citizens also had running water that flowed continuously through a nozzle. The great sporting arenas also had large and extensive systems of waste disposal, both for their patrons and to wash away the refuse from the gladiatorial contests. The great Colosseum, located in Rome, could be flooded so that mock naval battles could be conducted for the enjoyment of the spectators. Over time, Rome's water system proved to be a terrible liability. Roman engineers used lead pipes to move water throughout their cities. Prolonged consumption of water with high lead content poisoned their bodies and was one of the major factors that led to the decline of the Roman Empire.

Today, as in ancient times, water plays an essential role in the lives of every human being. Pollution, especially from toxic waste, threatens the world's supply of fresh water. If the nations of the world refuse to implement programs to protect this vital resource, the future of current civilization will be threatened.

RICHARD D. FITZGERALD

Further Reading

Kenoyer, Jonathan M. *Ancient Cities of The Indus Valley Civilization.* Karachi: Oxford University Press, 1998.

Postgate, J. N. *Early Mesopotamia: Society and Economy at the Dawn of History.* London: Routledge, 1994.

Romer, John. *People of the Nile: Everyday Life in Ancient Egypt.* New York: Crown, 1992.

White, K. D. *Greek and Roman Technology.* Ithaca: Cornell University Press, 1984.

Architecture and Engineering on the Indian Subcontinent

Overview

By 1000 B.C. India and China had both developed civilizations that were independent of and would ultimately outlive those of their neighbors in Egypt and Mesopotamia. Objects of daily life found in villages in India today, such as the bullock cart and potter's wheel, are identical to those used thousands of years ago and testify to the continuity of Indian life despite a wave of invasions over millenia. Religion is also part of that continuity: it has consistently formed the basis of the social structure in the country. That reality is reflected in the art and architecture of India, which are a genuine expression of its civilization.

Background

The earliest Indian art and architecture emerged from the valley of the Indus River around 2500 B.C. The best-known sites are Harappa, which was destroyed in the nineteenth century, and Mohenjo-Daro. Each city was fortified by citadels built on artificial oblong platforms large enough to include public buildings as part of the structure. In the cities themselves, houses, markets, and administrative buildings were arranged in a gridlike fashion. Dwelling houses were functional and plain, and ranged from two-room cottages to three-story palaces. Most dwellings had a central courtyard, surrounded by rooms for different purposes. The ground floor of an average house was 30 square feet (9 square m). The inner walls were coated with mud plaster, and the outer walls were made of plain brick. Residental quarters were varied by occupational groups. For example, at Mohenjo-Daro, workmen lived in parallel rows of two-room cottages. The great bath at Mohenjo-Daro was a rectangular bathing pool made of handsome brickwork sealed with bitumen to keep it water-tight. It could be drained in one corner. The bath also featured rows of small private chambers around the pool, and stairways leading down into the water.

Houses sometimes had indoor wells. Most had bathrooms that drained to sewers under the main streets. In fact, the drainage systems at Harappa and Mohenjo-Daro were state of the art until the arrival of Roman civilization. Burnt brick was used for construction throughout the Indus River valley, and the consistency in the size of the bricks suggests a standardized system of weights and measures.

The style of building was elegant but austere. The only architectural ornamentation was simple decorative brickwork. No monumental sculptures have been found, though the civilization produced a plethora of small objects such as toys with wheels, statuettes, and figurines. These objects, as well as bronze and copper implements, indicate a high degree of craftsmanship. The Indus Valley civilization thrived for a millenium, and then went into decay and disappeared around 1700 B.C. for reasons that are still unknown.

At the same time, Aryan invaders with superior military technology began migrations to India. The Aryans were a nomadic people, not taken to living in cities, and after the fall of Harappa and Mohenjo-Daro, the region became one of little villages with buildings of wood and reed. Aryans were skilled bronzesmiths, and their tools and weapons were superior to those of Indus Valley civilization. But their culture was oriented around warfare, and there are few traces of it from the period stretching from c. 1700 B.C. until Alexander the Great of Macedon (356-323 B.C.) crossed the Indus River in 325 B.C

Alexander did not linger in India, but his invasion had the effect of paving the way for the Maurya dynasty (c. 325-c. 183 B.C.). By this time, small towns and trade routes existed all over northern India. Buddhism, which arrived in India in the sixth century B.C., was a reaction against Hinduism, but coexisted alongside it. To proclaim his devotion to the Buddha, Emperor Asoka (d. 232 B.C.) erected edict pillars, tall monolithic columns that show mastery of stonework but serve no architectural purpose. He also dug wells at intervals along roads and set up rest houses for travelers.

The early Buddhist period is notable for the appearance of stupas, hemispherical mounds built to house the relics of the Buddha. The core of the stupa was fashioned of unburnt brick, and the outer face was constructed from burnt brick covered with a thick layer of plaster. At the top of the structure was an umbrella of wood or stone. The stupa was surrounded by a wooden fence that enclosed a path for people to walk around. Over time, Indian stupa architecture became increasingly ornate. Carved railings, terraces, and gateways appeared. The Stupa of Amaravati, completed c. A.D. 200, featured two promenades adorned with carved panels. In the north of India, stupas were taller in proportion to their bases, and often set on square platforms. One of the most famous stupas, which has been called a Buddhist wonder of the world, was the great tower raised by King Kanishka in Peshawar. According to the description of a Chinese traveler to the site, this monument incorporated many varieties of wood in a 13-story structure rising to a height of 700 feet (213 m). The platform was decorated with stucco images of the Buddha. The stupa was dominated by an iron mast that supported 13 gilded copper umbrellas. This feature proved to be the monument's undoing when it was hit by lightning.

The earliest freestanding religious building of which traces remain is a small round hall made of brick and wood that dates from the third century B.C. No temples survive previous to the Gupta period, but from that era forward, temples show a general pattern: small, with flat roofs and ornate pillars. The masonry is joined without mortar, which betrays a certain inexpertise of the builders. But by the sixth century, temple masonry was held together by iron dowels, and covered walks surrounded the buildings.

The Gupta dynasty (A.D. 320-600) oversaw the greatest cultural age in India. Architecture, sculpture, and painting all flourished, and their grandeur has not diminished with time. Chaitya halls, monastic sanctuaries hewn out of rock, evolved from simple structures to complexes of caves with elaborately carved facades and painted interiors. The most famous of these cave groups are the 27 caves at Ajanta, and those at Ellora, near Aurangabad. The cave temples at Elephanta, an island off Mumbai (formerly Bombay), contain exquisite sculptures.

The resurgence of the Indian empire under Harsha in 606 was the catalyst for another wave of construction, most notably in the capital city, Kanauj. Monumental stone architecture appeared only as Buddhism was beginning to die out in India. One example is the Pancha Rathas (c. 650) at Mahabalipuram, which are five small monolithic shrines cut out of live rock.

Impact

Art in the true sense of the word did not appear in India until the Maurya period in the third century B.C. The style of buildings and artefacts from the Indus Valley civilization cannot be said to show an aesthetic intent. Perhaps the intent existed, but will never be known, since after the cities vanished the culture was not picked up by the invading Aryans. Writing, for example, which appears on seals from Harappa and Mohenjo-Daro, disappeared and did not return until the middle of the first millennium B.C.

Of the intervening centuries, little remains until the stone-carving of the Mauryas. The edict pillars of Emperor Asoka were used to propagate official messages. Etched into their smooth surfaces were recommendations to the emperor's subjects that laid out a new social philosophy of dignity, religious tolerance, and nonviolence. The pillars, which were erected throughout north India, were a focus for political and social unity. Artistically, they represented the culmination of a native tradition of stonework.

In India, art and religion are synonymous. The purpose of art in India is to communicate great truths to humankind. Indian religious sculptures and paintings reveal the personality of the gods (Buddhist and Hindu texts state that the way to heaven is to make images). Nor is it possible to sort out art from religion from architecture. Architecture and sculpture are always complementary. The stupa represented a cosmic mountain. The temple was a model of the universe. Workers dedicated to the Indian temple relied on manuals of aesthetic procedure to guide them in architecture, sculpture, and painting.

The Gupta dynasty marked an important stage in Indian aesthetic development. For one thing, Indian artistic life came to maturity during this period, For another, the aesthetic conceptions of the Buddhists and Hindus began to diverge. For the first time, also, free-standing structures were constructed from lasting materials. The stone temples built during this period were unsurpassed up to the Muslim era. A particular kind of bell-shaped stupa spread throughout Southeast Asia.

Secular art in India was essentially unknown, which is a disadvantage in the sense that very little is known about the material life of the inhabitants of the subcontinent for a very long time. What is left, however, is a window into their minds. The gods and demons that figured in Hindu and Buddhist religious representations in the long ago past are the same that figure in village shrines all over India today. The ingenuity that went into the building of temples and other religious structures made them centers both for local worship as well as for pilgrimages, which in turn transformed the greatest of them into small, wealthy cities.

Technical achievement in India was not negligible. The Indus Valley civilization testified to advanced concepts of town planning, as well as water management and flood control. Such was the skill of Indian spinners and weavers that their silks and muslins were in demand in the Roman Empire. The monolithic columns of Mauryan times were carved from single blocks of stone that weighed up to 50 tons, and were polished and transported many hundreds of miles, all by a process has never been fully explained. Similarly, the Iron Pillar of Meharauli stands over 23 feet (7 m) high and is made of a single piece of iron. The craftsmen who fashioned it must have been extremely skilled metallurgists: it still shows no signs of rusting. Boats were used to carry goods and people along the great rivers. They also provided ferry services, since rivers crossed by major roads were not bridged. But oceangoing vessels were rarer. In the words of the scholar A. L. Basham, superstitions about sailing the sea made India "a nation of landlubbers."

GISELLE WEISS

Further Reading

Basham, A. L., ed. *A Cultural History of India*. Oxford: Clarendon Press, 1975.

Basham, A. L. *The Wonder That Was India*. New York: Grove Press, 1954.

Kramrisch, Stella. *The Hindu Temple*. 2 vols. Delhi: Motilal Banarsidas, 1976.

Rowland, Benjamin. *The Art and Architecture of India*. Melbourne: Penguin, 1953.

Zimmer, Heinrich. *The Art of Indian Asia*. 2 vols. Princeton: Princeton University Press, 1955.

The Impact of Mayan Architecture

Overview

The buildings left behind by the Maya inspire onlookers with a sense of awe and admiration. These buildings hold the secrets of Mayan religion, identity, and history. Known for its powerful features, Mayan architecture is not only aesthetically pleasing, but also technically accurate. Even though the thick jungle of Central America encroaches on these buildings, they still stand after thousands of years. The ruins themselves have been preserved by the vast and hostile jungle landscape, which has given archaeologists and scientists a chance to understand Mayan culture, political systems, and social and economic activities.

Background

Despite thorough study by scholars, much of Mayan civilization is still unknown today because much of the civilization's writing and texts were destroyed. The virtual elimination of Mayan writing forces experts to turn to buildings in hopes of understanding many aspects of Mayan culture that have disappeared. Even though many mysteries surround these great ruins, it is beyond a doubt that Mayan architecture is an art of original expression and remains unparalleled in history.

By 1000 B.C. the Maya began building villages in the mountainous regions of Mesoamerica. Their initial work essentially formed the templates for all subsequent societies in the region. Future generations looked to them in developing their own political and cultural views. The Maya used architecture as their starting point to express their beliefs and to create their civilization.

Through architecture, the Maya created intricate social institutions. The main focal point of each town became large pyramid-like structures. The people regarded these buildings as mountains rising out of the swamps and forests. The Maya gave these buildings meaning by molding plastered imagery of events, thus creating a pictorial history. The early Mayan civilization laid the foundation for most of the great architectural achievements to occur later.

The larger masonry structures are the most commonly remembered and explored examples of Mayan architecture, which include public buildings, palaces, temples, and ball courts. It is unlikely that the Maya had "professional" architects, rather a group of master builders separated tasks based on skill. For example, since they based building orientation on sacred beliefs, religious specialists became involved in the design and later dedication of the building. Common people provided the labor in order to fulfill their duty to the king or head of state.

The Maya considered houses and temples the center of the world, one for family and the other for the gods. Plazas and courtyards surrounded public facilities, which made up the operational spaces of the cities. Interior spaces were dark and small, especially in temples, since these places housed gods and their ancestors. Public access into these palaces was prohibited, but they were allowed to stand in the courtyards where many Mayan rituals and festivals were held. Mayan architecture is at the heart of many of these ceremonies. The buildings act as a stage and set the scene for the drama to unfold. In fact, courtyards also placed restrictions on where people could go. The Maya controlled movement by using various architectural designs, like constricted entrances, causeways, stairways, and other devices used to channel movement.

The Maya lived in and around their cities and towns in dense and permanent settlements. Just like in modern Mayan communities, houses were in clusters of two to six units centered around a patio. *Xanil nah,* or "thatched house," represents the oldest known example of Mayan architecture. These structures were built on slightly raised platforms. They adapted house construction to the tropical environment and gathered materials from the nearby forests. Today's Maya continue to construct their housing in a similar manner. In ancient times, they used termite-resistant wood for the frame and roof, palm leaves for the thatch, and strips of bark to tie everything together. Each house is similar in the fact that it is arranged in a single room. A three-stoned hearth served as the center of activity. Royal homes were similar in design, but used stone and a scale much larger in size, and were also supported on higher platforms.

The arrangement of the city was based on their view of the world. Building positions were perfectly aligned with symbolic meaning. The center of the city, or the center of their world, was represented by the palaces of the living ruler. To the north were tombs and shrines of their royal ancestors, and the ball court was perfectly positioned, since this ritualistic game combined past and present myths and legends. The sacred centers contained pyramid-like structures with temples on top and sculpted monuments to document the history of the ruling king and ancestors. The sacred precinct also housed administrative, religious, and residential complexes for royal lineage. The area around this precinct held

smaller-scale buildings that housed nonroyal, but wealthy lineages. The cities and towns sometimes had a causeway or *sak beh* that led from outer-lying zones into the center. These road systems are evidence of the degree of political authority contained in each city. The most extensive system of roads is at Coba, where a multitude of causeways extend over 60 miles (96.5 km) from the center, showing how powerful it must have been in its heyday.

Similar to the Egyptians, the Maya used a pyramid shape, only truncated, to construct their great temples. Most times, these buildings were only for show and represented sacred beliefs about the world below and the gods above. However, archaeologists have discovered cases in which these monumental pyramids served as tombs for great leaders. It seems that some of the tallest and most grandiose temples are found in ruins that marked earlier Mayan civilizations. Later periods in Mayan history never reached the gigantic scale their ancestors had once achieved.

Impact

While Mayan cities increased in population, partly due to their superior skills in agriculture, they rarely added additional buildings. Instead, the Maya used the technique of superimposition, where a new building is constructed on top of an older one. In most cases, after a building outlived its usefulness, it was partially or totally encased in a larger and taller structure. An unintended consequence of this architectural strategy is that modern-day archaeologists can study the site and find incredibly valuable information. Usually, the building concealed by the outer building is well preserved. Excavations into these superimposed works have allowed scientists to uncover the earliest dates of Mayan occupancy. Other discoveries have been made thanks to this technique, including the ability of experts to follow the lineage of leaders based on mosaics and plaque depictions and styles that would have eroded otherwise.

Maya architecture has had a profound influence on the region. Many features of Mayan architecture appear throughout Central America today. The use of color, open space, and texture provide ideas and guidelines that are evident in contemporary architecture. For example, in order to create open spaces in buildings, the Maya relied on courtyards, square buildings, and quadrangles. Included in this open space design was the use of platforms, creating a variety of shapes, sizes, and levels. Respect for the

Sculpture of a Mayan god from Chichén Itzá. *(Corbis Corporation. Reproduced with permission.)*

Mexican landscape has always been a constant. Building color and texture are natural looking in order to compliment and blend in with the environment. These styles are quite evident in Central America today.

Thanks to archaeologists, the legacy of the Maya is revealed through their architectural ideas that provides present-day architects with a lesson on using the environment as a friendly tool to enhance a building, not hinder it. The Maya utilized the tropical forests that surrounded them to improve their lives. By building extensive canals, reservoirs, and raised fields, the Maya produced an abundant supply of food. Even today, modern farmers in remote regions of Central America are taught the agricultural methods of the Maya for their own harvesting. In fact, the Maya were so successful in produc-

ing food that they soon created overpopulated cities, which in turn demanded a greater turnover of food production. Sometimes the environment could no longer keep up with the demand for harvesting and this created episodes of severe malnutrition. However, the Maya improved the domestication of maize and also developed an extensive list of beans, pumpkins, vanilla, manioc, chili peppers, and, most importantly, the creation of chocolate.

The descendants of the Maya react toward their climate and landscape the same way their ancestors did. The environment has not changed for them and the land provides the same building materials as in the past. Mayan artisans and architects pride themselves on a high level of craftsmanship and continue to keep many of their customs and traditions alive, even though modern buildings surround them. This pride links today's Maya with their ancestors and keeps them in direct contact with an ancient world that once served as one of mankind's greatest civilizations.

KATHERINE BATCHELOR

Further Reading

Andrews, G. F. *Maya Cities: Placemaking and Urbanization.* Norman: University of Oklahoma Press, 1975.

Hammond, Norman. *Ancient Maya Civilization.* New Brunswick, NJ: Rutgers University Press, 1988.

Sabloff, J. A. *The New Archaeology and the Ancient Maya.* New York: W.H. Freeman, 1994.

Sharer, Robert J. *Daily Life in Maya Civilization.* Westport, CT: Greenwood Press, 1996.

The Great Wall of China

Overview

What is commonly referred to as the Great Wall of China is actually four great walls rather than a single, continuous wall. The oldest section of one of the four Great Walls of China, was begun in 221 B.C., not long after China was unified into an empire out of a loose configuration of feudal states. The most famous early wall construction is attributed to the first Chinese emperor, Qin Shihuangdi. Scholars generally credit him with the restoration, repair, and occasional destruction of earlier walls, and ordering new construction to create a structure to protect China's northern frontiers against attack by nomadic people. Historians continue to debate the form these fortifications took. Though records mention the *chang-cheng* (long wall) of Emperor Shihuangdi, no reliable historical accounts indicate the length of the Qin wall or the exact route they followed.

Background

Wall building, around houses and settlements and along political frontiers, began in China more than 3,000 years ago. The first walls were probably between households, marking an important stage in the evolution of the traditional Chinese home. Next came walls around villages and towns. Earthen walls surrounded some prehistoric villages, and there are visible remains of a 4.3-mi (7-km) wall that is still more than 29.5 feet (9 m) high. These durable walls were constructed by the technique of pounded layers of earth alternated with stones and twigs inside wooden frames. In the period before the Qin dynasty, when political power was divided effectively among the rulers of feudal kingdom, these earthen walls were used to build state border walls.

During the Warring States period (403-221 B.C.) before China was unified, feudal states fought for control of the area constituting most of modern China. Although most of these segments of wall are impossible to locate today, some of them were reused during the construction of later walls. These early walls, however, are not usually thought of as being a Great Wall.

In 214 B.C., to secure the northern frontiers, Qin Shihuangdi, ordered his general Meng Tian, to mobilize all the able bodied subjects in the country to link up all the walls erected by the feudal states. This wall became a permanent barrier separating the agricultural Han Chinese to the south and the nomadic horse-mounted herdsman to the north. According to historical records, the Great Wall of Qin Shihuangdi was completed in about 12 years by the 300,000-person army, conscripted labor of nearly 500,000 peasants, and an unspecified number of convicted criminals. These workers were subjected to great hardships. Dressed only in rags, they endured cold, heat, hunger, exhaustion, and often cruel supervisors.

The wall was built across rugged terrain that included streams, rivers, mountains, and desert landscapes. Local earth, stone, timber, and bricks were the primary materials used to build the wall during the Qin dynasty. Although some stones were quarried to build walls in the mountains, tamped earthen walls—a traditional Chinese building technique—were most common in the flatter terrain and desert areas. Posts or boards were fixed on both sides of the wall, and earth and small stones were used as fill between the boards. This process was repeated layer upon layer, and the wall slowly rose 4 in (10 cm) at a time. This earthen fill was rammed into a solid layer by wooden hammers. In recent years archeological work has found evidence that most of the early walls were built in this manner. One example reveals that the wall was composed of bundled twigs, approximately 6 in (15 cm) thick, alternated with thin layers of coarse clay or gravel.

Building materials were transported on the backs of people or with carrying poles. A complex system of trails accommodated pack goats and donkeys bringing food and materials. Materials were often passed from hand to hand; builders stood in line from the foot of a section of wall and then building materials were passed from person to person. This method was safer and more efficient especially with sections on narrow mountain trails. Handcarts were used to transport materials on flat ground or gentle slopes. Large heavy rocks were carried using wooden rods and levers. Ropes were slung across deep ravines and valleys to move basketfuls of building materials. Some of the surviving members of the construction crews settled into some of the agricultural areas opened up after the construction was completed.

Impact

Most scholars suggest that the practice of extensive wall building, along with many other aspects of Chinese life, was instituted in the third century B.C. when the first unified Chinese state came into existence. This unification was effected when one of the earlier states, the Qin, defeated its rivals and the king adopted the newly created title of *huang-ti* (emperor) of the Qin dynasty (221-207 B.C.).

The first emperor proceeded to embark on a series of drastic reforms and massive public works projects to consolidate his rule. Along with a network of roads leading from the capital city, his laborers linked protective walls to deter raiding nomads into a defense system known as a *wan-li-ch'ang-ch'eng* or "ten-thousand *li* long wall." From the beginning, the construction of the first Great Wall of Qin was linked with the emperor's developing dynasty.

From 230-221 B.C. the warring rivals, the Han, Zhao, Wei, Yen Chou, and Qi, were crushed and the kingdom of Shihuangdi expanded eastward and northwards. Warfare during this time became larger in scale and ruthless with the development of the iron sword and crossbow. The technique of casting less expensive iron rather than bronze resulted in the production of swords with sturdy iron blades, pre-

ceding those invented and produced in the West by nearly 1,000 years. The crossbow, which fired arrows up to 250 yds (228.6 m) with accuracy, gave the Qin army an important edge in subduing enemies. By 221 B.C. Qin Shihuangdi had united nearly all of China.

The Qin unification ushered in a series of major changes that included the construction of the wall. The Great Wall as a project entailed an enormous investment of human and material resources. A single state government with strong centralized control was instituted. Law was standardized and a strong authoritarian rule was imposed. Also at issue was how the Qin dynasty (and later ones as well) would deal with the threat of nomadic people. Nomads were people who followed a pastoral way of life, subsisting on their flocks and moving with the seasons from one place to another. Equipped with horses and weapons, they held a great advantage over the settled Chinese people. The decision to allocate massive amounts of human and material resources to the construction of the wall was in part a strategy to maintain Chinese integrity at the borders.

Agricultural production along the Great Wall developed quickly. The once barren land was turned into a flourishing agricultural zone with irrigation and the use of the traction plow. Weights and measures, as well as coinage and writing, were standardized. Wide highways were also built to allow construction of the Wall. Some of these roads were themselves feats of engineering. They were raised in places where flooding was likely to occur and bridges allowed for the efficient transport of goods and hundreds of thousands of laborers.

The first of the Great Walls is an example of two conventional Chinese ideas: that a defensive system should be built where the terrain makes access difficult, and that locally available material should be used. The natural terrain—mountains and deserts—was fully utilized to make the structure of the wall both useful and practical. The first walls were designed to deter warriors wielding swords, spears, and bows and arrows. However, despite the defensive nature of the Wall, it did not become a barrier to cultural, political, and economic interchange. Goods, people, and ideas traveled back and forth at different times and places. For example, advanced metallurgical techniques, innovative farming methods, horses, camels, and music infiltrated Chinese culture over time.

The impact of the first Great Wall continues to be researched and discussed. The imperial system initiated during the Qin dynasty set a pattern that was to develop over the next two millennia. Steeped in myth and legend the history of the Great Wall of Qin became a rich legacy and a blueprint for subsequent generations of Chinese people.

Though much of this first Great Wall has disappeared as a result of centuries of natural and human-inflicted damage, remains of compressed earth, sand, and stones can be seen. The second "ten-thousand *li* long wall" was built during the Han Dynasty, the third was built by the Jin Dynasty that made peace with the Mongol invaders, and the fourth was built by the Ming Dynasty beginning in 1368. This series of

walls has become China's best-known monument and national symbol, and embodies some of most innovative and ingenious ideas of any people in the world.

LESLIE HUTCHINSON

Further Reading

Fryer, Jonathan. *The Great Wall of China*. London: New English Library, 1975.

Waldron, Arthur. *The Great Wall of China: From History to Myth*. Cambridge and New York: Cambridge University Press, 1990.

Zewen, Luo, et al. *The Great Wall*. New York: McGraw-Hill, 1981.

Cities of Ancient America

Overview

Many centuries before the rise of the Aztec and Inca civilizations most familiar to modern students of the pre-Columbian New World, the Americas were home to a number of highly advanced peoples. The Olmec, Maya, and other groups in Central America, as well as the Chavín of the Andes, were civilized in the truest sense of the word—in other words, they built cities. The splendor of their achievements becomes all the more impressive when considered in light of their technological limitations.

Background

Whereas the Olmec of Mesoamerica (archaeologists' term for ancient Central America) dwelled in a land of steaming rain forests and lush, vegetation-covered mountains, the Andean home of the Chavín peoples far to the south was rocky and dry. Despite these differences in environment, however, the two groups had much in common, not least of which was time: the Olmec flourished between c. 1200 and c. 100 B.C., and the Chavín culture during the period from c. 1000–c. 400 B.C.

Among the most interesting features common to the Olmec and Chavín was the practice of building pyramids. Certainly it is a mystery why ancient pyramid-building seems to have taken place primarily in Egypt and in the Americas far across the ocean, though it should be noted that the pyramids of the Americas were

places of worship rather than burial chambers. Equally perplexing, however, is the adoption of the pyramid by the Olmec and Chavín, who were not only separated by wide distance, but completely unaware of one another.

Another curious aspect of ancient American civilizations is their relatively low level of technological development in comparison to their achievements as builders. Most notable was their ignorance of the wheel. While it appears that the Egyptian pyramid-builders also had no knowledge of the wheel, they were working some 1,500 years before their American counterparts—and furthermore, archaeologists have found wheeled toys at various sites in Mesoamerica. Why the Olmec did not adapt the wheel to more practical usesemains a mystery.

Furthermore, pre-Columbian America was virtually without domesticated animals. (The sole exception was the llama, domesticated late in the game by the Inca; however, the llama could only carry light burdens—hardly the equivalent of a draft horse or ox.) Nor did ancient Americans possess sophisticated tools. The Chavín became highly accomplished in the art of fashioning objects from gold, but it appears that no civilization in the Americas entered the Bronze Age before c. A.D. 1200. Metal was chiefly for decoration, as with gold jewelry; tools, on the other hand, were of stone. Thus as one contemplates the pyramids of Mesoamerica, it is amazing to consider that they were built by peoples living quite literally in the Stone Age.

An Olmec carved head. *(Springer/Bettmann-Corbis. Reproduced with permission.)*

Both the Olmec and the Chavín civilizations grew out of agricultural systems that began developing in about 3500 B.C. At the hub of these agrarian civilizations were cities, though the earliest of these were not cities as that term is commonly understood among modern people: rather, they were ceremonial centers to which worshippers made pilgrimages. These first American cities had a static, planned character that distinguished them sharply from contemporary metropolises of the Old World such as Babylon. Whereas the latter had sprung up organically, and within its sprawl encompassed numerous functions, the ceremonial centers of America were almost entirely the province of priests and rulers.

Impact

The principal Olmec ceremonial centers were San Lorenzo, La Venta, and Tres Zapotes, all located in the Isthmus of Tehuantepec, the narrowest part of what is today southern Mexico. (The cities' Spanish names were obviously adopted later; archaeologists do not know the names the Olmec used for these centers.) La

Venta and Tres Zapotes were on more or less the same latitude, near the Gulf coast on the north, whereas San Lorenzo lay to the south, and the three cities together formed a downward-pointing triangle with sides about 100 miles (160 kilometers) long.

Founded in about 1300 B.C., San Lorenzo was built on a plateau atop a mountain—but the plateau was a manmade one, representing thousands of hours' labor and oceans of human sweat. The ability of their unnamed ruler to command this feat indicates the level of organization in Olmec society: only a nation with a strong central government can successfully call on its people for such massive undertakings. Further evidence of the highly organized Olmec system can be found in the massive public works projects, including drainage systems, water storage ponds, and stone pavements, uncovered by archaeologists. San Lorenzo was a city of houses built in the shape of mounds: at one point, there were some 200 of these house-mounds, home to about 1,000 priests and royalty. But just as suburbs surround modern cities, thousands more people—farmers, mostly—lived in surrounding areas.

By 900 B.C. San Lorenzo had declined, and was replaced by La Venta. Both cities were built on salt domes, large deposits of rock salt underneath the earth. But whereas San Lorenzo was primarily a ceremonial center, La Venta apparently also served functions typical of any city, housing tradesmen and people of other professions. In some ways, it was a model for the much more splendid city of Teotihuacán that would follow it: La Venta was built on a grid pattern, as Teotihuacán would later be; and just as Teotihuacán was dominated by the Pyramid of the Sun, La Venta had a main pyramid about 100 feet (about 30.5 meters) tall.

All around the Olmec were other cultures, the remains of which have been found at archaeological sites throughout Central America. The most prominent of these were the group whose presence in Mesoamerica spanned the periods designated as "ancient" and "medieval" among European civilizations: the Maya. They had developed ceremonial centers of their own as early as 2000 B.C., and by about 300 B.C. the Maya populated parts of what is now Guatemala, Honduras, and El Salvador before moving into what is now Mexico.

The modern-day Mexican state of Chiapas on the Guatemalan border, where the majority of the people still speak some version of the Mayan language, contains an archaeological site at Izapa, which may have been a ceremonial center between 1500 and 800 B.C. It is possible that Izapa preserved traditions of the Olmec that later became part of the Mayan culture, including the cult of the rain god. At the outset of the period coeval with the European Middle Ages, a number of Mayan cities, beginning with Tikal in what is now northern Guatemala, began to flourish.

In addition to the Maya, there were the Zapotec, who lived in what is now the Mexican state of Oaxaca and established the first true city (as opposed to a ceremonial center) in Mesoamerica, Monte Albán. By A.D. 200 it had become a dominant urban center, containing some 30,000 people, and it survived until A.D. 800. Yet as great as Monte Albán was, there was another city even greater: Teotihuacán in the Valley of Mexico, near the future site of the Aztec capital at Tenochtitlán, as well as the modern Mexican capital, Mexico City.

Built in about A.D. 100, Teotihuacán was the first true metropolis in the Western Hemisphere, and within 500 years, it would grow to become the sixth-largest city in the entire world. Teotihuacán, which appears to have been a planned city, covered some 8 square miles (20.5 square kilometers), a vast area for an ancient city. It had a population of between 125,000 and 200,000, an astounding figure for its time. (Because of sewage and sanitation problems, among other difficulties, ancient cities were seldom larger than small towns in modern times.)

Like Rome, Teotihuacán was a meeting-place for many cultures, and it appears that people from all over Mesoamerica lived in apartment-like buildings. The "skyscrapers" of Teotihuacán were its pyramids, the most significant of which was the Pyramid of the Sun. The latter stood on the city's main street, which the Aztec later dubbed the Avenue of the Dead. Other great temples lined the avenue, which ended at the Pyramid of the Moon.

Teotihuacán survived until about A.D. 750, when it began to decline rapidly. Archaeologists have suggested several possible reasons for its downfall, including a fire that engulfed much of the city. The fire may in turn have been the result of organized action, either by rebels or outside invaders such as the warlike Toltecs, then on the rise. On the other hand, the end of Teotihuacán may have come because its great population depleted natural resources and created sanitation problems that resulted in widespread disease.

A pre-Aztec pyramid at Teotihuacán. *(Kevin Schafer/Corbis. Reproduced with permission).*

About 2,500 miles (4,000 kilometers) south of the Olmec lived the people of the Chavín culture, around the border areas of modern Peru and Ecuador. The term *Chavín* refers to Chavín de Huántar, a ceremonial center that developed in what is now north-central Peru about 1200 B.C. Like the ceremonial centers of Mesoamerica, Chavín de Huántar was a city of pyramids and platforms, including a large structure dubbed the "Great Pyramid" by archaeologists.

Chavín de Huántar was about 1.5 miles (2.4 kilometers) across, and contained a "Great Plaza" in the southeast. To the northwest was the court and temple of the Lanzon, a stone idol representing the supreme deity worshipped at Chavín. Like the Mesoamericans, the people of Chavín revered the jaguar; hence there were also the Stairs of the Jaguars leading down to the Great Plaza from the Great Pyramid.

As with other ceremonial centers, Chavín de Huántar had a small resident population—probably no more than 1,000 people—with thousands more (presumably farmers and laborers who served the priests and rulers) living in surrounding areas. Between 400 and 300 B.C., Chavín de Huántar entered a period of decline, and eventually a less advanced group built a village over the site. Yet its memory lived on to inspire the Inca, just as Teotihuacán did the Maya and Aztec.

Numerous other peoples in the Andes, including the Moche, Paracas, and Nazca (who created the mysterious "Nazca Lines" between 200 B.C. and A.D. 600), were influenced by Chavín culture. Another impressive site near Chavín de Huántar is Tiahuanaco, in the Andean highlands of what is now Bolivia. It was to the Andes what Teotihuacán was to Mesoamerica: a great city, much more than a ceremonial center, which served as a focal point for the peoples all around it. Like Teotihuacán, it was the site of impressive achievements in architecture and engineering, including the massive Gateway of the Sun, cut from a single stone. Yet one thing distinguished Tiahuanaco from Teotihuacán or virtually any other major city, then or now: its altitude. At 13,125 feet (2.5 miles or 4 kilometers), Tiahuanaco was without a doubt the highest major city in history. Though it flourished between 200 B.C. and A.D. 600, Tiahuanaco continued to exert an influence over an area from southern Peru to northern Argentina until about A.D. 1100.

The latter date also marks the approximate founding of Cuzco, today the oldest continually inhabited city of the New World. Located in what is now Peru, Cuzco became the capital of the Inca Empire, which grew over the next 300 years and began to flourish in the mid-fifteenth century. Simultaneous with this Andean renaissance, but independent of it, was the emergence of an Aztec urban culture centered around Tenochtitlán. Both civilizations, which flowered in the brief period that remained for the Americas prior to the arrival of the Spanish, owed an immeasurable debt to their ancient forebears in the cities of the jungles and mountains.

JUDSON KNIGHT

Further Reading

Books

Cotterell, Arthur, ed. *The Encyclopedia of American Civilizations.* New York: Mayflower Books, 1980.

Leonard, Jonathan Norton. *Ancient America.* Alexandria, VA: Time-Life Books, 1967.

Internet Sites

Mesoweb: An Exploration of Mesoamerican Cultures. http:// www.mesoweb.com (November 14, 2000).

"New World Civilizations." http://www.emayzine.com/ lectures/classical%20maya.html (November 14, 2000).

"Teotihuacán." http://www.du.edu/~blynett/Teotihuacán. html (November 14, 2000).

The Development of Dyes by the "Purple People," the Phoenicians

Overview

Though the Phoenicians were among the most influential peoples of ancient times, becoming merchants and explorers who settled the western Mediterranean and beyond, there was—strictly speaking—no such place as "Phoenicia." Rather, the homeland of the Phoenicians was a coastal strip centered on what is today Lebanon, a chain of city-states dominated by Tyre and Sidon. As for the name *Phoenicia,* it comes from the Greek *Phoinike,* which shares roots with the word *phoenix*—a term that to the ancient Greeks connoted blood-red or purple. The latter was the color of a natural dye developed by the Phoenicians, who became so closely associated with it that their name reflected the fact.

Background

Today virtually all clothing dye comes from synthetic sources, but this is very much a latter-day development; prior to the mid-nineteenth century, all fabric coloring came from nature. So too did textiles, the first examples of which—found in the Judean desert—date to the seventh millennium B.C. Flax, hemp, rush, palm, and papyrus all became material for clothing in the Near East during the period from c. 6000-c. 5000 B.C., and in the centuries that followed, the peoples of the region began using wool and other animal fibers.

The first Near Eastern reference to dyes comes from c. 3000 B.C. (Chinese sources on the subject are even older), and the coloring agent

mentioned was madder. This is a plant that grows in northern Africa and southwestern Asia, and which contains in its roots coloring that varies from a pinkish to a brownish red. The oldest known garment colored with madder is an item of red linen from the tomb of Egypt's "boy king" Tutankhamen (d. 1323 B.C.)

Probably the next type of dye to appear, around 2500 B.C., was indigo. Though the ingredient indigoten can be found in a number of plants, its principal source is the legume *Indigofera tinctoria,* which probably originated in the Indian subcontinent (hence its name) and spread westward. Using sodium hydrosulfite, a caustic soda, to ferment the leaves of the indigo plant, dye-makers extracted a bluish paste that they then processed into cakes and ground finely. Examples of indigo-dyed garments have been found in excavations at Egyptian Thebes.

Impact

To the east of Egypt, facing the Mediterranean from the east, was a narrow strip of land just 200 miles (322 kilometers) long and 30 miles (48 kilometers) wide—much smaller, in fact, than modern-day Lebanon. This was the homeland of the Phoenicians, a Semitic people related to the Canaanites of the Old Testament. In the Bible they were called Sidonians, a reference to one of their leading cities, but in Greek literature—where their earliest mention can be found in the writings of Homer—they are identified with the name by which they are known today.

The Phoenicians are almost unique among ancient peoples in that they did not maintain an army or attempt to conquer other peoples; rather, their focus was on trade, which they expanded primarily by means of sea routes. Geography suited them for this endeavor. Though their soil was not bad for agriculture, the mountain ranges to the east meant that the available area for raising crops or animals was limited. Furthermore, the country's famed cedars were ideal for shipbuilding.

The first major Phoenician city-state, Tyre, was established in about 2000 B.C., and over the next centuries other port cities such as Sidon, Byblos, Tripolis, and Berytus emerged. Egypt invaded in c. 1800 B.C., and maintained control over the land for about four centuries, until the Phoenicians took advantage of Egyptian involvement in a war with the Hittites of Asia Minor to establish their own independence. The region truly came into its own as a trading power after about 1200 B.C., when attacks by the Sea Peoples—a mysterious nation that disappeared from history as suddenly as it appeared—broke the power of Ugarit, a Syrian port that had dominated trade in the Levant up to that time.

One of several distinguishing characteristics of the Phoenicians was the ability of their craftsmen: thus when Israel's King Solomon (r. c. 960-922 B.C.) was building his temple in Jerusalem, he brought in Phoenician workers. The Bible also indicates that Phoenicians were talented at working with bronze, and extensive evidence exists of Phoenician carving and glassmaking (they may in fact have been the first to make glass). But long before these appeared, there were the colored garments for which the Phoenicians later were named—garments that numbered among the first major Phoenician export products.

Apparently the Phoenicians adapted existing technology to color cloths blue with indigo, whereas red came from the kermes, a parasitic insect that lives in oak trees. (The term *crimson* comes ultimately from the Arabic word for this tiny creature, *qirmiz*.) But though the Greek word for the Phoenicians suggests the color red, in fact the most famous of all Phoenician-produced colors was purple, or more properly Tyrian purple.

In producing both red and purple, the Phoenicians went a step beyond vegetable dyes to produce colors from animal life. Purple came from the murex or *Murex brandaris,* a variety of mollusk found in the Mediterranean. The Minoans in c. 2500 B.C. had been the first to use murex for making dyes, but the Phoenicians greatly expanded on the practice—as was evident from the many heaps of murex shells found by modern archaeologists at Sidon.

Each murex produced just two drops of dye, and to make a single gram (0.035 ounces) of coloring required between 10,000 and 20,000 murex. The mollusk was thus worth more than its weight in gold, and garments colored in Tyrian purple were extremely expensive. Hence the idea of "royal purple"—the concept that because of its value, the color was to be worn only by royalty. Also the basis for the Phoenicians' trading empire, Phoenician mariners began searching for beds of the precious murex shell all across the Mediterranean.

During the period from c. 900-c. 600 B.C., the Phoenicians established a number of overseas colonies, which though they began partly as sites for gathering murex, in time also became warehouses for storing goods, as well as trading posts for ongoing business with the local peoples. Far across the Mediterranean, Phoenician traders founded their most important colony at Carthage in what is now Tunisia, as well as cities on the islands of Sicily and Sardinia off the coast of Italy.

They also established cities on the European continent, including Marseilles in France and the Spanish cities of Barcelona, Cadiz, Malaga, and Algeciras. Further away, at the edge of the known world, were what the Phoenicians called "the tin islands": Britain, as well as the region of Britanny on the northwest coast of France. The Phoenicians brought in purple cloth and traded it with the locals for tin, essential for making bronze. Thus it could be said that the Phoenicians' purple garments were the ultimate cause behind their wide-ranging expeditions, which in turn influenced the spread of their greatest contribution, the alphabet.

The Phoenicians, however, ultimately became victims of superpower conflicts in the region. Assyria had begun threatening Phoenician lands as early as 868 B.C., and the Assyrian monarchs Tiglath-Pileser III (r. 745-727 B.C.) and Sennacherib (r. 704-681 B.C.) led successful assaults on the city-states. Later, when Babylonia replaced Assyria as the dominant empire, Nebuchadnezzar II (r. 605-562 B.C.) crushed Tyre.

Still later, as part of the Persian Empire, the Phoenician fleet helped wage war on the Greeks during the Persian Wars (499-449 B.C.) More than a century after that, the armies of Alexander the Great (356-323 B.C.) conquered the Persians' empire, and Phoenicia passed into Greek hands in 333 B.C. Like much of the Middle East, it then fell under the Seleucid Empire before becoming part of the Roman province of Syria in 64 B.C.

As for dye-making, the ancient world saw developments in the use of safflower for red and yellow vegetable dyes, the insect dye lac in red Persian carpets, and a blue dye called techelet from the sea creature known as the chilazon. By c. A.D. 1300 a new purple dye known as archil, the product of lichen species, had taken the place of murex as a source of purple. Then in 1856 an 18-year-old English student named William Henry Perkin (1838-1907) produced the world's first synthetic dye, a tar-like black solution that when applied to silk produced mauve—that is, a light purple.

JUDSON KNIGHT

Further Reading

Books

Barber, Elizabeth J. Wayland. *Prehistoric Textiles.* Princeton, NJ: Princeton University Press, 1991.

Odijk, Pamela. *The Phoenicians.* Englewood Cliffs, NJ: Silver Burdett, 1989.

Internet Sites

"The Ancient Phoenicians." *St. Maron Parish of Cleveland.* http://www.stmaron-clev.org/phoenicians.htm (November 15, 2000).

A Bequest Unearthed, Phoenicia. http://phoenicia.org (November 15, 2000).

"Guide to Dyes." *Rugnotes.* http://www.rugnotes.com/discussions/zz9948.htm (November 15, 2000).

Metallurgy through the Ages

Overview

Over a period of thousands of years, humans learned to identify, extract, blend, and shape metals into tools, ornaments, and weapons. The ability of metals to alter the wealth, power, and culture of societies is so profound that the Bronze Age and the Iron Age label distinct eras in human development. Metallurgy makes the current Information Age possible and continues to shape our lives.

Background

Metals have shaped history—magnifying our efforts, providing leisure time, and creating empires—because metals allow us to shape our environment like no other materials. Ironically, the first recognized metal, gold, is unchanging and nearly useless. Gold exists in an almost pure state in nature. It does not rust or corrode, and, undoubtedly, it gleamed out from rocks or streambeds, catching the attention of humans in prehistoric times. Gold can be easily shaped, but it is so soft that it cannot be used for weapons or tools. The most useful material of this early time was stone, so this period is not known as the Golden Age, but the Stone Age. The only metal artifacts of this age are beautiful ornaments and simple utensils, such as cups and bowls. Still, gold did teach some fundamental principles of metallurgy that would become useful in later

times: discovery (finding and recognizing a metal in nature), concentration (in the case of gold, by cold hammering smaller pieces into larger pieces), and shaping (working the metal into a desired form).

It was copper, beginning in about 4000 B.C., that allowed humans to extend the techniques of metallurgy. Smelting, the use of heat to extract metal from ores, may have been discovered accidentally by potters. Kilns are hot enough to form of copper if the malachite and other copper-containing minerals are present during the firing process. Copper is too brittle to be cold hammered, but it could be hot hammered into sheets. Concentrating copper would have required the melting together of smaller pieces. Copper is a relatively soft metal, but it can be cast into tools and weapons. Copper became the starting point for the invention of alloys. This might have been helped out by natural contamination, mistakes (such as confusion caused by the similarity of the flames from copper and arsenic), or scarcity of ores. Whatever the source, it led to the creation of bronze, the metal that ended the Stone Age, in about 3000 B.C.

The first bronze was arsenic-based, but true bronze, an alloy of tin and copper, can be traced to the Sumerians in 2500 B.C. It was initially made by smelting different ores together, rather than by combining pure metals. Bronze is much harder than copper. It was widely adopted and made into weapons, tools, such as axes and scythes, and ornaments.

The dominance of bronze ended with the production of iron, a harder and stronger material. Iron began to replace bronze in about 1200 B.C. Iron oxide was used as flux in the smelting of copper to help the metal agglomerate. As temperatures in kilns were raised to accommodate new ores, this flux would leave iron residues. The first smelting of iron probably occurred in Anatolia, part of modern Turkey, in 2000 B.C. But pure iron is fragile, and the first uses of iron were generally ornamental. A breakthrough occurred with the development of coking, which allowed melting at lower temperatures and provided a harder, more durable version of the metal (really a carbon-iron alloy, steel). Iron smelting appears to have developed independently in both China and sub-Saharan Africa. In fact, there is evidence of smelting near the African Great Lakes as far back as 800 B.C. This technology began spreading throughout sub-Saharan Africa in about A.D. 100 with the migration of Bantu-speaking tribes, and continued until about A.D. 1000.

The working of iron began with wrought iron, which was simply sequential hot hammering, quenching (quick cooling in water to change the crystal structure), and reheating (annealing) of "blooms" (spongy, impure globules of iron). The result was strong, tough, and workable steel. The Iron Age lasted over 1,000 years in Europe, and Iron Age cultures were still dominant in some areas of Africa into the nineteenth century.

Impact

The chief value of metals is derived from their physical properties. Metals are malleable and can be melted together so small amounts can be combined and worked or cast into useful forms. Metals are hard, strong, and can be flexible while resisting permanent distortion, so they can be used for shields, blades, and springs. Different metals and nonmetals can be combined as alloys, blending and transforming their characteristics to customize materials to specific tasks. In recent times, other properties of metals—their ability to conduct electricity, their biological roles (such as in enzymes), their properties with light (paints), their radioactive properties (uranium)—have kept them at the leading edge of our technology and have made the principles of metallurgy discovered in ancient times even more essential.

Metals provide exquisite control over the use of energy. A blade concentrates and directs forces that can be used to plow a field, shave a man's face, or kill an enemy in battle. A spring stores and redistributes mechanical energy. A pipe directs the flow of material. Metal technology also

THE THREE AGES SYSTEM

~

The system of three ages provides historians with a yardstick for measuring ancient and prehistoric levels of technological development according to the materials from which a society makes it tools. Although years are assigned to certain ages, these dates reflect the time when the most advanced civilizations—primarily those of the Near East, India, and China—evolved to the next level of development. Technological evolution was much slower in areas where environmental conditions forced the populace to maintain a subsistence lifestyle.

The Stone Age is divided into Paleolithic or Old Stone Age, which roughly corresponds to the geological Pleistocene Age (1.8 million-10,000 years ago); the Mesolithic or Middle Stone Age, from the end of the last Ice Age to between 8,000 and 6,000 years ago; and the Neolithic or New Stone Age, which began thereafter. Dates for the last of these vary widely: the cultures of the Americas, for instance, did not enter the Neolithic Age until c. 1500 B.C., by which time the Near East had long since entered the Bronze Age. The latter has been divided into an Early Bronze Age (c. 3300-1950 B.C.), Middle Bronze Age (1950-1539 B.C.), and Late Bronze Age (1539-c. 1200 B.C.) Again, these dates are for the Near East: the cultures of the Americas did not begin to use bronze tools until c. A.D. 1100. Finally, the Iron Age is divided into Iron Age I (1200-950 B.C.) and Iron Age II (950-586 B.C.)

Technological development does not always reflect advancement in other areas: thus the Nok people of what is now Nigeria developed ironworking in about 1000 B.C., but lacked a written language or cities. The Aztecs, by contrast, had writing, cities, sophisticated engineering methods, and a highly organized society—but they never entered the Iron Age.

JUDSON KNIGHT

provided the first skills in transformation of materials that led to the development of chemistry. The alchemist's hope of changing lead into gold has been surpassed by our control of materials at the atomic scale, control that reaches back to the lessons of purification, recombination, and qualitative analysis that came from metallurgy. When the concept of chemical elements arrived, metals helped populate the first periodic tables.

A curious side effect of the development of metal alloys was the discovery of the principle of buoyancy. Because it was possible to mix silver with gold, the king of Syracuse asked Archimedes (287?-212 B.C.) to find out if his new crown was really pure gold. Archimedes found that he could determine the volume of the crown by displacing water. A similar volume of pure gold should weigh the same as the crown, so any discrepancy would point to the use of an alloy. The famous cry of "Eureka!" indicated not just that Archimedes had found an answer to a royal problem, but that he had discovered an important principle of physics.

The development of metallurgy had a profound effect upon the environment and the relationship between humans and nature. Wherever iron was introduced, deforestation and an increase in agriculture followed. Mining operations leached acids and toxic minerals, including mercury and arsenic, into nearby water. Waste products fouled the land and the air. The smelting of lead in 150 B.C. Rome produced clouds of toxic gas so extensive that a record of the air pollution is evident today in ice deposits in Greenland.

Metals are of such social and historical significance that two eras are named for them, the Bronze Age and the Iron Age. As an alloy, bronze was the first truly artificial material. With a wide range of characteristics that could be controlled, bronze was used for tools, utensils, and uniquely expressionistic ornaments. Bronze also made the sword possible, the first specialized tool for combat. Before the Bronze Age, warfare was informal and disorganized. With the introduction of bronze, artisans who created weapons and defensive armaments (including shields) came to be. Campaigns of conquest became possible and fortifications were built to defend newly arising cities, trade routes, and the sources of tin and copper ores. Bronze was so versatile and central to economies that, even after effective production methods for iron were developed, it took centuries for the new metal to supplant bronze.

Eventually, iron replaced wood, flint, and stone, as well as bronze. Its use was wider and more intensive than bronze, extending and revolutionizing agriculture and putting high-quality weapons into the hands of large masses of people. Iron resources changed trade routes. In particular, trade between northern Europe and the people of the Mediterranean withered, making the regions more culturally distinct. Iron also forged connections between tribes, and it is in the Iron Age that the roots of most modern European nations are found. Iron enabled large-scale migration, often driven by the march of powerful armies. While the bronze sword was a stabbing tool, the iron sword was a slashing tool, making equestrian warfare possible and allowing extended, large-scale battles. Iron also improved the use and durability of wheels, adding chariots to combat. The first tires were hot bands of iron wrapped around wood that shrunk upon cooling for a tight fit.

Settlements became more permanent during the Iron Age. Both the increased size of human societies and the need for defense led to new roles. For the first time, there is evidence of significant stratification across different cultures, with a well-fed class that did not do extensive, hard labor, and those with more limited diets who regularly took on the backbreaking work. The Iron Age was an age of kings and heroes, and this is reflected in the poetry and religion of the times.

PETER J. ANDREWS

Further Reading

Asimov, Isaac. *Isaac Asimov's Biographical Encyclopedia of Science & Technology.* New York: Doubleday, 1976.

Bisson, Michael S., et al. *Ancient African Metallurgy: The Socio-Cultural Context.* Walnut Creek, CA: Altamira Press, 2000.

Collis, John. *The European Iron Age.* New York: Routledge, 1997.

Ramage, Andrew, and P. T. Craddock. *King Croesus' Gold: Excavations at Sardis and the History of Gold Refining.* Cambridge, MA: Harvard University Press, 2000.

Treister, Michail Yu. *The Role of Metals in Ancient Greek History.* Boston, MA: Brill Academic Publishers, 1997.

The Development of Glassmaking
in the Ancient World

Overview

Glass is an inorganic solid substance. It is usually translucent, hard, and resists natural elements. While glass occurs naturally, it is one of the most important and oldest manufactured materials in the world. Civilization could not exist as it does today without glass. Glass has important uses in science, industry, work, home, play, and art. A world without glass is virtually inconceivable. In ancient times, glass was used in practical and decorative objects, from which its role in society has evolved into the important status it holds today. Glass is currently used in such diverse applications as housewares, clothing, building construction, and telecommunications.

There is much debate regarding the origin of manmade glass, and the exact date it was first manufactured cannot be accurately assessed. Early man used natural glass, such as obsidian, for making sharp tools used for cutting and hunting. It is believed that the first objects manufactured entirely from glass originated in Mesopotamia around 2500 B.C. While most of the artifacts from that time consisted of beads, some of the earliest glassware, in the form of sculpted vessels, still survives into modern times and is speculated to come from Mesopotamia. It has been surmised that they were used for oils and cosmetics of nobility. Glassmaking techniques eventually radiated from Mesopotamia into other geographical areas, bringing the technology to new regions of the world.

Small glass vessels that were sculpted directly from blocks of glass have been found with other Egyptian artifacts, indicating that they had this technology as much as 4,000 years ago. A glass bottle bearing the sign of Thutmose III, a pharaoh of the eighteenth dynasty of Egypt, is on display in the British Museum in London. The Egyptians, who set the early standard in glassmaking, also used glazes of glass to embellish objects made of other material. In time, glassmaking skills spread throughout the known world and became important to many societies.

Background

Perhaps the first notable advancement in glassmaking technology was that of the millefiori process for making open beakers and shallow dishes. The process originated around 100 B.C. in Alexandria and consisted of a shaped core over which sections of colored glass were overlaid. Later, an outer mold was placed over the glass to maintain the shape during the baking process. The pieces of glass fused together in the oven and the glass was then ground smooth. The finished project was both beautiful and practical.

Before the turn of the millennium, the Phoenicians introduced the new technique of blowing glass. They used a form of glass that had a consistency that was well suited for this type of work. In this process, a blowing iron was used to shape and mold molten glass. This was an iron tube about 39-79 in (1-2 m) long consisting of a mouthpiece at one end and a wider knob for holding the soft glass at the other. The craftsman loaded molten glass at one end and rolled it into shape on a hardened surface called a marver. The softened glass could also be blown from the opposite end into a mold or blown freely in the air. Subsequent trials of molding and reheating would allow the craftsman to shape the glass into the desired configuration. He also made use of a solid iron rod, called the pontil, that was used to help sculpt the glass. More complex features, such as a handle, could also be added whenever the craftsman desired.

The technique of blow molding spread rapidly through the known world. Skilled craftsman would readily go wherever they thought there was a suitable market for their skills. Glass engraving took root in Italy where they made cameo glass. This technique involves grinding through an opaque white outside layer to an inner darker region, which leaves a silhouette. The most famous example of this exacting technique is the Portland vase now on display in the British Museum, London. It was the Romans, however, who took the lead in the glassmaking industry.

The Romans used the glass blowing procedure for shaping glass, which made it possible to manufacture low cost, high quality decorative glassware. The Romans were also the first to produce a glass that was relatively clear and free of most impurities. Glass objects were then available to almost all strata of society. They made various objects such as bowls, bottles, and lamps. The Roman artisans took their craft very seriously and their work became the world stan-

Ancient Roman glass vessels. *(Peter Harholdt/Corbis. Reproduced with permission.)*

dard. Glassmaking became such a lucrative field in Rome that all glassmakers paid heavy taxes.

Early forms of glass consisted of three major components: lime, silica, and soda. Any impurities within the mix would cause the glass to become opaque or colored. Both the Romans and Egyptians probably mixed sand, ground seashells, and hardwood ash as sources of silica, lime, and soda, respectively. In order to color the glass, they added various metallic oxides. For example, copper was used to make green or ruby colored glass. It is noteworthy that these techniques depend on precise measurement of the metallic oxides and the early glassmakers were remarkably consistent in their colors and tints.

As successful as the Romans were in many glassmaking areas, they were never successful in producing flat slabs of glass, such as those now used in windows. They could not achieve a transparent pane without an enormous amount of labor to polish and grind the glass. The difficulty in reproducing adequate panes of glass eventually lead to the widespread use of the stained glass window.

The fall of the Roman Empire lead to a decline in glassmaking craftsmanship in the Western world, but the industry continued to thrive in the Near East. Outstanding examples of both highly technical and artistic forms of glass were made in this part of the world through the Renaissance.

Another important advancement in technology, which did not occur until the latter stages of antiquity, was the invention of the bellows. Bellows are mechanical devices that increase the air pressure within the device so that it can expel a jet of air. They usually consist of a hinged container with flexible sides that expand the volume of the container to draw air in and then compresses the volume to expel air out. These devices are used to stoke a fire, which results in an increased speed of combustion so the fire burns hotter.

Impact

The introduction of manufactured glass into society has been beneficial to mankind. We now rely heavily on glass in modern times. In the form of windows, glass has the obvious dual qualities of letting light into a room while protecting that room from extreme weather at the same time. Glass is used in lights, televisions, mirrors, optics, and telecommunications. In many works of art, it is a necessary component. Glass is extremely durable, although subject to breakage with sharp forces. It is nonporous, does not retain odors, and can be completely sterilized. It is absolutely essential to our modern way of life. Those early glassmakers are owed a debt of gratitude for the constant development of their craft.

The principle reasons behind the widespread use of glass is its flexibility in terms of use and manufacture, and the low cost at which it can be produced. Raw materials for the process are inexpensive and abundant so that most objects can be mass produced at a reasonable cost.

The most significant advancement in glassmaking technology was that of blow molding, which had a tremendous impact on society and could be considered one of the greatest technical innovations in history. This enabled humans to mold molten glass into virtually any desired size and shape. Glass ranges in size from extremely small fiber optics (less than 1/100,000 of a meter) to the extremely large reflecting mirror of the Hale telescope (more than 16.4 feet [5 m]).

Glassblowing made possible new commercial applications of glass and resulted in the creation of high quality pieces of art. Much of our modern glassblowing techniques rely on technology that had already been developed by the A.D. 300. Glassblowing gave craftsman remarkable control over their work, yet at the same time yielded an infinite variety of potential shapes and sizes. These pieces could be made at a relatively low cost and helped to set the course for our heavy reliance on an inexpensive, but practical commodity.

The immediate social impact of improved glassmaking was that glass goods, previously available to only the upper class, could be obtained by virtually everyone. Important, useful household items such as bottles and lamps became a mainstay of every home. It also provided those societies that developed and exported glass, an important commodity to trade, with great economic benefits as well. Glass had a tremendous positive influence on earlier societies, but has special relevance in modern times because it is incorporated into all facets of society. It is truly one of the most important manufactured products.

JAMES J. HOFFMANN

Further Reading

Corning Museum of Glass. *A Survey of Glassmaking from Ancient Egypt to the Present.* Chicago: University of Chicago Press, 1977.

Dodsworth, R. *Glass & Glassmaking.* New York: State Mutual Book & Periodical Service, 1990.

McCray, Patrick, ed. *The Prehistory & History of Glassmaking Technology.* Westerville: American Ceramic Society, 1998.

Oppenheim, A. L. *Glass & Glassmaking in Ancient Mesopotamia.* Corning, NY: Corning Museum of Glass, 1988.

Lighting the Ancient World

Overview

Until the nineteenth century—ironically, on the eve of the light bulb's invention—methods of lighting remained more or less unchanged since earliest antiquity. Three forms of lighting existed, in order of their appearance: torches, lamps, and candles, all of which used animal fat or, in the case of lamps in the most advanced ancient societies, vegetable oil. Thus people thousands of years ago rolled back the darkness, not only of night, but of remote places far from the Sun.

Background

In the characteristic abode of prehistoric man, the cave, light remained a necessity at all hours, because typically the Sun's illumination did not penetrate the rocky depths of these homes. Though popular belief pictures fire and the wheel as more or less simultaneous discoveries—give or take a few thousand years—in fact

the wheel only appeared during historic times, whereas man's use of fire stretches back into the earliest recesses of unwritten history.

Among the relatively more recent uses for fire is cooking; but even when humans still gnawed the raw flesh of animals, they required the warmth and light fire provided. Though warmth would seem somewhat more essential to human sustenance than light, in fact it is likely that both functions emerged at about the same time. Once prehistoric man began using fire for warmth, it would have been a relatively short time before these early ancestors comprehended the power of fire to drive out both darkness and the fierce creatures that came with it.

One major step in prehistoric development was the fashioning of portable lighting technology in the form of torches or rudimentary lamps. Torches were probably made by binding together resinous material from trees, though to some extent this is supposition, since the wood material has not survived. By contrast, many hundreds of prehistoric stone lamps have endured.

Paleolithic humans typically used as lamps either stones with natural depressions, or soft rocks—for example, soapstone or steatite—into which they carved depressions by using harder material. Most of the many hundreds of lamps found by archaeologists at sites in southwestern France are made of either limestone or sandstone. The former was a particularly good choice, since it conducts heat poorly; by contrast, lamps made of sandstone, a good conductor of heat, usually had carved handles to protect the hands of the user.

In addition to stone lamps, cave art at La Moute in France shows pear-shaped lamps made from the heads and horns of an ibex, a large wild goat plentiful in the region at that time. It should be noted in this regard that the very existence of prehistoric art, of which the most famous examples may be found in the Lascaux caves of southern France, illustrates the way that artificial lighting changed the world even in those early years. These splendid frescoes, deep in the recesses of caves and impossibly far from the Sun's light, would never have existed if the peoples of prehistoric times had not developed a reliable means of lighting their caves.

Impact

The history of lighting is generally divided into four periods, each of which overlap and that together illustrate the slow pace of change in illumination technology. First was the Primitive, a period that encompasses that of the torches and lamps of prehistoric human beings—though in fact French peasants continued to use the same lighting methods depicted on nearby cave paintings until World War I.

The two more recent stages are Medieval, which saw the development of metal lamps, and the Modern or Invention stage. The latter began with the creation of the glass lantern chimney by Leonardo da Vinci (1452-1519) in 1490, culminated with Thomas Edison's (1847-1931) first practical incandescent bulb in 1879, and continues today. Between Primitive and Medieval, however, was the world of ancient Greece and Rome, the Classical stage, that marks the high point of lighting in antiquity. Earlier ancient civilizations, such as that of Egypt, belong to the Primitive era in lighting—before the relatively widespread adoption of the candle and of vegetable oil as fuel.

It is important to recognize that, while the ancient Egyptians were infinitely more advanced than prehistoric peoples in their demonstrated ability with written language, building, and political organization, in some ways they were still living in the Stone Age. Indeed, Egypt during the Old Kingdom (c. 2650- c. 2150 B.C.), the era in which the pyramids were built, was quite literally just out beyond the Stone Age, since the metalworking of the Bronze Age was in its early stages.

With the emergence of metal tools in the Bronze Age and even more so the Iron Age, which began c. 1200, came the development of the cresset, a bronze basket of wrought iron into which resinous material—pine knots and rich pieces of hardwood—could be placed. This was the case in parts of Europe and other regions heavy with tree growth; but the Egyptians, possessing few trees, had to burn animal greases, and instead of a wrought-iron basket, the typical light fixture in a pharaoh's palace was a wrought-iron bowl. (As would be the case centuries later with the Greeks and Romans, the material from which lamps were made—metal for the wealthy, clay for the poor—served to distinguish classes.)

At various times, the fat of seals, horses, cattle, and fish was used to fuel lamps. (Whale oil, by contrast, entered widespread use only during the nineteenth century.) Primitive humans sometimes lit entire animals—for example, the storm petrel, a bird heavy in fat—to provide light. Even without such cruel excesses, however, animal fat made for a smoky, dangerous, foul-smelling fire.

Though archaeologists in France have found a 20,000-year-old lamp with vegetable-fiber residue inside, the use of vegetable oils for lighting did not take hold until Greek and especially Roman times. The favored variety of fuel among Romans was olive oil with a little salt that dried the oil and helped make the light brighter. Animal oils remained in use, however, among the poor, whose homes often reeked with the odor of castor oil or fish oil. Because virtually all fuels came from edible sources, times of famine usually meant times of darkness as well.

As with the use of vegetable oils, the development of the candle dates to earliest antiquity, but the ancient use of candles only became common—that is, among the richest citizens—in Rome. In its use of animal fat, the candle seems a return to an earlier stage, but the hardened tallow of candles made for a much more stable, relatively safer fuel than the oil of lamps.

Common to candles and lamps was the wick, made of fibers that burned slowly. In a

An ancient oil lamp. *(Bettmann/Corbis. Reproduced with permission.)*

lamp, the wick draws in the liquid fuel, which becomes gas as it burned, and the burning carbon at the end of the wick produces light. With a candle, the heat from the flame at the end of the wick liquefies wax near the wick's base. Capillary action draws the liquid wax upward, where the heat vaporizes it, and the combustion of the vapor produces light.

Even as the wealthiest Romans burned candles or vegetable oil in bronze lamps, and the poorest lit their homes with fish oil in lamps of clay or terra cotta, soldiers and others in need of portable lighting continued to use torches of resinous wood. This was also the case in classical Greece, as one can surmise from a reference in Thucydides (c. 471-401 B.C.) to the use of a torch in burning down the temple of Hera at Argos.

Etruscan tomb paintings at Orvieto, Italy, depict candles, and indeed a piece of candle from the first century A.D. was found by archaeologists at the French town of Vaison. Notable references to candles in classical writing include a description of candle-making by Pliny (c. A.D. 23-79), as well as these lines form *On the City of Rome* by Juvenal (fl. first century A.D.): "...led home only by the Moon / or a small candle, whose wick I tend with care...."

References to lamps, of course, are far more plentiful, as are examples of lamps found in archaeological digs—for instance, at Pompeii, where 90 decorated lamps were preserved by the eruption of Mt. Vesuvius in A.D. 79. Juvenal and Pliny both wrote about lamps, the latter noting that "the wicks made from fibers of the castor plant give a brilliantly clear flame, but the oil burns with a dull light because it is much too thick." Furthermore, there are the many passages in the New Testament that mention lamps: perhaps the most well-known reference to lighting in ancient literature is Jesus's admonition that "no one lights a lamp and hides it in a jar...." (Luke 8:16, NIV).

JUDSON KNIGHT

Further Reading

Books

Faraday, Michael. *The Chemical History of a Candle* (reprint of 1861 volume). Atlanta: Cherokee, 1993.

Forbes, R. J. *Studies in Ancient Technology.* Leiden, Netherlands: E. J. Brill, 1955.

Phillips, Gordon. *Seven Centuries of Light: The Tallow Chandlers Company.* Cambridge, England: Granta Editions, 1999.

Internet Sites

"An Appreciation of Early Lighting." *Cir-Kit Concepts.* http://www.cir-kitconcepts.com/EarlyLighting.html (November 16, 2000).

McElreath, Elizabeth F., and Regina Webster. "Artificial Light in Ancient Rome." *University of North Carolina.* http://www.unc.edu/courses/rometech/public/content/

arts_and_crafts/Libba_McElreath/artificial_light_in_r ome.html (November 16, 2000).

Pressley, Benjamin. "Conquering the Darkness: Primitive Lighting Methods." http://www.hollowtop.com/spt_ html/lighting.htm (November 16, 2000).

The Calendar Takes Shape in Mesopotamia

Overview

The calendar used today in the West has its roots in the system developed by the astronomers of Mesopotamia—and particularly the Mesopotamian civilization of Babylonia—during the period from the third to first millennium before the Christian era. Other civilizations created their own calendars with varying degrees of accuracy, but it is from Mesopotamia that the concept of the year, month, and day each gained their most consistent and lasting definition. A fourth means of marking time, the week, may also be traced (if perhaps indirectly) to Babylonia.

Background

Practically from the beginning of recorded time, men recognized that a year lasts about 360 days, a number still reflected in the use of a 360° circle among mathematicians and astronomers today. This may also have influenced the Babylonian adaptation of a sexagesimal, or base-60, number system (as opposed to the base-10 system used by Westerners today) in c. 2700 B.C.

The figure of 360 represented a mean, or close to one, between the length of the lunar and solar calendars. As its name implies, a lunar calendar is based on the Moon's revolutions around Earth, of which there are 12 during a solar year. With an average of 29.53 days per synodic month—*synodic* being a term that refers to the conjunction of celestial bodies, in this case the Moon and the Sun—the lunar calendar lasts about 354.37 days. The adoption of a lunar calendar makes sense on a short-term basis, but over a longer period it soon gets out of phases with the seasons. This explains why today the only major lunar calendar in use is that of the Muslim Middle East, a region that experiences little seasonal variation in climate.

Problems with the lunar calendar influenced the development of a second form, the luni-solar calendar. According to this system, most years consist of 12 months, but every few years it becomes necessary to insert a thirteenth month—a process known as intercalation—to keep the calendar in phase with the seasons. The Chinese calendar in use throughout much of East Asia today, as well as the Jewish or Hebrew calendar, are surviving examples of the luni-solar method. The remainder of the world, however, uses a solar calendar that has its origins in Rome. Yet elements of the Roman, Jewish, and Islamic calendars can be traced to foundations established by the astronomers of Babylonia and other Mesopotamian civilizations.

Impact

Babylonian achievements in astronomy resulted from an interest in astrology, and indeed Mesopotamian star-gazing always had a religious component. These astrologers were perhaps the first to equate heavenly bodies with deities: thus the Moon was called Sin, a goddess first worshipped by the Sumerians. In this vein it is interesting to note the close relation between early Babylonian nomenclature for heavenly bodies, and the terms later used by the Romans.

The latter, for instance, associated the Sun with Apollo, who drove across the sky in a fiery chariot; in Babylon it was Shamash who performed the same function, and his became the Babylonian name for the Sun. This similarity may also be found in the names for planets still used today: Venus was a fertility goddess, as was Ishtar, the name used by the Babylonians for that planet. Likewise Marduk was king of the gods, and the Babylonians applied this name to the largest of the solar system's planets, known today as Jupiter.

These facts establish two key points: later astronomers' indebtedness to the Babylonians, and the close relationship between religion and the beginnings of astronomy. But the calendar also had numerous down-to-earth applications, of course, reflected in Mesopotamian methods for dividing the year. Early astronomers of the region simply divided the solar year into two seasons, roughly equivalent to spring-summer and fall-winter. Because Assyria was further north, it became natural to recognize a third season, and still further north, in the Hittite civilization of Anatolia (modern Turkey), astronomers divided the year into four seasons that reflected the cycles of planting and harvesting.

Then there was the idea of the month, which began at the first sign of the new Moon. This practice of reckoning months had already become common by the third millennium B.C., but month names were not standardized. Each city had its own name for months, and sometimes several names, and by the twenty-seventh century B.C. Sumerians began reckoning dates according to the ruler in power. This is a practice familiar to anyone who has read the Old Testament, which is full of passages that begin "On the ____ day of the ____ month in the ____ year of the reign of ____."

Driven by the practical need for a luni-solar year that encompassed the entire agricultural cycle, Sumerian scribes in about 2400 B.C. first adopted a 360-day year consisting of 12 30-day months. According to this system, the financial (what modern people would call "fiscal") year began two months after barley-cutting, when it was time to begin settling accounts. The barley harvest itself marked the beginning of the agricultural year, and since the Sumerians associated good harvests with good leadership both in a political and religious sense, it was a natural step to make this the beginning of the royal year as well.

Thus at the beginning of the year, the ruler offered the first fruits of the harvest to the gods, ensuring their continued favor. Political considerations also influenced the naming of years, which were not numbered except—as noted above—in terms of the ruler in power. Thus if something notable happened during the "____ year of ____'s reign," it also became, for instance, "the year in which ____ built the temple of Inana." By the seventeenth century B.C., however, the Babylonians had standardized year nomenclature, counting regnal years in the form later adapted by the authors of the Bible.

Earlier, in the eighteenth century B.C., the Babylonians under Hammurabi (r. 1792-1750 B.C.) standardized the lunar calendar that had been in use among various Mesopotamian civilizations for about four centuries. The Babylonian year began in the spring on the first day of the month of Nisanu, and after the seventeenth century B.C., the period between the time when a ruler assumed power and Nisanu 1 was referred to as "the beginning of the kingship of ____."

The adoption of the lunar calendar eventually resulted in the need for an intercalated month. This, too, had been a feature of the earliest lunar calendars of the twenty-first century B.C., but its implementation had been rather erratic. Each of the various Sumerian city-states used their own intercalation systems, resulting in enormous confusion. The foundation of multinational empires by the Babylonians, Assyrians, and still later the Persians helped lead to standardization of this system through directives from above. By about 380 B.C., Persian emperors had ensured that the lunar and solar calendars were more or less aligned.

The Babylonian system that came to prevail throughout the Near East consisted of 12 basic months: Nisanu, Ayaru, Simanu, Du'uzu, Abu, Ululu, Tashritu, Arakhsamna, Kislimu, Tebetu, Shabatu, and Adaru. Every 19 solar years, or 235 lunar months, marked an entire luni-solar cycle, which required intercalation in years 3, 6, 8, 11, 14, 17, and 19. In all but the seventeenth year, the added month was called Adaru II, but on the seventeenth year it was Ululu II.

This system may seem highly complicated to a modern observer, but it worked for many centuries, and the Israelites' exposure to Babylonian culture during the Captivity (587-539 B.C.) ensured its lasting impact on the Hebrew calendar. The Bible in turn reflects the Mesopotamian and particularly Babylonian influence on a number of particulars—including, as perhaps shown in the opening chapters of Genesis, the concept of the week. It is not clear, however, whether the modern idea of a week has its roots in the Near East or in a separate system that evolved in Rome.

There, an eight-day market cycle had long prevailed, and by the second century B.C. this became a seven-day cycle with days referring to the gods and their ruling planets: Saturn, the Sun, the Moon, Mars, Mercury, Jupiter, and Venus. The week-day names used by the French and other speakers of Romance languages still reflect the Roman influence, whereas Germanic languages such as English use a combination of Roman and Norse terms. Thus Tiu, Woden, Thor, and Freya replaced Mars, Mercury, Jupiter, and Venus to lend their names to Tuesday, Wednesday, Thursday, and Friday respectively.

Likewise the modern form of the solar calendar more clearly shows the impact of Roman civilization, from whence comes the present system of months, as well as the designation of January 1 as the first day of the year. Yet long before Rome, the Babylonians established the basic idea of a 365-day year divided into 12 months of approximately 30 days apiece, and Roman calendar-makers built on a Mesopotamian foundation established centuries before.

JUDSON KNIGHT

Further Reading

Books

Moss, Carol. *Science in Ancient Mesopotamia.* New York: F. Watts, 1998.

Neugebauer, Otto. *The Exact Sciences in Antiquity.* New York: Dover Press, 1968.

Internet Sites

"The Babylonian Calendar." http://ourworld.compuserve.com/homepages/khagen/Babylon.html (December 3, 2000).

"Calendars." http://www.freisian.com/calendar.htm (December 3, 2000).

Harper, David. *A Brief History of the Calendar.* http://www.obliquity.com/calendar (December 3, 2000).

The First Clocks

Overview

Water clocks and sundials were the first artificial measures of time. They allowed people to see time in an abstract way, apart from nature, and also helped create an objective, shared view of time that facilitated social cooperation. Combined with other measurements, such as those of space and weight, timekeeping devices ultimately provided the basis for science and contributed to new ways of understanding and controlling nature.

Background

The first timepiece was probably a stick in the ground whose shadow showed the progress of the sun across the sky. Building on this principle were shadow clocks, or *gnomon* (Greek for "pointer"), which date from around 3500 B.C. This simple measurement, however, was soon insufficient. As humans established settlements and their societies grew, a common estimate of time became necessary for religious observances and to prompt or coordinate tasks, such as the time animals needed to be milked.

Sundials, which assign a numerical value to the position of a shadow cast by the Sun, were a further refinement in time measurement. Although first used simply to determine the local noon (the point during the day when the Sun is highest), early sundials improved upon shadow clocks by dividing the day into twelve equal periods. The sundial's pointer, however, is still called a *gnomon.*

The first hemispheric sundial, the kind familiar to us today, is attributed to the Chaldean astronomer Berrosus sometime around 300 B.C. It was little more than a bowl-like indentation carved into a cube of stone or wood. A pointer stood in the center of the bowl, creating the shadow that traveled arcs of varying lengths (to compensate for seasonal variations) that were carved into the surface. Each arc was divided into 12 hours. (These were called "temporary" or "temporal" hours because varied in length from summer to winter.) In 30 B.C., the Roman engineer Vitruvius described three types of sundials that were in common use. The most sophisticated took into account the varying length of days throughout the year, and could be adjusted according to the length of the noon shadow, which is shortest in the summer and longest in the fall. Dials were etched onto cones or inside bowls to increase their precision.

Of course, sundials require sunlight, making them unusable in interior rooms, on cloudy days, or at night (although a device known as a *merkhet,* used as early as 600 B.C., measured the night hours by tracking stars). Hourglasses, which use the regular, controlled action of gravity, were another way to measure time. Sand flowing through a narrow opening is a simple, effective way to measure small, specific units of time. They work anywhere, at any time, and units of time are independent of the Sun's variable motion. The hours of a water clock (or *clepsydra,* Greek for "water thief") were similarly fixed. However, people weren't used to hourglass or water clock time, whose hours were the same length in every season. A key development in water clocks was the ability to emulate the sundial and vary the hours of each day.

The basic mechanism of a water clock is the regular emptying or filling of a marked vessel by steadily dripping water. The oldest-known model was found in Amenhotep's (1353-36 B.C.) tomb. More sophisticated versions used floats as indicators, siphons to automatically recharge the source vessel, ringing bells, hour hands that turn, and even elaborate gears (at least as early as 270 B.C.). From the beginning, of course, they made a distinctive regular dripping sound that would evolve into the ticking of our own clocks.

Impact

With the sundial and the water clock, the essentials of timekeeping—a regular, repeated process, a means of tracking that process, and a way to display the results—had been introduced to human culture. This had profound social consequences. Timekeeping devices facilitated co-operation, providing appointed times for religious observances, work and community activities. They enabled bureaucracies and, in Athens, became a measure of fairness, allocating time for debates and court proceedings. Until the advent of the pendulum clock, sundials in particular were equated with time itself and were found not just in public areas, but in houses, baths, temples and even tombs. Portable sundials were carried as late as the seventeenth century. King Charles I of Great Britain left his pocket sundial to his son on the eve of his execution in 1649.

One of the most powerful expressions of the social significance of timing devices was the Tower of the Winds in the Agora, the main marketplace in Athens. Built in the first century B.C., a sundial was fixed on each of side of this octagonal tower. It also indicated wind direction, displayed the season of the year and the astrological date, and contained an elaborate 24-hour water clock.

For all the benefit and efficiency sundials and clocks brought, there was a dark side. The dictatorship of the clock was lamented as early as the second century B.C. in a poem written by the Roman poet and playwright Plautus. He asks the gods to confound the inventor of the sundial and "confound him, too, who in this place set up a sundial to cut and hack my days so wretchedly." Clocks told Plautus when to eat, even though his stomach is a better guide. Over 2,000 years later, Mark Twain also reflected on the artificial rule of the clock when he observed "Man is the only animal who goes to bed when he's not sleepy and gets up when he is."

Besides marking time, sundials also revealed nature in new ways. People had long known that the length of days changed throughout the year. With the sundial, they realized how much.

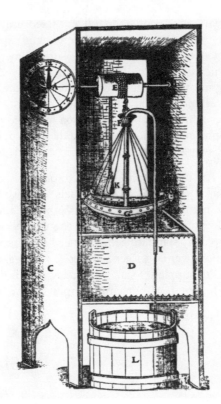

A drawing of a Roman water clock. *(Bettmann/Corbis. Reproduced with permission.)*

These variables in the length of hours and the changing angle of the Sun throughout the year gave them clues about the true nature of the world, which eventually led to a deeper understanding of the heavens. Sundials also encouraged a more careful study of natural phenomena and helped develop ever-more-precise measurement. This, in turn, helped develop a quantitative, scientific approach to knowledge. In fact, timekeeping, combined with other measurements, such as those of volume and mass, ultimately provided the basis for science in the modern sense. Galileo himself used a mercury clepsydra to time the motion of falling bodies. Ultimately, the measurement of time also made navigation and exploration more practical, expanding our view of the world.

Today, sundials give us a gentler view of time, revealing a connection between humans and nature. They are found in parks, forums, and gardens, where they provide a quiet reminder of the past and an aesthetically pleasing sense of order.

Water clocks were the first mechanical devices, the precursors of automation and the applied use of energy. They allowed ancient scientists to understand the regulation, transfer, and use of energy, and also helped establish the essential concept of standardized, fixed units of time that allowed repeated, consistent experimentation. The very idea of measuring time mechanically led to clocks and watches driven by weights and springs, which led to still more advanced mechanical and electrical devices. Ultimately, this mechanical view even informed philosophy and religion, providing the concept of a "clockwork" universe.

The use of timing devices to organize society and synchronize individuals' activity has only intensified over the ages. With the coming of the railroad, the hours of the day became completely standardized. The need for synchronization over larger geographies caused regional times to supplant local ones. Today, time is digital. It ties together our computers and our economies worldwide, with "24–7" the proud boast of every dotcom.

PETER J. ANDREWS

Further Reading

Asimov, Isaac. *Isaac Asimov's Biographical Encyclopedia of Science & Technology*. New York: Doubleday and Co., 1976.

Barnett, Jo Ellen. *Time's Pendulum: From Sundials to Atomic Clocks, the Fascinating History of Timekeeping and How Our Discoveries Changed the World*. Chestnut Hill, Mass.: Harvest Books, 1999.

Landes, David S. *Revolution in Time: Clocks and the Making of the Modern World*. Cambridge, Mass.: Harvard University Press, 2000.

Rhor, R.J., and Gabriel Godin. *Sundials : History, Theory, and Practice*. Mineola, NY: Dover Pubns, 1996.

Slave Labor

Overview

Through most of human history societies have used slavery as a form of labor. The ancient Greeks, Native American tribes, the Roman Empire, the ancient Egyptians, Nazi Germany, and even European nations of the Renaissance and the Enlightenment were slave-owning societies. In fact, it is only very recently in human history that slavery has become illegal in most of the world. The questions to ask are why slavery has been so widespread through history and how this affected the societies that used and owned these slaves.

Background

Slavery is one of the most reviled institutions in human history. Slaves are owned, they have no personal freedom, and are often asked to do menial or backbreaking work with no pay other than their food and shelter. Slavery has also been with humanity throughout recorded history and, likely, much earlier. It was largely accepted by the Christian, Muslim, and Jewish religions, and, in some societies, over half the population consisted of slaves.

People entered into slavery in myriad ways. Most frequently, slaves were captured enemy soldiers or enemy populations. In the minds of the conquering soldiers, this was appropriate— by enslaving enemy soldiers, they could not be formed into another army, and they could help to work the fields of those killed in the wars. To a large extent, the Roman armies were possible only because of the presence of slave labor that freed Roman landowners for military duty.

People also entered slavery through slave merchants. In many societies, captured citizens and soldiers would be sold to slave traders instead of being pressed into labor by the victors. This mechanism fed much of the African slave trade and was, in fact, the primary method for finding slaves to work in the New World.

Finally, some sold themselves or their family members into slavery to pay debts. If a family was deeply in debt with no hope of paying it off, it was possible to sell a child, a spouse, or even oneself into slavery to pay off these debts. In some cases, this at least gave the slave a place to sleep and meals to eat, albeit at the cost of personal freedom.

Typically, slaves were used to perform tasks that were laborious, distasteful, of that simply would not have been possible using only paid or volunteer labor. Thus, the massive constructions of the Inca and Maya were almost certainly built with slave labor, and of course the agriculture of the American South and throughout the Caribbean was built on the backs of slaves, mostly from Africa.

Slave labor provided much of the power in Egyptian society. *(Archive Photos. Reproduced with permission.)*

Impact

Although slavery is most often associated with the American South, it has been a nearly universal institution in most human societies. Although slaves everywhere were largely looked upon as being subhuman, in many societies slaves had the option of manumission, that is, eventually becoming free. Slavery also impacted societies in many respects, and these impacts were not always the same for all slave societies. In particular, slave labor made possible national feats not otherwise realistic, such as large construction projects or waging large-scale wars. Slavery also served as a source of income for some poor societies, and in some cases, slavery supported a large leisure class. Finally, slavery could not help but be a reflection on the values of the societies that owned them, for any society that condoned the involuntary servitude of large numbers of people seems less likely to value human life or to give equal rights to all its citizens.

Slavery made possible the wars of Rome, the massive constructions of many societies, and the intellectual freedom for which ancient Greece is justly famed. As noted above, Roman and Greek armies captured enemy soldiers and civilians during their frequent wars, and these captives were put to work. In Rome, they farmed fields, freeing Roman citizens for other work, particularly for fighting in the wars that expanded and defended the empire. Through Rome's centuries of growth and imperialism, slave labor was an essential factor in the equation that led to Rome's domination of most of the known world. In addition to their agricultural role, slaves also helped build the famed Roman roads, many of Rome's monuments and monumental buildings, and to conduct many of the other tasks of the empire. In fact, there is some evidence that Rome began to falter only after freeing many of its slaves, though it is doubtful that freeing its slaves was solely responsible for this. There is also strong evidence that contributing to Rome's fall was the expansion of empire, leading to a reduction in the flow of slaves and a concomitant reduction in cheap labor.

Although Greece is better known for intellectual rather than political dominance, the Greeks also built a substantial political empire and fought many wars against Persia and other rivals. These wars provided slaves, who were put to work doing a great deal of the manual labor necessary to sustain their society. Where Rome's energies went into expansion, Greece placed a high premium on the intellect. Slave labor made possible the free time necessary to build the incredible intellectual achievements for which Greece is still famed. As with the Roman roads, there is little doubt that the Greek intellectual tradition would be much less remarkable without the indirect benefits of slavery.

Other societies made similar use of slave labor. The Chinese used slaves for agriculture and construction, and many Native American tribes used slaves to till their fields. African tribes used slaves internally as well as selling them to Arab merchants, the Inca used slave labor to build their roads and temples, and the Maya also used slaves for agriculture, construction, and as human sacrifices.

Several societies have used slavery for direct economic gain. Of these, the African tribes are perhaps the best known. For centuries, many tribes sold slaves to Arab traders. At first, many of these slaves were used in the courts of the Arab empire, and most of the slaves sold were captured in wars or raids with neighboring tribes. This dynamic changed, however, with the growth of agriculture in the American South and the increasing demand for slave labor to help this expansion. In addition, the colonization and exploitation of the Caribbean and Latin America demanded more slave labor, to the extent that many of these islands and nations are now populated largely with the descendents of former slaves. This huge demand led to higher prices for slaves that, in turn, led to increased pressure by Arab slave traders on the tribes supplying them. Over time, some tribes began waging war solely for the purpose of obtaining slaves, and some even sold their own members into slavery. By the time the North American slave trade was outlawed, many African tribes had been decimated by the slave trade, changing much of the power and political structure in those parts of Africa.

It is interesting to note that, before the slave trade began, many African tribes would take only women and children as slaves, executing the men who were viewed as a possible threat. Thus, ironically, the demand for male slaves in the New World may have saved the lives of many men who would otherwise have been killed. This cannot, however, be seen as justifying the slave trade in any way. It must also be noted that the Africans were not the only society to sell slaves; they were simply the most recent and the best known. Similarly, the American colonies and the slave-holding society in the American South was not the only society to purchase slaves; they were simply the best documented. Both cases are just among the more recent in a long line of societies based on slavery.

Slavery had more impact on societies than simply providing labor and revenue. Slavery reflected on the values of the society, and the nature of slavery changed between societies. Although slavery was never a blessing, in some societies it was far less pernicious than in others.

In ancient China, for example, it was not uncommon for people to sell themselves or family members into slavery if they were overwhelmed with debt. In this case, the slaves were apparently viewed with somewhat less disdain than in those societies in which slaves were captured or purchased abroad. In addition, Chinese slaves could, and often did, purchase their freedom through work, and many slaves reentered "mainstream" society. Similarly, many Roman and Greek slaves could purchase their freedom and seem to have been integrated into society with little discrimination.

In fact, it is mostly those societies that enslaved members of other races in which discrimination seems to have been important. Europeans who were enslaved by other Europeans, African slaves in Africa, and Asian slaves in the Orient all seem to have had a great deal of mobility in moving from slave status to freedom. However, African slaves held by Arab courts and African slaves in the American colonies were subject to much more discrimination and more attempts were made to disenfranchise them following their freedom. Part of this might have been because slaves of another race are so obviously different. For example, a Chinese slaveowner may, perhaps, have more easily sympathized with his slaves because he could picture himself in the slaves' place. When the face you see looks like yours or your family's, perhaps despotism is more difficult. On the other hand, no Southern slaveholder could see himself in the face of an African slave. This may have made is easier to deny the slave's humanity and the equality, and it certainly made it easier to discriminate against former slaves who had either escaped or been freed.

This notwithstanding, condoning and relying on slavery to build a society reflects certain societal values that, to most of us today, seem almost incomprehensible. It is hard for many of us to understand how any society can place so little value on humanity that it would allow the ownership of fellow human beings. In some cases, this lack of comprehension reflects modern values that are being applied to ancient societies. To some extent, life *was* less valuable then. In the words of one philosopher, life was "nasty, brutish, and short," and slavery was simply seen as another unpleasant aspect of an unpleasant life. In some cases, too, slavery was seen as preferable to a life of poverty, starvation, and disease. Slaves had no freedom and were worked brutally, but they were at least alive, fed, and housed. This does not condone slavery, but it does help to explain why, in the ancient world, it was not viewed with the same degree of revulsion we have today. Perhaps, in a sense, the values of the societies that owned slaves can be placed in this perspective.

Finally, it must be noted that, in modern times, the economics of slavery have been shown to be unfavorable. There has been some argument that, even without the Civil War, the American slave-holding states would have had to stop the practice because of its economic inefficiency. There is some evidence that reliance on slave labor contributed to the South's reluctance to invest in agricultural machinery and, as a result, contributed to the industrial and economic inefficiency that led to the South's defeat in the American Civil War. There is also compelling evidence that, in spite of the difference in direct costs, slavery is inherently less efficient than the paid labor of free workers, who are more likely to work more willingly. Nevertheless, projecting these trends into the distant past, when technology, society, and economics were completely different, may not prove particularly useful.

P. ANDREW KARAM

Further Reading

Rodriguez, Junius P. *Chronology of World Slavery*. Santa Barbara, CA: ABC-CLIO, 1999.

Archimedes and the Simple Machines That Moved the World

Overview

"Give me a place to stand," Archimedes is said to have promised, "and I will move the world." In this perhaps apocryphal quote, the Greek mathematician, scientist, and inventor was discussing the principle of the lever and fulcrum, but he could very well have been describing his whole career. In addition to his mathematical studies and his work on buoyancy, Archimedes contributed to knowledge concerning at least three of the five simple machines—winch, pulley, lever, wedge, and screw—known to antiquity. His studies greatly enhanced knowledge concerning the way things work, and his practical applications remain vital today; thus he is aptly named the "father of experimental science."

Background

Born in the Greek town of Syracuse in Sicily, Archimedes (287?-212 B.C.) was related to one of that city's kings, Hieron II (308?-216 B.C.). Son of an astronomer named Phidias, he went to Alexandria in around 250 B.C. to study under Conon and other mathematicians who had been disciples of Euclid (330?-260? B.C.). He later returned to his hometown, where he lived the remainder of his life.

Though he contributed greatly to understanding of the lever, screw, and pulley,

Archimedes did not invent any of these machines. Of these three, the lever is perhaps the oldest, having been used in some form for centuries prior to his writings on the subject. Actually, the more proper name for this simple machine is "lever and fulcrum," since the lever depends on the fulcrum as a pivot. The simplest example of this machine in operation would be the use of a crowbar (lever) balanced on a block of wood (fulcrum), which greatly increases the lifting ability of the operator.

Levers appeared as early as 5000 B.C. in the form of a simple balance scale, and within a few thousand years workers in the Near East and India were using a crane-like lever called the *shaduf* to lift containers of water. Archimedes's contribution lay in his explanation of the lever's properties, and in his broadened application of the device. Similarly, he used the screw principle to improve on the shaduf and other rudimentary pumping devices.

The shaduf, first used in Mesopotamia in about 3000 B.C., consisted of a long wooden lever that pivoted on two upright posts. At one end of the lever was a counterweight, and at the other a pole with a bucket attached. The operator pushed down on the pole to fill the bucket with water, then used the counterweight to assist in lifting the bucket. By about 500 B.C., other

The Archimedean screw lifted water from one level to another. *(Bettmann/Corbis. Reproduced with permission.)*

water-lifting devices, such as the water wheel, had come into use.

Another water-lifting device was a bucket chain using a pulley, which is believed to have provided the means of watering the Hanging Gardens of Babylon. Archimedes, for his part, applied the screw principle to the pump, and greatly improved the use of the pulley for lifting. The pulley, too, was ancient in origin: though the first crane device dates to about 1000 B.C., pictorial evidence suggests that pulleys may have been in use as early as the ninth millennium B.C.

Impact

Returning now to the topic of the lever, it should be noted that Archimedes was first of all a mathematician and physicist, and secondarily an inventor. Not only was that his role in history, but that was how he saw himself: like virtually all great thinkers of the Greek and Roman worlds, he viewed the role of the practical scientist as on a level with that of the artisan—and since most artisans were slaves, he considered applied science as something infinitely less noble than pure science. This is of course an irony, given his great contributions to applied science, but it is also essential

to understanding his work on the lever and other machines. In each case, his practical contributions sprang from a theoretical explanation.

Where the lever was concerned, Archimedes explained the underlying ratios of force, load, and distance from the fulcrum point, and provided a law governing the use of levers. In Archimedes's formulation, the effort arm was equal to the distance from the fulcrum to the point of applied effort, and the load arm equal to the distance from the fulcrum to the center of the load weight. Thus established, effort multiplied by the length of the effort arm is equal to the load multiplied by the length of the load arm—meaning that the longer the effort end, the less the force required to raise the load. Simply put, if one is trying to lift a particularly heavy stone, it is best to use a longer crow bar, and to place the fulcrum as close as possible to the stone or load.

Some three centuries after Archimedes, Hero of Alexandria (first century A.D.) expanded on his laws concerning levers. Then in 1743 John Wyatt (1700-1766) introduced the idea of the compound lever, in which two or more levers work together to further reduce effort—a principle illustrated in the operation of the nail clipper. Physicists have also applied Archimedes's laws on the operation of levers to situations in which the fulcrum rests beyond the load (as with the wheelbarrow, whose wheel serves as the fulcrum), or beyond the effort (as with tongs, in which the elbow joint serves as fulcrum).

With regard to the screw, Archimedes provided a theoretical underpinning, in this case with a formula for a simple spiral, and translated this into the highly practical Archimedes screw, a device for lifting water. The invention consists of a metal pipe in a corkscrew shape that draws water upward as it revolves. It proved particularly useful for lifting water from the hold of a ship, though in many countries today it remains in use as a simple pump for drawing water out of the ground.

Some historians maintain that Archimedes did not invent the screw-type pump, but rather saw an example of it in Egypt. In any case, he developed a practical version of the device, and it soon gained application throughout the ancient world. Archaeologists discovered a screw-driven olive press in the ruins of Pompeii, destroyed by the eruption of Mount Vesuvius in A.D. 79, and Hero later mentioned the use of a screw-type machine in his *Mechanica*. Certainly the screw is a widely used device in modern times, and though its invention cannot be attributed to Archimedes, it is certain that he influenced the broadening of its applications. Thus in 1838, when the Swedish-American engineer John Ericsson (1803-1899) demonstrated the use of a screw-driven ship's propeller, he did so on a craft he named the *Archimedes*.

Again, in the case of the pulley, Archimedes improved on an established form of technology by providing a theoretical explanation. He showed that a pulley, which may be defined as any wheel supporting a rope or other form of cable for the transferring of motion and energy, operates according to much the same principle as a lever—that is, the pulley provides the operator with a mechanical advantage by reducing the effort required to move the object.

A single pulley provides little mechanical advantage, but by about 400 B.C. the Greeks had put to use compound pulleys, or ones that contained several wheels. Again, Archimedes perfected the existing technology, creating the first fully realized block-and-tackle system using compound pulleys and cranes. This he demonstrated, according to one story, by moving a fully loaded ship single-handedly while remaining seated some distance away. In the late modern era, compound pulley systems would find application in such everyday devices as elevators and escalators.

Archimedes's studies in fluid mechanics gave rise to the most famous story associated with him. It was said that while trying to weigh the gold in the king's crown, Archimedes discovered the principle of buoyancy: when an object is placed in water, it loses exactly as much weight as the weight of the water it has displaced. Supposedly he made his discovery in the bath, and was so excited that he ran naked through the streets of Syracuse shouting "Eureka!" or "I have found it." Again, the story itself may be apocryphal, but the application is very real: thanks to Archimedes's principle, shipbuilders understood that a boat should have a large enough volume to displace enough water to balance its weight.

In the realm of mathematics, Archimedes developed the first reliable figure for π, and in his work with curved surfaces used a method similar to calculus, which would only be developed some 2,000 years later by Isaac Newton (1642-1727) and Gottfried Wilhelm Leibniz (1646-1716). As an astronomer, he developed an incredibly accurate self-moving model of the Sun, Moon, and constellations, which even showed eclipses in a time-lapse manner. The model used a system of screws and pulleys to move the globes at various speeds and on different courses. In addition, he conducted important studies on gravity, balance, and equilibrium that grew out of his work with levers.

During the Second Punic War (218-201 B.C.), Archimedes worked as a military engineer for Syracuse in its efforts against the Romans and either invented or improved upon a device that would remain one of the most important forms of warfare technology for almost two millennia: the catapult. He is also said to have created a set of lenses that, using the light of the Sun, could set ships on fire at a distance. But Archimedes may have been a bit too successful in his wartime efforts: he was killed by a Roman soldier, no doubt as retribution, when Rome took Syracuse.

Archimedes remains one of the towering figures both of pure and applied science. He developed the three-step process of trial and experimentation that helped form the basis for scientific work in subsequent centuries: first, that principles continue to work even with large changes in the size of application; second, that mechanical toys and laboratory experiments may yield practical applications; and third, that a rational, step-by-step logic must be applied in solving mechanical problems and designing equipment. In so doing, he created the machines that transformed the world, and his impact remains powerful today.

JUDSON KNIGHT

Further Reading

Bendick, Jeanne. *Archimedes and the Door of Science*. Warsaw, ND: Bethlehem Books, 1995.

Lafferty, Peter. *Archimedes*. New York: Bookwright, 1991.

Stein, Sherman K. *Archimedes: What Did He Do Besides Cry Eureka?* Washington, DC: Mathematical Association of America, 1999.

The Chinese Invent the Magnetic Compass

Overview

Sometime before the fourth century B.C. the Chinese noticed that certain minerals, properly prepared, always pointed to the south. The mineral was magnetite, commonly called lodestone, and it was made into what the Chinese called a "south pointer." For many centuries, these were used primarily for mystical purposes. By the eleventh century A.D., the Chinese had turned their south pointer into a magnetic compass, and a century later this device spread to both Europe and the Islamic world. The magnetic compass would eventually find its most significant use as an aide to naval navigation during the late Middle Ages.

Background

The Earth has a planetary magnetic field, a fact known by every Boy Scout, hiker, and sailor. Like most planets, the geomagnetic field is roughly aligned with the planet's rotational axis with the result that magnetic north points towards the North Star, above the North Pole in the sky.

Due to the physics of magnetic fields, any magnetized objects are attracted to each other. Any magnetized object has what is called polarity; it has a north and a south magnetic pole and the magnetic field lines of force go through space between the two poles. Opposite poles are attracted to one another, so a south magnetic pole is attracted to a north magnetic pole and vice versa. Anyone playing with magnets witnesses this when they notice that, with one orientation the magnets stick together while, flipped, they repel one another.

Compasses are no different. The simplest compass is merely a magnet that is allowed to hang or float freely so that it can align itself with the Earth's magnetic field. The compass' south pole will try to connect with the Earth's north magnetic pole and so it will swing to the north if allowed, while the north pole of the compass will swing to the south. So, if we take a magnet shaped like a needle, the needle will align itself with the Earth's magnetic field and will point north and south.

Sometime before the fourth century B.C. an unknown Chinese person noticed that the mineral magnetite is naturally magnetic. How this was noticed may never be known, but it can be surmised that the magnetite was spontaneously attracted to another piece of magnetite or to an iron implement. In any event, this property is one of the defining characteristics of magnetite, and large pieces of pure magnetite are quite magnetic. Some time later (again, it is not certain when), it was noticed that pieces of magnetite would spontaneously line themselves up to point to north and south and that implements made of magnetite (such as needles) would do this, too. Special conditions were required; specifically, the magnetite had to be freely suspended on a silk thread or floating on a card in water. In some cases, the magnetite was hammered flat and floated on the water itself, staying afloat with surface tension. There is no doubt that, by the fourth century B.C. this phenomenon was so well know as to be taken quite for granted in surviving Chinese texts from that time.

At first, the Chinese formed the magnetite into shapes that could be used to point south (the Chinese preference). These shapes included needles, fish, turtles, and even spoons. In fact, one of the most common shapes was the spoon, resembling the Big Dipper, whose "pointer stars" also point to the north. Still later, the Chinese also discovered that magnetite could be used to magnetize steel and that steel needles could also be used in a compass.

Impact

For over 1,000 years the compass was used primarily to help align houses and other buildings in accordance with the Chinese practices thought to bring luck and positive energy. It was thought that lining buildings up in exact directions would allow energy to flow more easily into a house, bringing good fortune to those living and working within. It may have been thought that the forces acting to align a compass with north and south were the same the builders were trying to harness, but this is only speculation. However, it is certain that the first use to which magnetic compasses were put was geomancy.

These first devices were shaped like spoons, formed from pure magnetite with the handle being the north magnetic pole of the mineral. When set on a bronze platter marked with the compass directions, the spoon would slowly turn so the handle would point directly south, giving builders an accurate alignment for a new building.

Over time, the Chinese army took notice of this device and realized it could also be used to help soldiers maintain their bearings on cloudy days. In fact, this use of lodestone was not new. The first definite mention of a compass in world literature—written in the fourth century B.C., possibly by the philosopher Su Ch'in—notes, "When the people of Cheng go out to collect jade, they carry a south-pointer with them so as not to lose their way." However, this use seems to have escaped the notice of the military until several centuries later. The earliest mention of this use is in a book written in A.D. 1044 by Tseng Kung-Liang, but it is noted with such familiarity that it was almost certainly a well-known technique dating back to much earlier times. Similarly, the first use of a compass in marine navigation is noted in a book written in the tenth century A.D. but, again, this is noted as a common fact and not as a novelty. This also suggests a use dating back for many years rather than being a more recent innovation.

In these uses, the compass was not revolutionary, but was a huge improvement over existing tools. Its impact was not that it made navigation possible or even reliable, but that navigation was made more dependable under all conditions. No longer would a ship have to wonder through days of cloudy weather about its direction of travel; instead, the captain had simply to float a magnetized needle to determine his direction of travel. It must also be remembered that, at that time, ships rarely sailed out of sight of land, so most navigation consisted of simply looking for familiar landmarks. This is why the compass was more of a navigational aid rather than a revolutionary device. It was not until open-ocean sailing began in the fifteenth century that the magnetic compass really came into its own. When ships routinely sailed for days or weeks out of sight of land, the magnetic compass was indispensable and, in fact, such sailing likely would not have become routine without such devices.

The increasing use of compasses also had a somewhat wider, less obvious impact than its adoption for geomancy and navigation. In their quest to make better and more reliable compasses, the Chinese began experimenting in a number of areas. This led them to better steel-making techniques, and also to make some important observations about the physics of magnetized materials that were several centuries ahead of the rest of the world.

With magnetized metals in particular, the Chinese noticed that iron could be made magnetic by stroking it with a lodestone. This induced magnetism, however, faded rapidly with time and required renewal. Steel, however, stays magnetic longer, and the Chinese began experimenting with different "recipes" to try to find a steel that would stay magnetized as long as possible. What they found was that steels of higher purity and with a higher carbon content made the best magnets and, incidentally, were also harder and more durable than low-carbon steels.

In other experiments, the Chinese also noted that, when heated to red-hot, steels would lose their induced magnetism. This temperature and its significance is now called the Curie point, in honor of Pierre Curie (1859-1906), who explained it just about a century ago. It must be noted that the Chinese only observed this phenomenon and did not understand it. However, notice it they did, and they made use of their observation quite frequently.

At the same time, the Chinese noticed that steels would become spontaneously magnetized when they cooled down below the Curie point. This was shortly followed by noting that a steel needle could be aligned north-south during cooling and, by so doing, it turned into a permanently magnetized needle, suitable for use as a compass. What happens is that, when cooling through the Curie point, the needle picks up and "locks in" Earth's magnetic field. This becomes the needle's magnetic field, and lets it be used as a compass afterwards. Again, it must be emphasized that the Chinese did not understand the physics behind this phenomenon, and it is likely that they made no effort to do so. Even if they had, their understanding of science at that time simply was not up to the task and their explanations would likely have been more like magic or superstition than what we consider science. However, this is, to some degree, immaterial because they were the first to notice the phenomenon and to make use of it. Science usually proceeds by trying to explain an observation; without the initial observation there is nothing to try to explain, and science is not advanced. Therefore, the Chinese observations were as important to the progress of science as were later explanations of these effects.

In summary, it can be said with a fair degree of certainty that the Chinese were the first to notice the magnetic and "south-pointing" properties of magnetized materials. Although they first used these properties for geomancy, they also put them to use for direction-finding, which became the most significant use of such materials nearly 2,000 years later. The use of magnetic materials also spurred improvements in Chinese metallurgy and, later, caused Chinese scientists to make some interesting physical observations that further advanced the future manufacture of compasses and other magnetic devices.

P. ANDREW KARAM

Further Reading

Temple, Robert. *The Genius of China: 3000 Years of Science, Discovery, and Invention.* New York: Simon & Schuster, 1986.

Needham, Joseph. *Science and Civilization in China.* Multi-volume series. Cambridge: Cambridge University Press, 1954-[2000].

Development of Seagoing Vessels in the Ancient World

Overview

The first seagoing vessels were developed by the Egyptians, followed by other peoples living on the Mediterranean Sea. These peoples developed seagoing vessels when they decided to venture out onto the seas in order to trade, conquer other lands, find new resources, and or simply to see what was beyond the next island. Seagoing travel was not possible until several inventions refined early shipbuilding techniques to make ships stronger and more seaworthy. From then on, commerce and contact expanded between the peoples of the Mediterranean, then throughout Europe and North Africa, and this commerce and contact eventually led to an exploration and understanding of the ancient world.

Background

The first vehicle to take to the water with a person aboard was probably a log, and the water crossed probably a river or a lake. Woven baskets lined with tar were sometimes used to carry food, clothing, and even babies across bodies of water. When desire and curiosity drove people to take longer trips on the water, they hollowed out logs and lashed them together. Early humans who lived where wood was scarce created usable craft by making sacks of animal skins, inflating them with air, then joining them together to make a raft. Clay pots were used the same way.

Waterborne vessels were used as early as 4000 B.C. From the remains of clay tablets and containers, historians know that the Mesopotamians (c. 3500-c. 500 B.C.) built boats of reeds coated with tar or of wood frames supported by inflated skins. These boats floated downstream and were used as floating stores from which food and goods were sold to people along the route. When they reached

the end of the journey, the travelers took the boat apart, threw away the reeds, and sold the wood frames. They then loaded the skins onto pack animals and made their way back upstream on foot. These vehicles were used only on inland waterways, lakes, or rivers.

The earliest historical evidence of boats is from Egypt. From studying Egyptian art, scholars know that the early Egyptians built rafts from bundles of reeds that grew in abundance in the Nile Valley. These early boats were used only for transport along the Nile River. Archaeologists now have concrete evidence of early Egyptian boatbuilding techniques. In 2000, they excavated a boat at a royal burial site in Abydos, about 280 miles (450 km) south of Cairo. Other remains at the site indicate the boat was associated with pharaohs of the first dynasty, beginning about 3000 B.C. It is about 75 feet (25 m) long, 7-10 feet (about 2-3 m) wide at its widest, and less than 3 feet (1 m) deep, and appears to have been built from the outside in, rather than by a later technique of starting with an internal frame. It was made of thick wooden planks lashed together with rope fed through mortises. The boatbuilders filled the seams between the planks with bundles of reeds to make the boat watertight. Judging by the length of the boat, it probably would have been propelled by as many as 30 rowers. Although experts believe it is an actual, rather than symbolic, boat, it was probably intended for use by a pharaoh in the afterlife.

Before any craft could be put out to sea, people had to design boats that could stand up to longer voyages on rougher water. The most important invention was a hull made of wood planking. The planked hull may have been a refinement of the dugout canoe, strengthened for sea travel with long planks of wood held snug

A cargo boat from ancient Egypt. *(Archivo Iconigrafico, S.A./Corbis. Reproduced with permission.)*

against the sides of the dugout with reeds, cord, or twine. Wooden planks give a vessel stability and weight, and allow men to carry more goods and to propel the vehicle with oars. Two techniques of constructing wood-plank hulls developed in the first millennium B.C. Northern Europeans favored clinker construction, in which the hulls were built of overlapping timbers. In the Mediterranean, the boats were carvel-built; the planks in the hull were joined along their lengths to form a smooth surface.

The other important invention was a sail that allowed the boats to be propelled by wind. Sails could be made of fronds, woven reeds, or animal skins. Egyptians created the first-known examples of sails around 3000 B.C., first for traveling on the Nile and later for trips into the Mediterranean. Egyptian boats commonly featured oars in addition to sails, because they were traveling on the narrow channel of the Nile where winds were variable.

One motive for going to sea was to conquer other lands to acquire more territory or more power. Another motive was to obtain new sources for goods and to trade with nearby people and settlements for food, resources, and luxury items. The function of the ships determined their design. Long, narrow, rowed ships were used in warfare, because these ships needed speed and maneuverability, as well as room for carrying large numbers of fighting men. Vessels used for trading needed to have room for as much cargo as possible, but carried a small crew, so they had round bottoms. Trading ships were also taller than warships to prevent high seas from swamping the cargo.

Impact

The Egyptians put to sea as early as the third millennium B.C., sailing larger wooden versions of their reed boats. Most of their voyages were to Crete, and the boats built by the Minoans of Crete used Egyptian boatbuilding techniques. The Minoans were the first truly seafaring people. By around 2000 B.C., they had war galleys (vessels propelled primarily by oars) driven by many oarsmen on each side of the hull. The ships were probably built of planking, an Egyptian technique.

Phoenician seafarers dominated the eastern Mediterranean for about three centuries, beginning from about 1100 B.C. They knew Egyptian shipbuilding techniques and copied Minoan styles. Their galleys were long and narrow with high stern posts and low bows to which a heavy, pointed ram was attached. The ships were rowed by two tiers of oarsmen and controlled by a single steering oar and a single pole mast with a square sail.

The early Greek ships were long and narrow, with several tiers of oarsmen. The first ships were called penteconters and had about 25 oarsmen on each side of the ship. Penteconters were used in exploring and for communicating with outlying colonies. They were also used for raiding, and carried soldiers who then went ashore or boarded other ships to fight an enemy. As the Greeks spread commerce, they eventually needed protection at sea. The first galley built primarily for fighting was the unireme (from the Latin word *remus,* meaning "oar"). In the eighth century B.C., the Greeks developed the bireme, a ship with two banks of oars (the number of banks is the number of rowers on each oar or bank of oars, not the number of layers of oarsmen).

As Greek power increased in the sixth and fifth century B.C., they developed their most effective ancient warship, the trireme. The trireme had a single mast with a broad, rectangular sail that could be furled, and was manned by about 200 officers, seaman, and oarsmen. It had three tiers of oarsmen on each side of the boat, up to 85 a side. The deck was open in the center, only partially covering the area below to give rowers air and space. Triremes were graceful, maneuverable, and fast. Their main armament was a ram covered with bronze, which was designed to pierce the hulls of enemy warships. The triremes of the Athenian fleet defeated the Persians at Salamis in 480 B.C. and continued as the backbone of the Greek fleet from that time. Although effective in warfare, they had little room for provisions and usually put into shore at night to take on fresh supplies. In spite of this, they helped to create the Athenian Empire, spread Greek culture around the Mediterranean, and establish Greek colonies in Italy, North Africa, and western Turkey.

Triremes reached their highest point of development in the fifth century B.C. From that time, ships with ever-increasing numbers of banks were built. In the middle of the fourth century, the Athenians built quadriremes (four banks), followed by quinqueremes (five banks). Late in the fourth century and early in the third, the rulers of Macedonia built 18-banked ships crewed by 1,800 men. Rulers in Egypt built ships with 20 and 30 banks, but Ptolemy III topped them all with a 40-bank ship which would have required 4,000 rowers, if it had actually put to sea.

When the breakup of the empire of Alexander the Great in the late fourth century B.C. ended Greek naval domination, sea power developed in other parts of the Mediterranean. By about 300 B.C., Carthage, a Phoenician city on the north coast of Africa, became the foremost naval power in the area. Also at this time, the city-state of Rome began to expand south through Italy. When Rome collided with Carthage, the result was the First Punic War, which began in 264 B.C. Rome was not a naval power, but the conflict with Carthage convinced the Romans that they needed a navy. Using shipbuilders from defeated Greek city-states and knowledge they gained from Carthage, the Romans built a fleet of triremes and quinqueremes. Although these early ships were large, the Romans later developed the liburnian, a light, fast unireme with two banks of oars, which they used for escorting merchant ships and fighting pirates. Liburnians eventually became the most common form of Roman warship. The Romans defeated the Carthaginians in 184 B.C., taking control of the Mediterranean Sea and dominating it for several centuries.

The development of vessels that could travel the oceans made possible the dissemination of cultures, knowledge, and goods between the peoples of the Mediterranean, then Europe and North Africa and eventually led to man's exploration and understanding of the world. The conflicts that resulted from this were essential, if unhappy, steps in the evolution of the ancient world.

LYNDALL BAKER LANDAUER

Further Reading

Casson, Lionel. *Ships and Seafaring in Ancient Times.* Austin, TX: University of Texas Press, 1996.

Gould, Richard A. *Archaeology and the Social History of Ships*. Cambridge, MA: Cambridge University Press, 2000.

Landels, J. G. *Engineering in The Ancient World*. Berkeley, CA: University of California Press, 1978.

Ships and Seamanship in the Ancient World. Princeton, NJ: Princeton University Press, 1971.

Starr, Chester. *The Influence of Sea Power on Ancient History*. New York: Oxford University Press, 1989.

The Royal Road of Persia

Overview

In about 3500 B.C., a 1,500-mi (2,414-km) long road running from the Persian capital of Susa to the Aegean Sea came into use. Not necessarily a road as understood by modern usage, this was more a track worn into the soil that was used in a routine, if not an organized, way for over 2,000 years. Eventually organized by the Assyrians, it served the Persian kings for centuries, and was later used by Alexander the Great (356-323 B.C.) to convey his troops, ironically helping Alexander to conquer the Persian Empire. At the time, the Persian Royal Road was not unlike other roads built in Egypt, Greece, and Babylon, all of which were, though somewhat limited in scope, served to link parts of an empire. However, it was not until the Roman Empire that roads were brought to their logical conclusion in a system of integrated, well-engineered and well-constructed highways that linked all parts of the empire. In this, the Persians prefigured and inspired the Romans, who built on the success of their predecessors.

Background

The world's first roads were scarcely roads as we would recognize them. Primarily trails that, over centuries of use, became worn into the ground, early roads were largely used because they happened to be the shortest or fastest routes between two cities.

The first long-distance road was what later became the Persian Royal Road, running about 1,500 miles (2,414 km) from Susa (the ancient Persian capital) through Anatolia (now Turkey) to the Aegean Sea. At about the same time, the Indus civilization built paved streets in many of their cities, although these were not long-distance roads of the same scope. Other civilizations also seem to have built roads but, again, they were of a lesser scale, running mostly within or between nearby cities.

By about 1500 B.C., the Persian Royal Road had been organized and placed into regular use by the Persian kings. It was put in use for mail, trade, and for the military, uses that were later mirrored by the Roman Empire in its extensive road network. It is said that, with a system of relay stations and fresh horses, a messenger could travel the 1,500 miles (2,414 km) length of this road in just nine days, although normal travel time was closer to three months. Like the Roman roads, the Royal Road helped to link the ends of a large empire. Unlike Rome, most of the Persian Empire was lightly populated and the network of roads was not nearly as extensive as those that were to follow. In addition, being built in a dryer climate, there was little need for the extensive engineering that characterized the Roman roads.

Impact

The primary function of the Royal Road was to facilitate communication from the emperor to his distant subjects. In this, the impact was clearly to make it possible to administer an empire that, at that time, was geographically among the largest in the world. The Royal Road helped make the Persian Empire possible. In addition, the Royal Road demonstrated to contemporary civilizations the utility and value of such a road, and in so doing, it helped to inspire similar projects in other empires. This process culminated in the Roman roads. Finally, and ironically, the Royal Road proved the undoing of the Persian Empire, giving Alexander the Great and his armies rapid access to critical parts of ancient Persia and making his conquest all the easier.

The Persian Empire was one of the Middle East's first large empires. Ruled by the King of Kings in his capitals of Susa, Persepolis, and Ecbatana, Persia was divided into at least 20 provinces called satrapies. These satrapies were ruled by satraps, and all were in constant communication with the king. The empire was protected by a standing imperial army of at least 10,000 troops, augmented by local troops from each province. Some of these troops were permanently stationed on the frontier, and they were reinforced by others as necessary.

All of the Persian civil and military administration depended on the Royal Road. As with the later Roman Empire, the king and his government utilized a postal system to carry orders and information to the provinces and the frontier while ferrying information and requests for assistance back to the capital.

Unlike the earliest version of the road, the Persian kings made improvements, turning it into an all-weather byway that was usable in all seasons, including the relatively short rainy season. In fact, in a phrase often attributed to Benjamin Franklin (1706-1790), Herodotus (484?-420? B.C.) noted that the Persian royal messengers were stopped by "neither snow, nor rain, nor heat, nor gloom of night." The road, the dedication of the messengers, and the construction of relay stations located about a day's travel apart let messengers travel at a rapid pace, and gave the Persians somewhat of a diplomatic and military advantage over their neighbors, whose orders and armies responded much more slowly.

As with the Roman Empire of later centuries, the Royal Road was essential to the creation of what was, in its time, the largest empire on Earth. As noted above, the ability to communicate and move troops rapidly as needed gave the Persians a decided advantage over their rivals. In addition, this road made it possible for the emperor to hear the grievances of his subjects quickly, and let him move to resolve disputes in the provinces before they could fester and become disruptive. This, combined with a remarkably tolerant attitude towards the religions and practices of subject people helped the Persian Empire to grow and contributed to its remarkable stability for many hundreds of years.

The value of the Royal Road did not go unnoticed, and in the centuries following its construction, others tried to imitate it. The Greeks built some roads, although not as extensive because their empires were usually smaller than that of Persia. The Egyptians also built roads, though these were at first used primarily for moving building materials for the pyramids and other monuments. Some of the Egyptian roads predated those of Persia and cannot, of course, have been inspired by the Royal Road, but others seem to have been constructed following the Persian model, and are more likely to have been influenced by the Persian's success. Other civilizations building roads that may have been in-

fluenced in whole or in part by Persia included imperial China, India, and possibly Crete (although there is evidence that the Cretan roads may have been developed independently).

In spite of these minor influences, however, it is possible that the most significant culture inspired by the Royal Road was Rome. The Romans had, of course heard of the Royal Road and realized early on that a road system was essential for carrying out the business of the empire. However, the Romans also realized their needs differed from those of Persia, driven by geographic and climate differences. So, instead of simply copying the Persian system, the Romans took from the Persians the basic concept of an improved roadway, melded it with the civil engineering and skilled construction practices of the Cretans, Egyptians, and Babylonians, and added to it their concept of a network of roads connecting all parts of a sprawling empire. The result was, until the construction of the American interstate highway system, the greatest network of highways in history. In a very real sense, the Royal Road not only made the Persian Empire possible, it also made the Roman Empire possible.

Finally, in one of history's ironies, the Royal Road also made possible Persia's fall from power. Alexander the Great, in his wars of expansion, stumbled across the Royal Road with his armies. In this case, he had already defeated the Persian border armies, and he then used the Royal Road to rapidly move his troops into the heart of the Persian Empire. Quickly reaching Persepolis, Alexander's army sacked and burned the Persian capital and then moved on to defeat more Persian armies until receiving the surrender of the Persian emperor. This completed, Alexander went on to more conquests before finally stopping in India, having conquered virtually all of the world known to the Greeks of that era.

Alexander's use of the Royal Road demonstrated that virtually any strategic advantage can be used as a weapon for either side, if only an attacking enemy can put it to their use. In this case, by seizing the Royal Road, Alexander was able to turn the Persian's strategic road to his advantage because he could now move a superior army very rapidly, reaching Persian cities before they could establish an adequate defense. In this manner, the same road that helped the older empires expand and defend their nation then turned into an instrument that, in the hands of their enemy, helped to dismantle everything that had been built so laboriously over the course of centuries.

This lesson, apparently, was not fully appreciated by the Romans. A thousand years later, under attack by the barbarians from outside the empire, Roman border troops were defeated. The victorious barbarian armies were then able to proceed quickly through the Roman Empire along Roman military roads, attacking Roman garrisons and cities before they could establish their defenses. This culminated in the sack of Rome, the eventual fall of the Roman Empire, and ushered in the "Dark Age" of medieval Europe.

P. ANDREW KARAM

Further Reading

Curtis, John. *Ancient Persia.* Cambridge, MA: Harvard University Press, 1990.

Green, Peter. *The Greco-Persian Wars.* Berkeley: University of California Press, 1998.

Olmstead, Arthur T. *History of the Persian Empire.* Chicago: University of Chicago Press, 1959.

The Building of Canals in the Ancient World

Overview

By far the most impressive and well-known canals of the modern world are those in Panama and the Suez. The former, completed in 1903, connects the Pacific and Atlantic oceans, fulfilling a dream of several centuries. But the Suez Canal, which for the first time opened up the route between the Mediterranean and Red seas in 1869, represented the culmination of literally thousands of years' effort. Among the earliest canals were ones whereby builders in Egypt attempted to connect the Nile and the Red Sea; but these were far from the only major canal-building projects of the ancient and early medieval world. Indeed, the period before A.D. 700 saw the building of the world's longest canal in China, a waterway aptly known as the Grand Canal.

Background

Canals are manmade waterways built for irrigation, drainage, water supply, or—and this is most often the case in modern times—transportation. As early as 5000 B.C., more than two millennia before the unification of Egypt and the beginnings of what is typically recognized as Egyptian civilization, the first canals appeared in Egypt. Spurred by the drying of lands along the Nile, the early Egyptians began constructing dikes and irrigation canals, developments that greatly enhanced the livability of the area and helped lead to the establishment of cities.

Further east, in the river valleys of the Tigris and Euphrates that spawned the Mesopotamian civilizations, villagers began building primitive irrigation canals and ditches. This occurred at about the same time as the Egyptian projects, and likewise signaled a major step in the development of a civilization, as the peoples of Sumer banded together for the purpose of building these channels.

Historians commonly date the beginnings of true civilization—complete with agriculture, a settled way of life, an ordered system of government, writing, and cities—to about 3500 B.C. in Sumer, and slightly later in Egypt. The first Egyptian monarch of which anything is known was Menes, who united the kingdoms of Upper and Lower Egypt in about 3100 B.C., and it is said that he too undertook canal-building projects. Certainly in societies built around rivers, as those of Egypt and Sumer were, the development of irrigation for surrounding lands was a priority. Thus in Sumer in about 3000 B.C. the people of the lower Euphrates valley were using basic tools such as the lever for the construction of canals.

For the Egyptians, one of the facts of life along the Nile was the existence of cataracts, or rapids, that interrupt the smooth flow of the river. Just above the site of modern-day Khartoum, Sudan, is the Sixth Cataract, and as the Nile snakes gradually northward, it passes through several more of these rapids, each numbered in descending order. The First Cataract lies near the modern city of Aswan, and it was above this point that Egyptian civilization developed. Yet by the Sixth Dynasty (2350-2150 B.C.), the Egyptians had begun building transport canals to bypass the First Cataract. Once again there was a Mesopotamian parallel, when in about 2200 B.C. the building of the Shatt-el-hai Canal linked the Tigris and Euphrates rivers.

Impact

Egypt's pharaoh Senusret II (1842-1836 B.C.) had floodgates built along part of the Nile Valley in order to reclaim valuable farmland, and his son Senusret III (1836-1817 B.C.) ordered the First Cataract cleared. This removed one of Egypt's natural barriers, and for that reason he had fortresses built to protect the country from invasion by Nubians or Kushites to the south.

Yet Egypt and Mesopotamia were not the only lands in which canal-building took place during the period from 2000 to 600 B.C. Far away in Peru were the city-states of Layzón and Agua Tapada, influential in the development of the Chavín culture (1000-400 B.C.). Greatest among the legacies of the Layzón and Agua Tapada was the Cumbemayo Canal, which—more than 3,000 years before the building of the Panama Canal—linked waters that drained into the Pacific with those of the Atlantic watershed.

China, too, had a number of canals in ancient times. There the needs were different than in Egypt and Mesopotamia, due to a differing geographical configuration. Despite its large size, China has only one major coast, on the east. Though its major cities developed along the country's eastern portion, they tended to be river ports rather than seaports—yet most rivers in China run east to west, whereas the line of cities ran north to south. Thus the Chou Dynasty (1027-246 B.C.) saw numerous attempts to link the great Yangtze and Yellow rivers, and these efforts expanded greatly under the harsh Ch'in Dynasty (221-207 B.C.)

The latter is best remembered for another public-works project, the Great Wall, but this was far from the only large undertaking overseen by the autocrat Ch'in Shih-huang-ti (259?-210 B.C.). Decades before he seized control of the entire country in 246 B.C., his predecessors in the state of Ch'in had forced an army of slave-laborers to build a canal joining the Ching and Lo rivers. Once Shih-huang-ti unified China, making himself emperor, he brought Ch'in work-gang methods to bear on the country as a whole, creating a vast nationwide system of roads and canals to keep his armies supplied.

In the Near East during this time, a number of empires had risen and fallen. Babylonia reached its peak under Nebuchadnezzar II (630?-562 B.C.), renowned as the builder of Babylon's famous Hanging Gardens. These required an extensive network of irrigation canals to keep their trees and flowers watered, and indeed the construction of canals throughout the city was a hallmark of Babylon's magnificence. Yet in about 600 B.C. these life-giving canals yielded an unexpected product: malaria, dysentery, and other plagues bred by flies and mosquitoes living in stagnant canal-water.

It was not the plague, however, but the armies of Cyrus the Great (585?-529 B.C.), that brought an end to Babylonia with the Persian invasion of the capital city in 539 B.C. Cyrus would be succeeded by his son Cambyses II (reigned 529-522 B.C.), who added Egypt to the growing Persian Empire in 525 B.C. Cambyses's successor, Darius (550-486 B.C.), would in turn undertake the most impressive canal-building project up to that time.

According to the Greek historian Herodotus (484?-420? B.C.), Egypt's pharaoh Necho II (reigned 610-595) had begun a project to connect the Nile with the Red Sea; but after the Persian conquest of Egypt, it was Darius who completed the waterway. Herodotus reported that the canal was "four days' voyage in length, and it was dug wide enough for two triremes [fighting ships] to move in it rowed abreast." He also claimed that more than 120,000 workers died in its construction, but this was probably an exaggeration.

Some 50 miles (80 km) long, the canal was as much as 148 feet (45 m) wide and 10 feet (3 m) deep. Archaeologists in 1886 discovered the first of four stelae, or pillars, along the path of the ancient canal. Inscribed on these is the announcement of Darius: "I am Persian; from Persia I seized Egypt; I gave orders to dig this canal from a river by name Nile which flows in Egypt, to the sea which goes from Persia. Afterward this canal was dug; thus I had ordered and ships went from Egypt through this canal to Persia. Thus was my desire."

It appears that as early as the New Kingdom era (1539-1070 B.C.), the Egyptians dug a canal from the Nile via the Wadi Tumilat to the Red Sea. However, that channel had long since been covered by sand, and a similar fate would attend Darius's ambitious project. It had to be re-excavated by Ptolemy I (366?-283? B.C.), the first Greek ruler of Egypt, and by the time of the last, Cleopatra (69-30 B.C.), parts of it had been completely blocked by sand. Later, the Roman emperor Trajan (53-117) had it cleaned out, after which time it was called "Trajan's River;" but as the Roman Empire declined, so did the Nile-Red Sea canal.

In the years that followed, a number of Roman rulers devoted themselves to canal projects, but by that point they were more concerned with the upkeep of existing waterways—including Roman ones, dating to about 100 B.C., in what is now France—rather than to new constructions. Between the fall of the Western Roman Empire and the rise of Islam, there were only two significant powers in the Near East, the Eastern Roman or Byzantine Empire, and the Sasanian Empire of Persia, whose Khosrow I (reigned A.D. 531-579) built and restored a number of irrigation canals. The most important canal-building project of the early medieval era, however, occurred in China.

With the rise of the Sui Dynasty (589-618), history seemed to repeat itself. Like the Ch'in some 800 years before, the Sui were a short-lived, highly autocratic ruling house of immeasurable importance to the country's history, and, also like the Ch'in, they put into place massive building projects undertaken by thousands of slave laborers. Yang Ti (580-618) ordered the construction of numerous canals, most notably the Grand Canal, which ran for some 1,100 miles (1,760 km) and connected the Yangtze River in the south with the Yellow River in the north. Not only did this make possible much greater centralization of authority in the vast nation, it encouraged commerce and influenced a major shift in population, from the heart of Chinese civilization in the north to the rice-growing lands of the south.

Like the Great Wall, the Grand Canal would be improved many times during the centuries that followed, but in the West canal-building came to a halt with the confusion brought on by the end of Roman rule. Europeans took hope from the brief state of order created by the rule of Charlemagne (742-814), and contemplated the deceptively simple-seeming project of connecting the Main and Danube rivers at their closest approach; but this proved much harder than it seemed, and in any case the peoples of Western Europe during that era were concerned chiefly with mere survival.

Only in the twelfth century and afterward, with projects such as the Naviglio Grande in Italy, did canal-building in Europe resume. The Chinese completed a 700-mi (1,110-km) branch of the Grand Canal in 1293, and introduced the crucial development of the canal lock. But by the time France built its famous Canal du Midi in 1681, Western nations had taken a decisive lead in the building of manmade waterways. This would culminate in the French completion of Suez, and the American project in Panama a few years later. In 1992, 12 centuries after Charlemagne, the Main-Danube Canal was completed.

JUDSON KNIGHT

Further Reading

Oxlade, Chris. *Canals*. Chicago: Heinemann Library, 2000.

Woods, Michael, and Mary B. Woods. *Ancient Transportation: From Camels to Canals*. Minneapolis: Rinestone Press, 2000.

Roman Roads: Building, Linking, and Defending the Empire

Overview

The Roman Empire was, until the peak of the British Empire, the mightiest empire the world had ever known. At its peak it dominated virtually all of Europe, part of Africa, and most of the Middle East. One of the factors that made it possible to build, administer, and defend the Roman Empire was its intricate system of roads, which were so well designed and well constructed that they were still in use a millennium after they were first built. As the saying went, "All roads lead to Rome," and it was upon its roads that the Roman Empire extended and controlled its vast expanses.

Background

Roads have existed in some form for nearly 4,000 years. They were mostly used for trade and, in general, were no more than frequently followed paths with some sort of improvements at river crossings, swamps, and other difficult stretches. In some cases, branches and logs were laid on the ground to ease walking or horse-riding, but little more than this was standard. Different cultures made their own unique contributions to road building: the Egyptians were master surveyors, the Greeks excelled at masonry, the Etruscans developed cement-making and paving, and the Cretans were also skilled at paving. The Roman contribution was twofold: they first built drainage ditches alongside their early roads to help maintain them in passable condition in any weather conditions, and they recognized the advances of others. This second contribution was the most important; the Romans were not above borrowing technology from others, and they were the first to incorporate all the technological innovations noted above into a single network of roads. By doing so, and adding their own innovations as time went on, the Romans were able to construct a system of roads that remained unequalled for centuries.

The first Roman road was the Via Appia (the Appian Way) built in about 334 B.C. In the following few centuries, over 53,000 miles (85,295 km) of Roman roads ran to all corners of their empire. Twenty-nine of these roads were military roads, designed to rapidly convey the Roman legions to the frontier for offense or defense. And there is no doubt that these roads were a strategic advantage that helped Rome to build and to hold its empire. In fact, the Roman road system constituted the world's first integrated highway system.

Perhaps the chief innovation, however, lay in the roads' design, in particular the military roads. Made to last for centuries, the roads were usually wide, well drained, and built of several layers of rock, gravel, and concrete. In fact, not only did the roads allow travel at up to 75 miles (121 km) per day, but they lasted for over a millennium and served as Europe's roads during the Renaissance.

Impact

The Roman system of highways was instrumental in shaping the destiny of the Roman Empire. They also set new standards for road design and technology, and they were to serve Europe for centuries after the fall of the Roman Empire.

First, and most importantly, in many ways the Roman roads *were* the empire. The Romans understood that controlling a far-flung empire depended on rapid and efficient travel, and it was to that end that they built their highway system. Not only did this facilitate trade throughout the empire, the roads also made rapid communications possible, and they carried the Roman armies to trouble spots with dispatch.

Earlier roads, such as the Amber Road, could more accurately be considered routes—there were improvements in difficult spots, but they were generally simply paths or more-trav-

eled areas linking these improvements. Travel was usually slow, and many travelers became bogged down in bad weather. Travelers took months or years to reach their destinations, and messages took as long. For an empire the size of Rome's, this was not acceptable.

What the Romans did was to speed this travel. In a sense, the roads were the first "information superhighway," moving at walking speeds rather than electronic speeds. Nonetheless, it was the most rapid transportation available. This came into play primarily in the area of administering the empire, particularly by means of the Roman *cursus publicus,* or postal system.

With today's rapid communications, it is difficult to conceive of a time in which information traveled at the speed of a walking man or a walking horse. In such a world, where a good day's travel was usually less than 20 miles (32.2 km), carrying on the business of empire could be a slow matter. In the case of Rome, the business of empire included waging war, negotiating treaties, conducting diplomacy, sending orders to army generals, collecting taxes, distributing imperial decrees, receiving reports from emissaries abroad, and more. To have all of these vital functions and communications take place at a walking pace could be nearly intolerable.

The construction of high-quality highways was a tremendous advantage, for it sped up the pace of communications considerably. For example, a courier on a Roman road could travel at speeds of up to 75 miles (121 km) daily. Although it may be tempting to decry a rate of 75 miles (121 km) per day as still being slow, especially compared to travel today, making such a comparison is not appropriate because at that time there were no airplanes or automobiles. Instead, consider today's exploration of the Solar System. Today's space probes take months to reach Mars and years to reach the outer planets. These travel times are comparable to those needed to travel around the Roman Empire in the absence of the Roman highway system. Now, consider how revolutionary we would view a new spacecraft propulsion system that would let us reach Mars in a week and the outer planets in a month or so. What would this do to the way we view our solar system? What a revolution to be able to send people easily to visit or colonize Mars or to study Jupiter! The advent of the Roman roads opened up travel through the empire in a similar way.

Of course, the same roads that led from Rome also led *to* Rome, and information was

Remains of a Roman road in the ruins at Carthage. *(Bernard and Catherine Desjeux/Corbis. Reproduced with permission.)*

carried in both directions. In most cases, this simply facilitated receiving news and taxes from distant parts of the empire. However, Christianity spread along these roads, too, traveling easily to Rome and throughout the empire. There is, of course, no way to know how far, or how fast Christianity would have spread in the absence of these roads, but there can be little doubt that much of the new religion's early success was owed to the speed with which its adherents could spread their message through the empire.

In addition to the administrative advantages, the Roman roads were used as a military weapon. Just as the roads could be used to speed information to the far corners of the empire, they could also be used to speed armies to defend the borders, or to expand the borders through further conquest. In addition, the roads were easily used to provision armies, to send reinforcements to the battlefield, to deliver orders to the generals, or to bring reports from the front back to Rome.

This lesson was not lost on history, even during the twentieth century. Hitler used the German autobahn to move his troops to the front, North Vietnamese troops moved equipment on the Ho Chi Min Trail, and even the American interstate highway system was built with an eye towards moving troops rapidly. In fact, this highway system was designed to pro-

vide a number of straight stretches long enough to serve as emergency runways for warplanes, giving it dual wartime functionality. Incidentally, Swedish and Swiss highways are also designed for similar uses during wartime.

From the standpoint of civil engineering, Roman roads were similarly impressive. The typical Roman road was flanked by drainage ditches that helped to keep it dry during heavy rains or during the Spring snowmelt. Soil taken from the ditches was used to elevate the roadbed at least three feet above the surrounding terrain, further helping to keep the road dry. On top of that were spread layers of gravel, sand, concrete, and paving stones. The entire road might be as much as 4.5 feet (1.4 m) thick. All of this made the roads smooth, dry, and exceptionally durable. These roads might be as wide as 15 feet (4.6 m), allowing two-way horse and chariot traffic, and they would be heavily crowned (i.e., raised in the center to let rain drain to the side). On the sides were curbs up to 2 feet (0.6 m) wide, and auxiliary side lanes up to 7 feet (2.1 m) wide on either side of the road. The total width might be as much as 35 feet (10.7 m) across on a fully developed, heavily traveled road. The roads also usually ran straight across swamps, plains, lakes, ravines, and (as much as possible) mountains.

This construction led to the longest lasting impact of the Roman roads—their continued use for centuries after the fall of the Roman Empire. Even a millennium after Rome fell, the roads remained in use, in many cases as the only decent roads in parts of Europe. Thus, as late as the Renaissance, roads that were 1,000 years old were still in use, facilitating trade and travel, and serving Rome's successor states just as they had served Roman legions and couriers centuries before.

Roman roads were one of the major tools of the Roman Empire. They helped Rome build, maintain, and administer its empire, and they served future nations equally well. Early Christians used these roads to spread their gospel through the ancient world, and later generations moved pilgrims, armies, and goods. Whatever their faults, the Romans built well and built to last. They set the standard, too, for future generations of architects and civil engineers to aspire to.

P. ANDREW KARAM

Further Reading

Claridge, Amanda, Judith Toms, and Tony Cubberley. *Rome: An Oxford Archaeological Guide to Rome.* Oxford and New York: Oxford University Press, 1998.

Edward Gibbon. *The Decline and Fall of the Roman Empire.* 3 vols. New York: Modern Library, 1995.

Johnston, David E. *An Illustrated History of Roman Roads in Britain.* Bourne End, England: Spurbooks, 1979.

Laurence, Ray. *The Roads of Roman Italy: Mobility and Cultural Change.* London and New York: Routledge, 1999.

Nardo, Don. *Roman Roads and Aqueducts.* San Diego, CA: Lucent Books, 2001.

Writing Preserves Knowledge and Memory

Overview

The development of writing took place more than once, in different places and at different times. The earliest was about 5,500 years ago in Mesopotamia, the ancient Fertile Crescent of the Middle East, home of the first farmers and city builders as well as the first scribes. Writing began as idealized pictures and simple symbolic notations such as slashes and dots. Eventually, alphabets provided the flexibility to represent any sounds in the spoken language, facilitating the transmission of abstract ideas. Writing was arguably humanity's most important invention, because it provided the means to record and pass along knowledge between people separated in time and space.

Background

Before writing could be conceived, humans first needed spoken language. Anthropologists are unsure when our remote ancestors first developed the physiological capability to speak, and the abstract reasoning necessary to use verbal symbols to convey meaning. Whether the Neanderthals of 100,000 years ago had this ability, for example, is still in question. However, it is fairly well established that by 40,000 years ago, *Homo sapiens* had evolved the ability to express thoughts through language.

It was perhaps natural, then, that about the same time early humans learned to express ideas in words, the first evidence of representational art appeared. The pictures we find on walls of caves, such as the beautiful 22,000-year-old paintings at Lascaux, France, may have been meant to record an event, serve some religious purpose, or simply decorate the neighborhood. Whatever their primary purpose, they were clearly intended to represent the beauty of the animals they depict, and they do that for us still.

Archaeologists specializing in prehistory have found other steps on the path to writing: pieces of bone from as early as 30,000 B.C. that are marked with either groups of slashes or dots. These artifacts have been found all over the world. Some are controversial, with critics arguing that the marks could have been left by the teeth of carnivores. But in other cases, the bones are clearly tally sticks, each keeping count of some quantity important to their makers, perhaps animals killed in a hunt, days in the lunar cycle, or even a score in a game. Similar tally sticks were used by some European farmers until about 100 years ago, and by aboriginal Australians into the second half of the twentieth century. Knotted strings, such as those used by the Inca in South America, also served record-keeping purposes.

While these early numeric records were important precursors to writing, the information they conveyed was still quite limited. They were essentially mnemonic devices. The person who kept the tally stick may have known he hunted

Egyptian hieroglyphics carved in stone. *(Archive Photos. Reproduced with permission.)*

antelope, so that all he had to record, if such knowledge was considered necessary, was the number he bagged. To him, five strokes meant five antelopes. To us, they just mean five. Five what? We cannot know for sure. The context is gone, along with the memory of the maker.

A richer variety of information could be conveyed by pictures. In a relatively dense urban civilization such as the ancient Sumerians built in the Fertile Crescent of Mesopotamia, these became simplified into representational symbols called *pictograms* by about 3500 B.C. The Sumerian pictograms were, as far as we know, the world's first writing.

The earliest Sumerian pictograms, carved onto clay tablets, represented concrete, easily recognizable, familiar objects that were the common currency of daily life. The first known inscriptions are agricultural records, listing sacks of grain and heads of cattle. Over time, the signs were combined to signify more complex ideas. Most signs had more than one possible meaning. For example, the sign for the human foot could also mean "walk" or "stand up." About 600 signs were regularly used.

The ancient Egyptians began using pictograms called *hieroglyphs* in about 3000 B.C. They viewed writing as a gift of the god Thoth, although they may actually have gotten the idea from the Sumerians. In any case, their script appears to be an independent invention. They developed their own symbols, rather than adopting those of the Sumerians, and they did not press them into clay tablets. Instead, they carved their hieroglyphs in stone, and painted or drew them on pottery. They also had *papyrus,* which resembled paper. Made from fibers of plants that flourished along the Nile, papyrus could be rolled into scrolls for easy transportation and storage. Parchment, made from animal skins, was similarly portable. However, these writing materials were expensive to process, and so they did not replace stone and pottery for everyday use.

The next key step was the use of *rebus writing.* The rebus, a construction familiar to puzzle enthusiasts, consists of a series of simple pictures, each denoting not the object it represents, but the sound of its name. The result is a type of visual pun. For example, a picture of an eye followed by one of a can would represent "I can." In this example, you can see how pictures of concrete objects are used to represent concepts that would otherwise be more difficult to express in a pictogram, that of the "self," and of "being able to do something." Thus writing, in taking a step from the pictographic toward the *phonetic* (based on sounds) gained the ability to express more abstract ideas.

In Sumer, by about 2800 B.C., pictographic script had been simplified into *cuneiform* (wedge-shaped) symbols incised in clay with the edge of a stylus. At the same time, some of the symbols were being pressed into double duty. Words that sounded alike in the Sumerian language, such as the words for "water" and "in," began to be expressed by the same sign. Special symbols called *determinatives* were sometimes used to resolve

ambiguity about whether a sign was to be understood phonetically or as a pictogram.

This rebus writing concept made the cuneiform script very powerful, and it propagated across the Middle East, remaining in use for more than 2,000 years. The Elamites, for example, who lived about 200 miles (322 km) east of the Sumerians in what is now Iran, had independently invented a pictographic script that still puzzles scholars, but adopted cuneiform instead.

Waves of Semitic tribes, nomads from the Arabian desert who spoke languages related to today's Hebrew and Arabic, were spreading across the region at about this time. In Mesopotamia, Semitic kingdoms including Assyria and Babylonia arose and supplanted the Sumerians, ruling for about 1,800 years. In these kingdoms, the cuneiform script was adapted to the different words and sounds of the Semitic languages.

Further west in Asia Minor, the area that is now Turkey was ruled between the twentieth and thirteenth centuries B.C. by the Hittites. They spoke an Indo-European language and had independently developed their own hieroglyphic writing that was used mainly for ceremonial purposes. The Hittites adopted the cuneiform script for ordinary activities, proving that the Sumerian script was adapted to handle the sounds, vocabulary, and structure of yet another language family.

Writing arose in several areas outside the Middle East as well, perhaps as early as 3000 B.C. The peoples of the Indus Valley, in present-day Pakistan, had some contact with Sumer, but their script used different symbols. Unfortunately, they seem to have written mainly on perishable materials like wood or leather. Only a few inscriptions have survived on seals and monuments.

Around the same time, writing was also invented in the Yellow River Valley of China. Unlike the various civilizations in and around the Middle East and the Mediterranean who seem to have developed their own systems of writing after learning about it from their neighbors, the development of writing in China was probably truly independent. It is not impossible that there was some contact with the earlier literate societies of the Middle East, located in the western fringe of Asia, but the 4,500-mile (7,242 km) distance across mountains and desert makes it unlikely. Chinese writing materials, mainly bamboo and silk, were indigenous. Their characters were unique, and have remained essentially unaltered since their inception. Thus the Chinese script, with its approximately 1,000 basic signs, is the oldest in continuous use.

The inhabitants of Crete may have gotten the idea of writing from their trading partners, particularly the Egyptians. However, the script they devised, written from left to right and based on individual syllables, was unique. A form of it, called Linear B, was adopted by the Mycenaean Greeks in about 1600 B.C. and used to keep palace records and inventories.

It was back in the Middle East, though, with its long history of writing and its trend toward phonetic representations, that the alphabet arose. It emerged among the Semitic groups who had settled near the eastern shores of the Mediterranean, in today's Israel, Lebanon, Jordan, and Syria. This region was called "Canaan" in the Semitic languages, and its people were later called "Phoenicians" by the Greeks.

While many of the Canaanites were farmers, their territory was not endowed with the rich soil of the Nile Delta or the Fertile Crescent. However, they did have olive and cedar trees, wool from their sheep, and a marine mollusk called the murex snail that provided a coveted purple dye. They also had a number of sheltered harbors, and a location at the crossroads of civilization. As a result, they became traders and merchants, and built a string of important cities.

Among these was the trading port of Ugarit, in what is now northwestern Syria. There, scholars have found both typical cuneiform inscriptions and also the earliest known alphabetic script, dating from the fourteenth century B.C. The 30 characters, which resemble cuneiform symbols, correspond to sounds in the Semitic languages. Among the inscriptions were stories with striking resemblances to Biblical themes.

Something even more important was found at Ugarit: the first known alphabetical listing. This consisted of a table of the symbols in a standard order, along with cuneiform signs for the syllables *a, be, ga*...in essentially the same order they would appear in the Phoenician and later Semitic alphabets. The Ugaritic script, however, is unlike others used by the Canaanites, and does not appear again after the city was sacked and burned in about 1200 B.C.

A syllabic script was used in the Phoenician city of Byblos between about 2100 and 1300 B.C. It had about 80 characters based loosely on Egyptian hieroglyphics. In the Sinai peninsula, another group of Semites, workers in Egyptian copper and turquoise mines, used a script with about 27 characters, some resembling hieroglyphs, in about the fifteenth century B.C. It seems likely that the Semites in each case were using reduced sets of Egyptian symbols to represent the sounds of their own language. Scholars theorize that the Sinaitic Semites gave these symbols the names of some of the most common objects in their world, such as ox, house, camel and door: *alef, beth, gimel,* and *daleth.* These names are still used in the Hebrew alphabet today.

Whether it descended directly from the Ugaritic script, the script of the Sinaitic Semites, the Byblos pseudohieroglyphs, or some other Semitic script, the Phoenician alphabet, with its 22 letters, appeared in about 1200 B.C. In the Middle East, it evolved into Aramaic script, from which is derived square Hebrew, the alphabet adopted by the Jewish people. Aramaic essentially supplanted a script called early Hebrew, although the older script continued to be used by the Samaritans, a people closely related to the Jews. The earliest known Hebrew texts date from around 1000 B.C. The Arabic, Persian, and Indian scripts are all later descendants of Aramaic.

The Phoenician alphabet accompanied traders around the Mediterranean from the port of Byblos, from which came the Greek word for book, *biblios,* and the word *bible.* By about 800 B.C., the alphabet had been adopted by the Greeks. They wrote it from left to right like the old Linear B, added vowels (which are generally omitted from Semitic scripts), and made other changes in order to accommodate their Indo-European language. However, they kept the names of many of the letters, and their *alpha, beta, gamma,* and *delta* are clear reminders of their alphabet's Semitic roots.

The Greek alphabet was the forerunner of the Cyrillic alphabet now used in Slavic countries; it was carried there by Christian missionaries from Constantinople. Another descendant of the Greek script was Etruscan, used by the early inhabitants of Rome. The Etruscans were evicted by conquerors from the plain of Latium, and the Etruscan script disappeared.

Scholars differ as to whether the Latin alphabet used by the ancient Romans was adapted from the Etruscan or directly from the Greek. The earliest known Latin inscription, "Manius made me for Numerius," was found on a cloak pin dating from the seventh century B.C. As the Roman Empire spread across Western Europe, so did the Latin script. Today it is the most widely used alphabet in the world, and it is the alphabet in which this article is written.

Impact

Pictographic forms of writing are relatively difficult to learn, because of the number and complexity of their symbols. By contrast, alphabetic systems use only a few dozen symbols. Since these symbols are strung together to represent sounds, even new or unfamiliar words could be spelled, and even a word that was spelled incorrectly was usually recognizable.

These characteristics encouraged the rapid geographic spread of literacy along with knowledge of the Phoenician alphabet and its derivative scripts. While hieroglyphs and cuneiform script held sway, say around 1600 B.C., literacy was concentrated in the Nile Delta, the Fertile Crescent, and the Indus and Yellow River Valleys. By 400 B.C., the entire Middle East and the Mediterranean basin were literate. In addition, an alphabetic system made writing more accessible to the general population within each society, although the overall literacy rate was still low.

Since the alphabet made writing easier, it was used more often. Hieroglyphs and cuneiform were often used for ceremonial purposes, such as monuments and for praising the deeds of the kings who employed the scribes. They were also employed for practical functions such as keeping inventories and other records. Less frequently, scholars find snippets of poetry, stories, and letters between family members and friends.

One important piece of literature preserved in cuneiform fragments is the Sumerian *Epic of Gilgamesh*. It includes an account of an ancient flood similar to the one that appears in the Bible, and other legends resembling some of the Greek myths. The *Book of the Dead* is probably the best-known ancient Egyptian literary work. These types of texts are extremely valuable in helping us to understand the civilizations and worldviews of the people who came before us. With the alphabet, such texts increased in number.

Some material that had been passed down orally for generation upon generation began to be recorded in writing. In about 1000 B.C., Jewish scribes began writing down a collection of creation stories, religious laws, and oral history. These became the five earliest books of the Hebrew scriptures, called the Torah, and were followed by collections of prophetic teachings,

proverbs, and other texts. Together they comprise what Jews call the *Tanach* (a Hebrew acronym for "Torah, Prophets and Writings"), known to Christians as the Old Testament.

The New Testament was written in Greek within a few hundred years of the life of Jesus. It includes the four gospels (accounts of the life, teachings, and death of Jesus) and Paul's letters to fledging Christian communities around the Mediterranean. Clearly the spread of the new religion would have been much slower without the benefit of writing.

Writing was also important in the spread of Islam. Its central text, the Koran, is revered by Muslims as "the writing of Allah," just as devout Jews and Christians regard their Scriptures as divinely authored or inspired. Since Muslim law forbids the depiction of God or the prophet Mohammed, and some sects prohibit picturing any living being, Arabic calligraphy has evolved into a highly decorative art form.

Although these "founding texts" of important civilizations were religious in nature, writing is not important simply to propagate religions, but to exchange ideas of all kinds. Learning from someone once required proximity: a seat around a campfire, in a market square, or in a schoolroom. With writing, a person's words and thoughts can be transmitted to others far away.

Written words retain their power even after the writer is long gone, thus conferring a sort of immortality. A twenty-first century reader can go to any library and be in the company of Plato, Shakespeare, Jefferson, or Einstein. But not only the words of the famous have been preserved. Writing enables us to hear the voice of an Egyptian scribe from 4,000 years ago:

> *A man has perished and his body has become earth. All his relatives have crumbled to dust. It is writing that makes him remembered.*

SHERRI CHASIN CALVO

Further Reading

Claiborne, Robert. *The Birth of Writing*. New York: Time-Life Books, 1974.

Illich, Ivan, and Barry Sanders. *ABC: The Alphabetization of the Popular Mind*. San Francisco: North Point Press, 1988.

Jackson, Donald. *The Story of Writing*. New York: Taplinger, 1981.

Jean, Georges. *Writing: the Story of Alphabets and Scripts*. New York: Harry N. Abrams, 1992.

The Development of Writing Materials: 2000 B.C. to A.D. 699

Overview

Today we associate communication technology with high-speed presses and digital computers. However, they are only the most recent of the many ingenious methods people have devised to record and preserve information. The earliest writing surfaces were the walls of caves, where as many as 30,000 years ago images thought to be narratives of hunts or artifacts of archaic spiritual rites were drawn with charcoal and clay slip (a mixture of clay and water). During the next 25,000 years, calendars and inventories were carved into bone and rock. Business and legal transactions, religious texts, and other documents were etched or painted on the walls of public buildings and tombs.

The first materials produced specifically for writing emerged around the fourth millennium B.C., with a corresponding shift from the visual shorthand called *pictographs* to alphabetic scripts. Texts were either etched with a stylus or drawn with inks made by mixing powdered charcoal, ground insects, plants, or earth pigments with water. By the second millennium B.C., a variety of writing surfaces adapted from local natural resources were in wide use throughout the expanding empires of Asia and the Middle East.

Background

Carved stone tables have preserved many early forms of writing. One such artifact, the Code of Hammurabi (a second millenium B.C. Babylonian king), is a record of the rules of law and punishment, property rights, and the duties of family members "to promote the welfare of the people" and "to cause justice to prevail...." The Rosetta Stone (ca. 200 B.C.), named after the area of the Nile Delta where it was unearthed (the town of Rosetta [Rashid]), is a pronouncement of divine favor by the priests of Memphis on King Ptolemy V Epiphanes (205-180 B.C.) in honor of the ninth anniversary of his ascension to the throne. The message, repeated in Greek, Egyptian demotic (cursive) script, and hieroglyphics was etched into a black basalt stone. The Rosetta Stone's triple-tiered message became the key to deciphering hieroglyphics in the nineteenth century. Runic script, possibly evolved

from the Etruscan alphabet of northern Italy, was carved into stone (as well as bone and ivory) by Germanic peoples of Northern Europe, Britain, Scandinavia, and Iceland around the second century A.D. Rune stones were thought to be active vessels for the powerful words carved into them, rather than historical documents.

Clay, available in abundance throughout the Middle East, was used to create portable documents beginning in the fourth millennium B.C. A sharp stylus was pressed into the soft surface of wet clay tablets which were then dried in the sun. Those containing legal and business transactions were often wrapped in thin slabs of wet clay producing a reverse copy of the original. These "envelopes" were labeled to identify their contents, and sometimes marked to identify the parties involved. These dried packets could then be "archived" (cataloged and stored) for future reference. The widespread use of clay tablets coincides with the introduction of *cuneiform,* a name derived from the Latin *cuneus* (wedge) and *forma* (form). The angular writing system, attributed to the Sumerians who invaded Mesopotamia around 3500 B.C., may have become popular among expanding administrations, since angular marks could be created more quickly and easily than the curved lines of pictographs. Approximately 15,000 cuneiform tablet documents created by the Hittites survive from c. 1900-1200 B.C. Gradually scribes also began recording medical theories and scientific observations on clay tablets. One tablet documents a sighting of Halley's comet between September 22 and 28 in 164 B.C. Philosophical, historical, and mythical works were also preserved in clay. One, a 12- tablet version of the Babylonian epic poem *Gilgamesh* survives from c. 1600-1000 B.C.

A less permanent but still versatile recording material was the wax tablet, an ancient version of our "scratch paper." Soft wax in a hollow wood frame was used to record temporary data and was then recycled. Single wax tablets were used in early Mesopotamia, Greece, and Etruria. In the classical period, Greek students used them to practice their lessons. By the first century B.C., the Greeks and Romans were using multiple wax tablets tied together called *codices* from the Latin *codex,* meaning "wood." The term derives from an earlier practice of making writing sheets from birch or alder saplings. Beginning around 300 B.C., wood was also used by Mesoamerican Indians, who pounded the inner bark of *Ficus* trees and covered it with a thin lime-based plaster to create folded screen books.

Between c. 3100-2900 B.C., the Egyptians began making writing material from *Cyperus papyrus*, a tall, triangular-shaped reed that, though now extinct, grew abundantly along the banks of the Nile river. Our word "paper" comes from "papyrus," meaning "that which belongs to the house" (referring to the ancient Egyptian administration). It was processed by removing the rind, then splitting the soft, inner pith into strips and pounding them into two perpendicular layers to create a two-ply sheet. The pages were washed, dried, and glued together into rolls and wound around a spindle to prevent bending. Many documents were up to 35 feet (10.5 m) long; a few survive that measure over 100 feet (30.5 m). Writing was usually done on the inner, horizontal side, though some papyri fragments have writing on both sides. The polished surface was washable, making it one of the earliest recyclables: It was not only reused for writing, but for "cartonnage" or wrapping in the mummification process.

The earliest surviving papyrus scroll dates to c. 2400 B.C. Among the most significant Egyptian papyri were copies of the *Book of the Dead,* which were buried with people of distinction to insure their successful journey to the afterlife. By c. 650 B.C. the first rolls of papyrus arrived in Greece, though most papyri that have survived come from the drier climates of the Middle East. Papyrus was the primary writing surface among the Greeks and Romans from around the third century B.C. until after the Arab conquest of Egypt in A.D. 641. Codices (bound books) made of papyrus were introduced in Rome in the first century A.D. Papyrus was used throughout the Middle East until around the eleventh century A.D. when competition from the cheaper rag paper and the overexploitation of the papyrus beds ended its production permanently.

Tanned leather was also widely used for writing throughout the ancient world. Many of the Dead Sea Scrolls (thought to have been written by an ascetic community around second century B.C.) were written on a thin, whitish leather. Leather was used for documents in the Roman empire as late as the first century A.D., but was gradually replaced by the thinner, more versatile vellum and parchment. Technically speaking, vellum is made from the skins of young animals, and parchment from the skins of adult animals, although the terms have come to be used interchangeably. They were produced by soaking the skins in lime, then stretching them across a frame to dry. The hair was scraped away, and the surface smoothed with a pumice stone. Despite their labor-intensive preparation, vellum and parchment were less expensive than imported papyrus and silk, since they could be produced locally from domestic animals. Although they were initially used to create scrolls, eventually the skins were cut into large, uniform, oblong sheets, then folded over and stitched into the codex or book form we recognize today. Many medieval religious books and secular manuscripts—like the c. A.D. 1000 copy of *Beowulf* (an Anglo-Saxon epic poem)—were produced in this way. The decline of parchment books by the fifteenth century was due to the compatibility of cheaper rag paper with the printing press.

The Chinese were writing on bamboo strips formed into a roll by the fifth century B.C. This practice was eventually replaced by rolls of silk, a material cultivated for clothing as early as the third millennium B.C. The cocoon fibers of the caterpillar *Bombyx mori* were baked to kill the worms, then dipped in hot water to loosen the filaments. The threads, each hundreds of yards long, were wound around a spindle and woven into long sheets. Both official documents and paintings were created on silk scrolls in China for many centuries. Silk production was introduced to Korea by Chinese immigrants in c. 200 B.C. It spread to India by A.D. 300, and to the Byzantine Empire by the sixth century A.D. However, silk was always expensive even when produced locally. Consequently artisans developed other methods of producing flexible writing surfaces.

The invention of rag papermaking has been attributed to Ts'ai Lun, who ran the Chinese imperial workshop during the late Han dynasty (202 B.C.-A.D. 220). He is supposed to have demonstrated his version of paper to the emperor around. A.D. 104-105, although papermaking may have began in China 200 years earlier.

It was made by soaking hemp in water, pulverizing the fibers with a mallet, and spreading the resultant pulp into a mold made of a coarse-grained cloth stretched across a bamboo frame. As the water dripped through the mold, the intertwined fibers dried into sheets. Ts'ai Lun may also have incorporated other materials like mulberry and rags into the mixture. Over time, the

Cuneiform on clay tablets. *(Archive Photos. Reproduced with permission.)*

Chinese made other improvements, including the addition of starch for sizing and a yellow dye to repel insects. Production time was improved after the invention of a mold cover made from thin strips of rounded bamboo tied together that made it easier to release individual sheets and refill the mold immediately.

Papermaking contributed to the refinement of block printing in China. By the second century A.D., paper was used to make copies of original stone documents like the Confucian classics. The paper was pressed onto the stones, then covered with a black ink, producing a reverse image. This inspired the less costly and more portable method of making copies by printing from wood blocks, an innovation that would also be used in the West until the invention of moveable type. Papermaking technology reached Korea by the sixth century A.D., where the formula was modified to include fibers such as rice, straw, seaweed, and rattan. The spread of papermaking to Japan in the seventh century A.D. is attributed to a Korean monk named Doncho. It was subsequently introduced into India by Arabs who had learned papermaking from captive Chinese craftsmen in the eighth century.

Impact

Modern communications technology owes as much to papermaking as it does to the invention of moveable type in the fifteenth century and the digital revolution of the late twentieth century. To a significant degree, religious devotion is responsible for its widespread production. The Buddhist practice of making multiple copies of sacred texts and prayers was a key factor in the initial expansion of the papermaking industry in China. Of the approximately 15,000 books discovered in the Cave of the Thousand Buddhas in 1906, most were made of paper. Although papyrus was used by the Vatican until the eleventh century, between A.D. 1150-1390 papermaking had spread through Christian Europe. One factor in Gutenberg's pursuit of a career in printing was the opportunity to tap into the market for religious works and the thousands of indulgences sold by the Church. Multiple copies could be made quickly and inexpensively on paper with the printing press; with his invention of moveable type, the press could be reset relatively quickly. Knowing that he could sell few expensive parchment books, Gutenberg printed about 80% of his 200-copy run of the Bible on paper.

Inexpensive paper also made possible the distribution of political and often revolutionary books, pamphlets, and posters. Although the originals of important manuscripts continued to be printed on expensive vellum into the nineteenth century, paper was essential to every major bureaucracy in Europe by the Reformation. The production of cheap paper by steam

Funerary papyrus showing an offering by the deceased to the god Horus. *(Archivo Iconigrafico, S.A./Corbis. Reproduced with permission.)*

power in the nineteenth century fed the growing demand for popular reading material like newspapers, magazines, children's books, and novels. Inexpensive scientific, religious, and school texts played an important role in the education and religious life of the working and middle classes.

From ancient times individuals were charged with organizing and conserving documents. Libraries already existed in China and Sumer and Akkad in the second millennium B.C. The Assyrian King Ashurbanipal (668-627 B.C.) assembled a clay tablet library at Nineveh during his tenure. By the third century B.C., the preser-

vation of philosophy, history, poetry and other literature was considered so important that King Ptolemy I hired the Greek orator Demetrios Phalereus to collect all the works of the world for his library at Alexandria. Like today's digital backup disks, many surviving documents are contemporary copies of originals, created to preserve important information, and with good reason: the thousands of texts in the Alexandrian library twice fell victim to fires set by invaders.

Unfortunately written work has always been vulnerable to destruction by insects, vermin, air pollution, flood, fire, and accidental or deliberate acts. Many biblical works, carefully copied for preservation on parchment in monasteries throughout medieval Europe were destroyed by fire, as well as by mice and insects who ate their way through the parchment, ink, and glue. Using paper to copy documents has created even greater problems for archivists in recent years. Many paper documents printed on low-rag-content chemically processed paper deteriorated even more quickly than parchment books. Until the revival of acid-free papers and storage containers in the late twentieth century, many valuable documents were lost to overhandling, poor storage, petroleum-based inks, and even their own chemical makeup.

Today, old documents are being saved through digital scanning. The creation of electronic texts may also help slow the depletion of trees used for paper pulp, so that they may avoid the fate of Egypt's papyrus plant.

LISA NOCKS

Further Reading

Allman, William F. "The Dawn of Creativity." *U. S. News and World Report.* (May 20, 1996): 53-58.

Ashmore, Wendy, and Robert J. Sharer. *Discovering Our Past: A Brief Introduction to Archeology.* Mountain View, CA.: Mayfield, 1988.

Ceran, C. W. *Gods, Graves and Scholars.* 2nd. rev. ed. New York: Vintage-Random House, 1986.

Dawson, Raymond. *The Chinese Experience.* London: Phoenix, 1978.

Duke University Special Collections Library. Duke Papyrus Archive. http://scriptorium.lib.duke.edu/papyrus/.

Eisenstein, Elizabeth. *The Printing Revolution in Early Modern Europe.* New York/Cambridge: Cambridge University Press, 1983.

Institute of Paper Science and Technology. "The Invention of Paper: The Birth of Papermaking." www.ipst.edu/amp/museum_invention_paper.htm.

Olmert, Michael. *The Smithsonian Book of Books.* Washington, D.C.: Smithsonian, 1992.

Posner, Ernst. *Archives in the Ancient World.* Cambridge, MA.: Harvard University Press, 1972.

Sullivan, Michael. *The Arts of China.* Berkeley: University of California Press, 1999.

University Libraries, University of Iowa. Keeping Our Word: Preserving Information across the Ages. "The First Books." www.lib.uiowa.edu/ref/exhibit/book1.htm.

University Libraries, University of Iowa. Keeping Our Word: Preserving Information across the Ages. "Vellum." http://www.lib.uiowa.edu/ref/exhibit/vellum.htm.

White, J. E. Manchip. *Ancient Egypt: Its Culture and Heritage.* New York: Dover, 1970.

The Development of Libraries in the Ancient World

Overview

Libraries are institutions designed to preserve records, written material, legends, and literature. They preserve the history of time and place as well as the intellectual activity, discoveries, and innovative ideas within a culture. The first libraries of the western world were collections of literature, commentaries, records, and speculations on the way the world worked. Many of these institutions also encouraged scientific investigation, new ideas, and innovative methods of understanding the world.

Background

Libraries were inconceivable until writing was invented between 5,500 and 6,000 years ago in Mesopotamia and Egypt. Other scripts were invented by the Minoans on Crete 5,000 years ago, the Hittites in Anatolia (modern Turkey) about 4,000 years ago, and in China about 3,500 years ago.

The Sumerians in Mesopotamia developed the first writing system. Along the riverbanks, they found both clay and reeds. Pressing the end of a reed into wet clay made a distinct mark which remained after the clay dried. Sumerian writing, called cuneiform, was wedge shaped because the reeds were roughly three sided. Egyptian writing, called hieroglyphics, was done with a reed stylus which was dipped in ink. The stylus was then pressed onto a flat sheet made from papyrus, which grew in marshes along the Nile River. To make the papyrus sheet, the stalk was peeled, cut into strips, and pressed flat to form long scrolls of writing material. Both cuneiform and hieroglyphics developed from pictures that soon evolved into symbols as scribes refined the language. As the number of records increased, the need arose for storage places where they could be preserved and made available for use.

Little is known about the earliest libraries, and few have survived in any form. Some of the written works they contained deteriorated because they had been recorded on perishable surfaces, some libraries were destroyed by conquerors, and others fell into disuse when no one was left who could read the material.

The first and largest library of which there are tangible remains was in Nineveh, the capital of Assyria (an empire in what is now northern Iraq and southeastern Turkey), which existed from about 5000 to 612 B.C. The last ruler of Assyria was Ashurbanipal, who was the most powerful man in the world of the seventh century B.C. A fierce warrior, Ashurbanipal ruled Babylon, Assyria, Persia, and Egypt. He was a scholar and a patron of the arts and built a great library at his palace in Nineveh. He instructed his subjects to collect texts from all parts of his realm. Eventually the library held tablets detailing the history and culture of ancient Mesopotamia as well as what was known of chemistry, botany, mathematics, and cosmology. We know little of the activities that surrounded the library at Nineveh, but they probably centered on gathering, copying, translating, and reproducing the material that was available. Fourteen years after Ashurbanipal died, Nineveh was sacked and the library destroyed.

There had been two libraries in Egypt, one in Amarna in the fourteenth century B.C. and another at Thebes, but nothing of these libraries remains. The most important library of the ancient world was in Alexandria, a city in the Nile delta that had been founded by Alexander the Great (356-323 B.C.) in 332 B.C. The library was part of an institution of learning called the Alexandria Museum, which was established and supported by the rulers of Egypt beginning in the third century B.C. The purpose of the museum was to teach and to do scientific research, and the library was created to support this effort.

Much is known about the activities of the library at Alexandria. The librarians purchased scrolls from the private libraries of scholars and collectors in Athens and other cities, and copied and stored them at Alexandria. The library is believed to have contained a copy of every existing scroll in the Mediterranean area, and all scrolls were available to scholars associated with the museum. At its peak, at least 100 scholars worked, did research, or taught at the museum at one time. Some did original work in a sort of early research laboratory, some did original research, others wrote commentaries on the works of other scholars.

Many of the brightest minds of the ancient world worked or studied at the museum and library in Alexandria. Erasistratus, a Greek who lived from 325 to 250 B.C., became the assistant to Herophilus, the founder of the school of anatomy in Alexandria. Herophilus was one of the first anatomists to conduct postmortems (examinations of the body after death). Erasistratus is credited with being first to distinguish between motor and sensory nerves. He also traced the veins and arteries to the heart and named the trachea and the tricuspid valve in the heart. No works by either man have survived, but their ideas endured in fragments cited by others.

Euclid, the Greek mathematician called the "Father of Geometry," taught at the Alexandria Museum around 300 B.C. The ruler of Egypt, Ptolemy I, invited him to come there to work. Euclid's teaching affected many scholars in the ancient world. The *Elements,* of which Euclid wrote most sections, has been of greater influence on scientific thinking in its field than any other publication. The *Elements,* compiles and arranges all the knowledge about the various fields of mathematics that existed during Euclid's time, especially geometry, and includes his famed proof of the Pythagorean theorem.

Archimedes, the most original mathematical thinker of the ancient world, was also an engineer, inventor, and physicist. Archimedes is thought to have studied at the Alexandria Museum around 260 B.C., although he did most of his later work in his hometown of Syracuse on the island of Sicily. Archimedes discovered basic laws of hydrostatics (the study of fluids), engineering, and mathematics and invented many devices that continued in use for years, including the Archimedian screw used to raise water.

Eratosthenes, the first man known to have calculated Earth's circumference, was the librarian at Alexandria from about 250 B.C. He knew Earth was round and that the Sun cast almost no shadow at the equator at noon. To calculate the polar circumference of Earth, he placed two sticks at a measured distance apart north and south and calibrated the difference in the angle of the shadow cast by the Sun at each location at the same time. His result was amazingly close to the actual figure. He also devised a method of

A hall in the Alexandrian Library. *(The Stock Montage. Reproduced with permission.)*

finding prime numbers which is called the "sieve of Eratosthenes."

Ptolemy, the greatest astronomer of the second century A.D., also studied at Alexandria where he did work in mathematics and geography. His astronomical ideas were gathered and published in a work called the *Almagest*, which was accepted as the final authority on astronomy until the sixteenth century.

One of the most interesting scholars associated with the library at Alexandria was Hypatia, the daughter of a Greek mathematician and final director of the museum. Born in A.D. 370, Hypatia was the first woman to make a contribution to the development of mathematics. She studied mathematics and lectured on it as well as on philosophy. None of her works remain, but she is mentioned in later works. She was a leader in

new ideas of philosophy and wrote extensive commentaries on math, although there is no evidence that she did original research. As a pagan philosopher, she antagonized members of a fanatical Christian sect and was murdered by a mob in 415. Hypatia was instrumental in preserving the ancient works on mathematics and philosophy that still exist.

The museum and library buildings at Alexandria were destroyed in the civil war that occurred in the late third century A.D. Another branch of the library, located in the Temple of Sarapis, was destroyed by Christians in 391.

Impact

The collection and preservation of works of literature and history in the great libraries of ancient times was a great service for later ages, as these works became the basis of our knowledge of vanished cultures.

When the city of Nineveh was destroyed, Ashurbanipal's library was buried in the rubble and its location lost. However, when the library was rediscovered in the 1850s, many clay tablets found in the remains of this library were still readable because the clay had been fired by the burning of the city. Some of these tablets contain ancient codes of law, including the code written by Hammurabi in the eighteenth century B.C. About 20,700 surviving tablets and fragments were taken to England, and some are on display in the British Museum in London. These clay tablets provide modern scholars with most of what is known of the science, history, and literature of Babylon and Assyria. If it were not for the library of Ashurbanipal, we would know very little of the Assyrian's knowledge of the movements of the Sun, Moon, planets, and stars. In addition, important Mesopotamian epics such as the story of Gilgamesh would not have survived.

From the variety of those scholars who studied, taught, or worked at the museum and library at Alexandria, it is clear that these institutions were of vital importance to the learning and culture of the ancient world for centuries, and that they spread learning to all areas of the Mediterranean. Many books written by these scholars became influential references about specific scientific disciplines. Without the museum and library at Alexandria, we would know much less today about the world from which our culture and sciences evolved.

LYNDALL BAKER LANDAUER

Further Reading

Clagett, Marshall. *Greek Science in Antiquity*. New York: MacMillan Publishing Co, 1955.

Frankfurt, Henry. *The Birth of Civilization in the Near East*. Bloomington, IN: Indiana University Press, 1950.

Woolley, Leonard. *The Sumerians*. New York: W. W. Norton & Co., Inc. 1965.

The Development of Block Printing in China

Overview

Many centuries before the invention of the printing press in Europe, the Chinese developed a form of printing using carved wooden blocks. Two earlier Chinese inventions, paper and ink, paved the way for block printing; so too did the practice of using carved seals, which dates to early Mesopotamian civilizations. As for block printing, it too had appeared outside China, where textile-makers used it for making patterns on cloth; but in China during the seventh century A.D., the technique of printing large quantities of text with blocks first came to fruition. In time this would spawn an innovation that, when adapted in the West, would literally transform society: movable-type printing.

Background

Long before paper and printing was the invention of writing itself, which seems to have come about independently in Sumer, Egypt, the Indus Valley, and China about 6,000 years ago. This was one of the signal developments toward the beginnings of civilization itself, since the trans-

mission of ideas is essential to the propagation of learning. The Egyptians carved their hieroglyphs into stone, but the Sumerians, lacking an abundance of stone in their homeland, instead used clay blocks. Not only did they write on clay using a stylus, but in time the scribes of the earliest Mesopotamian civilizations began to use carved seals to repeat certain images—in particular, the "signature" of a ruler.

In China, the written word first appeared on bones or shells, and as technology developed, stone and later bronzeware became the preferred medium. The seal made its first appearance in the Chou Dynasty (1027-246 B.C.), when rulers and nobles commissioned artisans to carve them from precious jade or even rhinoceros horn, as well as copper. Using a rudimentary form of ink, the seal would then be impressed on a variety of materials. As for ink itself, it may have originally come from various animal and vegetable substances, but in time the Chinese discovered a more stable material. The latter came from the black excretions, such as creosote, left by burning wood and oil in lamps; later, when this innovation passed to the West, it would be incorrectly called "India ink."

Thus were born the four elements of written communication: the text itself (writing); the material on which the words or symbols were impressed (that is, clay and all the other forerunners of paper); the medium for making the text or symbols visible (ink); and the technology of transmitting ink to tablet: the seal. During the Chou Dynasty, the second of these elements evolved into more usable forms, including silk and flat strips of bamboo or wood, which when sewn together made a type of scroll.

The earliest Chinese scrolls or "books" tended to be unwieldy: it was said that when a learned man of the Han Dynasty (207 B.C.-A.D. 220) presented a series of written suggestions to the emperor, he did so on some 3,000 bamboo strips that required two strong men to carry them. Thus the invention of paper by Ts'ai Lun or Tsai-lung (c. 48-118) in c. A.D. 105 represented a significant innovation. This early paper was made of hemp, bark fibers, cloth, and even fishing nets, and over time Chinese paper-makers perfected the process. Eventually paper made its way to Southeast Asia and Korea, and by the mid- eighth century A.D. it appeared among the Arab civilizations of the Near East.

Meanwhile, the technology of making seals had continued to develop as well. There were two types of seals: relief or intaglio. The former involved carving an impression by removing all material aside from that which formed a negative or reverse image of the symbol one wished to imprint, whereas the latter entailed the opposite process—that is, carving the negative image *into* the material. The former could be used for making black impressions on a white background, whereas with the latter, all the area around the character was covered in ink while the character itself remained white.

Impact

Improvements in seal-making technology, combined with the development of paper, had paved the way for the creation of block printing. Then in the seventh century came the last crucial element: the need to mass-produce texts. This arose among Buddhist monks, who required numerous copies of their sutras, or sacred writings. So great was their need to spread information, in fact, that demand exceeded their own ability to produce copies by hand.

The solution proved to be block-printing, which involved a process that improved upon the earlier seal imprints. To create texts with block print, a monk wrote the material to be copied in ink on a sheet of fine paper. Then he coated a block of wood with rice paste, and carefully attached the written side of the sheet to the wood block. The consistency of the paste was such that, when used properly, the paper did not stick to the block—but the inked characters did. This left a negative image on the wood, enabling an engraver to cut away the areas without ink so that the text itself stood out in relief.

Next the printer used a paintbrush to ink the carved wooden block, and while the ink was still wet, spread a sheet of paper over it. He then rubbed the back of the sheet with a brush, causing the ink to imprint on the paper. Due to the need for a strong impression on the paper, it was necessary to print on only one side; otherwise the text would have been impossible to read. Despite this drawback, as well as the fact that carving the wood block required painstaking work, the new method represented a vast improvement over earlier forms of transmitting text. Instead of requiring the work of numerous monks over months or years, the production of a single text could take place over a matter of weeks, and the resulting work could be distributed to hundreds or thousands of monks. Thus by 1000, the Buddhists had printed all their scriptures, an effort that required 130,000 wood blocks and took 12 years to complete.

Thanks to the new technology, the printed word spread rapidly throughout the Buddhist world, a fact illustrated by the geographical locations of three texts variously cited as the world's first printed documents. The first of these, though most likely printed in China between 704 and 751, was a scroll discovered in Korea. Another oft-cited candidate is a Japanese text, dated to c. 764-770, commissioned by Empress Koken or Shotoku (718-770). Then there is the oldest full "book," the Diamond Sutra, later discovered in China's Gansu Province. Apparently made in 868, the Diamond Sutra consists of seven sheets of paper, which together form a scroll 16 inches (41 cm) long and 12 inches (30 cm) wide.

During the Sung Dynasty (960-1279), printing in China expanded greatly. In addition to the project of printing all Buddhist texts was an undertaking by the Imperial Academy, which commissioned the creation of some 100,000 engraved blocks for printing both sutras and Chinese histories. Then in c. 1045, a blacksmith and alchemist named Pi Sheng (fl. 1030s-1040s) developed a process even better than block-printing: movable type.

Thanks to Pi Sheng, printers did not have to carve out a new block of wood every time they wanted to print something; instead, the printer had at his disposal precast pieces of type. Using baked clay, Pi Sheng created type that he placed in an iron frame lined with warm wax. He pressed down on this with a board until the surface was perfectly flat, and after the wax cooled, he used the tray of letters to print pages. Three centuries later, on orders from the ruler Tsai-Tung (fl. c. 1390), Korean engravers developed type from bronze, which represented a vast improvement over clay because it was more durable and less brittle.

During these centuries of progress in the East, Western Europe lagged far behind: indeed, Europeans remained ignorant even of paper until the fourteenth century. Prior to that time, monks had used parchment, derived from animal skins, and instead of printing from blocks, they laboriously copied texts by hand. This in turn had numerous social implications: because books required enormous effort to produce, they were precious, and far beyond the reach of the common people. Without easy access to written material, the populace was largely illiterate, and virtually all learning lay in the hands of the Church. Furthermore, the inability to mass-produce the written word and the lack of multiple copies of many texts meant that literally thousands of writings from the ancient world, destroyed by barbarian invaders, were lost forever.

Yet in one of history's great ironic twists, the West rapidly caught up to the East and soon surpassed it. Block-printing technology apparently made its way westward at the hands of Mongol invaders, and combined with the adoption of paper in the fourteenth century helped spawn a minor revolution in information. In fact there was a brief period when block printing flourished in the West, a time that saw the production of texts such as the *Bois Protat,* dated to c. 1380, which depicts the crucifixion of Christ.

For several centuries afterward, Europeans would use a form of block printing, the woodcut, to provide inexpensive illustrations; but by then society had long since been transformed by the movable-type press. The latter was the invention of Johannes Gutenberg (c. 1395-1468), who—independent, and most likely ignorant of, Chinese advances—developed his own printing press in about 1450. As a result of this innovation, literacy spread rapidly, fueling the fires of the Reformation and other movements that completely changed the cultural fabric of Europe.

Movable type would never have the same impact in the East that it did in the West, a fact that springs from differences in the form of most Eastern and Western societies' written languages. European languages use alphabets, with just a handful of characters, making it easy for a printer to use movable type. By contrast, Chinese—as well as Korean, Japanese, and most languages of East Asia—uses pictograms, phonograms, and ideograms to represent words or syllables. Chinese is particularly complex in this regard, having more than 30,000 characters, which meant that a printer using movable type had to sort through endless trays of precast blocks. Thus the movable-type press never really caught on in the East, and printers there largely continued to use, and improve upon, the block-printing technology developed by Chinese Buddhist monks in the seventh century.

JUDSON KNIGHT

Further Reading

Books

McDonald, T. David. *The Technological Transformation of China.* Washington, D.C.: National Defense University Press, 1990.

Ross, Frank Xavier, and Michael Goodman. *Oracle Bones, Stars, and Wheelbarrows: Ancient Chinese Science and Technology.* Boston: Houghton Mifflin, 1993.

Steffens, Bradley. *Printing Press: Ideas into Type.* San Diego, CA: Lucent Books, 1990.

Internet Sites

"Historical Stories." http://china.tyfo.com/int/literature/history/200092lit-story2.htm (December 3, 2000).

"Woodblocks for Printmaking." *ANU Forestry.* http://www.anu.edu.au/Forestry/wood/nwfp/woodblock/woodblock.html (December 3, 2000).

The Early History of Cartography

Overview

A recent survey noted that nearly one-third of the people living in the United States could not distinguish North from South on a map. These results are somewhat surprising in light of the fact that maps have been an integral part of human society for over 5,000 years. Mapmaking is one of the oldest forms of communication and has taken on many different forms and functions throughout the course of history. Nearly every conceivable material has been used to produce maps, including stone, clay, skins, parchment, and even snow. Maps are generated in attempt to help people navigate better and to give us a clearer understanding of our world and surroundings.

The art and science of graphically or pictorially representing a geographic region is called cartography. These representations are usually made on a flat surface and are referred to as maps or charts. In addition, they may have non-geographical representations on them to indicate cultural areas, political precincts, natural phenomena, and many other categories. Cartography is an ancient discipline that dates back to the time of recorded history. The first maps were believed to be illustrations of prime hunting and fishing territories.

The oldest known map is Babylonian in origin and dates from about 2300 B.C. In addition, various pictorials of land features have been found with Egyptian artifacts from around the same period. It is notable that both of the areas depicted were river valleys and that knowing the intricacies of the geography would provide vital information for sustainable communities. Maps from later periods show plans for the construction of canals, roads, and places of worship. These were the predecessors of modern city planning and engineering maps.

While cosmography, the science of mapping the shape of the entire known world, was not heavily practiced until the time of the ancient Greeks, archeological studies in Iraq have uncovered a map dating from 1000 B.C. that shows Earth as a concentric circle with Babylon at the center of it, surrounded on all sides by water. However, there is little additional evidence that the Egyptians or Babylonians attempted to depict the entire planet and their place within it. Rather, their cartographic endeavors focused on pursuits of a more practical nature. They were more concerned with mapping fertile areas, regions with exceptional game, or their own boundaries. It would not be until the Greek philosopher-geographers began to speculate on the nature and shape of Earth that mapmakers attempted to map not only their surroundings, but also the entire world.

Background

The Greeks made the greatest early contributions to mapmaking through their systematic scientific pursuit of geography. This was somewhat motivated by need because they lacked productive soil for agriculture. This need led to colonization and the establishment of trade, primarily through navigable sea routes that needed to be charted. The city of Miletus was considered to be the center of cartographic knowledge and speculation around 600 B.C.

Hecataeus (sixth-fifth century B.C.) produced the first known geography book in about 500 B.C. In it he speculated that the world was a flat disk surrounded by a great ocean. This book was later modified and expanded by the great historian Herodotus (484?-420? B.C.). His significant contributions included a reference to the idea that the Phoenicians circumnavigated Africa nearly 2,000 years before Vasco da Gama (1460?-1524). He added a significant amount of new information regarding the geography of the known world and even extended into the realm of the unknown by predicting the natural features of unfamiliar lands. Herodotus also questioned the idea that the Earth was a flat disk and proposed several different theories regarding its actual shape, including one that supported the theory that Pythagoras (580?-500 B.C.) put forward that the world was a sphere.

By 350 it was commonly accepted by Greek scholars that Earth was in fact a sphere. Aristotle (384-322 B.C.) strongly argued this point and presented six lines of reasoning to demonstrate the world was shaped like an orb. Nearly all subsequent cartographers generally accepted this idea.

Dicaearchus of Messina, who was a follower of Aristotle, made significant contributions in this area. He was the first cartographer to place reference lines on world maps. He placed one running in an east-west direction and passing through Gibraltar and Rhodes. This had a significant influence on others and eventually led to the development of longitude and latitude.

The next significant Greek figure in cartography was Eratosthenes of Cyrene (276?-194? B.C.), the first person to reasonably estimate Earth's circumference. He realized that the Sun was very distant from Earth and that by using the angle by which the sunrays fell on two different cities of a known distance apart, he could estimate Earth's circumference. Eratosthenes made other contributions as well. He improved upon the reference lines of Dicaearchus and provided much scientific thought in other areas.

Hipparchus was a contemporary of Eratosthenes and was often quite critical of his work. The major contribution of Hipparchus was to apply rigorous mathematical principles to the field of cartography. He used trigonometry to help determine locations on Earth's surface and extended the reference lines of Dicaearchus to specify longitude and latitude as it is still used today. He attempted to measure latitude by using the ratio of the longest to shortest day for a particular area. Hipparchus was also the first person to partition the known inhabited world into climatic zones on a map.

Certainly the greatest and most important cartographer from the ancient world was Ptolemy (A.D. 100?-170?). Ptolemy was a great scientist who wrote one of the most influential scientific works of all time, *The Guide to Geography*. This eight-volume work dealt with the basic principles of map and globe construction, locations of various cities, theories of mathematical geography, and instructions for preparing maps of the worlds. Interestingly enough, this work had little initial influence, as it was largely forgotten and not rediscovered until 1,400 years later. The maps and directions were often crude approximations from discussions with travelers, but they were accurate enough to show relative locations and direction. The work, however, provided a summary of all of the knowledge of geography at that time. Ptolemy also contributed to the mathematical aspect of cartography as well as many other areas of science.

After Ptolemy, there seems to be stagnation in the science of mapmaking. While the Romans constructed maps of their extensive road systems, they had little use for the mathematical principles that the Greeks had extolled. Additionally, most of the ancient records had been lost or destroyed, so they had little influence until some were discovered over 1,000 years later.

Impact

Interestingly enough, there was little progress in the field of cartography immediately after Ptolemy. In fact, it was not until the late fifteenth century, when copies of his maps were published in an atlas, that the field saw renewed activity. Because of this, it can be surmised that the early cartographers exerted little impact, at least initially. Certainly each important predecessor in the field of cartography influenced those who came later, but the real impact of these people is not readily seen on society until the late fifteenth century, when explorers began to study the early maps of Ptolemy. Because of a constant exchange of ideas in ancient Greece, successive improvements in cartography were made quickly. It was less than 100 years from the introduction of the idea that maps could contain reference lines to the incorporation of longitude and latitude. As a matter of fact, this is largely the same system that exists today.

The early explorers such as Christopher Columbus (1451-1506), Ferdinand Magellan (1480?-1521), and Americus Vespucci (1454-1512) used Ptolemy's map as a guide for their voyages. While the map was as accurate as it could be for the time that it was produced, it was still grossly inadequate in many areas. For instance, it greatly exaggerated the combined size of Europe and Asia, while underestimating the size of Earth. This was a critical mistake that strengthened the idea that Columbus could reach Asia by traveling westward and actually led him to underestimate the distance to Asia as he set out across the Atlantic on his first voyage. Ptolemy's influence even extended into the Southern hemisphere, where his idea that a great southern continent existed was perpetuated for many years. It was not until 1775 and many voyages when James Cook (1728-1779) demonstrated that it did not exist. Thus, Ptolemy helped to spur the age of exploration.

This increase in exploration set up a domino effect in human society. Exploration encouraged many improvements in technology, which further aided mapmaking. These improvements included the development of the principles of navigation and improvements in the instruments for these purposes. This age of discovery brought cultures together. These interactions sometimes had a positive effect, such as an exchange of commerce and ideas, however, it also had tragic effects for some cultures.

The ancient cartographers also had considerable influence in emphasizing for future generations the need to have mathematical map constructions, as opposed to more abstract and philosophical aspects of geography. Modern scientists have made significant strides in this area, including the development of satellite technology. Yet as advanced as we have become, we still need to emphasize basic map-reading skills in our society.

JAMES J. HOFFMANN

Further Reading

Crone, G.R. *Maps & Their Makers: An Introduction to the History of Cartography.* North Haven: Shoe String Press, 1978.

Goss, John. *The Mapmaker's Art: An Illustrated History of Cartography.* Skokie: Rand McNally, 1993.

Wilford, John. *The Mapmakers.* New York: Alfred A. Knopf, 2000.

Biographical Sketches

Imhotep
2667?–2648? B.C.
Egyptian Vizier, Chief Priest, and Architect

Imhotep was an Egyptian official who served the third-dynasty pharaoh Djoser (r. 2630-2611 B.C.) as vizier, chief priest of the sun god, and chief architect. A commoner, Imhotep rose through the ranks at court to become so respected as a sage, architect, and healer that he was later deified and worshipped as a god. Today he is best known for building the Step Pyramid, one of the world's earliest stone monuments and the first pyramid in Egypt.

As Djoser's chief advisor, Imhotep was assigned the important task of building the pharaoh's tomb at Saqqara. At first Imhotep planned to build the traditional square mastaba tomb, but through a series of changes his plan evolved into Egypt's first pyramid, which he built in stages like stairs. Not only did Imhotep build his pharaoh a symbolic "stairway to heaven," but he built it to last forever, from stone instead of the more traditional mud brick. Because working with stone is much different than working with mud brick, Imhotep had to develop new building techniques so that the pyramid would not collapse under its own weight. When it was finished, the pyramid rose in six stages to a height of about 200 feet (60 m). The chamber for the pharaoh's body was dug deep into the rock under the pyramid, along with 3.5 miles (5.5 km)of shafts, tunnels, galleries, and storage rooms.

Imhotep's vision for Djoser's tomb did not end with the Step Pyramid, but extended to a huge, complex surrounding the pyramid-chapels, tombs, shrines, terraces, courtyards, life-size statues, and subterranean passageways, all built of stone. When it was finished, a stone wall 34-feet (10 m) high surrounded the complex of buildings, covering an area of about 37 acres (1.5 sq km). All the stone surfaces in the complex displayed various hand-carved decorations, including buttresses and recesses, fluted columns, papyrus-shaped capitols, and pictorial wall reliefs. Imhotep's achievement at Saqqara involved more than architecture. The whole complex was constructed to express Imhotep's vision of king and country. Egyptologists think that the complex was as large as a large town of the time.

It is clear that by building the pyramid complex, Imhotep was attempting to give material expression to the spiritual ideals of the Egyptians while providing Djoser with a model city to rule in the afterlife. Nothing on this scale had ever been attempted before, and the political implications of Imhotep's achievement are almost as important as the work itself. Only a very strong central authority could hope to muster, organize, support, and finance the labor involved in such an undertaking. Although the step pyramid was adopted as the standard type of tomb for hundreds of years, the vast complexity of the Step Pyramid complex was not repeated. The incredible organization necessary to build Djoser's complex anticipated the political structures required to build the great pyramids of the fourth dynasty (c. 2597-2475 B.C.).

In addition to his talents as an architect, Imhotep was famous in antiquity for his wisdom and his skill as a doctor. Ancient Egyptians attributed the earliest "wisdom texts" to him, although none of them has survived. Although there is no contemporary evidence that Imhotep was a doctor, he was invoked as a healer in twelfth dynasty (1938-c. 1756 B.C.) inscriptions and was worshipped as a god of medicine, possibly as early as the nineteenth dynasty (1292-c. 1190 B.C.). Later in the Ptolemaic period (310-330 B.C.), the Greeks equated Imhotep with their god of healing, Asclepios, and Ptolemy VIII built a shrine to him. The cult was still active during the first century A.D. when two Roman emperors, Tiberius and Claudius, praised Imhotep in the inscriptions on the walls of their Egyptian temples.

SARAH C. MELVILLE

Appius Claudius Caecus
340?-273? B.C.
Roman Statesman

Appius Claudius Caecus was born as a member of the noble class of Rome in about 340 B.C. He was one of the first noteworthy personalities in early Roman history and had a significant impact on society. While he is best known for initiating the construction of the famous Roman road, the Appian Way, he was also influential in other areas. Appius was an outstanding statesmen and politician. He was responsible for the first aqueduct in Rome, provided much legal expertise and reform of the legal practice, and remained an influential member of Roman society throughout his life.

Little is known of Appius's early life. The first accounts of him are in regard to his being elected censor of Rome in 312 B.C. As a member of the noble class, Appius was quite interested in political reform. He began a program of reform intended to give groups such as urban artisans and commercial interests full political rights. This, in turn, would give them a greater voice in government. To accomplish this goal, Appius admitted the sons of freedmen into the Senate and redistributed landless citizens among the basic political units.

Another example of his interest in the lower classes was his contribution to the legal field. He helped to write a publication outlining the methods of legal practice and published a list of court days for the public so they could make use of the legal system. He also wrote less technical works, most of which have been lost.

Appius was popular with the masses, as he was among only a handful of patricians who actively sought to increase their rights. There is much speculation as to why this member of nobility fought so hard for a class of which he was not a member. The explanations range from philanthropy to an attempt to break the power of the new nobility to accusations that he was a dema-

gogue obsessed with creating a new base of power with him at the center. However, his power was fleeting, as many his reforms were soon withdrawn and some of the freedmen he had franchised had been denied places in the Senate.

Certainly the lasting legacies of Appius were his construction projects. He built the first aqueduct (Aqua Appia) in Rome, which brought water from the Sabine Hills. He also initiated the construction of the great military and commercial road (Via Appia) between Rome and Capua. He was such an integral part of these projects that they were named after him. This was a unique honor during those times. The Appian Way was originally only 132 miles (212.5 km), but was extended another 230 miles (370 km) over the next 60 years. The Appian Way was so important to the Roman Empire that during that time only a curator of high rank administered it. The road was an engineering marvel, consisting of stone and mortar that lasted for centuries.

With advancing age, Appius Claudius began to suffer from a typical affliction at that time, blindness. In fact his surname, Caecus, means "the blind." Even in this state, Appius was still a viable statesman and leader. When Rome was considering making a treaty with their enemy, Pyrrhus, and subsequently giving up large portions of southern Italy, Appius delivered a passionate and eloquent speech urging the rejection of the proposal. His plea convinced the Senate and eventually Pyrrhus was driven from Italy. Few others had such a dramatic influence on early Roman society.

JAMES J. HOFFMANN

Callicrates
fl. fifth century B.C.
Greek Architect

Although virtually nothing is known about his personal life, Greek architect Callicrates is credited as one of the codesigners (with Ictinus) of the Parthenon in Athens, part of the larger Acropolis. Built for Pericles, a prominent political leader, the building is an example of the Greek attempt at perfect orderliness. The Parthenon was constructed from roughly 447-438 B.C.

During the long reign of Pericles in the fifth century B.C., Athens experienced a golden age. Pericles's most lasting contribution, however, was a beautifying campaign that displayed the power and authority of Athens to the rest of the world. He believed the glory of Athens should be visible to all.

Pericles placed Phidias (a famous sculptor) in charge of the building program for the Acropolis. Sitting high above the city on a rocky plateau, the Parthenon was its crowning jewel. For the temple's design, Phidias turned to Ictinus (considered the leading architect of the time), while Callicrates is most often listed as "master-builder" or "master of works." In this respect, Callicrates served primarily as Ictinus's contractor and technical director.

Some scholars have suggested that Callicrates might have served as the official city architect of Athens. Others have speculated that he was more concerned with the technical and managerial aspects of architecture. If the latter were true, then he would have led the construction aspects of the Parthenon, including the supervision of the workers, but would not have been responsible for the aesthetic elements.

The Parthenon became the spiritual center of Athens. While many Greek temples served as places of worship for many gods, the Parthenon was dedicated to Athena, the god of creativity and wisdom. The building could be viewed from all parts of the city and from the harbor, which displayed the city's power to passing ships.

The Parthenon looks simple and straightforward, a columned rectangle, but on closer inspection, its simplicity becomes an illusion. Many lines in the Parthenon look either straight or tapering, but in actuality are neither. Ictinus and Callicrates used Attic marble to construct the building, making it the first to be made completely of marble, including the ceiling tiles. The designers also adhered to strict rules of proportion throughout the structure.

Outside the Parthenon stood eight columns along the ends, with 17 on the sides. Inside, the Parthenon revealed itself as a place of worship. Phidias created an enormous wooden statue of Athena that stood 40 feet (12 m) high. He covered the statue in gold and ivory—ivory for the skin and gold for the clothing. Unfortunately, the statue was stolen and whisked off to Constantinople in the fifth century A.D. and later destroyed by fire. All that remains of the statue is its image on coins and small copies made from marble.

The sculptures inside the Parthenon also represent the highest form of art in ancient Greece. Phidias designed the works, but employed numerous other sculptors to complete the great number of pieces. A frieze of reliefs covered more than 500 feet (152 m) on the pediment. The fact that more than 420 feet (128 m) of the frieze still remains attests to the craftsmanship and skill of the sculptors.

Callicrates is also recognized as the designer of the Temple of Athena Nike, on the Acropolis, which was commissioned in 449 B.C., after a peace treaty was signed with Persia. The temple was designed with pentelic marble and was much smaller than the grandiose Parthenon. After many delays, construction began on the project in 427 B.C. and ended in 424 B.C.

In addition, scholars believe Callicrates designed a small Ionic temple on the banks of the Ilissos River in Athens and a temple in honor of Apollo on the island of Delos. Renowned classics professor Rhys Carpenter, in his book *The Architects of the Parthenon*, proposes that Callicrates was also responsible for the Hephaesteum, the temple of Poseidon at Sunion, the temple of Ares at Acharnae, and the temple at Rhamnous.

Despite the tourists who flock to see the Parthenon today in modern Greece, the building has been through some tumult. In A.D. 393 it was turned into a Christian church and the statue to Athena was changed to one for Mary. A few centuries later, the Parthenon became a mosque. In 1687 the building suffered its worst fate when used as a storage place for Turkish gunpowder. A Venetian soldier deliberately fired into the building and blew out its side. The Parthenon, however, survived and after hundreds of years of study is being restored.

BOB BATCHELOR

Callinicus of Heliopolis
600?-700?
Syrian Architect and Inventor

Callinicus (also spelled Kallinikos) was born in the Syrian city of Heliopolis sometime in the seventh century. Little is known regarding his life, and if it were not for one significant invention, he would have been just one of the billions of forgotten people in history. Callinicus, an architect by trade, is credited with the invention of Greek fire. This military weapon was a highly combustible material that, once explosively expelled from a cannon-like barrel onto enemy troops, ships, or buildings, was nearly impossible to extinguish. It was a secret weapon of the Byzantine Greeks and was instrumental in many of their military conquests.

Around the time Callinicus was born, there was significant hostility between the Arab and Byzantine empires. This would eventually spill over to his hometown of Heliopolis in Syria, where Callinicus was an architect and inventor. In order to flee from the advancing onslaught, he escaped from Syria and made his way to Constantinople. Callinicus was still concerned about the advancing Arabs, however, as he had evacuated his city just months before the battle of Yamuk. He was concerned that not only would the Arabs capture his beloved homeland, but possibly his newfound country as well. Thus, it seems that this Jewish refugee began to experiment with various combinations of chemicals to develop a weapon that would help defend against the Arabs. Callinicus eventually hit upon a specific combination of materials that was so insidious and effective that it helped to change the course of history. He had invented a weapon known as Greek fire, and delivered its formula to the Byzantine emperor.

The exact composition of Greek fire is unknown. It was kept as a closely guarded state secret with only the emperor and the Callinicus family who manufactured it knowing the formula. Despite the fact that the exact makeup of Greek fire still eludes our detection even today, it is assumed that it was a mixture of naphtha, pitch, sulfur, possibly saltpeter, and possibly some other unknown ingredients. When exposed to air, the mixture spontaneously burst into fire and could not be extinguished with water. In fact, this substance would burn even when submerged in water. There were few known substances that could extinguish it, with sand and urine being the two most common.

In order to use this effectively, the Romans developed a large siphon that served as a propellant, which was mounted on the hull of the ship and operated in similar fashion to a syringe. From there the Greek fire could be propelled onto a ship to wreak havoc with the enemy. Another great advantage was that it rarely backfired on its user. This weapon was quite effective and gave the Roman navy a distinct advantage.

Greek fire was first successfully used by the Byzantine fleet against the invading Arabs in the Battle of Cyzicus, off the coast of Constantinople, in 673. It was a weapon that gave the user such a decisive tactical edge that its introduction into the warfare of its time was comparable in its demoralizing influence to the introduction of nuclear weapons in modern times. Historical sources derived from Roman, Greek, and Arab writers agree that it surpassed all other incendiary weapons of the day in both its physical and psychological dominance. Thus, the dominance and longevity of the Roman fleet owes a great deal to Callinicus and his closely guarded secret.

JAMES J. HOFFMANN

Flavius Magnus Aurelius Cassiodorus
490?-585?
Roman Statesman and Historian

Flavius Magnus Aurelius Cassiodorus was born in Scylletium, Bruttium, in the Kingdom of the Ostrogoths (in modern Italy) in around 490. Cassiodorus was a statesman, historian, and monk and is credited with saving Roman culture from impending barbarism.

Cassiodorus was born as the son of a governor during the period of the Ostrogothic kings in Italy. He served as an apprentice to his father until he became a statesman in his own right. In 507 he was appointed quaestor, which was followed by an appointment to consul in 514. In 526 he became the chief of the civil service. He attained his last political post in 533 when he was selected as praetorian prefect. As a statesman under the Ostrogoths, Cassiodorus was interested in public education and developing a sound infrastructure to support it. He was relatively successful in this endeavor and the practices of ancient education continued to survive under barbarian rule. During his time in office, Cassiodorus had a relatively uneventful career that did not foreshadow the great impact he was to have on history. His zeal for public education and his desire to preserve it would continue to be the focus of his life. However, it was not until he retired from political office in 540 that he began his life's most important work in earnest.

After Cassiodorus retired, he became a monk and founded the monastery named Vivarium. The major goal of the monastery was to keep Roman culture alive and to perpetuate it through the ages. While Cassiodorus was neither a great author nor scholar, he provided the impetus for the maintenance and reproduction of cultural texts from Rome. Cassiodorus collected all types of manuscripts and instructed his monks to transcribe these works. It is noteworthy that they copied not only Christian texts, but also works that were considered to be pagan. This practice was significant because it influenced others to do the same, thereby preserving a great many of the ancient works that would not have been saved without this process. This practice was used as a model for other monasteries in the coming centuries. The significance of this cannot be overestimated because, if such a practice had not been undertaken, much of the wisdom and philosophy of antiquity would have been lost in the disintegrating Roman Empire.

The actual works of Cassiodorus fall into two distinct categories. He wrote extensively on historical and political topics, including summations of his edicts when in office. He also wrote texts that were concerned with theology, such as *De anima*, in which he discusses life after death and the soul. His most influential text, translated as *Institutes of Divine and Secular Literature,* was written for his monks and seems to be intended as a guide for learning. The first section discusses the study of the scriptures, while the second is an encyclopedia. The latter section was widely read during the Middle Ages and gave an overview of the liberal arts. The format of this book also served as a guide for encyclopedic works for many centuries.

Through his desire for public education, Cassiodorus effectively saved a large portion of the culture of Rome from total loss. Through his own writing and the obligatory writings of his monks, he helped influence others to do the same. Cassiodorus would surely be pleased by the result of his labor, which had a much more far-reaching impact than even he could have imagined.

JAMES J. HOFFMANN

Ch'in Shih-huang-ti

259-210 B.C.

Chinese Emperor

Ch'in Shih-huang-ti was a formidable figure in ancient Chinese history. As the emperor of the Ch'in Dynasty he established the parameters of dynastic rule that all others would follow for the next 2,000 years. During his rule, he unified much of China with his aggressive government, which was based on the teachings of Legalism. In fact, the name for China derives from the Ch'in dynasty. It was under Ch'in Shih-huang-ti that much of the Great Wall of China was built, as well as a huge burial compound now known as the Ch'in tomb.

Between 771 and 221 B.C., China consisted of numerous independent states found mostly in the north. Each state was fighting for control of land, in what was known as the "Warring States Period." The Ch'in, a small state in the northern regions of the Wei River Valley, gained power in the wake of this period. Ch'in Shih-huang-ti, who was first known as Ch'eng, was made king of the Ch'in state in 246 B.C. at the age of 13 while his father was held hostage in the state of Chao. Not originally intended for the throne, his mother, guided by financial motives, worked to put him there. Until Ch'in Shih-huang-ti came of age in 238 B.C., the government was run by his mother. Upon taking control of the throne he executed his mother's lover, who had joined the opposition, and exiled his mother for her role in the disobedience.

The Ch'in dynasty began in 256 B.C., but it was not to achieve its greatest power until years later, when under the advice of Li Ssu and Chao Kao, his advisors, Ch'in Shih-huang-ti began a mission to unify all the northern states under his rule. It was then that he took the title "Ch'in Shih-huang-ti," or "The First Sovereign Emperor of the Ch'in." Ch'in Shih-huang-ti then formed a government that was based on the ideals and principles of Legalism, as taught to him by his advisors. Legalism held that people were essentially selfish and base and needed a strong central government with strict rules and harsh punishments in order to function as a society. At the center of the new government were the emperor and his ministers. A harsh, sometimes cruel, autocratic rule was the result, replacing the old feudal system of aristocracy and nobility. Other schools of thought and philosophy were outlawed, especially Confucianism. Many of its teachers were executed and their books burned. By 221 B.C. Ch'in Shih-huang-ti had conquered his rival states and unified China. In an effort to reinforce the idea of a unified China, Ch'in Shih-huang-ti instituted a program to standardize the Chinese language, as well as measurements for width and length, and a series of roads and canals were built to converge on the capital city of Xianyang.

To protect his state from a Hunnish tribe of people to the north known as the Hsiung Nu, Ch'in Shih-huang-ti embarked on an amazing effort to connect the walls and fortresses, created during the Warring States Period, to protect his kingdom. The result was the Great Wall of China. It spanned, not including its many branches, 4,160 miles (6,700 km), and is one of the largest manmade features on Earth. Construction started under General Meng T'ien in 214 B.C., and lasted 10 years. Another structure of astounding proportions built under Ch'in Shih-huang-ti was a massive burial compound, known as the Ch'in Tomb. It was discovered by archaeologists in 1974, near the present-day city of Xiam. The tomb, encompassing 20 square miles (50 sq km), was a huge subterranean complex, landscaped to resemble a low, wooded mountain. In the chamber, 6,000 life-sized terracotta soldiers were found in battle formation, and in adjoining chambers thousands of smaller figurines were found. A stable of skeletonized horses was discovered and the remains of bronze gilded chariots accompanied them. Valuable gems, jade carving of trees and animals, as well as silks were also unearthed. The emperor's actual burial tomb has yet to be excavated. It was purported that it took 700,000 men more than 36 years to complete.

In the later years of his life, Ch'in Shih-huang-ti survived three assassination attempts and weathered the constant threat of revolt. When he came to power, Ch'in Shih-huang-ti claimed his government would last 10,000 years, but in fact it collapsed only four years after his death in 210 B.C., and was replaced by the Han Dynasty. Ch'in Shih-huang-ti and the Ch'in dynasty were regarded as evil aberrations, but the fact remains that the Ch'in dynasty served as the basis for all subsequent dynasties. The power achieved by the Ch'in in such a short time continues to stupefy historians. The Great Wall and Ch'in Tomb stand as testament to that great power.

KYLA MASLANIEC

Ctesibius of Alexandria
300?-200? B.C.
Greek Physicist and Inventor

Ctesibius (also spelled Ktesibios) was a Greek physicist and inventor who was probably born in Alexandria sometime around 300 B.C.. He was the first of many Greeks to become part of the great ancient engineering tradition in Alexandria, Egypt, and, as such, was the influential predecessor to many subsequent inventors. While he was an enthusiastic and prolific inventor, he is most famous for two particular inventions. The first was an improvement of the clepsydra or water clock, by which time was kept with dripping water maintained at a constant rate. The second invention was the hydraulis or water organ, a mechanized device in which air was forced by water through organ pipes to produce sounds.

Like many of the significant individuals from antiquity, very little is known about the life of Ctesibius. There are no direct sources dealing with his life and times, but a cloudy sketch of his life can be pieced together from various historians. Ctesibius was born the son of a barber and started out in the same career. One of his first inventions was an occupationally related counterweighted mirror.

The device involved a mirror, placed at the end of a tubular pole, and a lead counterweight of the exact same weight, placed at the other end. This setup would allow the mirror to be adjusted readily to accommodate the height of each patron. Ctesibius also noticed that when he moved the mirror, the weight bounced up and down while making a strange whistling noise. He understood this noise to be the air escaping from the tube and wondered if this principle could be utilized to make music. He began to think about the power of both air and water and made use of them in his inventions.

Ctesibius has been credited with building singing statues, pumps, water clocks, and the world's first keyboard instrument. His improvement of the water clock resulted in a time-recording device whose accuracy would not be surpassed for over 1,500 years.

Modern society is to a large extent dictated by time. We need and rely on accurate devices to measure the passage of time. This, however, is a relatively recent phenomenon. In the past, the forces of nature, rather than raw time, governed people's lives. In addition, the technology needed to measure accurate time was not well understood. During Ctesibius's day, water clocks were used to measure the time that defendants were able to speak when on trial. It was a simple device much like an hourglass, but with water rather than sand as the timekeeping medium. Water was placed into a jar with hole in the bottom, and when the water ran out, so did the defendant's time. Ctesibius realized that as the volume of water changed, so did the time, so he improved on the design by adding two other containers. The first feed into the jar to keep it at a constant level and the second had a float with a pointer that could accurately measure the number of drips. In this fashion, Ctesibius invented a timing device that remained the primary model until into the fourteenth century, when falling weights replaced falling water.

Ctesibius has also been credited with the invention of the organ. He recognized that water displaced air in a bucket and used that principle to keep the pressure high in the organ even when the pump was on the recovery cycle. This gave his organ a continuous sound, which could be changed by selecting different operating valves. It is unfortunate that our knowledge of Ctesibius is severely fragmented and second-hand. He was obviously a mechanical genius who influenced his peers and certainly left a much greater legacy than he is given credit for.

JAMES J. HOFFMANN

Dionysius Exiguus
500?-560?
Scythian Theologian, Mathematician, and Astronomer

Dionysius Exiguus, a Roman theologian, mathematician, and astronomer, is known for inventing a Christian calendar that was later incorporated into the currently used Gregorian calendar. In addition to providing calculations for determining the date of Easter, Dionysius's calendar is notable for establishing its starting point from the birth of Jesus Christ, thus introducing the designations B.C. and A.D.

Little is known of Dionysius's early life. It is documented, however, that he arrived in Rome about the time of the death, in 496, of Pope Gelasius I, who had summoned him to organize the official archives of the church hierarchy. The Roman Empire was in ruins, and the city of Rome itself was dilapidated and nearly deserted.

A trained mathematician and astronomer, as well as a master theologian, Dionysius spent his days working at a complex now known as the Vatican. He wrote church canons and spent a great deal of time thinking about how to make order of time itself. He worked as a church scholar for many years and in 525, at the request of Pope John I, he began work on the calculations that would become the basis of the Gregorian calendar centuries later. At the time, the Julian calendar, devised by Julius Caesar (100-44 B.C.), was used as the working calendar of the Church. One of the most problematic calculations for these Roman timekeepers was figuring out where the holy days fell, especially Easter.

Pope John I asked Dionysius to calculate the dates upon which future Easters would fall. The Church had adopted a formula some 200 years earlier stating that Easter should fall on the first Sunday after the first full moon following the spring equinox. Dionysius carefully studied the positions of the Moon and Sun and was able to produce a chart noting the dates of upcoming Easters, beginning in the year 532. This compilation was received with some controversy, but was accepted. At that time, most people used the year designated as either 1285, dated from the founding of Rome, or 248, based on a calendar that began with the first year of the reign of Emperor Diocletian.

Dionysius defended his decision, suggesting that his calculation was taken from the reputed birth date of Jesus Christ. He "preferred to count and denote the years from the incarnation of our Lord, in order to make the foundation of our hope better known." The new Easter charts reflected this preference, which began with anno Domini nostri Jesu Christi DXXXII, or A.D. 532. Dionysius's system used the designations A.D. (after Christ's birth) and B.C. (before Christ). However, Dionysius miscalculated Christ's birth as A.D. 1, as there was no zero value in the Roman numeral system, instead of 4 or 5 B.C. (before Christ) as is now generally accepted. The sixteenth-century pope Gregory III later adapted and incorporated the Julian calendar, with its problem of loosing entire days over time, into what became known as the Gregorian calendar. This calendar is the most popular and widely used in the world today.

In addition to his skill as a mathematician and astronomer, Dionysius was highly regarded as a theologian. He is credited with writing a collection of 401 ecclesiastical canons or scriptures that would become important historical documents about the early years of Christianity. He is also responsible for compiling an important collection of decrees written by Pope Siricius (344?-399) and Pope Anastasius II (?-498). He translated many works from Greek into Latin, thus preserving many important documents for later scholars.

LESLIE HUTCHINSON

Eupalinus
fl. sixth century B.C.
Greek Architect and Engineer

In the sixth century B.C. Eupalinus designed and constructed a tunnel that brought water from a source outside the capital city of Samos down to its citizens. Deemed one of the seven wonders of the ancient world, the Eupalinion Tunnel passes from the top of a hill and goes under the earth more than 65 feet (19.8 m) to a water tank in the town. Amazingly, the water still flows the same way today.

Hardly any information survives about Eupalinus. According to the great science fiction writer Isaac Asimov, who also loved and studied ancient Greece, the Greeks "emphasized abstract thought and paid little attention to their own record as practical engineers." Thus, little remains about men like Eupalinus. What is known about Eupalinus is that he was an architect/engineer from Megara. However, his famous work took place on Samos, an island on the eastern Aegean Sea, only a mile from the coast of Asia Minor. The historian Herodotus (c. 484-c. 420 B.C.) believed that the three greatest engineering feats in ancient Greece all occurred on Samos: the great temple to honor the goddess Hera, the Eupalinion Tunnel, and the amazing sea wall that protects its harbor.

Ancient Greece set the standard for public works, including community waterworks. Leaders realized that such public works helped boost the economy and led to better sanitation among the people. Often, tyrannical leaders would focus on public works to win favor with the citizenry. Bringing water into the cities was a monumental task, and many great thinkers spent their lives grappling with the issue.

The Greeks relied mainly on aqueducts and bridges to bring water into the cities. They favored these devices because they thought water could only be transported downward or on a straight path. Thus, they would build aqueducts through mountains or bridges to pass over valleys.

Eupalinus began work on the tunnel in Samos on the command of Polycrates, the tyrant governor of the region. Interestingly, after Eupalinus designed the tunnel, digging began from both ends simultaneously, which led it to be called the "two-mouthed tunnel" by Herodotus. It is unclear exactly how many people actually worked in the tunnel at one time, but estimates range from at least two and perhaps as many as 15. When the two sides met, they were only off by 15 feet (4.5 m). Scholars believe that the teams were comprised of slaves who carved the tunnel out of the rock using hammers and chisels.

In modern times, a German team studied the tunnel and found details that made its construction even more impressive. At one point, the workers had to deviate from the plan due to unstable soil. Even with this turn, they still found their way to the workers coming from the other end.

Since the aqueduct brought fresh drinking water into the city, it had to be completely lined with stone. Clay pipes delivered the water through a trench dug in the floor. After 10 years of work, the tunnel was completed. It measured more than 3,000 feet (915 m) long and approximately 6 feet (1.8 m) in diameter. Dug through the rock that makes up Mt. Castro, the tunnel is located less than 100 feet (30.5 m) above sea level.

The people of Samos attempted to use the aqueduct in 1882, but were unsuccessful. Nearly a century later, from 1971 to 1973, the German Archaeological Institute of Athens began uncovering the tunnel. Today, the tunnel is visited by flocks of tourist who revel in the natural beauty of Samos.

BOB BATCHELOR

Hadrian
76-138
Roman Emperor

During his more than 20 years as emperor, Hadrian traveled throughout Rome's vast empire, ensuring the well being of its citizens, building its defenses, and overseeing great public works projects, including construction of the wall in southern Britain that bears his name.

Hadrian, whose full Latin name is Publius Aelius Hadrianus, was believed to have been born in his family's homeland, Italica, which is now in southern Spain. His father died when he was just 10 years old, and he went to live with his cousin, Ulpius Trajanus. Hadrian returned to Italica five years later and received his military training, but he remained there for only a few years before moving to Rome and beginning his ascent to power. He served as military tribune with three Roman legions in the provinces of Upper and Lower Moesia.

In 97 he was summoned to Gaul to convey congratulations to the newly designated emperor Trajan. Hadrian gained the favor of Lucius Licinius Sura, the man responsible for Trajan's power, and earned the trust of Trajan's wife, Plotina. In the year 100, Hadrian married Trajan's grandniece, Vibia Sabina. Two years later, Trajan appointed him to the command of the First Legion, and called on Hadrian to assist him in fighting the Dacian war.

The emperor's young protégé rose to the praetorship in 106, earned the position of governor of Lower Pannonia a year later, then attained the coveted post of consul in 108. Unfortunately, Sura died and powers opposed to Sura, Plotina, and Hadrian took over Trajan's court, stalling Hadrian's rise to power for nearly 10 years. It was not until 117, when he was put in charge of Trajan's army in Syria during the Parthian wars, that he returned to public service. On August 9 of that year, Hadrian learned that Trajan had adopted him. Two days later, Trajan's death was reported, and Hadrian succeeded the elder emperor.

Hadrian set out to return to Italy, but before he could assume his new position, the prefect of the Praetorian Guard, Acilius Attianus, ordered the execution of four dissidents in Rome to assure the safety and stability of Hadrian's regime. This act made the public suspicious of their new emperor, and when he arrived, Hadrian had to regain his people's favor, which he did by committing great acts of generosity and sponsoring elaborate gladiatorial games.

Hadrian remained in Rome for three years before setting out on a lengthy journey throughout the Roman Empire. He began in Gaul, establishing order within his armies there, before continuing on to Britain in 122. Over the course of the next three years, he also visited Spain, the Balkans, and Asia Minor. He returned to Rome in 125, but just three years later set out again, this time venturing to North Africa and traveling as far as Egypt.

Over the course of his rule, Hadrian's artistic and architectural patronage was well renowned. During his visit to Britain in 122, he directed the construction of Hadrian's Wall to mark the boundary of Rome's empire and to protect Roman citizens living there. The 73-mile (117.5-km) long wall took six years to construct, and stretched from Wallsend-on-Tyne in the east to Bowness-on-Solway in the west. In Rome, he oversaw the construction of bridges, roads, aqueducts, and temples. He also built a grand villa for himself in Tivoli, outside Rome, oversaw construction of the Temple of Rome and Venus, and rebuilt the fire-ravaged Parthenon.

Hadrian made his final journey abroad in 134, to quell a Jewish revolt in Judaea. In 138 the aging emperor chose for his successor 18-year-old Annius Versus, who would later become Marcus Aurelius (121-180). Hadrian died after a prolonged illness at the seaside resort of Baiae.

STEPHANIE WATSON

Hero of Alexandria
First century A.D.
Greek Inventor and Mathematician

Hero (or Heron) of Alexandria was a prolific writer of mathematical and technical textbooks. His best known works are *Pneumonics*

and *Metrica*. Credited with the invention of an early form of steam engine, Hero created a number of technical devices, including the odometer, dioptra (surveying tool), and screw press.

Little is known about Hero and his life. In fact, the time during which he lived is subject to debate, with speculation ranging from 150 B.C. to A.D. 250. The most accurate estimate appears to be around A.D. 62. Even less is known about his personal life. Due to the number of books he wrote, and the content of these books, it has been suggested that he was appointed to the Museum or the University of Alexandria, where he probably taught mathematics, physics, pneumonics, and mechanics. Many of Hero's books were likely intended to serve as textbooks for his classes. What kind of person Hero was has also been the subject of debate. While some considered him to be incompetent and uneducated, simply copying the works of different scientists, others believed him a skilled mathematician and creative inventor.

Hero wrote many books, with *Pneumonics* being the longest and perhaps the most read. It was very popular during medieval times and during the Renaissance. The book outlined various pneumatic devices, and shared descriptions of how they worked. Most were no more than toys used for magic and amusement, and has led some scholars to believe he was not a serious scientist or inventor. Hero indicated that some of the inventions were his and that others were borrowed, but did not clarify which ones were actually his, giving the impression that he was merely collecting the knowledge of others. Most formed this opinion before some of Hero's works, such as *Metrica* and *Mechanics,* had been found. *Metrica,* his most important work on geometry, was lost until 1896 and contained formulas to compute the areas of things like triangles, cones, and pyramids. The area of the triangle is often attributed to Hero, but it is likely he borrowed it from Archimedes (287?-212 B.C.) or the Babylonians. *Mechanics* deals with machines, mechanical problems of daily life, and the construction of engines. Though these books have been criticized for their preoccupation with child-like toys and disorganization, they were likely used as textbooks. The attention to popular toys was probably employed to explain the principles of physics and pneumonics to students, and the lack of proper organization in his books may result from the fact that they were never completed. Other books by Hero include *Dioptra, Automata, Barulkos, Belopoiica, Catoprica, Definitiones, Geometrica, De mensuris,* and *Stereometrica.*

One of Hero's greatest achievements was the invention of the aeolipile, considered by many to be the first steam-powered engine. The plans for this machine are found in *Pneumatics.* Also described in *Pneumatics* were siphons, a fountain, a coin-operated machine, a fire engine, and other steam-powered machines. In *Dioptra,* Hero described the diopter or dioptra, a surveying instrument similar to the theodolite. Hero also displayed a familiarity with astronomy in a chapter of *Dioptra,* in which he described a method for finding the distance between Rome and Alexandria using a graphical formula based on the position of the stars. Another notable invention of Hero's was the screw press; at the time it was a new and more efficient way to extract juice from grapes and to extract oil from olives.

Hero's contribution to science was varied, though his tireless devotion to the collection of ideas and knowledge was significant in itself. Several of Hero's machines, such as the steam engine, are often cited as his most important contributions. While Hero did not invent the steam engine as we now know it, he did contribute to its eventual creation. The steam engine had a major impact on society, allowing physical, time-consuming labor to be completed by a machine, and freeing people to concentrate on other things, like exploration and discovery. The field of mathematics also benefited from Hero. His books chronicled the mathematical knowledge of his day and allowed others who came later to build on that work.

KYLA MASLANIEC

Ictinus

Fifth century B.C.

Greek Architect

Ictinus, a celebrated Greek architect, worked on such famous structures as the Parthenon on the Acropolis, the Temple of the Mysteries at Eleusis, and the Temple of Apollo Epicurius at Bassae. The sheer scale, as well as the artistic grace of the Parthenon, are testament to the skill of Ictinus and other Greek architects of the day.

The exact dates Ictinus's birth and death are not known, but it is clear that he lived during the fifth century B.C. Very little is known about his life, though much is known about his work. It thought that Ictinus was not Athenian, but rather from the western Peloponnese. During the golden age of art and architecture in Greece, Pericles commissioned Ictinus and Callicrates to

work, under the artistic vision of Phidias, on the Parthenon. The Parthenon, when completed, embodied all the refinements, civilization, and glory of ancient Greece and Athens.

The Parthenon was erected during the rule of Pericles in Athens and on the heels of a military victory over the invading Persians in 479 B.C. Pericles presided over the newly emerging democratic government and perhaps as part of a public relations campaign decided to rebuild the temples of the Acropolis, destroyed in war, in honor of the goddess Athena. The word "Parthenon" means "apartment of the virgin." Pericles enlisted Phidias, a famous sculptor and artist, to oversee the project, and Ictinus and Callicrates were chosen to actually design the Parthenon. Some have suggested that Ictinus and Callicrates were rivals not collaborators, while other sources claim Ictinus was the creative and artistic force behind the project and Callicrates played the role of engineer. Two years of difficult planning passed before construction began on the Parthenon in 447 B.C., during the Panathenaic Festival. The temple itself took less than 10 years to complete. The last stone was laid in 438 B.C., though work on the exterior of the temple continued until 432 B.C. The exterior Doric columns of the temple were 6 feet (2 m) in diameter and 34 feet (10.4 m) high. The temple itself was 101 feet (31 m) wide and 228 feet (69.5 m) long. There were three architectural styles used by the Greeks at that time: Doric, Ionic, and Corinthian. The Parthenon employed the use of the Doric, the most simple and severe of the styles.

Important to the design of the temple were the optical refinements used to make such an imposing structure more graceful and appealing. To the human eye straight lines appear to bulge or sag, but this optical illusion was counteracted in the design of the temple. Some of the methods employed by Ictinus were already in use by Greek architects, but these refinements reached new heights in the Parthenon. To that end, Ictinus and Callicrates made the Parthenon appear perfectly symmetrical when little of it was. The Parthenon was also famous for the artwork that it housed, including an enormous ivory and gold statue of Athena that stood 39 feet (12 m) high. Many of the sculptures in the Parthenon are now in museums or are missing, as in the case of the large statue of Athena.

The Epicurean Apollo, another notable architectural work by Ictinus, is one of the few nearly complete temples still standing. It was built in 420 B.C. on a 3,732-ft (1,131-m) rise near Phigalia, called Bassae. Ictinus was the main designer of the temple and incorporated all three orders of architectural style in the temple. It was a unique building—primitive, wild, and somewhat crude in appearance for the time. Perhaps in response to working under the artistic control of Phidias for so long on the Parthenon, Ictinus incorporated little sculpture in the Epicurean Apollo. Built in honor of Apollo, the temple harmonized the qualities of the temple with its natural environment. As scholars note, Ictinus seems to have deliberately built the temple in a wild and primitive style to reflect its wild surroundings. Ictinus also worked on the Temple of the Mysteries at Eleusis in around 430 B.C.

Despite the ravages of time, the Parthenon still symbolizes the strength and accomplishment of the society that built it. The aesthetic and emotional impact that the Parthenon has on those who see it now, as then, is incredible. Many of the artistic and architectural designs of the ancient Greeks, including those of Ictinus, are used in building designs today, and through the durability of these monuments modern society has learned much about the ancient Greeks.

KYLA MASLANIEC

Isidorius of Miletus
Sixth century A.D.
Turkish Architect and Engineer

Isidorius of Miletus was born in Turkey during the early sixth century. Along with Anthemios of Tralles, Isidorius designed and constructed the Church of the Holy Wisdom, or the Hagia Sophia, in Constantinople. This church, a magnificent example of Byzantine architecture and design, was built in 532-37 under the reign and personal direction of Emperor Justinian (483-565).

Little is known of Isidorius's early life. He was born in Miletus and presumably received his education and training in architecture and engineering in the city of Constantinople. In addition, he was a respected scholar and teacher, known for revising the work of Greek mathematician Archimedes (287?-212 B.C.) and writing a commentary of a book by Hero of Alexandria (first century A.D.), a mathematician who invented toys, a pneumatic pump, and a formula for expressing the area of a triangle. Isidorius invented a compass in order to study geometry and construct parabolas. Several of his students contributed to *Elements of Geometry of Euclid* and

commentaries on the work of Archimedes.

Isidorius is best known for his architectural and engineering collaboration with Anthemios of Tralles in the design and construction of the Hagia Sophia. Justinian commissioned this masterpiece of Byzantine architecture after a fire destroyed the first church in 532. Justinian was the driving force behind the architectural revival that built or reconstructed more than 30 churches in Constantinople. Byzantine churches reflected a wide variety of styles. The Hagia Sophia, a basilica type, incorporated vaulted arches and domes and a very elaborate and ornate interior.

The Hagia Sophia is the crown jewel of Byzantine architecture. Isidorius and Anthemius, under the directives of Justinian, designed, engineered, and built one of the most memorable buildings in the history of architecture. A central dome rises 185 feet (56.4 m) to give the interior of the church a spaciousness that mimics the feeling of being outdoors. This illusion of space was achieved by the use of pendentives, a new building form designed and used for the first time in the construction of the Hagia Sophia. Four pendentives in the shape of curved or spherical triangles support the rim and are in turn locked into the corners of a square, formed by four huge arches. This engineering style based on the use of pendentives became known as "hanging architecture." It gave the interior of the buildings an open ethereal quality and was incorporated into the exterior of the building with immense buttress towers. Walls covered with colorful mosaics and elaborate designs conceal the outside of the church.

The Hagia Sophia was built in the remarkably short period of five years. The innovative nature of the design and perhaps the speed of construction made the structure unstable. The first dome fell after an earthquake, and its replacement needed to be repaired again in the ninth and fourteenth centuries. Nearly all churches built during the next 1,400 years reflected the architectural and engineering work of Isidorius of Miletus.

LESLIE HUTCHINSON

Justinian I
483-565
Byzantine Emperor

The most famous of all the emperors of the Byzantine, or Eastern Roman Empire, was Justinian I. After becoming emperor, he em-

Justinian I, ruler of the Eastern Roman Empire from A.D. 526-565, preserved Roman law for future generations. (*Bettmann/Corbis. Reproduced with permission.*)

barked upon an extensive building program that produced many magnificent examples of early Byzantine architecture, including churches, aqueducts, and canals, throughout Constantinople. He commissioned the building of the Church of the Holy Wisdom, or the Hagia Sophia, which is the best known example of the Byzantine style of architecture. Many of his administrative programs as emperor have stood the test of time and have been integrated into modern policies.

Flavius Petrus Sabbatius Justinian was born in the year 483 to Slavic parents in a country along the eastern Adriatic coast. Little is known of his early years except that as a youth he was adopted by his uncle, Emperor Justin I, and was educated in Constantinople. In 527 Justin made him coruler of the empire. When his uncle died a few months later, Justinian became the sole emperor.

Justinian was known as a strong ruler and excellent administrator. When he assumed power the laws of the empire were in confusion. Many were out of date, many contradicted themselves, and different provinces had a different understanding of the laws. One of his most valuable contributions was his code of Roman law, which took all the laws from the Roman Empire and consolidated them into one uniform system. This became known as the *Codex Justini-*

anus and comprised the most logical and fairest system of law. In later centuries, when Europe began to develop into states, this code became the legal basis of the new governments. Today the laws of most European countries and the Roman Catholic Church show the influence of these laws compiled by Justinian I.

Under Justinian, the Eastern Roman Empire enjoyed its greatest glory. Financed by taxation, he used these funds to construct buildings in the capital city of Constantinople. The golden age of early Byzantine art and architecture blossomed under Emperor Justinian, who was a prolific builder and patron of the arts. Throughout his vast empire he authorized the building of forts and aqueducts, and the building or rebuilding of 30 churches. The most famous of these churches is the Church of the Holy Wisdom, or Hagia Sophia, in Constantinople. This church, designed and engineered by Isodorius of Melitus and Anthemius of Tralles, was a magnificent example of Byzantine architecture.

The Hagia Sophia was built in five years and incorporates a style of architecture known as "hanging architecture," which gives the church its ethereal quality. The domed structure uses pendentives, a new technique at the time of its construction, that support the dome on a square framework of four enormous arches. This engineering feat gives the structure a feeling of weightless stability and a visual sense of great spaciousness.

Justinan was also instrumental in the development of an art form known as mosaics. Mosaics were the favored medium for the internal decoration of the Hagia Sophia and other Byzantine churches. Mosaics were created by assembling small pieces of colored glass or enamels, occasionally overlaid with gold leaf, into pictures and designs. Spread over the walls and vaults of the churches, these mosaics created a luminous effect that complimented the mystic character of the Christian Church as well as embellishing the magnificence of the imperial court, presided over by emperor Justinian I.

LESLIE HUTCHINSON

Shotoku Taishi
574-622
Japanese Prince

Shotoku Taishi was born in Yamato, Japan, in 574. As the crown prince of Japan, he helped to shape Japanese culture and history in many aspects. Specifically, he was instrumental in the development of Japanese constitutional government, he opened cultural exchange with China, which had a tremendous impact on Japanese society, and he undertook important building projects, such as irrigation and building projects. Shotoku was also a prolific author, by which he influenced ideas about ethics, the system of government, and how history was recorded. He even influenced the hairstyles of those in both his own era and modern times.

Shotoku was born into the powerful Soga family, the second son of the short-reigned emperor Yomei. As a result of political upheaval, his aunt came to power and Shotoku was appointed as the crown prince and regent in 593. He remained in this position for nearly 40 years until his death in 622. Shotoku firmly believed that Chinese culture had significant things of value that Japan could extract for its own use. His initial and most influential act was to send envoys to China in order to facilitate cultural exchange. This was the first gesture of its kind in over 100 years and it opened up avenues for cultural, economic, and political exchange.

The infusion of Chinese culture into Japan had many positive effects. Once the cultural doors were open, scholars, monks, and skilled workers, such as artisans and craftsman, flooded into Japan and helped to bring about social, political, religious, and economic reforms. The Chinese calendar was adopted for use and support of both Buddhism and Confucianism was strongly encouraged. There was a flurry of construction of Buddhist temples, some of which still stand today. But the most important change came with the adoption of a Chinese style of governance.

Shotoku reorganized the court system using the Chinese model and instituted a system of ranks, which were identifiable by the color of the headgear associated with it. This was an important governmental change, as it helped to break free from the system of nepotism and introduced one based upon merit. His most important contribution, however, was the writing and adoption of a Chinese-style constitution in A.D 604.

The *Seventeen Article Constitution* is one of the most important documents in Japanese history. This constitution, authored by Shotoku, was intended to be held as a model for Japanese government, and it formed the philosophical basis of Japanese government for subsequent generations. It consisted of a set of instructions, aimed at the ruling class, concerning ethical concepts and the bureaucratic system. This constitution

was firmly entrenched in Confucian philosophy, although there are also a number of Buddhist elements. It expounds on the belief that there are three realms in the universe: Heaven, Man, and Earth. It further states that the general welfare of the people is the task of the emperor, who had been placed in authority by the will of Heaven. It also stressed following such virtues as harmony, regularity, and moral development.

Shotoku's influence was far reaching and of both a political and cultural nature. He even influenced hair designs, still worn today, that reflect traditional Japanese culture. He wore his hair pulled up to form a knot and bundled that knot on top of his head. With his reorganization of Japanese government and culture, Shotoku left Japan a well-defined central administrative system and a rich cultural legacy.

JAMES J. HOFFMANN

Theodoros of Samos
Sixth century B.C.
Greek Architect and Sculptor

Theodoros, a sixth-century B.C. architect from the Greek island of Samos, designed the third Ionian temple honoring the Greek goddess Hera. Theodoros was the son of Rhoikos of Samos, also an architect of the colossal temple. Although some scholars suggest Theodoros may have been the son of the sculptor Telekles, most regard Telekles as an additional son of Rhoikos, and the brother of Theodoros.

Theodoros and Rhoikos built the Temple of Hera at the Samian town of Herion, believed to be the goddess Hera's birthplace. Samos was an especially prosperous island during the Archaic era, as well as a center for engineering and the arts. Samos was ruled by the tyrant Polycrates, who commissioned the monumental building. Theodoros built the temple over the ruins of a previous prehistoric monument honoring Hera (mother of the gods) on such a massive scale (the structure had 104 columns, each rising 60 feet [18.3 m] high) that it became known as the "Labyrinth of Samos," named after the famous maze on Crete. Theodoros designed the temple according to the 10-part system, in which the field of vision is divided into 10 parts of 36° each. Using the geometry of Pythagoras (580?-500 B.C.), also from Samos, Theodoros worked with angles in proportion to his 36° standard, resulting in a design noted throughout the ancient Greek world for its symmetry and majesty.

The Temple of Hera's grand scale, as an architectural type, was without precedent in Greek temple architecture. Theodoros was the first to use the 10-part system of design, which became synonymous with Ionian architecture. The style endured more than 700 years, as medieval architects later applied similar principles of proportion in their designs of Gothic cathedrals. Ironically, Theodoros's Temple of Hera survived less than 100 years. Scholars credit its partial destruction with an attack by the Persians, an earthquake, or a sinking foundation.

By the mid-sixth century, word of the colossal temple at Hera had reached Ephesus, an Archaic cosmopolitan seaport. Not to be outdone by the Samians, the rival Ephesians began construction of a giant temple dedicated to the god Artemis. Theodoros provided supervision and technical advise to the two Cretan architects in charge of the temple. The monumental Temple of Artemis was built in 10 years (560-550 B.C.) in the Ionian style of Theodoros, with 127 white marble columns, each 65 feet (19.8 m) high, surrounding the interior. The temple was the largest of its time, and the first to be constructed entirely of marble. Fishermen on board boats approaching the harbor of Ephesius could see the massive white temple before land was visible. Theodoros's Temple of Artemis became recognized as one of the seven wonders of the ancient world.

Additionally, Theodoros and his brother Telekles were accomplished sculptors. Having spent time in Egypt, the brothers were said to have utilized Egyptian techniques of proportion so that, while Theodoros was in Ephesus and Telekles was in Samos, each could independently make half of the same statue and later join them perfectly. When the brothers constructed the statue of Pythian Apollo in this manner, it was said that the two halves corresponded so well that they appeared to have been made by the same person. Theodoros brought from Egypt to Greece the technique of smelting iron and pouring it into molds to make statuary castings. Later, Theodoros improved upon the technique to introduce the fusing and casting of bronze.

Throughout his life, Theodoros used his mathematical and scientific abilities in artistic pursuits. Besides architecture and sculpture, Theodoros also made an emerald signet for the Samian ruler Polycrates, wrote a treatise on the Temple of Hera, and invented artisans' tools, including the lathe.

BRENDA WILMOTH LERNER

Vitruvius
d. c. 25 B.C.
Roman Architect and Engineer

Vitruvius is best known as the author of *De architectura*, the first attempt at a comprehensive study of architectural practice. This manual dealt not only with building methods and materials but also sought to place architectural practice within the larger sphere of liberal arts. Though its influence on later Roman architecture was limited, *De architectura* was widely read during the Renaissance and became the authoritative work on classical architecture.

Few facts of Vitruvius's life are known, and his identity remains in question. Only his family name, Vitruvius, is known with certainty. There is good reason to think that his cognomen was Pollio, and he is often referred to today as Marcus Vitruvius Pollio. He worked in some capacity for Julius Caesar (100-44 B.C.) and was later employed as a military engineer by Octavian (63 B.C.-A.D. 14), the future Emperor Augustus. Upon retirement, Vitruvius came under royal patronage. His only known civil engineering project was the basilica at Fanum Fortunae (modern Fano, on the Adriatic coast of Italy).

The 10 books of *De architectura* cover a wide range of topics. Vitruvius begins with a discussion of the nature of architecture, claiming the province of the architect encompasses all tasks associated with the building of a city. Consequently, the branches of knowledge architects should be familiar with include arithmetic, drawing, geometry, optics, history, philosophy, literature, music, and medicine. He then discusses town planning—the division of land by laying out of streets, the distribution of sites for public and private buildings, the methods and materials appropriate to their design, and techniques for supplying water via conduits and aqueducts. Various matters of practical import are also dealt with in detail, including flooring, stucco-work, painting, and color schemes. The last few books of *De architectura* deal with a number of subjects that, though seemingly unrelated, fell well within the purview of the architectural engineer of antiquity. These included horology, especially with reference to sundials; and mechanics, particularly concerning the construction of ballista, siege engines, and other machines of war.

Previous practitioners of architecture and the arts had produced manuals of rhetoric. Not only is *De architectura* the only surviving such

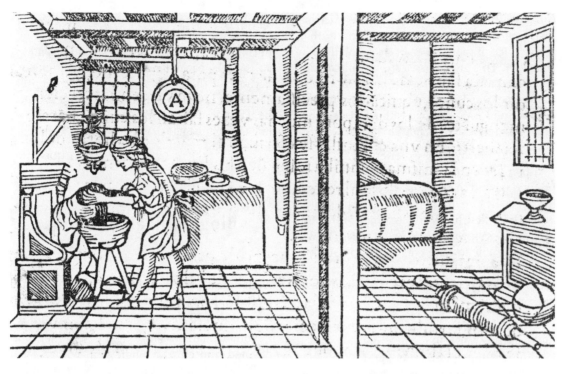

An illustration from Vitruvius's *De Architectura*. (*Archivo Iconigrafico, S.A./Corbis. Reproduced with permission.*)

work, it was also the first to attempt to systematically encompass the entire field of architectural theory and practice. However, instead of an overview and objective analysis of architectural history, Vitruvius drew selectively from the past to define and support his own methodology.

The defining characteristic of the Vitruvian method was its quantification of the principles and rules governing the design and construction of buildings. For example, Vitruvius sought to reduce temple planning to the application of rules governing the dimensions of the constituent parts in relation to the whole. Specifically, he developed a typology of rectangular temples. Seven different designs were included, each with rules governing the relationship of columns to walls. He developed other classificatory schemes as well, including one for private residences based on the style of atrium—Tuscan, Corinthian, tetrastyle, displuviate, and testudinate.

The best known and most significant of the Vitruvian classifications dealt with columns. Vitruvius identified three types: Doric, Ionic, and a variant of the Ionic known as the Corinthian. He explained the historical emergence of each before discussing their character. He found the relation of the base to height in the unadorned Doric suggestive of male strength and solidity, while the slenderness of the fluted Ionic, together with the volutes of its capital, was more suggestive of deli-

cate femininity. Thus, the choice of column determined the character of the building—Doric being more appropriate for a temple dedicated to bellicose Mars, Ionic for a temple to Diana.

STEPHEN D. NORTON

Wen-ti Sui
541-604
Chinese Emperor

Wen-Ti, the posthumous name given to Yang Chien, was the emperor of China from 581 to 604. The founder of the Sui dynasty, he is credited with the reunification and reorganization if China after centuries of unrest. He was well trained in military strategy, which he used to overthrow the government and assume the title of emperor. While in office, Wen-Ti was responsible for many reforms and building projects that had a significant impact on Chinese culture.

Wen-Ti was a member of an extremely powerful and influential family that typically held high offices in the governments of non-Chinese clans. These clans broke up much of southern China at that time. Wen-Ti was raised primarily by a Buddhist nun, but when he reached age 13 he attended a government sponsored school for the upper class. In school, he seemed more concerned with horsemanship and military exercises than composition and history. He joined the military service of the Yu-Wen of the Chou dynasty at age 14. This group had a strong military and conquered much of northern China. During that campaign, Wen-Ti performed well in a command function and was rewarded with marriage to one of the daughters of the Chou crown prince.

The Chou dynasty experienced an unsettled period during which the emperor prematurely died and the new crown prince was ineffective. This convinced Wen-Ti that he should overthrow the Chou's and assume control. Although it was a difficult battle, Wen-Ti was eventually able to seize control by way of superior organization and military skill. He assumed the imperial title in 581, and the Sui dynasty was established.

As emperor, Wen-Ti picked the best possible men to support him. His ultimate goal was to reunify China. In order to do this, he needed to supplant the current capital. He built a new capital and set into motion his grand design of centralizing the government and unifying a disjointed China under one common rule. This required many types of reforms. One major reform involved dismantling a system of bureaucratic nepotism, whereby government posts were awarded by heredity rather than performance, testing, and recommendations. At the same time, Wen-Ti planned the conquest of southern China. He launched an overwhelming assault by both land and water to take this region.

Wen-Ti's achievements consisted of much more than strengthening and reunifying the Chinese empire. He provided a means by which the government could be successfully administered. His lasting success was in the form of political and institutional reforms. He revised laws and rewrote the penal code, built up a tremendous infrastructure, and set up a system of checks and balances within his government. When the newly written laws were introduced, they were more lenient than the statutes they replaced and much effort was given to local education and enforcement of the new laws. The infrastructure was strengthened through many public works projects, including the construction of the Grand Canal, for example. When finished, it joined northern China with the Yangtze River. The central government was setup as a multi-level organization, administered by the emperor with the help of three central ministers. Each level answered to one above it, so a system of checks and balances was established.

Despite his prosperity and success, Wen-Ti did not seem to be happy. Although he had fulfilled nearly every goal, his family life was miserable and he became disenchanted with religion. Because of this, in 601 he demanded a public ceremony in honor of himself. Three years later, he fell ill and died. One of the most influential reigns in China had come to an end.

JAMES J. HOFFMANN

Biographical Mentions

Akhenaton (Amenophis IV or Neferkheprure' wa'enre')
r. c. 1356-1335 B.C.

Egyptian pharaoh and religious reformer who devoted himself to a single god, re-Herakhty. Akhenaton conceived of re-Herakhty as immanent in the sunlight streaming from the Aten or solar disk. He built the great city of Akhetaten—Horizon of the Aten—in honor of his god and instituted wide ranging social reforms throughout Egypt. Akhenaton was eventually deposed and his works execrated as heretical. Akhetaten was rediscovered in 1887 at modern-day Tell el Armana.

Amenemhet III
1842-1797 B.C.

Pharaoh of Egypt during the twelfth dynasty (1818-1770 B.C.). Amenemhet developed the agricultural potential of the Faiyum area southwest of Cairo by completing the large irrigation project begun by his predecessors, reclaiming 153,600 acres (62,200 hectares) of land for cultivation. His water regulation system included draining marshes and constructing an overflow canal. Amenemhet also exploited the resources of the Sinai desert, especially its turquoise deposits. His reign was the last long and prosperous reign of the twelfth dynasty.

Anarcharis of Scythia
fl. c. 592 B.C.

Scythian inventor credited with developing the first ship's anchor. Though the Scythians had settlements along the Black Sea, they are not typically regarded as a seafaring people. Nor was their civilization particularly advanced in comparison to those of their neighbors in Greece and Asia Minor, and these facts make Anarcharis's achievement all the more remarkable.

Anthemius of Tralles
fl. 532-537

Byzantine architect, engineer, and mathematician who, along with Isidorus of Miletus, was responsible for building the Hagia Sophia in Constantinople. Anthemius also wrote treatises on the focal property of parabolic mirrors, and on the possibility of using burning mirrors for military purposes.

Archytas of Tarentum
fl. c. 428-350 B.C.

Greek mathematician and philosopher who applied mathematical theory to the study of music, geometry, and astronomy. Archytas was born in Tarentum, an area of southern Italy that was, at the time, under Greek control. He was a follower of the philosopher Pythagoras, who had theorized that numbers could be used to understand virtually all phenomena. One of Archytas's greatest accomplishments was to duplicate the cube via the construction of a three-dimensional model. He also applied mathematical proportions to his study of pitch and scale in musical harmony. Archytas was also a great statesman, serving as commander in chief in Tarentum for seven years.

Lucius Junius Moderatus Columella
fl. first century A.D.

Roman agricultural writer. Born in Gades in Spain, he wrote, among other works, treatises on Roman agriculture. His most famous such work was *De Re Rustica* (On agriculture). All 12 books of this agricultural manual survive and provide a comprehensive and systematic treatment of the subject, particularly with regards to viticulture. It is also highly praised for its literary style. The owner of several estates, Columella drew on his own practical experience as well as the works of other classical authorities.

Dionysius the Elder
c. 430-367 B.C.

Greek tyrant of Syracuse who helped establish Hellenistic control in Sicily and southern Italy and developed the first catapult for warfare. Dionysius rose to power as tyrant in 405, after distinguishing himself in a war between the Greeks and Carthaginians of North Africa. In 399 B.C., while preparing for another battle against the Carthaginians, he invited Greek craftsmen to the island of Syracuse in Sicily to help him develop new weaponry. Their work resulted in the invention of more powerful ships, called *quinqueremes*, which had four or five banks of oars instead of the previous three banks, as well as the first catapults—machines able to hurl objects with deadly force. These innovations helped Dionysius lead the Greek army to victory against the Carthaginians in two conflicts, the first in 396 B.C. and the second in 392 B.C. He was finally defeated by Carthage during the third war (383-c. 375 B.C.).

Sextus Julius Frontinus
A.D. 35-c. 103

Roman governor of Britain water commissioner, and author of a book on the history and technical details of Rome's aqueducts. In A.D. 75. Frontinus succeeded Petillius Cerealis as Britain's governor. Approximately 22 years later, he was appointed superintendent of the aqueducts in Rome. His book *De aquis urbis Romae* (Concerning the waters of the city of rome) outlined the history and maintenance requirements of the aqueducts. He also wrote a book on the strategies of warfare titled *Strategematicon libri iii.*

Gudea
2141-2122 B.C.

Mesopotamian ruler noted as a temple-builder and patron of the arts. An *ensi* or governor, Gudea ruled the late Sumerian city-state of Lagash, and fostered a golden age of peace and prosperity despite attacks by the Gutians from the mountains in the north. Based on the many statuettes that exist from his reign, as well as numerous inscriptions, it appears that Gudea conducted an extensive building campaign. Due to the fact that most Sumerian architecture was of clay or mud and not stone, the buildings themselves are long gone; however, historians have noted signs of Gudea's impact on the religious life of the region.

Hammurabi
fl. c. eighteenth century B.C.

Emperor of Babylon who was the creator of the first known code of civil and criminal laws. He freed Babylon from Elam and expanded it into a powerful empire by conquering neighboring lands. In addition to being a capable military leader, he was an effective administrator, building cities, temples, and canals and promoting progress in agriculture. His legal system, the Hammurabi Code, was discovered in A.D. 1901, carved on the ruins of a monument.

Juvenal
c. 55-c. 127

Roman satirist whose *On the City of Rome* provides a richly detailed and highly revealing portrait of daily life in Rome. Seen through the eyes of a friend leaving the city for the simpler life of the country, the city is a bustling, lively—and dangerous—place. Among other things, the poem describes "carts clattering through the winding streets;" giant trees and blocks of marble going by in unsteady wagons; loose roof tiles and leaky jars that can fall from windows. The narrator watches a "long procession of servants and burning lamps" while he makes his way home lit by a candle (a sign of wealth). Later he comments disparagingly that "Iron is mainly used to fashion fetters, / So much so we risk a shortage of ploughshares...."

Khufu/Cheops
2551-2528 B.C.

Egyptian pharaoh during the fourth dynasty (c. 2575-c. 2465 B.C.) who built the Great Pyramid at Giza. This pyramid, the largest built by any pharaoh, was part of a funerary complex that included a mortuary temple, valley temple, causeway, seven boat pits, a satellite pyramid, and pyramids for three of Khufu's queens. Khufu's authority was so great that he was able to mobilize and support as many as 20,000 workers each year throughout his reign to work on his pyramid complex.

Ko Yu
fl. first century B.C.?

Semi-mythical Chinese inventor credited with developing the wheelbarrow. According to one ancient text, "Ko Yu built a wooden goat and rode away into the mountains on it." In fact this was a form of code intended to conceal the exact nature of the invention, which was so valuable for moving arms and materiel that the Han Dynasty armies kept its design secret. In time the wheelbarrow came to be known in China as the "wooden ox."

Lu Pan
fl. fifth century B.C.?

Chinese inventor credited with creating the first known kite. Lu Pan lived some time between the sixth century B.C.—some accounts describe him as a contemporary of Confucius (551-479 B.C.)—and the fourth century B.C. Known as "the mechanician of Lu," he reportedly created a glider in the shape of a magpie from bamboo, then became the first person ever to fly a kite.

Ma Chün
fl. c. 260

Chinese alchemist who created an early compass. Using differential gears, Ma Chün developed what he called a "south-pointing carriage." The design of his compass was probably much like that of the ones that have continued to be used in China communities for the divination art of *feng shui:* a spoon made of lodestone or magnetite ore on a cast bronze plate inscribed with symbols indicating the directions of various constellations. The name reflects early Chinese alchemists' mistaken belief that certain kinds of metal cause a compass to point southward.

Meng T'ien
fl. 220 B.C.

Chinese general who directed the construction of the Great Wall. The first Ch'in emperor, Shih Huang-ti, entrusted his general, Meng T'ien, to supervise the construction of a 3,000-mile-long (4,828-km-long) fortification designed to defend northern China from nomadic Asian tribesmen. Meng began construction of the wall in 221 B.C., utilizing the services of some 300,000 men. His section of the Wall reportedly took ten years to build, but the Wall was not completed until the Ming Dynasty, around A.D. 1500. Meng is also credited with inventing a type of harpsichord, the cheng, as well as a pen made of hair.

Mo-tzu
c. 470-c. 391 B.C.

Chinese philosopher, also known as Mo ti, who provided what may be the earliest account of a camera obscura. In about 400 B.C. Mo-tzu observed reflected light rays from an illuminated object passing through a pinhole into a otherwise completely dark room. He noted that these create a precisely inverted image of the original object. More than 2,200 years later, this principle would influence the development of the camera.

Pericles
495-429 B.C.

Greek statesman who led Athens to its democratic and cultural golden age, and directed construction of the Parthenon and Acropolis. Pericles rose to power as head of Athens' democratic party in 461 B.C. After the once feuding Greek city-states reached a truce in 451, Pericles worked to establish Athens as Greece's cultural and political center. He called for a massive building project, including the reconstruction of temples destroyed by the Persians, and the magnificent Acropolis and Parthenon were erected. Later, he extended Athenian settlements to accommodate the rapidly growing population, and constructed a third Long Wall to protect Athens and the port of Piraeus. In the late 430s, the Thirty Years' Peace with Sparta was at an end, and Pericles evacuated the countryside, calling his people within the Athens city walls. The city became crowded and unsanitary, and an ensuing plague decimated as much as one-third of the population. Pericles himself became ill and died in 429.

Philip II of Macedon
381-336 B.C.

Macedonian general who became king of Macedon and reorganized its army. Philip developed the deep phalanx, introduced the sarissa (18-foot-long, or 5.5-meter-long, spear), and increased the mobility of his army by reducing its baggage. He also reformed battlefield tactics by having cavalry units fight in close concert with infantry units. As a result, Philip subjugated Illyria, Thrace, and was made the supreme commander of Greece. He was assassinated before his planned invasion of Persia.

Pliny the Younger
c. A.D. 61-c. 113

Roman scholar and official whose published letters illustrated life during the Roman Empire. Following his father's death, Pliny the Younger was adopted by his uncle, the elder Pliny. He began practicing law at age 18, and eventually held several administrative positions, serving as praetor, consul, prefecture of the military and senatorial treasuries, and head of the drainage board in Rome. His collected correspondence, published in nine volumes between 100 and 109, carefully detailed the social and political life of the Roman Empire.

Priscian
fl. c. A.D. 500

Byzantine grammarian whose works include a long poem concerning the Roman weights and measures. Also useful for historical purposes are his Latin grammar textbooks, which include quotes from esteemed Greek and Roman thinkers and leaders—quotes that would otherwise have been lost. Priscian's 16-book *Institutiones grammaticae* became a classic grammar text during the Middle Ages.

Sesostris III
1878-1843 B.C.

Pharaoh of twelfth-dynasty Egypt who reigned from 1836 to 1818 B.C. Sesostris strengthened the central government and secured Egypt's southern border by building a series of forts on the Nubian frontier. He divided Egypt into administrative districts, each controlled by a vizier, a reform so effective that this form of central government functioned effectively for over a century. The 11 forts he had built in Nubia were strategically placed to protect the border and regulate trade.

Shih Lu
fl. c. 219 B.C.

Chinese engineer who designed the Miracle Canal, one of China's first notable inland waterways. Completed in 219 B.C., the canal was one of many projects undertaken during the Ch'in Dynasty (221-207 B.C.), which was most noted for the building of the Great Wall. Though it was only 20 miles (32 km) long, by connecting the Yangtze and Kan rivers, the canal made it possible for ships to travel inland from Canton in the south to the latitude of modern-day Beijing in the north.

Solomon
r. c. 962-922 B.C.

King of Israel who built massive fortifications, palaces, and a national temple in Jerusalem. Many of the technological details of his various construction projects, especially the temple, are given in the Bible (1 Kings 5-7). He was only 18 when he ascended the throne upon the death of his father, King David, and he reigned for 40 years. He is regarded as a philosopher and poet and is known for his wisdom and justice.

T'ai Tsung
599-649

Chinese emperor and cofounder of the T'ang Dynasty, noted for his reforms in agriculture and other aspects of Chinese life. Born Li Shih-min, in 626 he usurped his father, the first T'ang emperor Kao Tsu (r. 618-626). As emperor he expanded China's borders, greatly reformed the civil-service system, redistributed land, and patronized the arts and sciences. Under his rule, T'ang China was one of the most efficiently governed nations in the world, and the arts and sciences flourished.

Tao Yue
c. 608-c. 676

Semi-legendary Chinese inventor of porcelain. It is said that Tao Yue was born near the Yangtze River, whose "white clay" (kaolin) he added to other varieties of clay in order to develop porcelain. In the Chinese capital at Ch'ang-an or Xian, he sold the resulting creation as "artificial jade."

Ts'ai Lun
50-121

Chinese official credited with the invention of paper. In 105 Ts'ai Lun, a eunuch in the court of the Han Dynasty emperor, suggested creating paper out of tree bark, scraps of hemp, rags, and old fishnet. Not only was this far less expensive than the silk that provided the principal writing surface for documents of the time, the new material actually proved better suited to the task. (Furthermore, in a move that was about 1,900 years ahead of its time, Ts'ai Lun had figured out a way to recycle waste materials.) In honor of his achievement, the emperor granted him the title of marquess in 114.

Virgil (Publius Vergilius Maro)
70-19 B.C.

The foremost Roman poet of his day. Born in Mantua, he wrote, among other works, on agriculture in the *Georgics* (On farming). This didactic poem, written in hexameter verse, was published in 29 B.C. in four books. It was frequently cited by other classical writers, and the Roman agricultural authority Columella pays tribute to Virgil in his own work; however, it is to be noted more for its literary accomplishment than for its use as a farming manual.

Wu-ti
156-87 B.C.

Chinese Han Dynasty emperor noted for his reforms and expansion of the empire. Wu-ti replaced the power of nobles with that of Confucian civil servants, for whom he established a system of examinations (formalized c. 600) that would remain in use until modern times. He also issued an early form of banknote, and in 138 B.C. sent an official named Chang Ch'ien (c. 114 B.C.) westward on a diplomatic mission that resulted in the first Chinese contact with an outside civilization. As a result of Chinese exposure to Central Asia, Wu-ti became intrigued with the region's sturdy "Celestial Horses," which were brought back to China and bred.

Bibliography of Primary Sources

Cassiodorus. *Institutiones divinarum et saecularium litterarum* (English Translation, "Institutes of Divine and Secular Literature," sixth century A.D.). Provides a comprehensive overview of the liberal arts during its time. Apparently written as a guide for learning, the first section discusses the study of the scriptures, while the second is an encyclopedia. The latter section was widely read during the Middle Ages, and its format served as a guide for encyclopedic works for many centuries.

Epic of Gilgamesh. An important piece of Sumerian literature preserved in cuneiform fragments. It includes an account of an ancient flood similar to the one that appears in the Bible, and other legends resembling some of the Greek myths.

Hero of Alexandria. *Pneumatics*. c. first century A.D. Includes Hero's plans for the aeolipile, which he invented and is considered by many to be the first steam-powered engine. Also described in *Pneumatics* are siphons, a fountain, a coin-operated machine, a fire engine, and other steam-powered machines.

Juvenal. *On the City of Rome*. Second century A.D. Provides a richly detailed and highly revealing portrait of daily life in Rome. Seen through the eyes of a friend leaving the city for the simpler life of the country, the city is a bustling, lively—and dangerous—place. Among other things, the poem describes "carts clattering through the winding streets;" giant trees and blocks of marble going by in unsteady wagons; loose roof tiles and leaky jars that can fall from windows.

Philon of Byzantium. *Mechanics*. Third century B.C. Important collection of writings on ancient technology, particularly military devices such as siege engines and fortresses, and the art of defending and besieging towns. Consists of a total of nine books which, taken as a whole, summarize much of the world's knowledge about a variety of devices and technologies during Philon's time.

Vitruvius. *De architectura*. First century B.C. Considered the first attempt at a comprehensive study of architectural practice. This manual contains discussion of building methods and materials as well as an effort to place architectural practice within the larger sphere of liberal arts. Though its influence on later Roman architecture was limited, *De architectura* was widely read during the Renaissance and became the authoritative work on classical architecture.

JOSH LAUER

General Bibliography

Books

Agassi, Joseph. *The Continuing Revolution: A History of Physics from the Greeks to Einstein.* New York: McGraw-Hill, 1968.

Asimov, Isaac. *Adding a Dimension: Seventeen Essays on the History of Science.* Garden City, NY: Doubleday, 1964.

Benson, Don S. *Man and the Wheel.* London: Priory Press, 1973.

Boorstin, Daniel J. *The Discoverers.* New York: Random House, 1983.

Bowler, Peter J. *The Norton History of the Environmental Sciences.* New York: W. W. Norton, 1993.

Brock, W. H. *The Norton History of Chemistry.* New York: W. W. Norton, 1993.

Bruno, Leonard C. *Science and Technology Firsts.* Edited by Donna Olendorf, guest foreword by Daniel J. Boorstin. Detroit: Gale, 1997.

Bud, Robert, and Deborah Jean Warner, editors. *Instruments of Science: An Historical Encyclopedia.* New York: Garland, 1998.

Bynum, W. F., et al., editors. *Dictionary of the History of Science.* Princeton, NJ: Princeton University Press, 1981.

Carnegie Library of Pittsburgh. *Science and Technology Desk Reference: 1,500 Frequently Asked or Difficult-to-Answer Questions.* Detroit: Gale, 1993.

Crone, G. R. *Man the Explorer.* London: Priory Press, 1973.

De Camp, L. Sprague. *The Ancient Engineers.* Cambridge, MA: MIT Press, 1963.

De Groot, Jean. *Aristotle and Philoponus on Light.* New York: Garland, 1991.

Ellis, Keith. *Man and Measurement.* London: Priory Press, 1973.

Gershenson, Daniel E., and Daniel A. Greenberg. *Anaxagoras and the Birth of Scientific Method.* Introduction by Ernest Nagel. New York: Blaisdell Publishing Company, 1964.

Gibbs, Sharon L. *Greek and Roman Sundials.* New Haven, CT: Yale University Press, 1976.

Good, Gregory A., editor. *Sciences of the Earth: An Encyclopedia of Events, People, and Phenomena.* New York: Garland, 1998.

Grattan-Guiness, Ivor. *The Norton History of the Mathematical Sciences: The Rainbow of Mathematics.* New York: W. W. Norton, 1998.

Gregor, Arthur S. *A Short History of Science: Man's Conquest of Nature from Ancient Times to the Atomic Age.* New York: Macmillan, 1963.

Gullberg, Jan. *Mathematics: From the Birth of Numbers.* Technical illustrations by Pär Gullberg. New York: W. W. Norton, 1997.

Hellemans, Alexander, and Bryan Bunch. *The Timetables of Science: A Chronology of the Most Important People and Events in the History of Science.* New York: Simon and Schuster, 1988.

Hellyer, Brian. *Man the Timekeeper.* London: Priory Press, 1974.

Hodge, M. J. S. *Origins and Species: A Study of the Historical Sources of Darwinism and the Contexts of Some Other Accounts of Organic Diversity from Plato and Aristotle On.* New York: Garland, 1991.

Holmes, Edward, and Christopher Maynard. *Great Men of Science.* Edited by Jennifer L. Justice. New York: Warwick Press, 1979.

Hoskin, Michael. *The Cambridge Illustrated History of Astronomy.* New York: Cambridge Universit Press, 1997.

Lankford, John, editor. *History of Astronomy: An Encyclopedia.* New York: Garland, 1997.

Lewes, George Henry. *Aristotle: A Chapter from the History of Science, Including Analyses of Aristotle's Scientific Writings.* London: Smith, Elder, and Co., 1864.

Mayr, Otto, editor. *Philosophers and Machines.* New York: Science History Publications, 1976.

McGrath, Kimberley, editor. World of Scientific Discovery. 2nd ed. Detroit: Gale, 1999.

Mueller, Ian. *Coping with Mathematics: The Greek Way.* Chicago, IL: Morris Fishbein Center for the Study of the History of Science and Medicine, 1980.

Multhauf, Robert P. *The Origins of Chemistry.* New York: F. Watts, 1967.

Porter, Roy. *The Cambridge Illustrated History of Medicine.* New York: Cambridge University Press, 1996.

Sarton, George. *Hellenistic Science and Culture in the Last Three Centuries B.C.* New York: Dover Publications, 1993.

Sarton, George. *Introduction to the History of Science.* Huntington, NY: R. E. Krieger Publishing Company, 1975.

Singer, Charles. *Greek Biology and Greek Medicine.* New York: AMS Press, 1979.

Smith, Roger. *The Norton History of the Human Sciences.* New York: W. W. Norton, 1997.

Smith, Wesley D. *The Hippocratic Tradition.* Ithaca, NY: Cornell University Press, 1979.

Spangenburg, Ray. *The History of Science from the Ancient Greeks to the Scientific Revolution.* New York: Facts on File, 1993.

Stiffler, Lee Ann. *Science Rediscovered: A Daily Chronicle of Highlights in the History of Science.* Durham, NC: Carolina Academic Press, 1995.

Swerdlow, N. M. *Ancient Astronomy and Celestial Divination.* Cambridge, MA: MIT Press, 1999.

Temkin, Owsei. *Galenism: Rise and Decline of a Medical Philosophy.* Ithaca, NY: Cornell University Press, 1973.

Travers, Bridget, editor. *The Gale Encyclopedia of Science.* Detroit: Gale, 1996.

Whitehead, Alfred North. *Science and the Modern World: Lowell Lectures, 1925.* New York: The Free Press, 1953.

Young, Robyn V., editor. *Notable Mathematicians: From Ancient Times to the Present.* Detroit: Gale, 1998.

JUDSON KNIGHT

Index

*Numbers in bold refer to
main biographical entries*